Methods in Enzymology

Volume 177
NUCLEAR MAGNETIC RESONANCE
Part B
Structure and Mechanism

METHODS IN ENZYMOLOGY

EDITORS-IN-CHIEF

John N. Abelson   Melvin I. Simon

DIVISION OF BIOLOGY
CALIFORNIA INSTITUTE OF TECHNOLOGY
PASADENA, CALIFORNIA

FOUNDING EDITORS

Sidney P. Colowick and Nathan O. Kaplan

Methods in Enzymology

Volume 177

Nuclear Magnetic Resonance

Part B

Structure and Mechanism

EDITED BY

Norman J. Oppenheimer

DEPARTMENT OF PHARMACEUTICAL CHEMISTRY
SCHOOL OF PHARMACY
UNIVERSITY OF CALIFORNIA, SAN FRANCISCO
SAN FRANCISCO, CALIFORNIA

Thomas L. James

DEPARTMENT OF PHARMACEUTICAL CHEMISTRY
SCHOOL OF PHARMACY
UNIVERSITY OF CALIFORNIA, SAN FRANCISCO
SAN FRANCISCO, CALIFORNIA

ACADEMIC PRESS, INC.
Harcourt Brace Jovanovich, Publishers
San Diego   New York   Berkeley   Boston
London   Sydney   Tokyo   Toronto

ACADEMIC PRESS, INC.
San Diego, California 92101

United Kingdom Edition published by
ACADEMIC PRESS LIMITED
24-28 Oval Road, London NW1 7DX

LIBRARY OF CONGRESS CATALOG CARD NUMBER: 54-9110

ISBN   0-12-182078-5   (alk. paper)

Printed and bound by CPI Group (UK) Ltd, Croydon, CR0 4YY
Transferred to Digital Print 2011

Table of Contents

Section I. Enzyme Modifications for Nuclear Magnetic Resonance Studies

Section II. Protein Structure

v

Section III. Enzyme Mechanisms

Section IV. *In Vivo* Studies of Enzymatic Activity

Appendix

Contributors to Volume 177

Article numbers are in parentheses following the names of contributors.
Affiliations listed are current.

Russ B. Altman (11), *Stanford Magnetic Resonance Laboratory, Stanford University, Stanford, California 94305*

D. Eric Anderson (3), *Institute of Molecular Biology, University of Oregon, Eugene, Oregon 97403*

Lucia Banci (12), *Department of Chemistry, University of Florence, 50121 Florence, Italy*

Vladimir J. Basus (7), *Department of Pharmaceutical Chemistry, School of Pharmacy, University of California, San Francisco, San Francisco, California 94143*

Ivano Bertini (12), *Department of Chemistry, University of Florence, 50121 Florence, Italy*

Martin Billeter (8), *Institut für Molekularbiologie und Biophysik, Eidgenössische Technische Hochschule (ETH)-Hönggerberg, CH-8093 Zürich, Switzerland*

Philip H. Bolton (4, 14), *Hall-Atwater Laboratories, Department of Chemistry, Wesleyan University, Middletown, Connecticut 06457*

Jack S. Cohen (23), *Biophysical Pharmacology Section, Medicine Branch, National Cancer Institute, National Institutes of Health, Bethesda, Maryland 20892*

Sheila M. Cohen (22), *Department of Animal and Exploratory Drug Metabolism, Merck Institute for Therapeutic Research, Merck Sharp & Dohme Research Laboratories, Rahway, New Jersey 07065*

Frederick W. Dahlquist (3), *Institute of Molecular Biology, University of Oregon, Eugene, Oregon 97403*

Peter F. Daly (23), *Pittsburgh NMR Institute, Pittsburgh, Pennsylvania 15213*

Mark Dell'Acqua (4, 14), *Department of Chemistry and Biochemistry, University of Maryland, College Park, Maryland 20742*

J. T. Gerig (1), *Department of Chemistry, University of California, Santa Barbara, Santa Barbara, California 93106*

John A. Gerlt (4, 14), *Department of Chemistry and Biochemistry, University of Maryland, College Park, Maryland 20742*

David G. Gorenstein (15), *Department of Chemistry, Purdue University, West Lafayette, Indiana 47907*

Lynn Harpold (4, 14), *Department of Chemistry and Biochemistry, University of Maryland, College Park, Maryland 20742*

David W. Hibler (4, 14), *Department of Chemistry and Biochemistry, University of Maryland, College Park, Maryland 20742*

Oleg Jardetzky (11), *Stanford Magnetic Resonance Laboratory, Stanford University, Stanford, California 94305*

R. Kaptein (10), *Department of Organic Chemistry, University of Utrecht, 3584 CH Utrecht, The Netherlands*

Phyllis A. Kosen (5), *Department of Pharmaceutical Chemistry, School of Pharmacy, University of California, San Francisco, San Francisco, California 94143*

I. D. Kuntz (9), *Department of Pharmaceutical Chemistry, School of Pharmacy, University of California, San Francisco, San Francisco, California 94143*

David M. LeMaster (2), *Department of Biochemistry, Molecular Biology and Cell Biology, Northwestern University, Evanston, Illinois 60208*

Claudio Luchinat (12), *Institute of Agricultural Chemistry, University of Bologna, 40127 Bologna, Italy*

ROBBE C. LYON (23), *Laboratory of Metabolism and Molecular Biology, National Institute of Alcohol Abuse and Alcoholism, Rockville, Maryland 20854*

LAWRENCE P. MCINTOSH (3), *Institute of Molecular Biology, University of Oregon, Eugene, Oregon 97403*

ALBERT S. MILDVAN (17), *Department of Biological Chemistry, The Johns Hopkins University School of Medicine, Baltimore, Maryland 21205*

DAVID C. MUCHMORE (3), *Institute of Molecular Biology, University of Oregon, Eugene, Oregon 97403*

C. M. OSHIRO (9), *IBM Palo Alto Scientific Center, Palo Alto, California 94304*

TAYEBEH POURMOTABBED (4, 14), *Department of Chemistry and Biochemistry, University of Maryland, College Park, Maryland 20742*

B. D. NAGESWARA RAO (18), *Department of Physics, Indiana University-Purdue University at Indianapolis (IUPUI), Indianapolis, Indiana 46205*

JOHN M. RISLEY (19), *Department of Chemistry, The University of North Carolina at Charlotte, Charlotte, North Carolina 28223*

PAUL R. ROSEVEAR (17), *Department of Biochemistry and Molecular Biology, Medical School, The University of Texas, Health Science Center at Houston, Houston, Texas 77225*

CHRISTOPHER B. RUSSELL (3), *Institute of Molecular Biology, University of Oregon, Eugene, Oregon 97403*

CHARLES R. SANDERS II (16), *Department of Chemistry, Yale University, New Haven, Connecticut 06511*

R. M. SCHEEK (10), *Department of Physical Chemistry, University of Groningen, 9747 AG Groningen, The Netherlands*

J. F. THOMASON (9), *Department of Pharmaceutical Chemistry, School of Pharmacy, University of California, San Francisco, San Francisco, California 94143*

MING-DAW TSAI (16), *Department of Chemistry, The Ohio State University, Columbus, Ohio 43210*

ROBERT L. VAN ETTEN (19), *Department of Chemistry, Purdue University, West Lafayette, Indiana 47907*

W. F. VAN GUNSTEREN (10), *Department of Physical Chemistry, University of Groningen, 9747 AG Groningen, The Netherlands*

JOSEPH J. VILLAFRANCA (20, 21), *Department of Chemistry, The Pennsylvania State University, University Park, Pennsylvania 16802*

HANS J. VOGEL (13), *Department of Biological Sciences, The University of Calgary, Calgary, Alberta, Canada T2N 1N4*

JOYCE A. WILDE (4, 14), *Hall-Atwater Laboratories, Department of Chemistry, Wesleyan University, Middletown, Connecticut 06457*

KURT WÜTHRICH (6), *Institut für Molekularbiologie und Biophysik, Eidgenössische Technische Hochschule (ETH)-Hönggerberg HPM, CH-8093 Zürich, Switzerland*

Preface

NMR spectroscopy has undergone a remarkable transformation in the past decade which has few precedents in science. What was previously considered to be primarily an analytical tool for the study of small molecules has blossomed into one for investigating structure and dynamics, ranging from whole organisms to the atomic level. It is a field that is undergoing change at an astounding rate, with nearly continuous publication of powerful new techniques that further extend the range of experimental applicability. The applications of NMR spectroscopy to biochemistry, in general, and to enzymology, in particular, have also undergone parallel expansion. To date, however, there has not been an extended presentation in *Methods in Enzymology* of these applications.

In Volumes 176 and 177 we have attempted to serve two primary functions. The first is to rectify past omissions by providing a general background of modern NMR techniques, with a specific focus on NMR techniques that pertain to proteins and enzymology. The second is to provide a "snapshot" of the current state-of-the-art in NMR experimental techniques. Our overall goal is to provide information to enable the reader to understand a given technique, to evaluate its strengths and limitations, to decide which is the best approach, and, finally, to design an experiment using the chosen technique to solve a problem.

This volume covers protein modifications for NMR, including isotope labeling, techniques for protein structure determination, enzyme mechanism methodology, and means for examining enzyme activity *in vivo*. Volume 176 covers basic and advanced NMR techniques, including two-dimensional NMR, and methods for studying protein dynamics, including rate constants and molecular motions.

Although some techniques may be superseded by yet newer procedures and other techniques will certainly appear, it is our hope that the methods presented in these volumes will be of continuing value in the design and execution of NMR experiments for solving problems in protein and enzyme structure, function, and dynamics.

NORMAN J. OPPENHEIMER
THOMAS L. JAMES

METHODS IN ENZYMOLOGY

VOLUME 68. Recombinant DNA
Edited by RAY WU

VOLUME 69. Photosynthesis and Nitrogen Fixation (Part C)
Edited by ANTHONY SAN PIETRO

VOLUME 70. Immunochemical Techniques (Part A)
Edited by HELEN VAN VUNAKIS AND JOHN J. LANGONE

VOLUME 71. Lipids (Part C)
Edited by JOHN M. LOWENSTEIN

VOLUME 72. Lipids (Part D)
Edited by JOHN M. LOWENSTEIN

VOLUME 73. Immunochemical Techniques (Part B)
Edited by JOHN J. LANGONE AND HELEN VAN VUNAKIS

VOLUME 74. Immunochemical Techniques (Part C)
Edited by JOHN J. LANGONE AND HELEN VAN VUNAKIS

VOLUME 75. Cumulative Subject Index Volumes XXXI, XXXII, XXXIV–LX
Edited by EDWARD A. DENNIS AND MARTHA G. DENNIS

VOLUME 76. Hemoglobins
Edited by ERALDO ANTONINI, LUIGI ROSSI-BERNARDI, AND EMILIA CHIANCONE

VOLUME 77. Detoxication and Drug Metabolism
Edited by WILLIAM B. JAKOBY

VOLUME 78. Interferons (Part A)
Edited by SIDNEY PESTKA

VOLUME 79. Interferons (Part B)
Edited by SIDNEY PESTKA

VOLUME 80. Proteolytic Enzymes (Part C)
Edited by LASZLO LORAND

Section I

Enzyme Modifications for Nuclear Magnetic Resonance Studies

[1] Fluorine Nuclear Magnetic Resonance of Fluorinated Ligands

By J. T. GERIG

Nuclear magnetic resonance experiments can provide much information about ligand–protein complexes, including stereochemical details, rates of conformational change, and identification of interactions between atoms of the ligand with other atoms of the ligand or atoms of the protein. The fluorine nucleus has properties, including those listed below, that make attractive its use in studies of ligand binding to enzymes:

1. Fluorine is a spin-$\frac{1}{2}$ nuclide, is present in 100% natural abundance, and, among common nuclei, has a receptivity to NMR detection that is second only to that of the proton.[1]

2. Fluorine NMR is characterized by an extremely wide range of chemical shifts, with the shift being highly responsive not only to changes in chemical bonding within the structure that holds the fluorine nucleus but also to changes in the local environment that are the result of protein binding.

3. The van der Waals radius of covalent fluorine is about 0.14 nm,[2–4] only slightly larger than that of hydrogen, so that replacement of a single hydrogen by a fluorine in an enzyme is not expected to have a major structure-disrupting effect for reasons of steric bulk. It should be noted, however, that a group of fluorine atoms such as the trifluoromethyl group is collectively much larger than the corresponding all-proton system.

4. Perhaps most important for studies of biological systems is that with *in vitro* systems there are no fluorine background signals to interfere with the spectroscopy in the way that the water signal intrudes in proton NMR experiments. Moreover, *in vivo* concentrations of fluorine are sufficiently low that there are no background problems in these systems either.[5]

It should be noted at the outset that the advantages of NMR studies of fluorinated ligands interacting with enzymes are bought at a price. Al-

[1] R. K. Harris, "Nuclear Magnetic Resonance," p. 73. Pitman, London, 1983.
[2] L. Pauling, "The Nature of the Chemical Bond," 3rd ed., pp. 257–260. Cornell Univ. Press, Ithaca, New York, 1960.
[3] A. Bondi, *J. Phys. Chem.* **68**, 441 (1964).
[4] S. C. Nyberg and C. H. Faerman, *Acta Crystallogr., Sect. B: Struct. Sci.* **41**, 274 (1985).
[5] T. R. Nelson, F. D. Newman, I. M. Schiffer, J. D. Reith, and S. L. Cameron, *Magn. Reson. Imaging* **3**, 267 (1985).

though enzyme-induced fluorine chemical shift effects can be dramatically large for one accustomed to the more modest shift effects observed in proton or carbon-13 experiments, there is at present no way to reliably predict or interpret the direction of a protein-induced fluorine shift effect, much less its magnitude. A single fluorine spin attached to the ligand may not be much more sterically demanding than a proton at the same position, but the chemical properties of the carbon–fluorine bond are totally different from those of the corresponding carbon–hydrogen bond. The highly polarized C–F bond may alter local structure by changing solvation patterns, by dipole–dipole interactions with other groups, or by acting as a (weak) hydrogen bond acceptor, to mention just a few possibilities. Thus, there is a chance that fluorine placement in an enzyme–ligand complex will alter three-dimensional structure significantly relative to the corresponding nonfluorinated system, so that the NMR observations made may not be relevant to the behavior of a "natural" system. Some experimental evidence regarding the extent of such structural perturbations will be discussed later.

A final disadvantage is that the anisotropy of the fluorine chemical shift tensor is large. This has the practical consequence that contributions of the chemical shift anisotropy (CSA) relaxation mechanism to T_1 and T_2 relaxations become appreciable at high magnetic fields (corresponding to proton resonance at frequencies greater than 300 MHz), such that the sensitivity and resolution advantages one expects by operating at these high fields may not be realized. The CSA effect on T_2 relaxation means a rapidly increasing signal linewidth with increasing field, to the extent that spectra obtained at, say, 470 MHz may be inferior to those obtained at 282 MHz, both in terms of resolution (linewidths) and sensitivity.

The experimental aspects of getting fluorine spectra of enzyme-bound fluorinated ligands are no more daunting than those involved in obtaining proton spectra with the same systems, and have been discussed in an earlier chapter in this series.[6] For best sensitivity and line shape it is desirable to have a probe that has been optimized for fluorine observations. However, in many cases this is unnecessary and either a proton-observe probe or the decoupler coil in a broadband probe can be retuned to operate at the fluorine frequency. With any of these approaches it is critical that background signals from fluorine-containing materials in the probe be at a sufficiently low level. It is difficult enough to detect broad signals from an enzyme system without the uncertainties and irritations of competing signals from probe materials which can appear at intensities greater than those of interest!

[6] B. D. Sykes and W. E. Hull, this series, Vol. 49, p. 270.

The ability to decouple or produce radiofrequency (rf) pulses at the proton frequency while observing fluorine is becoming increasingly desirable in studies of biological systems; these experiments are best carried out in probes designed specifically for this purpose. Because the fluorine and proton frequencies are close to each other, special efforts are required to eliminate interferences from the proton channel while data accumulation of fluorine signals is in progress. These requirements are reasonably well-met by directional couplers and filters supplied by manufacturers as part of their fluorine-observe/proton-decouple systems.

We find the methods described by Fukishima and Roeder[7] for calibration of observe pulses and the method described by Thomas et al.[8] and others for calibration of decoupler pulses to be adequate. A useful sample for these calibrations is α-fluorocinnamic acid—the vinyl fluorine and proton define an AX system and the material is soluble in aqueous and organic solvents.

The observables in a solution-phase fluorine NMR experiment include T_1 (spin–lattice) relaxation times, T_2 (spin–spin) relaxation times, chemical shifts, coupling constants (if the linewidth of the signals involved will permit resolution of fine structure), and various Overhauser experiments. The T_1 determinations are best made with the inversion-recovery sequence using a composite 180° pulse and phase cycling of the 90° observe pulse.[9,10] A variety of methods can be used to estimate T_2, including simply measuring the linewidth and various spin-echo sequences. It should be noted that sequences involving formation of several echoes, such as the Meiboom–Gill experiment, are more difficult to set up and more subject to experimental errors than T_1 experiments.[11] For determination of the steady-state fluorine–proton nuclear Overhauser effect (NOE), we prefer an experiment in which data with and without the NOE are collected in alternate scans and stored separately, as this approach minimizes the effects of long-term drifts on the accuracy of the results. It is essential to include a $10T_1$ delay between scans in this kind of experiment, as discussed by Opella et al.[12] It should be recalled that for large molecules in which fluorine–proton dipolar interactions provide the only relaxation mechanism, the fluorine NOE, when protons are completely

[7] E. Fukishima and S. B. W. Roeder, "Experimental Pulse NMR," p. 434. Addison-Wesley, Reading, Massachusetts, 1981.
[8] D. M. Thomas, M. R. Bendall, D. T. Pegg, D. M. Doddrell, and J. Field, J. Magn. Reson. 42, 298 (1981).
[9] R. Freeman, S. P. Kempsell, and M. H. Levitt, J. Magn. Reson. 38, 453 (1980).
[10] J. D. Cutnell, H. E. Bleich, and J. A. Glasel, J. Magn. Reson. 21, 43 (1976).
[11] R. L. Vold, R. R. Vold, and H. E. Simon, J. Magn. Reson. 11, 283 (1973).
[12] S. J. Opella, D. J. Nelson, and O. Jardetzky, J. Chem. Phys. 64, 2533 (1976).

saturated, corresponds to a reduction of the signal intensity to -106% of its original value.

The discussion which follows will treat two types of fluorine NMR experiments, distinguished by whether the fluorinated moiety is covalently linked to the enzyme. A variety of experiments can be designed in which a fluorinated compound interacts with a site on the protein in the same manner that a substrate, inhibitor, or product might interact. The complexes so formed are capable of dissociating and the types of experiments that are feasible and their interpretation will depend on the rate of the dissociation. Experiments in which the fluorinated ligand is able to dissociate from the enzyme will be referred to as Type I experiments. In contrast, fluorinated molecules can be designed which will interact with a locus on the enzyme and, while bound, undergo a reaction with a group near the binding site such that the fluorinated species becomes permanently part of the covalent structure of the protein and is, therefore, unable to separate from the protein without breaking a chemical bond. These systems will be designated Type II systems. In the limit of very slow dissociation, methods for interpretation of results from Type I systems become identical to those used with Type II experiments.

Both Type I and Type II fluorine NMR experiments with proteins (mostly enzymes) have been reviewed, although these articles are now rather dated.[13,14]

Type I Systems

In the simplest case, Type I experiments can be described in terms of the equilibrium

$$\text{EF} \underset{k_a}{\overset{k_d}{\rightleftharpoons}} \text{E} + \text{F} \tag{1}$$

where E represents the enzyme and F represents the fluorinated species that interacts with the enzyme to give the complex EF. For a system at equilibrium, $k_d/k_a = K_D$, where K_D is the dissociation constant of the complex ($= [\text{E}][\text{F}]/[\text{EF}]$). The first-order dissociation rate constant k_d is given by $k_a K_D$ and, if one assumes that the rate of association is limited by diffusion such that k_a is about 10^{10} M^{-1} sec^{-1}, the dissociation rate constant for the complex can be estimated given knowledge of K_D. The lifetime (τ) of the EF complex is the reciprocal of this rate constant ($\tau = 1/k_d$).

[13] J. T. Gerig, *Biol. Magn. Reson.* **1**, 139 (1978).
[14] J. T. Gerig, *in* "Biomedical Aspects of Fluorine Chemistry" (R. Filler, ed.), pp. 163–189. Elsevier, New York, 1982.

Methods are available for measurement of exchange rates, including those based on analysis of complete line shapes, peak broadening, dependence of the apparent T_2 in a Carr–Purcell–Meiboom–Gill spin-echo experiment on the rate at which the refocusing 180° pulses are applied, and transfer of saturation or inversion (cf. Chapters 15 through 20).[15-19] If an exchange rate is very fast or very slow, these techniques are inappropriate for its determination, but under these conditions precise knowledge of the rate is not required for analysis of ligand binding data. It should be remembered that there may be certain combinations of relaxation times, chemical shifts, and exchange rates for which a fluorine signal will be so broad that it will disappear into the baseline.[20-22]

Equation (2) defines a two-site exchange problem, and all of the observables mentioned earlier are affected to some extent by the rate of the exchange process, even if the exchange process is slow enough to give separate signals for nuclei attached to the "free" and "bound" ligands. When the fraction of the fluorinated species that is protein bound is small and the exchange rate is reasonably rapid, the observed spin–lattice relaxation rate behavior is given by

$$\frac{1}{T_{1\,\text{obs}}} = \frac{1}{T_{1\,\text{f}}} + \left(\frac{1}{T_{1\,\text{b}}} - \frac{1}{T_{1\,\text{f}}}\right) x_\text{b} \tag{2}$$

where $T_{1\,\text{f}}$ and $T_{1\,\text{b}}$ correspond to the spin–lattice relaxation times of the free and bound forms of the ligand, respectively, and x_b is the mole fraction of the ligand that is enzyme bound. It can be seen that a plot of $1/T_{1\,\text{obs}}$ against x_b will be linear, with a slope giving $T_{1\,\text{b}}$ and an intercept of $1/T_{1\,\text{f}}$.[23] If exchange is slow, or the mole fractions of free and bound material are comparable, or $T_{1\,\text{f}}$ and $T_{1\,\text{b}}$ have very different magnitudes, then spin–lattice relaxation will be biexponential and a complete solution of the Bloch equations may be necessary for determination of $T_{1\,\text{f}}$ and $T_{1\,\text{b}}$.[23]

Spin–lattice relaxation times for spins in large molecules are expected to be a function of radiofrequency, although the exact form of this dependence will be defined by the details of molecular motions near the ob-

[15] J. I. Kaplan and G. Fraenkel, "NMR of Chemically Exchanging Systems." Academic Press, New York, 1980.
[16] J. Sandstrom, "Dynamic NMR Spectroscopy." Academic Press, New York, 1982.
[17] J. R. Alger and R. G. Shulman, Q. Rev. Biophys. 17, 83 (1984).
[18] J. P. Carver and R. E. Richards, J. Magn. Reson. 6, 89 (1972).
[19] J. T. Gerig and A. D. Stock, Org. Magn. Reson. 7, 249 (1975).
[20] J. L. Markeley, Acc. Chem. Res. 8, 70 (1975).
[21] J. Feeney, J. G. Batchelor, J. L. Albrand, and G. C. K. Roberts, J. Magn. Reson. 33, 519 (1979).
[22] J. L. Sudmeier, J. L. Evelhoch, and N. B. H. Jonsson, J. Magn. Reson. 40, 377 (1980).
[23] A. C. McLaughlin and J. S. Leigh, Jr., J. Magn. Reson. 9, 296 (1973).

served nucleus (cf. London [18], Vol. 176, this series).[24,25] T_1 as a function of the correlation time(s) describing these motions is usually double-valued and, when using experimental T_1 data to define correlations times, it is difficult to choose between the two solutions in the absence of a determination of the frequency dependence. (In this light it is distressing to see the demise of high-resolution instruments which can make fluorine observations at 56 and 94 MHz.) For the size of enzymes typically studied by fluorine NMR techniques, T_1 is found to increase with increasing radiofrequency.

The conditions required to give fast exchange averaging of the chemical shifts of the free and bound ligand are different from those needed to make Eq. (2) valid. Swift and Connick[26] have shown that for exchange between two sites with one of the components in large concentration excess over the other, the observed signal is shifted from its position when there is no exchange by an amount δ, given by Eq. (3), where $\Delta\nu$ is the chemical shift difference in hertz between the signals for the free and bound fluorine nuclei, T_{2b} is the transverse relaxation time for the bound spin, and k_d is defined in Eq. (1). Thus, the observed chemical shift effect will vary linearly with the mole fraction of bound material, but the coefficient of variation (S) in general will depend on T_{2b} and k_d in addition to the shift difference between the sites. Only when $k_d \gg \Delta\nu$ and $1/T_{2b}$ will the slope of a plot of δ against x_b give a reliable value for $\Delta\nu$. The slope of δ versus x_b plots will be radiofrequency dependent, since both $\Delta\nu$ and T_{2b} will change with variation of operating radiofrequency. Both parameters increase with increasing frequency and

$$\delta = \frac{k_d^2 \, \Delta\nu x_b}{\left(\dfrac{1}{T_{2b}} + k_d\right)^2 + (\Delta\nu)^2} \equiv Sx_b \tag{3}$$

the condition regarding the magnitude of k_d indicated above can become progressively more difficult to satisfy at higher magnetic fields; a system that may be in fast exchange at 94 MHz can enter the slow or intermediate exchange regime at 470 MHz. More complicated expressions are available when the experimental conditions are outside the range of applicability of Eq. (3)[27] and a complete line-shape simulation of the two-site exchange system may be necessary to obtain an accurate estimate of the

[24] D. Doddrell, V. Glushko, and A. Allerhand, *J. Chem. Phys.* **56**, 3683 (1972).

[25] R. E. London, *in* "Magnetic Resonance in Biology" (J. S. Cohen, ed.), Vol. 1, pp. 1–33. Wiley (Interscience), New York, 1980.

[26] T. J. Swift and R. E. Connick, *J. Chem. Phys.* **37**, 307 (1962).

[27] S. H. Smallcombe, B. Ault, and J. H. Richards, *J. Am. Chem. Soc.* **94**, 4585 (1972).

bound chemical shift (and T_{2b}). Programs for these simulations based on density-matrix considerations[28] or the Bloch equations[29] are easily implemented on a personal computer.

Methods for analysis of observed transverse relaxation data (T_2) obtained by spin-echo or line-shape methods for fluorine nuclei attached to a ligand that binds to an enzyme in a Type I process are basically the same as those used for determination of T_1 for the bound fluorine. However, this process is complicated because the exchange rate figures much more strongly in defining the observed T_2. Leigh has given a rigorous equation for T_{2b} and discussed various approximations that can be used to simplify its algebraic form.[23,30]

When the exchange rate places the system sufficiently close to the fast exchange limit variations of the observed chemical shift, T_1 or T_2 with the ligand and enzyme concentrations can be used to deduce the binding constant for equilibrium [Eq. (1)] as well as the values of these parameters for the enzyme-bound form of the ligand.[21,31] Complications such as association of the enzyme can be taken into account in such efforts.[32]

If there is more than one binding site for the ligand on the protein, the system is not a two-site exchange problem and evaluating the parameters which characterize each binding site becomes more complicated, if not intractable. This, unfortunately, is the situation if weak or nonspecific binding of the ligand is possible, and considerable experimentation may be necessary to identify that part of the observed NMR response that is due to binding at a specific site on an enzyme. One approach is to determine the NMR shift or relaxation parameters in the presence of a second nonfluorinated ligand that is known to bind specifically to the site of interest, the residual NMR effects being assigned to the interaction of the fluorinated ligand with the "weak" sites. However, interpreting these results is often problematic, since the nature of the competition of the two ligands for the "weak" sites is not clear.

Properly designed fluorine–proton NOE experiments can produce evidence regarding the conformation of a bound, flexible molecule containing a fluorine nucleus, provided that the bound conformation brings a proton of the ligand into close proximity of the fluorine. Protons of the enzyme may also be close enough to the fluorine of the ligand to produce a

[28] C. S. Johnson, Jr., *Adv. Magn. Reson.* **1**, 33 (1965).
[29] J. I. Kaplan and R. E. Carter, *J. Magn. Reson.* **33**, 437 (1979).
[30] J. S. Leigh, Jr., *J. Magn. Reson.* **4**, 308 (1971).
[31] O. Jardetzky and G. C. K. Roberts, "NMR in Molecular Biology," p. 328 *et seq.* Academic Press, New York, 1981.
[32] K. L. Gammon, S. H. Smallcombe, and J. H. Richards, *J. Am. Chem. Soc.* **94**, 4573 (1972).

NOE. Spin diffusion smears the specificity of fluorine–proton Overhauser effects in large molecules for the same reasons that proton–proton NOE specificity can be lost by this phenomenon,[33] and in this situation the most profitable NOE experiments are those that involve studies of the time dependence of generation of the NOE, either by following changes in the intensity of the fluorine signal after start of irradiation at a proton frequency, or by means of a two-dimensional experiment with a mixing time sufficiently short so that spin diffusion effects are attenuated.[34–36] Analysis of these experiments can provide estimates of the correlation times and internuclear distances as is detailed more fully elsewhere in this series (Borgias and James [9], Vol. 176).

Nuclear Overhauser effects in exchanging systems have been discussed by several authors.[37–39] Presuming fast exchange, irradiation of a ligand proton resonance produces an initial rate of change of the (averaged) fluorine signal intensity

$$\text{Rate} = \sigma_{H_fF}(1 - x_b) + \sigma_{H_bF} x_b \tag{4}$$

where σ is the cross-relaxation rate produced by a fluorine–proton interaction in the free or bound state and is a quantity that depends on the details of molecular motion and the distance between the fluorine spin and the proton in that state. If a proton signal of the liganded *protein* is the one irradiated, the initial rate of change of the fluorine signal intensity is $\sigma_{H_bF} x_b$.

We have found that a useful prelude to a two-dimensional fluorine–proton NOE experiment with Type I or II systems is determination of the time behavior of the fluorine intensity following a nonspecific 180° proton pulse. For systems larger than about 1 kDa the fluorine intensity will initially decrease, reaching a minimum, and, as spin–lattice relaxation comes into play, return to its equilibrium value. The time at the minimum and the shape of the curve before the minimum are indicative of the power of spin diffusion in the system. A mixing time appreciably shorter than this minimum time should give a two-dimensional NOE result that is relatively unaffected by spin diffusion. In crucial situations, it is still advisable to determine the mixing-time dependence of a peak appearing in the two-dimensional NOE in order to provide indications of whether it is

[33] A. Kalk and H. J. C. Berendsen, *J. Magn. Reson.* **24,** 343 (1976).
[34] K. Wüthrich, "NMR of Proteins and Nucleic Acids," p. 93. Cornell Univ. Press, Ithaca, New York, 1987.
[35] S. J. Hammond, *J. Chem. Soc., Chem. Commun.,* 712 (1984).
[36] M. Cairi and J. T. Gerig, *J. Magn. Reson.* **62,** 131 (1985).
[37] J. H. Noggle and R. E. Schirmer, "The Nuclear Overhauser Effect," pp. 77–95 and 126–166. Academic Press, New York, 1971.
[38] M. Borzo and G. E. Maciel, *J. Magn. Reson.* **43,** 175 (1981).
[39] G. M. Clore and A. M. Gronenborn, *J. Magn. Reson.* **53,** 423 (1983).

due to direct dipolar interactions or the result of aggressive spin diffusion (although see Borgias and James [9], Vol. 176, this series).

Application of the appropriate handling of the experimental data will produce for Type I systems two sets of data, parameters for the free (unbound) ligand[40] and the corresponding information for the bound ligand. In addition to the fluorine chemical shift of the ligand in the bound state, the spin–spin relaxation time for this fluorine, possibly some scalar coupling information, the spin–lattice relaxation time, fluorine–proton NOE data, and the rate of exchange of ligand between the free and the bound states may also be obtained.

Type II Systems

Exchange rate complications are absent in Type II experiments and one may determine by direct observation the relaxation parameters, enzyme-induced chemical shift, and Overhauser effects. Prior to interpretation of these data, knowledge of the stoichiometry of the reaction that linked the fluorinated group to the protein must be obtained and a demonstration of the specificity of the reaction for a site on the protein is also in order. (These are the requirements for any "reporter group" experiment.) Methods for design of reagents for Type II experiments and for determination of the stoichiometry and specificity of the modification reactions have been discussed.[41-43] Various mechanism based active-site-directed reactions would appear to be ideal for the generation of useful Type II systems.[44-46]

Design of the Fluorinated Ligand

Materials for Type I experiments with enzymes can be designed with the structures of the enzyme's substrate(s),[46,47] product(s), cofactors, and

---

[40] It should be noted that parameters obtained for the "free" ligand signal in the presence of enzyme may or may not be equal to those obtained in the absence of protein. Weak, nonspecific interactions with nondescript sites on the enzyme other than the active site may lead to relaxation effects and small protein-induced chemical shift effects.

[41] R. L. Lundblad and C. M. Noyes, "Chemical Reagents for Protein Modification," Vols. 1 and 2. CRC Press, Boca Raton, Florida, 1984.

[42] G. E. Means and R. E. Feeney, "Chemical Modification of Proteins." Holden-Day, San Francisco, California, 1971.

[43] J. Eyzaguirre (ed.), "Chemical Modification of Enzymes: Active Site Studies." Halsted, New York, 1987.

[44] C. Walsh, *Tetrahedron* **38,** 871 (1982).

[45] C. T. Walsh, *Adv. Enzymol.* **55,** 197 (1983).

[46] R. H. Abeles, *Drug Dev. Res.* **10,** 221 (1987).

[47] J. E. Leffler and E. Grunwald, "Rates and Equilibria in Organic Reactions," p. 173. Wiley, New York, 1963.

known inhibitors in mind. Substitution of one or more hydrogens in these structures by fluorine can be considered in light of the expected effects of this substitution on the stability, reactivity, or interactivity of the resultant molecule. A common target is the aromatic ring, where a hydrogen can be replaced by a single fluorine or a trifluoromethyl group, but monofluoro or difluoro substitution on an aliphatic side chain is also a possibility. The CF_3 group is an enticing substituent because the three equivalent fluorines give rise to a sharp signal that is not complicated by large proton–fluorine couplings and, because rotation about the C_3 axis of the group is rapid even when attached to a macromolecule, the signal stays reasonably sharp under conditions that extensively broaden other fluorine signals. However, the trifluoromethyl group is quite bulky and has a strong electron-withdrawing effect, comparable to a cyano group in aromatic systems.[47]

It is desirable to consider symmetry in design of the reporter molecule. For example, placement of a fluorine or trifluoromethyl group at a position ortho or meta to the group that will form the covalent bond with the enzyme, or will be a significant reference point in determining how the ligand will bind to the enzyme, may make it possible for the fluorinated molecule to bind to the enzyme in two ways which differ by a 180° rotation of the aromatic ring. Placement of the fluorinated substituent at the para position will vitiate this difficulty, since ring rotation will not change the position of the fluorine atom.

Table I indicates some fluorinated molecules that have been used in Type I and Type II experiments. The experiments summarized there, while certainly not exhausting the possibilities, suggest various approaches that have been used to produce fluorinated species that are appropriate for NMR experiments.

Doubtless because of the interest of pharmaceutical, agricultural, and polymer chemical firms in the often unusual effects of fluorine substitution on the properties of their products, there is a surprising variety of fluorinated organic compounds available from the standard supply houses; these compounds can serve as precursors to molecules useful in the types of experiments discussed above. There is also a very large literature relating to the synthesis of fluorine-containing structures.[48-50]

Interpretation of T_1 and T_2 Data

Bothner-By has introduced the concept of spin "islands," groups of spins within a protein that are sufficiently isolated from the other spins

[48] J. T. Welch, *Tetrahedron* **43**, 3123 (1987).
[49] D. Bergstrom, E. Romo, and P. Shum, *Nucleoside Nucleotides* **6**, 53 (1987).
[50] M. Schlosser, *Tetrahedron* **34**, 3 (1978).

TABLE I
FLUORINATED MOLECULES USED IN NMR EXPERIMENTS

Structure	Use	Reference(s)[a]
Type I experiments		

| | Pyridoxal analog | 1 |

| | Glucose analogs, useful in metabolism studies | 2, 3 |

| Guanosine·O-P-O-P-O-P-F | GTP analog | 4 |

| | Studies of tryptophan synthase | 5 |

| | Carbonate dehydratase inhibitor | 6 |

Type II experiments

| | Sulfhydryl reagent | 7 |

(continued)

TABLE I (continued)

Structure	Use	Reference(s)[a]
	Sulfhydryl reagent	8
	Sulfhydryl reagent	9
	Transition-state analog inhibitor of chymotrypsin	10
	Substrate analog, gives stable acyl chymotrypsin	11
	Inactivator of chymotrypsin	12
	Affords stable acyl chymotrypsin	13

[a] References: (1) Y. C. Chang, R. D. Scott, and D. J. Graves, *Biochemistry* **26,** 360 (1987); (2) B. A. Berkowitz and J. J. H. Ackerman, *Biophys. J.* **51,** 681 (1987); (3) I. L. Kwee, T. Nakada, and P. J. Card, *J. Neurochem.* **40,** 428 (1987); (4) O. Monasterio, *Biochemistry* **26,** 6099 (1987); (5) E. W. Miles, R. S. Phillips, H. J. C. Yeh, and L. A. Cohen, *Biochemistry* **25,** 4240 (1986); (6) J. T. Gerig and J. M. Moses, *J. Chem. Soc., Chem. Commun.,* 482 (1987); (7) L. E. Kay, J. M. Pascone, B. D. Sykes, and J. M. Shriver, *J. Biol. Chem.* **262,** 1984 (1987); (8) M. Brauer and B. D. Sykes, *Biochemistry* **25,** 2187 (1986); (9) J. P. Caradonna, E. W. Harlan, and R. H. Holm, *J. Am. Chem. Soc.* **108,** 7856 (1986); (10) B. Imperiali and R. H. Abeles, *Biochemistry* **25,** 3760 (1986); (11) J. T. Gerig and S. J. Hammond, *J. Am. Chem. Soc.* **106,** 8244 (1984); (12) H. Tsunematsu, H. Nishikawa, and L. J. Berliner, *J. Biochem. (Tokyo)* **96,** 349 (1984); (13) J. W. Amshey and M. L. Bender, *Arch. Biochem. Biophys.* **224,** 378 (1983).

such that the interactions within the group are of primary importance in defining their mutual T_1 and T_2 relaxation.[51] Hull and Sykes[52] have discussed the relaxation behavior expected for a fluorine nucleus within a protein. Flourine spins in proteins are surrounded by protons—protons attached to the ligand that also hold the fluorine, protons that are part of the structures of nearby amino acids, and, finally, protons that are part of or in exchange with the protons of the solvent. The effects of the last class of spins can be essentially eliminated by replacing the solvent with D_2O, but there will be some uncertainty generated regarding possible solvent isotope effects on protein structure.[53-57] In the simplest treatment of relaxation, the effects of spin coupling are neglected and the relaxation behavior of the spins within an island are described by a system of equations analogous to those discussed by Borgias and James [9] (Vol. 176, this series). In addition to contribution of the (dominant) fluorine–proton dipolar interactions to relaxation, it is necessary to include in such treatments the contribution that the CSA mechanism makes to the fluorine relaxation, if one is considering T_2 relaxation at any frequency and also T_1 relaxation at frequencies above 56 MHz. The analysis assumes a three-dimensional model for the spins in the island and a model for the dynamic behavior of the island as it resides in the enzyme.[58] The simplest model for the motion is one in which the collection of spins maintains fixed distances relative to one another as the entire collection tumbles isotropically in a way that can be characterized by the single-rotation correlation time τ_c. This model can be made more realistic by introduction of internal rotations and other motions that modulate the internuclear distances.[25,59-61]

Solution of the relaxation equations predict that T_1 relaxation of an n-spin system of fluorine and hydrogen nuclei should be an n-exponential process, but it is rare to observe nonexponential relaxation experimen-

[51] A. A. Bothner-By, in "Biological Applications of Magnetic Resonance" (R. G. Shulman, ed.), p. 177. Academic Press, New York, 1979. This chapter provides an excellent summary of NOE experiments with both Type I and Type II systems.
[52] W. E. Hull and B. D. Sykes, J. Chem. Phys. 63, 867 (1975).
[53] P. G. Pradhan and G. B. Nadkarni, Biochim. Biophys. Acta 615, 474 (1980).
[54] D. M. Franz and R. W. Voss, Jr., Fed. Proc., Fed. Am. Soc. Exp. Biol. 39, 363 (1980).
[55] C. L. Schauf and J. O. Bullock, Biophys. J. 30, 295 (1980).
[56] D. L. Erbes, R. H. Burris, and W. H. Orme-Johnson, Proc. Natl. Acad. Sci. U.S.A. 72, 4795 (1975).
[57] R. K. Gupta and A. S. Mildvan, J. Biol. Chem. 250, 246 (1975).
[58] J. T. Gerig, D. T. Loehr, K. F. S. Luk, and D. C. Roe, J. Am. Chem. Soc. 101, 7482 (1979).
[59] G. Lipari and A. Szabo, J. Am. Chem. Soc. 104, 4546 (1982).
[60] R. E. D. McClung and B. K. John, J. Magn. Reson. 50, 267 (1982).
[61] R. Rowan III, J. A. McCammon, and B. D. Sykes, J. Am. Chem. Soc. 96, 4773 (1974).

tally for fluorine that is bound to an enzyme. A logarithmic plot of fluorine T_1 data generally has a slope equivalent to the *initial* value for the slope expected if the fluorine is selectively inverted in the T_1 determination.

We and others have observed that models for the dynamics at a ligand binding site that are sufficient to explain T_1 and $^{19}F\{^1H\}$ NOE data often predict $1/T_2$ values or linewidths ($W_{1/2} = 1/\pi T_2$) that are significantly smaller than those determined experimentally.[62–65] The additional line broadening is presently thought to be the result of exchange between several microheterogeneous conformational states of the ligand–enzyme structure such that the fluorine nucleus experiences several rather different chemical shifts and these are not completely averaged by the exchange process that interconverts the microstates. Further investigation of this idea is warranted.

To further complicate things, it is possible that the dissociation process of the group F from the enzyme in Type I experiments may be more complex than is indicated by Eq. (1), perhaps because the process involves several enzyme-bound states that are appreciably populated and interconvertible.[66] What will then be observed in the (somehow averaged) fluorine resonance will reflect not only the properties of the free ligand and the ligand in its finally bound state, but also the properties of the intermediates.

Interpretation of Chemical Shift Data

Fluorine chemical shifts cover a very wide range, and a reaction of a fluorinated ligand with a group on an enzyme, such that there is a change in the covalent bond structure of the ligand, is usually made quite apparent by a dramatic change in the fluorine spectrum. However, simply interacting with a protein, with no change in covalent bonds, is sufficient to generate an appreciable change in shift. We will refer to this latter kind of shift as an enzyme- or protein-induced shift; such a shift must arise because of the change in environment experienced by the fluorine of the ligand as it leaves the aqueous phase and enters the enzyme binding site. The range of protein-induced fluorine shifts that have been observed to date is about 15 ppm. Usually this effect is downfield, sometimes by as much as 9 ppm, but there are examples where interaction with a protein produces a large upfield shift of a fluorine signal. A reliable interpretation

[62] B. J. Kimber, D. V. Griffiths, B. Birdsall, R. W. King, P. Scudder, J. Feeney, G. C. K. Roberts, and A. S. V. Burgen, *Biochemistry* **16**, 3492 (1977).

[63] J. F. M. Post, P. F. Cottam, V. Simplaceanu, and C. Ho, *J. Mol. Biol.* **179**, 729 (1984).

[64] J. T. Gerig and S. J. Hammond, *J. Am. Chem. Soc.* **106**, 8244 (1984).

[65] M. E. Ando, J. T. Gerig, and K. F. S. Luk, *Biochemistry* **25**, 4772 (1986).

[66] L. B. Dugad and J. T. Gerig, *Biochemistry* **27**, 4310–4316 (1987).

of these shifts could provide some insight into the interactions between fluorine of the ligand and structural features of the enzyme.

Several groups have discussed fluorine shielding effects in proteins.[67,68] The enzyme-induced shift effect can be the result of a change in electrostatic interactions, van der Waals interactions, electric fields generated by surrounding dipoles, and specific hydrogen bonding, or can arise from nearby magnetically anisotropic structures in the protein such as carbonyl groups and aromatic rings. The ring current effect from the side chains of phenylalanine, tyrosine, tryptophan, and histidine can perhaps be as large as ±2 ppm,[69] and is an important shift determinant of protein-induced *proton* shifts in proteins. Ring current effects (in parts per million) will be precisely the same for a fluorine nucleus at a given position in the tertiary structure of an enzyme as for a proton or other nucleus at the same location. The magnetic anisotropies of other groups in a protein are small and estimable by standard methods.[70] Collectively, the magnetic anisotropies of the structural components of proteins alone are not sufficient to account for the range of protein-induced fluorine shifts that have been observed.

While there is some evidence for covalent fluorine acting as a hydrogen bond acceptor, it is not clear how strong this interaction would be in a protein nor how large its effect on fluorine shifts might be. This interaction should be deshielding, but in studies of fluorinated compounds in hydroxylic solvents there have been no indications of shift effects large enough to be consistent with the magnitude of protein-induced shift effects. It is possible that the tertiary structure of a protein could enforce a hydrogen-bonding interaction to fluorine in some way to create a shift effect larger than that observed in solution, but this has not been demonstrated.

Theoretical considerations indicate that changes in van der Waals interactions that accompany the binding of a ligand to an enzyme can account for the experimental range of protein-induced fluorine shifts that have been observed.[71] These interactions are a strong function of internuclear distances, with nearest-neighbor interactions being the only ones of significance. If van der Waals interactions with fluorine are stronger within the enzyme–ligand complex than they are in the solvent, then a

[67] F. Millett and M. A. Raftery, *Biochem. Biophys. Res. Commun.* **47**, 625 (1972).
[68] B. J. Kimber, J. Feeney, G. C. K. Roberts, B. Birdsall, D. V. Griffiths, A. S. V. Burgen, and B. D. Sykes, *Nature (London)* **271**, 184 (1978).
[69] S. J. Perkins, *Biol. Magn. Reson.* **4**, 193 (1982).
[70] L. M. Jackman and S. Sternhell, "Applications of NMR Spectroscopy in Organic Chemistry," pp. 72–98. Academic Press, New York, 1969.
[71] F. H. A. Rummens, *NMR: Basic Princ. Prog.* **10**, 1 (1975).

downfield shift will be the result, while a weaker set of interactions in the bound state than are present in the solvent would produce an upfield shift. Certainly van der Waals interactions usually play a role in favoring the binding of a ligand, so the typically observed downfield fluorine shift on binding of a fluorinated ligand to a protein is consistent with this notion of the origin of the shift effect. It has been suggested in the literature that observation of an upfield fluorine shift in these situations is indicative of a ring current effect on the bound fluorine, but this conclusion cannot be supported solely by the direction of the shift. A quantitative treatment of van der Waals interactions in real (dynamic) molecules does not seem to be available, but investigations along these lines might provide a way to interpret at least roughly an observed protein-induced fluorine shift.

Consideration of electric field effects on fluorine shifts have produced mathematical results that are essentially identical in form to those that describe the van der Waals effect.[62] Thus, it would be difficult experimentally to determine the separate roles of these two effects in determining fluorine shifts.

Fluorine chemical shifts in enzyme systems may be expected to respond in some way to variation of solution variables, including temperature and pH. The ranges of these changes are usually considerably larger than those observed for proton shifts in the same kind of experiment, but this is not universally so. The direction of a fluorine shift change in a titration experiment may be opposite that observed in the corresponding proton titration,[67] pointing up again that the interactions responsible for shift changes in proton and fluorine spectroscopy are not the same.

Fluorine chemical shifts are temperature dependent, typically shifting linearly downfield 0.001–0.01 ppm/°C.[72] A temperature variation larger than this, or taking place in a significantly nonlinear manner, likely indicates the presence of a conformational change as the temperature is varied or the intrusion of an exchange process.

Fluorine chemical shifts are subject to a solvent isotope effect when the solvent is changed from H_2O to D_2O, typically 0.2–0.3 ppm upfield.[73,74] This effect has been used to indicate whether or not a fluorine within a protein–ligand complex is exposed to solvent[75]; an upfield shift of this magnitude when making this solvent change would indicate exposure of the enzyme-bound fluorine to solvent or solvent-exchangeable protons. Detection of the isotope shift effect may be challenging if the linewidth of

[72] C. J. Jameson, A. K. Jameson, and S. M. Cohen, *J. Chem. Phys.* **67**, 2771 (1977).
[73] I. B. Golovanov, V. Gagloev, I. A. Soboleva, and V. Smolyaninov, *J. Gen. Chem. USSR* (*Engl. Transl.*) **43**, 642 and 905 (1973).
[74] W. E. Hull and B. D. Sykes, *Biochemistry* **15**, 1535 (1976).
[75] P. E. Hansen, H. D. Dettman, and B. D. Sykes, *J. Magn. Reson.* **62**, 487 (1985).

the signal being observed is large (or difficult to interpret, if it is real) but is significantly less than the magnitude indicated earlier. It should also be kept in mind that the pK_a values of ionizable groups can be altered by change in the solvent isotopic composition and that the shift effect assigned to this change may actually be a titration shift. Finally, hydrophobic interactions (and presumably other interactions that define the binding energy for a ligand) are subject to an isotope effect[76] so that an observed shift change in a Type I experiment may be a reflection of a change in the fraction of ligand bound to the enzyme.

Flourine chemical shifts are also sensitive to substitution of protium by deuterium within the ligand.[77] Thus, deuteration of the ligand structure, perhaps in an effort to remove some fluorine relaxation pathway, can produce a shift of the fluorine resonance position.

Consequences of Fluorine Substitution

Alterations in reactivity in biological systems that result from fluorine substitution have been of continuing interest.[46,78] In addition to changes in electronic structure that fluorine substitution produces in a covalent molecule (inductive and resonance effects), with the attendant changes in reactivity, replacement of a carbon–hydrogen bond by a carbon–fluorine bond likely introduces a new set of nonbonded interactions, and these may have nonnegligible consequences on local and global protein structure. Murray-Rust et al.[78] have examined the crystal structures of a large number of fluorinated compounds, attempting to identify general interactions of covalent fluorine that could have structural implications in proteins. They concluded that substantial interactions of the C–F bond with alkali metal cations are detectable when such ions are present. Whether or not covalent fluorine can act as a hydrogen bond acceptor was also examined by Murray-Rust et al.[78]; it was concluded that the fluorine in a C–F bond could act as a weak proton acceptor in a hydrogen-bonding situation. Although such interactions apparently are energetically favorable, they usually form in the context of other, stronger hydrogen bonding and dipolar interactions and, thus, produce effects which are difficult to assign unambiguously to a specific interaction.

Replacement of C–H by C–F generally makes a molecule more hydrophobic and one can anticipate that fluorine substitution will increase the strength of hydrophobic interactions within an enzyme–ligand complex.

[76] D. Oakenfull and D. E. Fenwick, *Aust. J. Chem.* **28,** 715 (1975).
[77] J. B. Lambert and L. G. Greifenstein, *J. Am. Chem. Soc.* **95,** 6150 (1973).
[78] P. Murray-Rust, W. C. Stallings, C. T. Monti, R. K. Preston, and J. P. Glusker, *J. Am. Chem. Soc.* **105,** 3206 (1983).

For example, the Hansch π constant for a fluorophenyl ring is 0.15 larger than the value for the unsubstituted system, consistent with a somewhat greater hydrophobicity for the fluorinated structure.[79] All other effects neglected, this substitution would produce a decrease in the binding constant for a small molecule interacting hydrophobically with a protein.

Trifluoromethyl substitution strongly enhances lipophilicity, although the magnitude of the effect depends on the position of the CF_3 group in the molecule.[80]

Clearly fluorine substitution has the *potential* to change the structures of proteins or enzyme–ligand complexes, but are such changes detectable? Evidence on this point is slowly becoming available. 4-Fluorohistidine has been placed in the His-12 position of the S-peptide of ribonuclease and the structure of its complex with the S-protein examined crystallographically. At the limits of the resolution of the experiment (0.26 nm) there was no change in the tertiary structure of the protein, even though the fluorinated imidazole ring has a much lower pK_a than does the nonfluorinated analog.[81] RNase S' with a 4-fluorophenylalanine replacing Phe-8 has also been prepared. In this case there was no significant effect of the fluorine substitution on the binding or reactivity of substrate, although there was a significant enzyme-induced chemical shift effect.[82] Kimber *et al.*[62] have prepared an enzyme in which the five tryptophan residues were replaced by 5-fluorotryptophan. An unusual fluorine–fluorine spin coupling was observed between two of these, a result that could have obtained only if the fluorine nuclei were essentially within van der Waals contact of each other.[83] On the basis of initial X-ray results, Matthews[84] concluded that such a contact was inconsistent with the crystallographic results and, thus, the structure of the protein had been altered by the fluorine substitution. However, further refinement of the X-ray data[85] produced a structure in which fluorine–fluorine interactions were entirely consistent with the NMR observations.[86] Arseniev *et al.*[87] have shown that the appearance of the fluorine NMR spectrum of bacteriorhodopsin

[79] C. Hansch and A. Leo, "Substituent Constants for Correlation Analysis in Chemistry and Biology," p. 15. Wiley (Interscience), New York, 1979.

[80] N. Muller, *J. Pharm. Sci.* **75**, 987 (1986).

[81] H. C. Taylor, D. C. Richardson, J. S. Richardson, A. Wlodower, A. Komoriya, and I. M. Chaiken, *J. Mol. Biol.* **149**, 313 (1981).

[82] I. M. Chaiken, M. H. Freedman, J. R. Lyerla, Jr., and J. S. Cohen, *J. Biol. Chem.* **248**, 884 (1973).

[83] F. B. Mallory and C. W. Mallory, *J. Am. Chem. Soc.* **107**, 4816 (1985).

[84] D. A. Matthews, *Biochemistry* **18**, 1602 (1979).

[85] J. T. Bolin, D. J. Filman, D. A. Matthews, R. C. Hamlin, and J. Kraut, *J. Biol. Chem.* **257**, 13650 (1982).

into which varying levels of 5-fluorotryptophan had been incorporated was independent of the extent of incorporation of this fluorinated amino acid. Given the high sensitivity of fluorine shifts to microenvironment, this result suggests that fluorine substitution does not have an appreciable influence on the tertiary structure of this protein in solution. Finally, there are many systems involving Type I interaction, with an enzyme in which fluorine substitution seems not to have an effect on the thermodynamics of the interaction that is much different than that expected on the basis of the hydrophobic propensities of the fluorinated group.[88] Collectively, these observations indicate that placement of fluorine in an enzyme system may not produce significant structural perturbations.

However, there are also observations that point to a contrary conclusion. Both porcine and human leukocyte elastase exhibit high affinity for trifluoroacetylated peptides, an affinity that can be orders of magnitude greater than that for the corresponding acetylated materials.[89] NMR studies, interpreted in conjunction with X-ray results, suggested a specific binding site for the CF_3CO group, which is near the active site of the enzyme,[90] wherein fluorine–enzyme interactions develop that are not present when the corresponding CH_3CO derivatives bind. Interestingly, chloromethyl ketones derived from the CF_3CO peptide derivatives bind more strongly than do the corresponding CH_3CO compounds, but it is the latter that react most rapidly to give covalently modified enzymes,[91] not because of any electronic effects of the CF_3 group but because the complexes formed with the trifluoroacetylated inhibitors are nonproductive. That is, the orientation of the ligand at the active site that is forced by the fluorine–enzyme interactions puts the chloromethyl ketone group into an orientation that is unfavorable for subsequent covalent bond formation.

Semisynthetic analogs of horse heart cytochrome c in which Tyr-67 is replaced by either phenylalanine or 4-fluorophenylalanine have been prepared.[92] With the phenylalanine substituent, the protein retains 56% of its

[86] S. J. Hammond, National Institute of Magnetic Resonance, Mill Hill, London (unpublished observations).

[87] A. S. Arseniev, A. B. Kuryatov, V. I. Tsetlin, V. F. Bystrov, V. T. Ivanov, and Y. A. Ovchinnikov, *FEBS Lett.* **213**, 283 (1987).

[88] H. R. Bosshard and A. Berger, *Biochemistry* **13**, 266 (1974).

[89] P. Lestienne, J.-L. Dimicoli, and J. Bieth, *J. Biol. Chem.* **253**, 3459 (1978); see also P. Lestienne, J.-L. Dimicoli, and J. Bieth, *J. Biol. Chem.* **252**, 5931 (1977).

[90] J.-L. Dimicoli and J. Bieth, *Biochemistry* **16**, 5532 (1977).

[91] P. Lestienne, J.-L. Dimicoli, A. Renaud, and J. G. Bieth, *J. Biol. Chem.* **254**, 5219 (1979).

[92] A. K. Koul, G. F. Wasserman, and P. K. Warme, *Biochem. Biophys. Res. Commun.* **89**, 1253 (1979).

activity in an oxidase assay but with 4-fluorophenylalanine at this position activity is completely lost. Thus, there is a requirement for a "normal" aromatic ring in this position for electron transfer in cytochrome c, and the fluorinated analog, either because of its electronic nature or because of perturbation of local structure by the fluorine, is unable to support this process.

Binding of the ligand 3',5'-difluoromethotrexate to dihydrofolate reductase from *Lactobacillus casei* produced a complex that is presumably important to the cytotoxic activity of this material. The difluorobenzoyl ring of the inhibitor was found to have an appreciable barrier to rotation about its symmetry axis.[93] A number of interactions with amino acids of the enzyme were identified as probably being responsible for the barrier in this early work; further refinement of these conclusions became possible after structural studies of the complex by means of two-dimensional proton NMR techniques. These latter studies demonstrated that the difluorophenyl ring is likely tilted about 15° away from the orientation taken by the corresponding all-proton system.[94]

There are a number of signals that appear to high field of TMS in the ^{1}H NMR spectrum of human carbonate dehydratase I and II. The signals presumably arise from methyl groups that are juxtaposed to aromatic rings in the tertiary structure, and calculations show that several of the methyl groups in the active site of the enzyme could give signals in this region. When sulfonamide inhibitors bind to the enzyme they coordinate to the zinc atom at the active site; the upfield ^{1}H NMR spectrum of these complexes is different from the spectrum of the native enzyme, suggesting that a conformational change accompanies binding. The spectral changes brought about by binding of 4-fluorobenzenesulfonamide are different than those produced when benzenesulfonamide is bound.[66] Thus, a single fluorine substitution in this complex is sufficient to alter the structure of this enzyme–ligand complex in a detectable way. The extent of this change remains to be demonstrated.

No overarching conclusions regarding the effects of fluorine substitution in enzyme–ligand systems are possible at this time. There is sufficient cause, however, to be cautious when interpreting results. One can hope that in the future our abilities to predict the occurrence and impacts of interaction of the C–F group with proteins will improve, along with the ability to predict and interpret fluorine chemical shifts in these systems.

[93] G. M. Clore, A. M. Gronenborn, B. Birdsall, J. Feeney, and G. C. K. Roberts, *Biochem. J.* **217**, 659 (1984).
[94] S. J. Hammond, B. Birdsall, J. Feeney, M. S. Searle, G. C. K. Roberts, and H. T. A. Cheung, *Biochemistry* **26**, 8585 (1987).

Like all reporter group experiments, NMR studies of fluorinated ligands in enzyme structures offer a number of operational advantages and can produce results indicative of binding stoichiometry, ionization behavior, enzyme dynamics, and conformational changes. And, like all experimental results, analysis of observations made using this technique should be undertaken with awareness of those complications which preclude facile conclusions.

Acknowledgments

The comments made above have benefited from the collective wisdom and experience of my associates over the past 20 years. Their dedication, experimental skill, and insights are acknowledged with appreciation and gratitude.

[2] Deuteration in Protein Proton Magnetic Resonance

By DAVID M. LeMASTER

Deuteration techniques in protein NMR serve as a means of modifying ^{1}H spectra for improved resolution and assignment purposes as well as providing increased sensitivity for direct ^{2}H observation (see Keniry [19], Vol. 176, this series). There are three general patterns of deuterium labeling that in practice serve rather different functions: selective protonation in a deutero background, selective deuteration in a protio background, and random fractional deuteration in which all carbon-bound hydrogen positions have been uniformly enriched to an intermediate level with deuterium.

Selective Protonation

Selective protonation was first developed in the late 1960s as a means of attempting to observe individual resolved proton resonances in protein one-dimensional (1D) NMR spectra, which at natural abundance gave rise to rather featureless envelopes of severely overlapped aromatic and aliphatic resonances. Crespi et al.[1] approached the labeling process mainly by providing individual protonated amino acids to autotrophic algae grown on an otherwise perdeuterated medium in order to study

[1] H. L. Crespi, R. M. Rosenberg, and J. J. Katz, *Science* **161**, 795 (1968).

various small algal proteins. In contrast, Markley et al.[2] grew the hetero-trophic bacterium *Staphylococcus aureus* on deuterated algal protein hy-drolysate in D_2O supplemented with a small subset of protonated amino acids. In their subsequent study of the labeled staphylococcal nuclease, they succeeded in observing individual resolved proton resonances. The efforts of both of these groups have been reviewed in full earlier in this series.[3,4]

In addition to the elimination of overlapping resonances, generation of a deuterated background serves to improve the resolution of residual proton resonances for other reasons. The relaxation rate, and hence the natural linewidth, of protons in proteins is determined primarily by $^1H-^1H$ dipolar interaction. By diluting out the spatially adjacent protons with deuterium, the linewidths of the remaining protons will be decreased. This effect was anticipated by Crespi and co-workers[5] in the study of algal flavoprotein fully deuterated at all carbon-bound sites. The amide protons back-exchanged into this sample gave rise to resonances that were ap-proximately a factor of two narrower than was seen in the natural abun-dance sample. This made it possible to measure exchange rates for a large number of individual amide proton resonances in algal ferredoxin at 220 MHz.[6] Although the protonation experiments are not strictly selective in the sense used here, other studies have nevertheless made use of the decreased $^1H-^1H$ dipolar interaction obtained by general deuteration. Bösch et al.[7] observed that their NOESY data for micelle-bound glucagon were significantly improved by the use of deuterated detergent to reduce tertiary dipolar interactions (e.g., spin diffusion). Similarly, spectral edit-ing and decreased dipolar interaction can be obtained in ligand binding studies as observed in the binding of [^1H]mellitin to perdeuterated calmo-dulin.[8]

Selective Deuteration

Although in principle the technique of selective deuteration blurs into the technique of selective protonation as the level of labeling increases, in

[2] J. L. Markley, I. Putter, and O. Jardetzky, *Science* **161,** 1249 (1968).
[3] J. L. Markley, this series, Vol. 26, p. 605.
[4] H. L. Crespi and J. J. Katz, this series, Vol. 26, p. 627.
[5] H. L. Crespi, J. R. Norris, J. P. Bays, and J. J. Katz, *Ann. N.Y. Acad. Sci.* **222,** 800 (1973).
[6] H. L. Crespi, A. G. Kostka, and U. H. Smith, *Biochem. Biophys. Res. Commun.* **61,** 1407 (1974).
[7] C. Bösch, L. R. Brown, and K. Wüthrich, *Biochim. Biophys. Acta* **603,** 298 (1980).
[8] S. H. Seeholzer, M. Cohn, J. A. Putkey, A. R. Means, and H. L. Crespi, *Proc. Natl. Acad. Sci. U.S.A.* **83,** 3634 (1986).

practice the two techniques are generally distinct. In most cases selective protonation has been used for situations in which the resonances of interest are largely or entirely obscured by overlapping resonances, requiring the wholesale elimination of the background resonances in order to obtain adequate resolution. In contrast, selective deuteration has been generally used to provide assignment information when at least partial resolution is obtained in natural abundance spectra. A familiar example is that of the base-catalyzed deuterium exchange of histidine C-2 ring protons.[9] For proteins which undergo reversible acid denaturation, exchange of the C-2 ring proton of tryptophan is feasible as well.[10] As higher field instruments have become available, the aromatic regions of small to moderate-size proteins have become amenable to analysis by 1D techniques. Several studies have utilized deuterated[11,12] aromatic amino acids as a means of obtaining increased resolution and assignment information on the labeled protein samples. In favorable cases the use of selectively deuterated aromatics has resulted in observation and assignment of resonances in DNA–protein complexes of ~200–400 kDa[13] (G. C. King, personal communication). In a similar spirit the upfield methyl region has also been edited by selective deuteration.[14,15]

Although selective deuteration has most often been applied in cases in which the resonances of interest are observed as resolved peaks, the use of carefully matched samples of natural abundance and selectively deuterated proteins can yield difference spectra, as elegantly illustrated by McConnell and co-workers[16–20] in their combination of selective deuteration and spin labeling for the structural determination of antigen-binding sites in monoclonal immunoglobulin Fab fragments.

[9] D. H. Meadows, O. Jardetzky, R. M. Epand, H. H. Rüterjans, and H. A. Scheraga, *Proc. Natl. Acad. Sci. U.S.A.* **60**, 766 (1968).
[10] J. H. Bradbury and R. S. Norton, *Mol. Cell. Biochem.* **13**, 113 (1976).
[11] J. Feeney, G. C. K. Roberts, B. Birdsall, D. V. Griffiths, R. W. King, P. Scudder, and A. S. V. Burgen, *Proc. R. Soc. London, Ser. B* **196**, 267 (1977).
[12] K. S. Matthews, N. G. Wade-Jardetzky, M. Graber, W. W. Conover, and O. Jardetzky, *Biochim. Biophys. Acta* **490**, 534 (1977).
[13] G. C. King and J. E. Coleman, *Biophys. J.* **51**, 152a (1987).
[14] B. Birdsall, J. Feeney, D. V. Griffiths, S. Hammond, B. Kimber, R. W. King, G. C. K. Roberts, and M. Searle, *FEBS Lett.* **175**, 364 (1984).
[15] M. S. Searle, S. J. Hammond, B. Birdsall, G. C. K. Roberts, J. Feeney, R. W. King, and D. V. Griffiths, *FEBS Lett.* **194**, 165 (1986).
[16] J. Anglister, T. Frey, and H. M. McConnell, *Biochemistry* **23**, 1138 (1984).
[17] J. Anglister, T. Frey, and H. M. McConnell, *Biochemistry* **23**, 5372 (1984).
[18] T. Frey, J. Anglister, and H. M. McConnell, *Biochemistry* **23**, 6470 (1984).
[19] J. Anglister, T. Frey, and H. M. McConnell, *Nature (London)* **315**, 65 (1985).
[20] J. Anglister, M. W. Bond, T. Frey, D. Leahy, M. Levitt, H. M. McConnell, G. S. Rule, J. Tomasello, and M. Whittaker, *Biochemistry* **26**, 6058 (1987).

Selective Deuteration in Two-Dimensional NMR. The value of selective deuteration has increased significantly with the advent of two-dimensional (2D) techniques. It is now possible to resolve a great majority of the cross-peaks observed for small proteins. Selective deuteration provides a facile means of obtaining residue-type assignment as well as side-chain spin-coupling connectivity information. This technique has been used extensively in the study of the 108-residue protein *Escherichia coli* thioredoxin, for which two-thirds of the residues were assigned by residue type using both individual and multiresidue-type deuterated samples, as illustrated in Fig. 1.[21] Shown are phase-sensitive COSY spectra of the intra-residue α-amide coupling domain for [U-²H]aspartic acid-enriched (Fig. 1A) and [α-²H]serine-enriched (Fig. 1B) samples. The aspartic acid and asparagine cross-peaks observed in Fig. 1B are selectively eliminated in Fig. 1A, while the reverse is true for the serine cross-peaks. In addition, one of the two nonequivalent α-amide cross-peaks for the glycine residues is eliminated in the deuterated serine spectrum as a result of the stereospecific transhydroxymethylation which occurs during biosynthetic conversion of serine to glycine.

The use of deuteration to provide stereoselective assignments can be readily extended to the case of β-methylene protons, as has been carried out in a number of amino acid and peptide studies.[22-25] Stereoselective determination is equally difficult in the case of proteins in the absence of labeling techniques.[26] *Escherichia coli* thioredoxin samples have been produced in which one of the two β protons of the aspartic acid and asparagine residues has been selectively deuterated.[27] Figure 2 shows the β-amide NOESY spectral region of the two chirally deuterated samples as well as a reference spectrum. In addition to providing assignment information, the stereospecific labeling renders it feasible to determine the χ_1 side-chain dihedral angle around the αβ bond for nearly all labeled residues using coupling-constant and NOE intensity data. This should prove particularly crucial for structural analysis studies. Although NMR has been demonstrated to be capable of determining protein main-chain conformations to a moderate resolution, side-chain conformations are gener-

[21] D. M. LeMaster and F. M. Richards, *Biochemistry* **27**, 142 (1988).
[22] M. Kainosho and K. Ajisaka, *J. Am. Chem. Soc.* **97**, 5630 (1975).
[23] A. J. Fischman, D. H. Live, H. R. Wyssbrod, W. C. Agosta, and D. Cowburn, *J. Am. Chem. Soc.* **102**, 2533 (1980).
[24] J. Kobayashi and U. Nagai, *Biopolymers* **17**, 2265 (1978).
[25] H. Kessler, C. Griesinger, and K. Wagner, *J. Am. Chem. Soc.* **109**, 6927 (1987).
[26] S. G. Hyberts, W. Märki, and G. Wagner, *Eur. J. Biochem.* **164**, 625 (1987).
[27] D. M. LeMaster, *FEBS Lett.* **223**, 191 (1987).

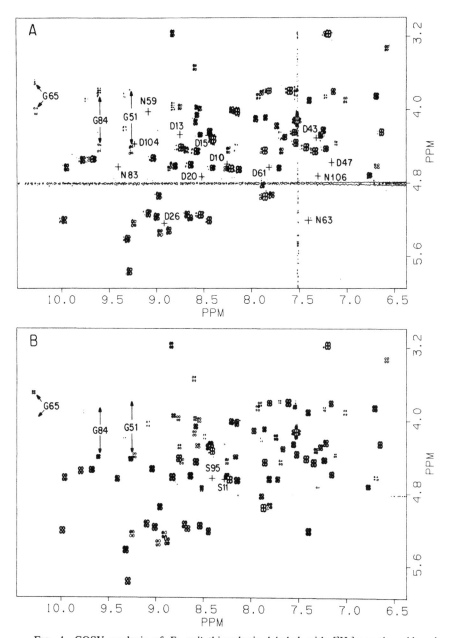

FIG. 1. COSY analysis of *E. coli* thioredoxin labeled with [²H₃]aspartic acid and [²Hα]serine. In the phase-sensitive spectra of the α-amide region, crosses mark positions in which cross-peaks have been eliminated by selective deuteration of aspartic acid (A) and serine (B). The peaks missing in A can be seen in B and vice versa. The glycine biosynthetically derived from [²Hα]serine is stereospecifically deuterated in the *R* position, resulting in elimination of passive coupling and hence signal enhancement of the *S* proton resonance. This effect is demonstrated for the downfield resonances associated with Glyc-51, -65, and -84. The arrows point to the Nα and Nα' cross-peaks in A; the altered intensities in the equivalent positions can be seen in B.

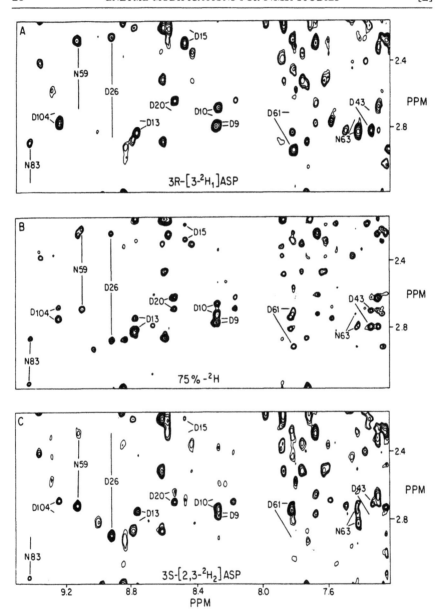

FIG. 2. Chiral β-deuteration editing of β-amide NOESY spectra. (A and C) A portion of the NOESY spectra of aspartic acid-labeled *E. coli* thioredoxin in which each of the β-hydrogen positions are selectively deuterated. The solid lines indicate the positions of the *pro-R* and *pro-S* intraresidue β-amide cross-peaks for the aspartic acid and asparagine residues as verified by reference to the random fractionally deuterated spectrum in B.

ally considerably less well determined.[28] The addition of the determination of the side-chain conformations based on deuteration, spin coupling, and NOE analysis offers the possibility of more closely approaching X-ray quality structural determinations by NMR.

Random Fractional Deuteration

By its nature, selective labeling results in selective information. As a result several separate samples must be prepared and analyzed in order to obtain information throughout a protein molecule. It would clearly be desirable to prepare a sample which offers some of the benefits of selective labeling while retaining the full complement of useful spectral information contained in the spectra of the natural abundance sample. This goal can be largely realized by the use of random fractional deuteration.

One of the major benefits of modern 2D NMR techniques is that for a number of suitable proteins of under 10 kDa most of the spectral data necessary for sequential assignment can be resolved under one set of sample conditions using various pulse experiments. In such a case, the value of isotopic labeling is largely limited to facilitating selected residue-type assignments and sequential connectivities. Unfortunately, this ceases to be the case for larger proteins. Problems of degeneracy in moderate-size proteins can be approached by observing the spectral effects of variations in sample conditions such as pH, temperature, and ligand binding. However, the resultant effects are generally haphazard in appearance and lack the directed nature required to resolve specific ambiguities in assignments.

The problem of spectral degeneracy becomes rapidly more severe as the molecular weight of the protein increases. The number of anticipated cross-peaks in 2D spectra increases roughly in proportion to the molecular weight. More important in terms of spectral resolution is the linewidth problem. The linewidth is linearly dependent on the transverse relaxation rate, which in turn is roughly proportional to the molecular weight of the molecule. In a 2D contour plot this effect enters in both dimensions, and thus the resolution depends on the square of the molecular weight. The increasing number of cross-peaks and the relaxation effect combine to give an overall dependence of resolution on approximately the cube of the molecular weight.

As noted earlier, the proton relaxation times are dominated by $^1H-^1H$ dipolar interaction. The benefit of narrower linewidths anticipated in

[28] G. Wagner, W. Braun, T. F. Havel, T. Schaumann, N. Gō, and K. Wüthrich, *J. Mol. Biol.* **196,** 611 (1987).

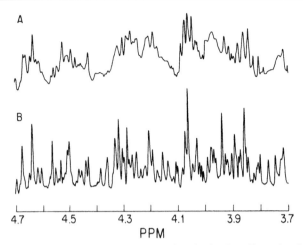

FIG. 3. ^{1}H NMR spectra of bacteriophage T4 thioredoxin. One-dimensional spectra for a natural abundance (A) and a 75% random fractionally deuterated (B) protein sample collected and processed under similar conditions. The increased resolution of the fractionally deuterated sample results from the longer relaxation times and hence narrower linewidths as well as reduced spin-coupling effects.

some of the earlier selective protonation experiments can be extended to all of the resonances by means of uniform deuterium dilution. The first attempt to examine this technique of random fractional deuteration was a brief 1D study of 90% U-^{2}H-labeled elongation factor TU from *E. coli*.[29] The observed linewidths of the resolved histidine resonances decreased by a factor of 5.5 following deuteration. Given that ^{2}H–^{1}H dipolar relaxation is 1/16 as efficient as the ^{1}H–^{1}H interaction, this value is quite close to that predicted from a 90%–10% labeling pattern.

Reduction of Passive Coupling by Random Fractional Deuteration. It has become feasible to prepare readily comparatively large amounts of random fractionally deuterated protein samples necessary to characterize the potential of the technique in terms of resolution and sensitivity. Figure 3 illustrates a portion of the 1D proton spectrum of the 87-residue protein bacteriophage T4 thioredoxin for both the natural abundance and the 75% U-^{2}H-labeled samples that were collected and processed under similar conditions. This spectral region predominantly contains H_α resonances, whose degeneracies have been noted to be an important limitation on assignment experiments. The clear enhancement of resolution observed in the deuterated sample is due not only to the narrower natural linewidths but also to the marked reduction in spin coupling.

[29] H. R. Kalbitzer, R. Leberman, and A. Wittinghofer, *FEBS Lett.* **180**, 40 (1985).

Despite the fact that spin coupling is not apparent in a standard 1D spectrum of the fractionally deuterated sample, its presence is readily observed in a 2D chemical shift correlation spectrum, as illustrated in Fig. 4B.[21] A phase-sensitive COSY spectrum of a portion of the $H_\alpha-H_\beta$ region of the 108-residue protein *E. coli* thioredoxin deuterated to 75% is shown. Each cross-peak is composed of a basic square quartet pattern in which the adjacent components are 180° out of phase. Projection of this pattern upon either axis yields the anticipated doublet from two coupled spins. In contrast, the corresponding cross-peaks from the natural abundance spectrum in Fig. 4A have a more complex component structure which represents the convolution of the basic quartet pattern with the coupling to other protons, so-called passive coupling, analogous to multiplet splitting in a standard 1D spectrum. The random fractionally deuterated sample results in a considerable improvement in resolution due to the combination of reduced passive coupling and narrower linewidths with little loss in sensitivity. Since both the α and β protons are reduced by a factor of four in concentration in the partially deuterated sample, the anticipated intensity of the cross-peak is reduced by a factor of 16. However, largely due to the longer relaxation times and hence narrower linewidths, the signal-to-noise ratio is only a factor of 2–3 less than that of the natural abundance spectrum. The relative sensitivity will become more favorable for larger proteins due to the antiphase character of the cross-peaks. When the natural linewidth becomes larger than the coupling constant, as is generally true of proteins above 10 kDa, significant mutual cancellation occurs between the cross-peak components.[30] With the narrower linewidths obtained upon deuteration, this effect should become important only in much larger proteins.

Reduced Spin Diffusion by Random Fractional Deuteration. In addition to narrower linewidths and reduced passive coupling, an important benefit of fractional deuteration is the effective suppression of spin diffusion, that is, tertiary and higher order spin-relaxation processes. In a NOE experiment, the initial buildup rate for a direct dipolar interaction is inversely proportional to the internuclear distance to the sixth power, which is the basis for using NOE data for structural analysis. However, as the longitudinal magnetization transferred from one nucleus builds up at a second one, it can in turn be transferred to a third nucleus. If the time allowed for transfer is sufficiently long in a simple 1D experiment, the net magnetization will become spread throughout much of the protein molecule.[31] Because this spin diffusion clearly negates the distance informa-

[30] D. Neuhaus, G. Wagner, M. Vasak, J. H. R. Kägi, and K. Wüthrich, *Eur. J. Biochem.* **151,** 257 (1985).
[31] A. Dubs, G. Wagner, and K. Wüthrich, *Biochim. Biophys. Acta* **577,** 177 (1979).

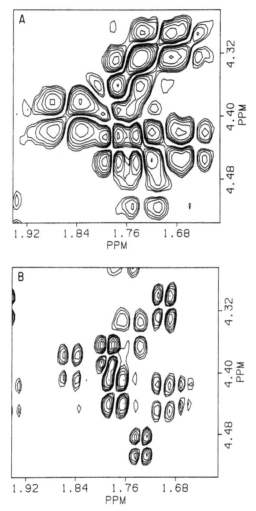

FIG. 4. Elimination of passive spin coupling by random fractional deuteration. (A) A section of the $\alpha\beta$ COSY region presented in phase-sensitive mode for a natural abundance *E. coli* thioredoxin sample. (B) The corresponding spectrum for the 75% U-^{2}H-labeled sample using identical acquisition and analysis parameters. The improved resolution observed in the fractionally deuterated sample is partially due to the longer T_2 relaxation time and hence narrower linewidth, but mainly it is a result of the fact that coupling from other protons to the two protons involved in the coherence transfer is largely eliminated due to the isotopic dilution.

tion that is potentially available, sufficiently short buildup or mixing times must be used so as to limit the effect, which in turn limits the sensitivity of the experiment.

By diluting out the tertiary spin interactions, random fractional deuteration strongly suppresses spin diffusion. As a result, longer mixing times can be used, thus, when combined with the narrower linewidths, yielding a compensation in sensitivity similar to that observed for the COSY spectra on partially deuterated samples. On 75% U-^{2}H-labeled *E. coli* thioredoxin, both NOESY and COSY spectra for cross-peaks between carbon-bound positions are two to three times less sensitive than those of natural abundance samples, whereas for cross-peaks between carbon-bound and amide protons (~95% ^{1}H) the spectra are of equivalent sensitivity.[21] Of particular value in sequential assignment work is the fact that the amide–amide NOESY spectra for the fractionally deuterated samples are markedly superior to those of the natural abundance samples. In this case both sites are essentially fully occupied with protons, yet the relaxation times of the amide protons of the fractionally deuterated samples are significantly larger, resulting in improved resolution as well as enhanced sensitivity. In the case of *E. coli* thioredoxin, over 80% of the sequential amide–amide connectivities are observed, including 50% of those in β strands for which the internuclear distances are greater than 4 Å and are only rarely seen in natural abundance spectra.[21]

In addition to its use in sequential assignment experiments, the suppression of spin diffusion should prove valuable in quantitative interpretation of NOE data. To date, distance information from NOESY experiments in macromolecules has generally been interpreted in rather qualitative terms. Tertiary spin interactions are largely responsible for the ambiguities limiting interpretation of NOE intensities (see Borgias and James [9], Vol. 176, this series). This is particularly severe in the case of methylene protons due to their efficient cross-relaxation. When combined with the previously described techniques for stereospecific assignment, random fractional deuteration should render feasible the quantitative interpretation of cross-peaks involving methylene protons. Since roughly 40% of all the resonances of a protein spectrum arises from methylene protons, on the order of 60% of all NOESY cross-peaks can be anticipated to involve methylene positions. The potential for obtaining more accurate buildup rates combined with stereospecific assignments clearly indicates the value of deuteration techniques as a means of providing structural information that in practice is inaccessible in data from natural abundance samples.

Variation in Percentage Deuteration. A majority of the data to date on random fractional deuteration have utilized a level of enrichment of 75%.

This value has been used in anticipation of largely suppressing passive coupling without severely sacrificing sensitivity. It would appear that neither aspect is particularly sensitive to the level of deuteration, so that perhaps deuteration anywhere within the range of 65–85% will provide qualitatively similar results. It seems doubtful that levels above 85% will prove useful in 2D experiments, unless perdeuteration is desired.

On the other hand, lower levels of deuteration should prove valuable. As in the case of the 1D spectrum, which gives the illusion of the elimination of spin coupling, the 2D COSY spectra seem to suggest that higher order spin interactions are absent. This is clearly not true. By reducing the level of deuteration to counteract loss in sensitivity, it should prove feasible to observe three and four spin interactions in moderately sized proteins with improved resolution as compared to natural abundance spectra. A complementary approach makes use of the fact that a significant component of the observed proton linewidth is generally due to interresidue dipolar interaction, and in the spirit of the previously discussed analysis of glucagon in deuterated micelles,[7] the study of samples with selected natural abundance side chains in a deuterated background will provide multiquantum spectra with improved resolution and sensitivity. Which of these two techniques will prove most promising is currently under investigation. In summary, it is clear that any currently envisioned 2D ^{1}H–^{1}H experiment that is anticipated to be practical in the study of moderate-size proteins can be significantly enhanced by the use of various deuteration techniques.

Synthesis of Deuterated Amino Acids

The chemical means of producing deuterated amino acids can be separated into the generally more facile exchange procedures, which start with the natural abundance amino acid sample, and *de novo* chemical synthesis. In addition, for amino acids possessing methylene positions, deuteration procedures can be separated according to whether the labeling is stereospecific.

Deuteration of the α and β Positions. Several techniques have been developed for hydrogen exchange of the α position. Perhaps the most generally useful techniques are acetic anhydride–sodium acetate-catalyzed[32] and alkaline pyridoxal-catalyzed exchanges.[33] Either of these techniques allows exchange for essentially all of the standard amino acids. The α-hydrogens of glycine can also be labeled stereospecifically.[34]

---

[32] D. A. Upson and V. J. Hruby, *J. Org. Chem.* **42,** 2329 (1977).
[33] S. W. Tenenbaum, T. H. Witherup, and E. H. Abbott, *Biochim. Biophys. Acta* **362,** 308 (1974).
[34] H. Ohrui, T. Misawa, and H. Meguro, *J. Org. Chem.* **50,** 3007 (1985).

In addition to its use in α exchange, at elevated temperatures near pH 5, pyridoxal catalyzes β exchange for the majority of the amino acids.[33,35] Unfortunately, amino acids with electron-withdrawing substituents on the β position, as well as histidine and tryptophan, react under these conditions[36] and hence require more involved syntheses. [3-^{2}H]Threonine can be synthesized via the reaction of [1-^{2}H]acetaldehyde with an activated glycine derivative.[37] In addition to the analogous reaction with deuterated formaldehyde,[38] reduction of ethyl acetamidocyanoacetate provides a facile route to β-deuterated serine.[39] Tryptophan synthase and tryptophanase serve to convert serine in the presence of indole[40] or appropriate thiols[41] into labeled tryptophan or cysteine, both of which are also readily produced by chemical means.[42,43] The constraint against electron-withdrawing groups at the β position even extends to aspartic acid, which decarboxylates under the standard pyridoxal exchange conditions. However, the β-hydrogens of aspartate can be exchanged in acid.[35]

Chiral β Deuteration. Most of the common amino acids possess β-methylene hydrogens that present the possibility of chiral β deuteration. In some cases this can be achieved on a practical synthetic scale by specific enzyme-catalyzed reactions. The chirally labeled aspartic acid samples referred to above were produced by the action of aspartase on ammonium fumarate.[44,45] One β-methylene position can be labeled by conducting the synthesis in ^{2}H$_2$O, while the other is labeled using [^{2}H$_2$]fumarate, which is easily synthesized.[46] Chirally labeled glutamate can be synthesized by the hydrogen exchange behavior of isocitrate dehydrogenase on 2-oxyglutarate, which in turn is converted to glutamate using glutamate dehydrogenase.[47,48] Both chiral derivatives can be produced by making use of preexchange of the β positions of 2-oxyglutarate in ^{2}H$_2$O.

A more generally applicable approach to chiral β deuteration involves the synthesis of α,β-dehydroamino acids followed by reduction with deu-

[35] D. M. LeMaster and F. M. Richards, *J. Labelled Compd. Radiopharm.* **19**, 639 (1982).
[36] E. H. Abbott and A. E. Martell, *J. Am. Chem. Soc.* **92**, 1754 (1970).
[37] D. H. G. Crout, M. V. M. Gregorio, U. S. Müller, S. Komatsubara, M. Kisumi, and I. Chibata, *Eur. J. Biochem.* **106**, 97 (1980).
[38] M. Levine and H. Tarver, *J. Biol. Chem.* **184**, 427 (1950).
[39] L. Berlinguet, *Can. J. Chem.* **33**, 1119 (1955).
[40] S. S. Yuan and A. M. Ajami, *Tetrahedron* **38**, 2051 (1982).
[41] N. Esaki, H. Tanaka, E. W. Miles, and K. Soda, *Agric. Biol. Chem.* **47**, 2861 (1983).
[42] H. R. Snyder and C. W. Smith, *J. Am. Chem. Soc.* **66**, 350 (1944).
[43] D. A. Upson and V. J. Hruby, *J. Org. Chem.* **41**, 1353 (1976).
[44] S. J. Field and D. W. Young, *J. Chem. Soc., Chem. Commun.*, 1163 (1979).
[45] K. H. Röhm and R. L. Van Etten, *J. Labelled Compd. Radiopharm.* **22**, 909 (1985).
[46] E. M. Richards, J. C. Tebby, R. S. Ward, and D. H. Williams, *J. Chem. Soc. C*, 1542 (1969).
[47] Z. B. Rose, *J. Biol. Chem.* **235**, 928 (1960).
[48] G. E. Lienhard and I. A. Rose, *Biochemistry* **3**, 185 (1964).

terium gas. The classic Erlenmeyer azlactone synthesis involves the reaction between an aromatic aldehyde and an activated glycine derivative followed by hydrolysis of the ring and reduction. This synthesis combined with D_2 reduction of the resultant α,β-dehydroamino acid has been carried out for all four of the standard aromatic amino acids.[49-52] A similar synthesis reaction has been successfully applied to chirally deuterated serine as well.[53] Unfortunately, for most aliphatic amino acids this synthesis is rather inefficient. However, the reaction between a cupric glycinate complex and an appropriate aldehyde followed by dehydration of the resultant β-hydroxyamino acid[54,55] offers a useful means of synthesizing aliphatic α,β-dehydroamino acids. In addition, there are several other alternative routes to the requisite α,β-dehydroamino acids.[56,57]

This synthetic approach gains considerable stimulus from the fact that much of the effort toward chiral synthesis of amino acids has utilized chiral reducing agents acting upon α,β-dehydroamino acids.[58] Although in this work the main focus is on the chirality at the α position, the cis addition generally observed in hydrogenation reactions imparts chirality on the β center as well. Some of the elegance of the chiral reduction is lost in cases in which samples with either chirality at the β position are desired, since either both the E and the Z isomers of the dehydroamino acid must be obtained or else one of the products must be racemized to effect conversion from the d to the l isomer at the α-carbon.

In a few cases, such as serine,[59] cysteine,[60] and glutamic acid,[61] chiral β deuteration has been obtained by entirely different chemical routes.

Specialized Deuteration Procedures for Side-Chain Positions. As the labeling pattern desired extends beyond the β-carbon perforce, the approaches become residue-type specific. The aromatic rings can be selec-

[49] G. W. Kirby and J. Michael, *J. Chem. Soc., Perkin Trans. 1*, 115 (1973).

[50] P. G. Strange, J. Staunton, H. R. Wiltshire, A. R. Battersby, K. R. Hanson, and E. A. Havir, *J. Chem. Soc., Perkin Trans. 1*, 2364 (1972).

[51] G. W. Kirby and M. J. Varley, *J. Chem. Soc., Chem. Commun.*, 833 (1974).

[52] A. R. Battersby, M. Nicoletti, J. Staunton, and R. Vleggaar, *J. Chem. Soc., Perkin Trans. 1*, 43 (1980).

[53] D. J. Aberhart and D. J. Russell, *J. Am. Chem. Soc.* **106**, 4902 (1984).

[54] L. Somekh and A. Shanzer, *J. Org. Chem.* **48**, 907 (1983).

[55] T. Kato, C. Higuchi, R. Mita, and T. Yamaguchi, Jpn. Patent 60,190,749 (1985) (*Chem. Abstr.* 104:109267j).

[56] R. Grigg and J. Kemp, *J. Chem. Soc., Chem. Commun.*, 125 (1977).

[57] A. J. Kolar and R. K. Olsen, *Synthesis*, 457 (1977).

[58] J. Martens, *Top. Curr. Chem.* **125**, 165 (1984).

[59] L. Slieker and S. J. Benkovic, *J. Labelled Compd. Radiopharm.* **19**, 647 (1982).

[60] J. E. Baldwin, R. M. Adlington, N. G. Robinson, and H. H. Ting, *J. Chem. Soc., Chem. Commun.*, 409 (1986).

[61] S. J. Field and D. W. Young, *J. Chem. Soc., Perkin Trans. 1*, 2387 (1983).

tively as well as fully deuterated by acid-catalyzed exchange.[62,63] In addition, Pt,[64] Pd,[65] and Raney Ni[66] can be used for both exchange and dehalogenation in order to obtain a more selective labeling pattern.

Procedures have been described that are suitable for the deuteration of the methyl positions of valine,[67] leucine,[68] threonine,[37] and methionine,[69] including chiral methyl labeling of valine[70] and leucine.[71,72] [4,5-^{2}H$_2$]Leucine[73] and [3,4-^{2}H$_2$]isoleucine[74] can be obtained by hydrogenation of the corresponding dehydroamino acid. In the case of isoleucine this results in the chiral labeling of the γ position. The γ position of methionine can be deuterated generally via base-catalyzed exchange of the S-methyl derivative[75] or stereoselectively by the action of cystathionine γ-synthase[76] (O-succinylhomoserine lyase).

The γ-hydrogens of glutamic acid can be exchanged in strong acid[35] and can be stereospecifically labeled.[77] When combined with the ability to label the α and β positions at will, this offers a practical route to the synthesis of labeled proline via the thiolation and reduction of pyroglutamic acid[78] as well as synthesis of labeled arginine via conversion to ornithine[79] and guanidylation. Several other approaches to proline labeling have been developed. Hydrogenation of 3,4-dehydroproline results in

[62] D. V. Griffiths, J. Feeney, G. C. K. Roberts, and A. S. V. Burgen, *Biochim. Biophys. Acta* **446**, 479 (1976).

[63] H. R. Matthews, K. S. Matthews, and S. J. Opella, *Biochim. Biophys. Acta* **497**, 1 (1977).

[64] R. S. Norton and J. H. Bradbury, *J. Catal.* **39**, 53 (1975).

[65] M. K. Anwer, R. A. Porter, and A. F. Spatola, *Int. J. Pept. Protein Res.* **30**, 489 (1987).

[66] R. C. Woodworth and C. M. Dobson, *FEBS Lett.* **101**, 329 (1979).

[67] T. W. Whaley, G. H. Daub, V. N. Kerr, T. A. Lyle, and E. S. Olson, *J. Labelled Compd. Radiopharm.* **16**, 809 (1979).

[68] S. S. Yuan and J. Foos, *J. Labelled Compd. Radiopharm.* **18**, 563 (1981).

[69] D. Dolphin and K. Endo, *Anal. Biochem.* **36**, 338 (1970).

[70] H. Kluender, F. C. Huang, A. Fritzberg, H. Schnoes, C. J. Sih, P. Fawcett, and E. P. Abraham, *J. Am. Chem. Soc.* **96**, 4054 (1974).

[71] R. Cardillo, C. Fuganti, D. Ghiringhelli, P. Grasselli, and G. Gatti, *J. Chem. Soc., Chem. Commun.*, 474 (1977).

[72] D. J. Aberhart and B. H. Weiller, *J. Labelled Compd. Radiopharm.* **20**, 663 (1983).

[73] A. J. Fischman, H. R. Wyssbrod, W. C. Agosta, and D. Cowburn, *J. Am. Chem. Soc.* **100**, 54 (1978).

[74] R. Cahill, D. H. G. Crout, M. V. M. Gregorio, M. B. Mitchell, and U. S. Müller, *J. Chem. Soc., Perkin Trans. 1*, 173 (1983).

[75] R. Wiesendanger, B. Martinoni, T. Boller, and D. Arigoni, *J. Chem. Soc., Chem. Commun.*, 238 (1986).

[76] M. N. T. Chang and C. T. Walsh, *J. Am. Chem. Soc.* **103**, 4921 (1981).

[77] C. Ducrocq, A. Righini-Tapie, R. Azerad, J. F. Green, P. A. Friedman, J. P. Beaucourt, and B. Rousseau, *J. Chem. Soc., Perkin Trans. 1*, 1323 (1986).

[78] A. Kleemann, J. Martens, and K. Drauz, *Chem.-Ztg.* **105**, 266 (1981).

[79] T. Itoh, *Bull. Chem. Soc. Jpn.* **36**, 25 (1963).

cis addition to the β and γ positions.[80] Chiral deuteration at the γ position is feasible via 4-hydroxyproline.[81] A fairly involved procedure serves to yield chiral deuteration at the δ position as well.[82] The fact that pyrrole carboxylic acid can be reduced to proline with Pd on carbon in [1-^{2}H]acetic acid (unpublished observation) provides an additional means of producing selective[83–85] as well as fully deuterated derivatives.

Enzymatic decarboxylation of diaminopimelic acid has been used to synthesize lysine chirally labeled in the ε position.[86] Presumably the pyridoxal-catalyzed β exchange followed by decarboxylation would result in labeling of the δ position as well. In addition, syntheses suitable for producing [4,5-^{2}H$_2$]-, [4,4,5,5-^{2}H$_4$]-, and [4,5,6,6-^{2}H$_4$]lysine have been reported.[87,88]

Biosynthetic Production of Uniformly Deuterated Amino Acids. In addition to various selectively deuterated amino acids, perdeuterated amino acids are also of use. Although to date virtually all experiments have used fully deuterated amino acids, the same procedures can equally well be applied to fractionally deuterated samples. Using the hydrogen exchange techniques discussed above, perdeuterated glycine, alanine, aspartic acid, glutamic acid, phenylalanine, tyrosine, and histidine can be produced. For the others, more involved chemical syntheses are necessary. A more general approach is by biosynthetic production via growth of microorganisms on deuterated media. Although much of the early work involved growth of algae, the use of heterotrophic bacteria such as *E. coli* grown on deuterated succinate offers numerous advantages in terms of ease and speed of production.[21,88] The labeled succinate can be generated by acetic acid–sodium acetate-catalyzed exchange[89] or by reduction of acetylene dicarboxylate esters with D$_2$ gas.[90] This holds for combined ^{2}H/^{15}N labeling as well. Only in the case in which ^{13}C is also desired do

[80] J. M. A. Al-Rawi, J. A. Elvidge, J. R. Jones, V. M. A. Chambers, and E. A. Evans, *J. Labelled Compd. Radiopharm.* **12**, 265 (1976).
[81] Y. Fujita, A. Gottlieb, B. Peterkofsky, S. Udenfriend, and B. Witkop, *J. Am. Chem. Soc.* **86**, 4709 (1964).
[82] P. Gramatica, P. Manitto, A. Manzocchi, and E. Santaniello, *J. Labelled Compd. Radiopharm.* **18**, 955 (1981).
[83] G. P. Bean and T. J. Wilkinson, *J. Chem. Soc., Perkin Trans. 2*, 72 (1978).
[84] P. E. Sonnet, *J. Med. Chem.* **15**, 97 (1972).
[85] D. J. Chadwick, J. Chambers, G. D. Meakins, and R. L. Snowden, *J. Chem. Soc., Perkin Trans. 1*, 201 (1973).
[86] Y. Asada, K. Tanizawa, S. Sawada, T. Suzuki, H. Misono, and K. Soda, *Biochemistry* **20**, 6881 (1981).
[87] L. Birkofer and K. Hempel, *Chem. Ber.* **93**, 2282 (1960).
[88] D. M. LeMaster and F. M. Richards, *Anal. Biochem.* **122**, 238 (1982).
[89] V. J. Stella, *J. Pharm. Sci.* **62**, 634 (1973).
[90] S. Borcic, M. Nikoletic, and D. E. Sunko, *J. Am. Chem. Soc.* **84**, 1615 (1962).

bacteria such as *E. coli* become impractical. Even in this case the hetero-
trophic methylotrophs that utilize methanol as a carbon source are an
efficient source of bulk protein, which has been used to produce both ^{2}H-
and ^{13}C-labeled amino acids (C. Unkefer, personal communication).

In order to use these labeled bacterial samples it is necessary to extract and hydrolyze the crude protein. In some cases the bulk hydrolysate
has been used directly in labeling experiments. Otherwise the constituent
amino acid must be separated. It is possible to separate 10 g of protein
hydrolysate into the individual purified amino acids on an ion-exchange
system composed of a carrier displacement column and an elution
column, each of roughly 500 ml of bed resin with approximately 10 liters
of eluent each.[88]

Production of Deuterated Proteins

Surely the most general approach to synthesis of isotopically enriched
proteins is by chemical means. This offers the possibility of overcoming
the major limitation of biosynthetic production, i.e., the inability to enrich
residues of the same type differentially throughout the protein sequence.
In addition, nonphysiological amino acids can be introduced, which has
proved very useful in peptide hormone studies. With the introduction of
"cassette" synthesis techniques,[91] production of small proteins has been
facilitated along with the ability to systematically introduce sequential
variations. Nevertheless, *de novo* protein chemical synthesis remains a
daunting enterprise. In certain cases enzymatically facilitated semisynthetic procedures have been successful, but the generality of this approach is fairly limited. Only in a few cases have these chemical approaches been used for the purpose of obtaining isotopically labeled
samples.[92,93]

To date and in the foreseeable future the method of choice is biosynthetic production. The application of molecular biological techniques to
yield increased synthesis levels and site-directed mutagenesis of proteins
has served to enhance the value of isotopic labeling as a means of studying their biophysical properties by NMR. Given the constraint of being
unable to differentially label within a residue type, the practical problem
centers on how specifically a given amino acid or a general labeling pattern can be incorporated into the cell protein. There are three related

[91] W. F. DeGrado and E. T. Kaiser, *J. Org. Chem.* **45**, 1295 (1980).
[92] R. Richarz, H. Tschesche, and K. Wüthrich, *Biochemistry* **19**, 5711 (1980).
[93] L. T. Hughes, J. S. Cohen, A. Szabo, C. H. Niu, and S. Matsuura, *Biochemistry* **23**, 4390 (1984).

factors to be considered: the organism in which the expression is carried out, the amino acid(s) to be incorporated, and what positions within the amino acid are to be labeled.

In many cases it is practical to force feed with large amounts of labeled amino acids in order to get efficient incorporation into the protein of interest. This was the method first used to produce deuterated staphylococcal nuclease[94] as well as several algal proteins.[95] This has also proved to be highly successful in labeling tissue culture cells as a means of producing selectively deuterated monoclonal Fab fragments. In these studies all four aromatic and six methyl-containing amino acids have been successfully incorporated.[16,18] More ambitious has been the growth of mice fed deuterated tryptophan[96] and quail hens fed a deuterated aromatic amino acid diet as a means of producing egg white lysozyme labeled to a level of 80%.[97]

Experiments requiring uniform levels of deuteration obtained via growth on D_2O-containing media are biologically more demanding. However, in addition to various bacteria and algal strains, yeast, fungi, and certain protozoa have been successfully cultured in fully deuterated media.[98] As might be relevant to random fractional deuteration experiments, higher plants have been adapted to up to 70% D_2O. Mammals are considerably more sensitive to D_2O toxicity, but the possibility remains that certain useful eukaryotic expression systems such as that of baculovirus[99] may prove tolerant to deuteration.

In using bacteria or presumably lower eukaryotic organisms such as yeast, the opportunity exists to use genetics to facilitate the labeling process. Generally speaking, it can be anticipated that the technique of force feeding will be least successful for the amino acids that are closest to the central pathways, such as glutamate, aspartate, and alanine. Nevertheless, in addition to the alanine H_β labeling of the tissue culture cells discussed above, 3-^{13}C- and ^{15}N-labeled alanine[100,101] have been success-

[94] I. Putter, A. Barreto, J. L. Markley, and O. Jardetzky, *Proc. Natl. Acad. Sci. U.S.A.* **64**, 1396 (1969).

[95] H. L. Crespi, U. Smith, L. Gajda, T. Tisue, and R. M. Ammeraal, *Biochim. Biophys. Acta* **256**, 611 (1972).

[96] P. Gettins and R. A. Dwek, *FEBS Lett.* **124**, 248 (1981).

[97] A. Brown-Mason, C. M. Dobson, and R. C. Woodworth, *J. Biol. Chem.* **256**, 1506 (1981).

[98] J. J. Katz, in "Thirty-Ninth Annual Priestley Lectures. I: Chemical and Biological Studies with Deuterium," p. 1. Penn. State Univ. Press, University Park, Pennsylvania, 1965.

[99] V. A. Luckow and M. D. Summers, *Biotechnology* **6**, 47 (1988).

[100] G. D. Henry, J. H. Weiner, and B. D. Sykes, *Biochemistry* **25**, 590 (1986).

[101] L. P. McIntosh, F. W. Dahlquist, and A. G. Redfield, *J. Biomol. Struct. Dyn.* **5**, 21 (1987).

fully incorporated into protein in *E. coli* for which a suitable alanine auxotroph has not been found.[102]

Although force feeding has proved useful in a number of cases, when feasible the use of auxotrophs offers several advantages. It is generally possible to avoid the addition of large excesses of labeled material commonly used in force-feeding experiments (however, note the cases of serine and threonine labeling in *E. coli* auxotrophs).[103] In addition, auxotrophs facilitate high levels of incorporation which, although often not essential for heteronuclear labeling, are often of major importance in deuteration and double-label experiments.

Selective Labeling of the α Position. The use of genetics becomes increasingly important when attention is turned to the position in the amino acid to be labeled, in particular if either the α-hydrogen or α-nitrogen is to be enriched. Aminotransferases often serve to scramble both of these positions, while the remainder of the amino acid positions are preserved intact during incorporation. Although ^{15}N labeling is technically beyond the scope of this review, the observations made on the selectivity of ^{15}N incorporation are directly relevant because the α-hydrogen position can only be more sensitive to biological exchange.

α-Hydrogen exchange was observed in the earliest NMR protein labeling experiments on force-fed algae.[1,104] Beyond these experiments comparatively little work has been done to quantitate the efficiency of α-hydrogen incorporation during forced feeding. In contrast, several studies have measured the efficiency of ^{15}N labeling as a function of overall media composition. Perhaps most valuable are the labeling studies of streptomyces subtilisin inhibitor by Kainosho and co-workers.[105,106] In this case they wished to observe ^{15}N splitting in the 1D spectra of the ^{13}C carbonyl resonance of the preceding residue in the sequence for ^{15}N-labeled alanine, glycine, cysteine, valine, and aspartic acid, and hence the efficiency of incorporation was of considerable importance. They observed levels of enrichment of at most 50%, and in the case of aspartic acid only 24%, despite considerable effort in examination of dependence on medium composition. In most cases of selective deuterium labeling high efficiency is essential for the subsequent NMR experiment. As the ^{15}N labeling experiments described suggest, α labeling is dependent on the use of genetics to provide both the auxotrophic and aminotransferase deficiencies in order to provide the desired efficiency.

[102] J. O. Falkinham, *J. Bacteriol.* **130**, 566 (1977).
[103] D. M. LeMaster and F. M. Richards, *Biochemistry* **24**, 7263 (1985).
[104] H. L. Crespi, H. F. Daboll, and J. J. Katz, *Biochim. Biophys. Acta* **200**, 26 (1970).
[105] M. Kainosho and T. Tsuji, *Biochemistry* **21**, 6273 (1982).
[106] M. Kainosho, H. Nagao, and T. Tsuji, *Biochemistry* **26**, 1068 (1987).

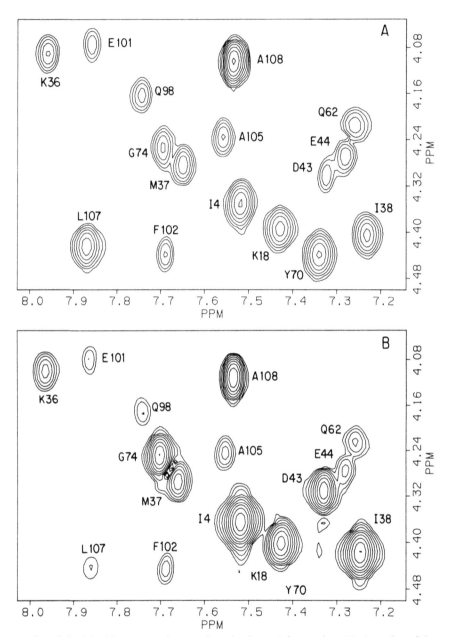

FIG. 5. Multiresidue-type assignment by selective α,β deuteration. (A) A portion of the α-amide region is displayed in magnitude mode for a random fractionally deuterated *E. coli* thioredoxin sample. A second sample was prepared in which selectively α,β-labeled amino acids are incorporated in a 75% ^{2}H background. [^{2}H$_\alpha$]Tyrosine and [^{2}H$_\alpha$,^{2}H$_\beta$]leucine result in the strong reduction of the corresponding cross-peaks (B) (Y70, L107), while [^{1}H$_\alpha$,^{2}H$_\beta$]aspartic acid, [^{1}H$_\alpha$,^{2}H$_\beta$]isoleucine, and [^{1}H$_\alpha$,^{1}H$_\beta$]serine (hence glycine) yield significantly stronger peaks (D43, I4, I38, G74) than are observed in the reference spectrum. The resonances of all other residue types in the region are the same within one contour level (1.4 scale factor) for the two spectra.

Escherichia coli is, of course, the obvious choice for such genetic manipulation. Auxotrophs have been found for all amino acids except alanine. Mutations in three glutamate-dependent general amino acid aminotransferases have been obtained.[107] A strain carrying all three lesions grows at the same rate as does the prototrophic parent in a deuterated minimal medium.[21] Introduction of additional auxotrophies into the aminotransferase-deficient strain has rendered feasible experiments demanding efficient label incorporations. This can be illustrated in a multi-residue-type labeling experiment in which several amino acids differentially labeled in the α and β positions are incorporated into a 75% ^{2}H randomly deuterated background. A portion of the α-amide region of a magnitude COSY spectrum of such a sample is presented in Fig. 5 along with a reference spectrum from a 75% U-^{2}H sample.[21] In this case residue-type assignment information can be obtained not only as a result of diminution of the cross-peak, as in the cases of Tyr-70 and Leu-107, but also by means of peak enhancement, as in the cases of isoleucine-4 and -38 as well as Asp-43, due to the 100% ^{1}H enrichment at the α positions of these β-deuterated amino acids. The additional complexities that arise in biosynthetic incorporation are manifest in consideration of the α-hydrogen labeling of the amino acids that are biosynthetically derived from aspartic acid. In addition to asparagine, threonine residues also show peak enhancement in the experiments described above, whereas lysine and methionine do not. Similarly for the serine family, both serine and glycine preserve the α-hydrogen label but cysteine and tryptophan do not. In these cases the increased lability of the α-hydrogen as compared to that expected for the α-nitrogen provides an additional means of differentiating multiresidue types in a single experiment.

In summary, a large number of deuteration patterns are accessible for all of the standard amino acids. When combined with the metabolic genetics and protein expression systems as available for *E. coli,* a truly extraordinary range of deuteration experiments is practical and can allow for detailed NMR studies that are not feasible using natural abundance samples.

[107] D. H. Gelfand and R. A. Steinberg, *J. Bacteriol.* **130,** 429 (1977).

[3] Expression and Nitrogen-15 Labeling of Proteins for Proton and Nitrogen-15 Nuclear Magnetic Resonance

By David C. Muchmore, Lawrence P. McIntosh,
Christopher B. Russell, D. Eric Anderson, and
Frederick W. Dahlquist

Introduction

Nitrogen-15 nuclear magnetic resonance is an established and rapidly developing technique for structural and dynamic investigations of biomolecules.[1-4] The large chemical shift dispersion of ^{15}N resonances is very useful for resolving individual peaks and, in some cases, the chemical shifts can be interpreted in terms of structure.[5] The low natural abundance (0.36%) and inherent insensitivity $[\gamma(^{15}N)/\gamma(^{1}H) = -0.101]$ of this spin-$\frac{1}{2}$ isotope have traditionally limited the use of ^{15}N NMR in the study of biological systems. However, the natural abundance of ^{15}N may be advantageous inasmuch as proteins and nucleic acids can be enriched by over 200-fold in this isotope. Selectively labeled sites are thus observed free of background signal from unlabeled sites.

The sensitivity of ^{15}N NMR is dramatically improved using modern polarization transfer techniques[6,7] or by indirect detection of the nucleus with ^{1}H NMR.[8] Several one- and two-dimensional NMR experiments have been developed to exploit the chemical shift dispersion and relaxation properties of ^{15}N for studies of biological molecules. Nitrogen-15 can also be utilized indirectly as a means to edit or filter complex ^{1}H NMR spectra of proteins and nucleic acids, such as nuclear Overhauser enhancement (NOE) spectra.[8-10]

In this chapter we will discuss the methods of biosynthetic enrichment of proteins with ^{15}N in backbone and side-chain groups, as exemplified by

[1] D. Gust, R. B. Moon, and J. D. Roberts, *Proc. Natl. Acad. Sci. U.S.A.* **72,** 4696 (1975).
[2] W. von Philipsborn and R. Mueller, *Angew. Chem.* **25,** 383 (1986).
[3] M. Witanowski, L. Stefaniak, and G. A. Webb, *Annu. Rep. NMR Spect.* **18,** 1 (1985).
[4] G. C. Levy and R. L. Lichter, "Nitrogen-15 Nuclear Magnetic Resonance Spectroscopy." Wiley, New York, 1979.
[5] W. M. Bachovchin and J. D. Roberts, *J. Am. Chem. Soc.* **100,** 8041 (1978).
[6] G. A. Morris and R. Freeman, *J. Am. Chem. Soc.* **101,** 760 (1979).
[7] D. M. Doddrell, D. T. Pegg, and M. R. Bendall, *J. Magn. Reson.* **48,** 323 (1982).
[8] R. H. Griffey and A. G. Redfield, *Q. Rev. Biophys.* **19,** 51 (1987).
[9] R. H. Griffey, M. A. Jarema, S. Kunz, P. R. Rosevear, and A. G. Redfield, *J. Am. Chem. Soc.* **107,** 711 (1985).
[10] L. P. McIntosh, F. W. Dahlquist, and A. G. Redfield, *J. Biomol. Struct. Dyn.* **5,** 21 (1987).

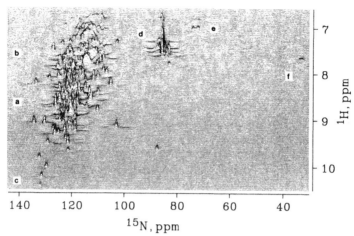

FIG. 1. The forbidden-echo (HMQC) spectrum of uniformly ^{15}N-labeled T4 lysozyme. The sample (250 μl) was ~2 mM protein in 100 mM sodium chloride, 100 mM sodium phosphate, and 10% D$_2$O, pH 4.9, at 24°. The peaks from the ^{15}N–^{1}H of the backbone (a and b) and side-chain amides (b), tryptophan indole ε_1N–H (c), arginine εN–H (d) and $\eta_{1,2}$N–H (e), and primary amines (f) are indicated. The glutamine and arginine side-chain peaks appear with distorted phase due to the more complex spin system of primary amides. The spectrum was recorded as described previously[8,11,41,74] using the 500-MHz spectrometer built by A. G. Redfield at Brandeis University. The data was processed using software written by A. G. Redfield and D. Hare (FTNMR).

our work with bacteriophage T4 lysozyme (Fig. 1). We will also outline the NMR experiments which we have used to study these labeled proteins.[10–13] We will focus on three necessary features for investigation of ^{15}N-labeled proteins. First, the protein must be rapidly and efficiently produced in milligram quantities. In the case of T4 lysozyme, this requires the development of controlled, high-level expression systems for synthesis of the protein by bacterial hosts. Second, the protein must be uniformly or selectively enriched in ^{15}N by growth of the bacteria on defined media supplemented with the isotope. This requires consideration of the metabolic pathways of bacteria and the use of bacteria with amino acid auxotrophies and aminotransferase deficiencies to ensure efficient and controlled incorporation of the ^{15}N labels. Third, new NMR experiments

[11] R. H. Griffey, R. E. Loomis, A. G. Redfield, and F. W. Dahlquist, *Biochemistry* **24**, 817 (1985).

[12] R. H. Griffey, A. G. Redfield, L. P. McIntosh, T. G. Oas, and F. W. Dahlquist, *J. Am. Chem. Soc.* **108**, 6816 (1986).

[13] L. P. McIntosh, R. H. Griffey, D. C. Muchmore, C. P. Nielson, A. G. Redfield, and F. W. Dahlquist, *Proc. Natl. Acad. Sci. U.S.A.* **84**, 1244 (1987).

have been developed which either directly or indirectly yield information regarding the ^{15}N nucleus or exploit the ^{15}N as a means to filter ^{1}H NMR spectra. The spectra produced by these experiments provide a means of obtaining information about the structural and dynamic properties of a protein that would be difficult or impossible to obtain by previously existing methods. These measurements are particularly suited to larger proteins, such as T4 lysozyme (18.7 kDa), for which resolution and assignment of proton NMR resonances become increasingly difficult.

Although we will restrict our discussion to ^{15}N, many of the labeling strategies and NMR experiments presented herein are applicable to other nuclei, such as ^{2}H, ^{13}C, and ^{113}Cd. These topics are further discussed in accompanying chapters within this volume.

Expression System

Our spectroscopic investigations of T4 lysozyme require that relatively large quantities of labeled wild-type and mutant proteins be easily produced. The expense of labeled amino acids also requires a highly efficient means of incorporation of labeled precursors into the protein. In theory, labeled T4 lysozyme ought to be available from a T4 phage infection of the relevant *Escherichia coli* strains. In practice, this has proved to be unsatisfactory because of modest yields and relatively complex purification protocols.[14-16] As a consequence, we designed a plasmid-based expression system which would meet our needs. We elected to use an *E. coli* bacterial host, as opposed to alternative microbial systems, since its metabolic pathways for amino acid synthesis and degradation have been well characterized and mutations that block many key steps are readily available.[17]

The criteria for the plasmid were centered on the need for a system suitable for efficient isotopic labeling of T4 lysozyme by expression of the lysozyme gene in a variety of *E. coli* host strains and for gene manipulation of the coding sequence of this protein. We required a controllable promoter to avoid the serious problem of selection for mutations in the vector, caused by continuous high-level synthesis of a gene product which is deleterious to the host cells, resulting in reduced expression levels or attenuated specific activity. Because we were particularly interested in the properties of thermolabile forms of the enzyme, we ruled out

[14] A. Tsugita, M. Inouye, E. Terzaghi, and G. Streisinger, *J. Biol. Chem.* **243**, 391 (1968).
[15] H. Jensen and K. Kleppe, *Eur. J. Biochem.* **26**, 305 (1972).
[16] B. Swewczyk, J. Kur, and A. Taylor, *FEBS Lett.* **139**, 97 (1982).
[17] B. J. Bachmann, *Microbiol. Rev.* **47**, 180 (1983).

temperature-inducible systems. To assist our labeling, we also needed maximum flexibility in the amino acid content of the medium, thus precluding use of amino acid regulation such as that found in expression systems based on the *trp* operon promoter.

We chose to construct a generalized high-level expression system which would put the lysozyme gene under the control of the *lacUV5* promoter on a plasmid containing the *lacI*^{q} gene, which codes for the *lac* repressor. This combination resulted in fairly high induced levels of lysozyme activity with almost nonexistent uninduced levels. We obtained increased expression by insertion of multiple copies of a *trp–lac* hybrid promoter[18] directly upstream from our gene. Our system also includes a transcription termination sequence directly downstream from the T4 lysozyme sequence which has been inserted into the vector. Details of the construction of this T4 lysozyme expression system are summarized below and are illustrated in Fig. 2.

Our initial attempt at high-level expression entailed insertion of the previously cloned T4 lysozyme gene[19] into the general vector, pCR043. This vector contained genes coding for two selectable markers, ampicillin and tetracycline resistance, in addition to the *lacUV5* promoter and the *lacI*^{q} gene. Eight residues of β-galactosidase were included in the fragment from which the promoter was derived. We used an *Hin*dIII site 130 bases downstream from the T4 lysozyme coding region and a *Cla*I site, derived from an *Ava*II site through use of molecular linkers, 70 bases upstream to insert the gene between *Cla*I and *Hin*dIII sites in pCR043. The resulting plasmid, pHSe3, led to isopropylthio-β-D-galactoside (IPTG)-dependent expression of lysozyme activity in *E. coli* hosts. The approximate levels of lysozyme production in plasmid-containing host colonies were monitored directly on nutrient agar using simple diffusion assays.[20]

The levels of T4 lysozyme expression by pHSe3 were relatively low. This was possibly due to interference by parts of the upstream T4 sequence with either the transcriptional or translational processes. By screening of a set of progressive *Bal*31 exonucleolytic deletions from the *Cla*I site we generated a plasmid which produced much higher levels of lysozyme activity. In this isolate, the T4 DNA had been deleted to within 15 bases upstream from the initiating ATG, and the remnant contained little more than the bacteriophage Shine–Dalgarno sequence. To further

[18] D. Russel and G. Bennet, *Gene* **20**, 231 (1982).
[19] J. E. Owen, D. W. Schultz, A. Taylor, and G. R. Smith, *J. Mol. Biol.* **165**, 229 (1983).
[20] G. Streisinger, F. Mukai, W. J. Dreyer, B. Miller, and S. Horiuchi, *Cold Spring Harbor Symp. Quant. Biol.* **26**, 25 (1961).

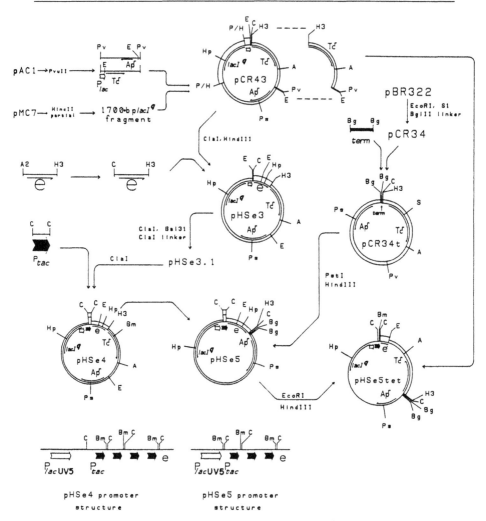

FIG. 2. Plasmid pCR43 was constructed from the 1700-bp fragment of pMC7 [M. P. Carlos, *Nature* (*London*) **274**, 762 (1979)] containing the *lacI*^{q} gene and the two *Pvu*II fragments of pAC1 [A. J. Carpousis and J. D. Gralla, *Biochemistry* **19**, 3245 (1980)] providing the *lacUV5* promoter (bottom, open arrow) as well as the ampicillin resistance (*Ap*^{r}) and tetracycline resistance (*Tc*^{r}) genes. The T4 lysozyme gene (*e*) was inserted downstream from the *lac* promoter after converting the *Ava*II site to a *Cla*II site, creating pHSe3. Treatment with *Bal*31 nuclease-deleted sequences between the promoter and the *e* gene to create pHSe3.1. Insertion of *tac* promoter cassettes (closed arrows) created the unstable pHSe4. Exchanging the *Hin*dIII–*Pst*I fragment with pCR34t created the stable pHSe5 which is the standard lysozyme expression vector for this study. Plasmid pCR34t is a derivative of pBR322 [F. Bolivar, R. L. Rodriguez, P. J. Greene, M. C. Betlach, H. L. Heynecker, H. W. Boyer, J. H. Crosa, and S. Falkow, *Gene* **2**, 95 (1977)], where the *Eco*RI site was converted

increase expression, we inserted a hybrid *trp–lac* promoter using molecular linkers. Fortuitously, four copies of this strong promoter were inserted between the *lac* fragment and the T4 fragment. Continued screening for .increased production resulted in isolation of a plasmid from which some portion of the *lac* operator, a short residual β-galactosidase fragment, and most of one copy of the hybrid promoter had been deleted. This plasmid, pHSe4.3, did exhibit some degree of instability, presumably due to unacceptably high levels of expression of the tetracycline resistance gene, and we found it expedient to delete this portion of the plasmid. This deletion resulted in readthrough from a strong constitutive T4 promoter, located just downstream from the lysozyme coding region, into the β-lactamase gene on the plasmid. The resulting overproduction of lactamase caused difficulties in use of ampicillin resistance as a selectable marker for the plasmid. Insertion of a transcription terminator downstream from the lysozyme gene alleviated this problem. The final plasmid, pHSe5, has reliably mediated high levels of T4 lysozyme production in a large number of host strains.

We wished to study a number of lysozyme variants which had been identified initially as phage T4 mutations. Since many of these lay in the carboxyl terminal of the peptide, encoded by DNA downstream from an *Eco*RI restriction site cutting in the codons for glycine-77 and isoleucine-78, we needed an unambiguous way in which to transfer this portion of gene into the plasmid. Thus, we inserted a DNA fragment with *Eco*RI and *Hin*dIII cohesive ends containing the tetracycline resistance gene from pBR322 in place of the back half of the lysozyme gene. Hosts to this plasmid were resistant to tetracycline and expressed no detectable lysozyme activity. When this vector, pHSe5$_{tet}$, has been cut with the above two enzymes and some other restriction enzyme with a unique site in the "spacer" tetracycline resistance gene and then ligated with unmodified T4 DNA[19] that also has been cut with these two enzymes, the transformants which produce lysozyme contain the altered lysozyme gene sequence of the mutant. Use of the same two restriction sites also facilitated movement into and out of M13 bacteriophage vectors used for DNA sequencing and for *in vitro* mutagenesis. We have subsequently incorpo-

---

to a *Bgl*II site, at which a *trp* operon transcription termination sequence (term) was inserted. Note that the promoter sequence underwent a spontaneous deletion during the construction of pHSe5 as shown at the bottom of the figure. Insertion of the *Eco*RI–*Hin*dIII Tc^r fragment into pHSe5 created pHSe5tet, which is useful in screening for reacquisition of the *Eco*RI–*Hin*dIII lysozyme fragment from mutant phages. The restriction sites are denoted by A, *Ava*I; A2, *Ava*II; Bm, *Bam*HI; Bg, *Bgl*II; C, *Cla*I; E, *Eco*RI; H3, *Hin*dIII; Hp, *Hpa*I; Ps, *Pst*I; Pv, *Pvu*II; S, *Sal*I; P/H, site of blunt end ligation of *Pvu*II/*Hin*cII sites.

rated an M13 origin of replication into versions of our vector in order to allow site-directed mutagenesis of the plasmid.

Subsequent to the completion of this phase of our work, another high-level vector for T4 lysozyme was reported.[21] In this case, the strategy involved tailoring the beginning of the structural gene in order to fit a preexisting successful translational start.

The plasmid series which we have devised differs only slightly from any number of other such vectors which have been reported. The important features are (1) a strong controllable promoter, (2) a gene needed for control of this promoter, (3) an appropriately situated ribosome binding site, and (4) a terminator downstream from the insert for prevention of unwanted messages. Many of the plasmids intended for general use also include a "polylinker" containing numerous restriction sites for convenient insertion of the gene of interest. It is important to stress that optimal levels of expression often require subtle manipulation of the structural gene to appropriately match the properties of the vector.

[15]N Labeling of Proteins

Using our T4 lysozyme expression system, we are able to rapidly produce tens of milligrams of the protein from liter-scale growths of host *E. coli*. This allows us to uniformly or selectively enrich residues in T4 lysozyme with ^{15}N by growing the host bacteria in defined media supplemented with the isotopic label. To maximize the efficiency of selective ^{15}N enrichment of the protein, to avoid isotopic dilution, and to control incorporation of label, we prepare T4 lysozyme using bacteria with lesions in appropriate amino acid metabolic pathways (Tables I, II, and III).

Uniform [15]N Labeling of Proteins

Bacteria grown on a minimal salt medium enriched with an ^{15}N source such as ammonia or glutamic acid will uniformly incorporate the isotope into the backbone and side-chain amide, amine, imidazole, indole, and guanidino groups of proteins[22,23] (Fig. 1). Ammonia is the preferred nitrogen source for *E. coli* and the only inorganic form of this element on which the bacteria can grow.[23] We have produced uniformly ^{15}N-labeled T4 lysozyme by expression of pHSe5 in the prototrophic *E. coli* strain 594 grown on a variation of M9 media supplemented with 0.5 to 1.0 g [15]NH$_4$Cl (>98% isotopic enrichment) per liter. The medium contains 6 g Na$_2$HPO$_4$,

[21] L. J. Perry, H. L. Heynecker, and R. Wetzel, *Gene* **38**, 259 (1985).

[22] B. Tyler, *Annu. Rev. Biochem.* **47**, 1127 (1978).

[23] F. C. Neidhardt (ed.), "*Escherichia coli* and *Salmonella typhimuria*: Cellular and Molecular Biology," Vol. 1. Amer. Soc. Microbiol., Washington, D.C., 1987.

TABLE I

Escherichia coli Amino Acid Metabolism[a]

Amino acid	Sole nitrogen source[b]	Key precursor	N source			Useful host genotype[c]
			α-NH₂	Side chain	α-N to end up	
Glu	+	α-Ketoglutarate; Gln	NH₃, Gln	—	All	gdhA, gltB (*aspC, avtA, ilvE, tyrB)
Gln	+	Glu	Glu	NH₃	All	glnA
Arg	+	Glu	Glu	Glu, Asp	—	argH
Pro	+	Glu	Glu	—	—	proC
Asp	+	Oxaloacetate	Glu	—	Glu, Asn, Lys, Met, Thr, NH₃	*aspC, tyrB
Asn	+	Asp	Asp	NH₃, Gln	—	asnA, asnB
Lys	—	Asp	Asp, Glu	Glu, Asp	—	lysA
Met	—	Asp	Asp	—	—	metC
Thr	—	Asp	Asp	—	Gly, NH₃	thrC
Ile	—	Thr	Glu	—	Glu	*ilvE
Leu	+/−	Pyruvate	Glu	—	Glu	*ilvE, tyrB
Val	—	Pyruvate	Glu, Ala	—	Glu, Ala	*ilvE, tyrB
Ser	+	3-Phosphoglycerate (Gly)	Glu, Gly	—	Gly, Cys, Trp, NH₃	serA
Gly	+	Ser (Thr)	Ser	—	Ser, NH₃	glyA
Cys	—	Ser	Ser	—	—	cysE
Trp	+/−	Chorismate, Ser	Ser	Gln	Ser	trpA,B
Phe	—	Chorismate	Glu	—	Glu	*aspC, ilvE, tyrB
Tyr	+	Chorismate	Glu	—	Glu	*aspC, tyrB
His	+/−	Adenine	Glu	Gln (adenine)	—	hisD
Ala	+	Pyruvate	Glu, Val	—	Glu, Val, NH₃	*avtA

[a] From Refs. 23, 42, and 43.

[b] Growth on the amino acid as the sole nitrogen source (+ indicates growth, +/− indicates possible growth). (From Ref. 22 and from H. Halvorson [Can. J. Microbiol. 18, 1647 (1972)].)

[c] From Ref. 17. Asterisk indicates an aminotransferase deficiency.

TABLE II
Escherichia coli AMINOTRANSFERASES

Aminotransferase[a]	Structural gene	Specificity	Comments
TrA (aspartate)	*aspC*	Glu ↔ Asp, Phe, Tyr	Nonrepressible[b]; highest affinity for Asp
TrD (aromatic)	*tyrB*	Glu ↔ Asp, Leu, Phe, Tyr	Repressible by Tyr[b]
TrB (branched chain)	*ilvE*	Glu ↔ Ile, Leu, Phe, Val	Repressible by Ile, Leu, and Val[c]
TrC (alanine–valine)	*avtA*	Ala ↔ Val	Inhibited by Leu, Ala, and α-amino-butyrate[d]
—	—	Glu ↔ Ala	Uncharacterized[d]

[a] From Ref. 47.
[b] From Ref. 48.
[c] From Ref. 49.
[d] From Refs. 50–52.

3 g KH_2PO_4, 0.5 g NaCl, 0.25 g $MgSO_2 \cdot 7H_2O$, 0.014 g $CaCl_2 \cdot 2H_2O$, 0.005 g $FeCl_3 \cdot 6H_2O$, 0.001 g thiamine, and 10 g glucose per liter. The yield of T4 lysozyme is typically 10–25 mg per liter of media or 10–20% of that obtained with broth. Similar protocols have been reported for uniform ^{15}N labeling of whole cells,[24,25] proteins,[26–28] and nucleic acids.[29,30]

Selective ^{15}N Labeling of Proteins by Residue Type

The backbone and side-chain functional groups in a protein often can be distinguished by their ^{15}N and ^{1}H chemical shifts[1,3] (Fig. 1). Unfortunately, only rarely can resonances be assigned to particular residues on the basis of chemical shift alone.[31] One solution is to label a protein at a single defined site; except for the case of an amino acid which occurs only once in the protein, this requires an approach such as chemical modification of the selected residue. A feasible compromise is selective ^{15}N enrichment of the side chains or peptide backbone of a protein by residue type by growing the host bacteria on a defined medium supplemented with

[24] A. Lapidot and C. S. Irving, *Proc. Natl. Acad. Sci. U.S.A.* **74,** 1988 (1977).
[25] A. Lapidot and C. S. Irving, *J. Am. Chem. Soc.* **99,** 5488 (1977).
[26] T. A. Cross and S. J. Opella, *J. Mol. Biol.* **182,** 367 (1985).
[27] G. M. Smith, L. P. Yu, and D. J. Domingues, *Biochemistry* **26,** 2202 (1987).
[28] P. Leighton and P. Lu, *Biochemistry* **26,** 7262 (1987).
[29] T. L. James, J. L. James, and A. Lapidot, *J. Am. Chem. Soc.* **103,** 6748 (1981).
[30] M. Kime, *FEBS Lett.* **175,** 259 (1984).
[31] H. R. Kricheldorf, *Pure Appl. Chem.* **54,** 467 (1982).

TABLE III
SELECTIVE LABELING OF T4 LYSOZYME WITH ^{15}N-LABELED AMINO ACIDS

Residue labeled	[α-^{15}N]Amino acid	Relevant host genotype[a]	Quantity of amino acid/liter	Comments
Ala	DL-Alanine	avtA aspC ilvE tyrB[b]	600 mg	—
Arg, Glx	L-Glutamate	avtA aspC ilvE tyrB[b]	380 mg	No Arg added; Arg labeled at αN and εN
Asp	L-Aspartate	asnA asnB aspC tyrB	100 mg	Thr labeled 20%
Asx	L-Aspartate	aspC ilvE tyrB[b]	200 mg	200 mg Asn did not prevent incorporation at Asn; 25% Thr labeled
Glx	L-Glutamate	avtA aspC ilvE tyrB[b]	550 mg	25% Ala labeled
Gly	Glycine	glyA	250–375 mg	—
Gly, Ser	Glycine	serA	500 mg	No serine added
Ile	L-Isoleucine	aspC ilvE tyrB[b]	50 mg	L-Alloisoleucine present but no effect
Leu	L-Leucine	aspC ilvE tyrB[b]	60 mg	—
Lys	L-Lysine · HCl	lysA	160 mg	—
Met	L-Methionine	metC	150 mg	—
Phe	L-Phenylalanine	pheA; aspC ilvE tyrB[b]	75 mg; 50–75 mg	Loss of >40% label
Thr	L-Threonine	thr	150 mg	—
Trp	L-Tryptophan	trp	35 mg	
Tyr	L-Tyrosine	avtA aspC ilvE tyrB[b]	35 mg	—
Val	L-Valine	aspC ilvE tyrB[b]	60–100 mg	—

[a] From Ref. 17.
[b] Bacterial strain DL39 (Ref. 46) and DL39 avtA::Tn10.

one or more ^{15}N-labeled amino acids or amino acid precursors. Several proteins, including hemoglobin,[32] α-lytic protease,[5,33] bacteriophage fd and Pf1 coat proteins,[26,34,35] subtilisin inhibitor,[36] T4 lysozyme,[11,37] E. coli thioredoxin,[38] phage P22 c2 repressor,[39] lambda cro repressor,[28] and

[32] A. Lapidot, C. S. Irving, and Z. Malik, J. Am. Chem. Soc. 98, 632 (1976).
[33] W. M. Bachovchin, Proc. Natl. Acad. Sci. U.S.A. 82, 7948 (1985).
[34] M. J. Bogusky, P. Tsang, S. J. Opella, Biophys. Biochem. Res. Commun. 127, 540 (1985).
[35] R. A. Schiksnis, M. J. Bogusky, P. Tsang, and S. J. Opella, Biochemistry 26, 1373 (1987).
[36] M. Kainosho and T. Tsuji, Biochemistry 21, 6273 (1982).
[37] F. W. Dahlquist, R. H. Griffey, L. P. McIntosh, D. C. Muchmore, T. G. Oas, and A. G. Redfield, in "Synthesis and Applications of Isotopically Labeled Compounds" (R. R. Muccino, ed.), p. 533. Elsevier, Amsterdam, 1985.
[38] D. M. LeMaster and F. M. Richards, Biochemistry 24, 7263 (1985).
[39] H. Senn, A. Eugster, G. Otting, F. Suter, and K. Wüthrich, Eur. Biophys. J. 14, 301 (1987).

N-*ras* p21 (S. Campbell-Burk and A. G. Redfield, unpublished observations), have been selectively labeled with one or more of the following ^{15}N-labeled amino acids: Ala, Arg, Asp, Asx, Gly, Glx, His, Ile, Leu, Lys, Met, Phe, Pro, Ser, Thr, Trp, Tyr, and Val. Analogously, tRNA has been selectively ^{15}N labeled by nucleoside type.[40,41]

Biosynthetic enrichment of selected amino acids in a protein with ^{15}N requires efficient incorporation of ^{15}N precursors into the protein in a predictable fashion. Two major concerns are isotopic dilution and incorporation of the ^{15}N label into undesired residues. Isotopic dilution results in a decrease in the expected ^{15}N enrichment of a residue, due either to transaminase activity which exchanges the ^{15}N label with ^{14}N from other nitrogen-containing molecules or to endogenous synthesis of unlabeled amino acids by the bacteria. In many cases this simply reduces the enrichment of selected sites relative to the background ^{15}N in the sample and hence decreases the effective quantity of labeled protein available for an NMR experiment. However, isotopic dilution can be very detrimental for experiments which rely on the simultaneous incorporation of multiple isotopic labels into a protein.[13] Incorporation of label into undesired sites results from the metabolic conversion of the ^{15}N-labeled compound to amino acids other than those originally intended to be labeled. "Scrambling" of the ^{15}N label to many amino acid types through the general nitrogen pool of the bacteria is generally not a problem when using media supplemented with most or all amino acids. However, undesired isotopic enrichment of one or a few residue types due to specific conversion of the exogenously supplied ^{15}N-labeled amino acid to a metabolic derivative, or due to aminotransferase-catalyzed nitrogen exchange, warrants serious concern. This particularly holds for experiments, such as those leading to the assignment of NMR resonances, which absolutely require ^{15}N labeling of residues in a predictable fashion. Isotopic dilution and incorporation of label into undesired sites can be limited or prevented by repression of bacterial amino acid metabolism using defined media supplemented with essentially all amino acids and by use of bacterial hosts with lesions blocking the appropriate steps of amino acid synthesis and degradation.

Repression and Inhibition of Amino Acid Metabolism. Amino acid biosynthesis in bacteria is regulated at the level of enzymatic activity and at the level of gene expression.[23] Endogenous synthesis of unlabeled amino acids and aminotransferase-catalyzed nitrogen exchange often can be restricted by supplying the host bacteria with sufficient quantities of amino acid nutrients. Therefore, the simplest approach to selective bio-

[40] R. H. Griffey, C. D. Poulter, Z. Yamaizumi, S. Nishimura, and B. L. Hawkins, *J. Am. Chem. Soc.* **105**, 143 (1983).
[41] S. Roy, M. Z. Papastravos, V. Sanchez, and A. G. Redfield, *Biochemistry* **23**, 4395 (1984).

synthetic labeling of proteins is to use defined growth media containing excess unlabeled amino acids. Such media are described below. The extent to which this approach will prevent isotopic dilution or incorporation of label at undesired sites depends on many factors, including the amino acid in question and the growth characteristics of the bacterial host.

In the cases of the amino acids which lie at the end of anabolic pathways (Table I), we and others have not observed any misincorporation of ^{15}N into specific residues, for example, via the general nitrogen pool. In our initial work, we had found reasonable success in labeling the phenylalanine sites in T4 lysozyme using a strain auxotrophic for phenylalanine (*pheA*) grown under conditions which repress the expression of some of the endogenous aminotransferases.[11] Similarly, Senn *et al.*[39] did not observe loss of ^{15}N when labeling with P22 *c2* repressor using prototrophic *E. coli* grown with [^{15}N]leucine in a defined medium. However, we and others[36] have not found this to be a generally useful approach, particularly when cells are exposed to the labeled amino acid for growth periods of several hours, as isotopic dilution of ^{15}N-labeled amino acids and conversion of the labeled amino acids to metabolic derivatives have been observed.

Although repression of endogenous amino acid synthesis with supplied nutrients may be sufficient for ^{15}N enrichment of proteins with some amino acids, our experience with T4 lysozyme demonstrates that use of bacterial hosts with lesions in their amino acid metabolism is a more efficient and reliable approach for selective labeling. In addition to controlling the labeling, this allows lower amounts of labeled amino acids in the defined growth media, since continuous high levels for repression are not needed. This also permits the cells to be incubated for longer periods after induction without problems arising from endogenous synthesis or transaminase activity.

Bacterial Strains Auxotrophic for Amino Acids. The metabolic pathways for nitrogen assimilation and amino acid biosynthesis in bacteria are extensively discussed in several reviews and texts.[22,23,42,43] In Table I, a summary of the amino acid metabolism of *E. coli* is given. Bacteria with lesions in these pathways are a powerful tool for selectively incorporating isotopic labels into an expressed protein.[44] Isotope dilution and label incorporation can be controlled using bacterial strains auxotrophic for the appropriate amino acids. Perhaps more importantly, the metabolic pathways of a bacteria can be exploited to direct ^{15}N from a precursor metabo-

[42] H. E. Umbarger, *Annu. Rev. Biochem.* **47**, 533 (1978).
[43] D. A. Bender, "Amino Acid Metabolism," 2nd ed. Wiley, New York, 1985.
[44] D. M. LeMaster and J. E. Cronan, Jr., *J. Biol. Chem.* **257**, 1224 (1982).

lite to one or more amino acids in a controlled fashion. *Escherichia coli* strains with defined lesions in amino acids biosynthesis are often readily available from laboratory stocks[17] or can be easily constructed by standard techniques such as generalized transduction.[45] We have had little problem in expressing the lysozyme vector in a variety of *E. coli* strains.

In Table III, a summary of the bacterial strains used to produce selectively T4 lysozyme samples is given. In the cases of Gly, Lys, Met, Ser, Thr, and Trp, which are not substrates for aminotransferases, we have used bacteria specifically auxotrophic for these amino acids. In the cases of Asx, Glx, Ile, Leu, Phe, Tyr, and Val, bacteria with auxotrophies due to aminotransferase deficiencies were used to label T4 lysozyme with ^{15}N; this approach is discussed below.

Loss or misincorporation of an isotopic label is a serious concern when growing bacteria with ^{15}N-labeled Glu, Gln, Asp, and possibly Ala, Gly, Ser and Thr. In these cases, the metabolic conversion of the amino acid to its derivatives as well as dilution of the amino acid by endogenous synthesis or transaminase activity are potential problems. Bacteria with lesions in both anabolic and catabolic pathways producing and utilizing these amino acids are useful if not necessary for obtaining ^{15}N-enriched proteins at these residues.

Difficulty in incorporation of [^{15}N]glutamate into proteins is expected due to the central role of this amino acid in nitrogen and amino acid metabolism.[22,23] Somewhat surprisingly, we produced T4 lysozyme labeled at both glutamate and glutamine residues and only 25% at alanine residues by expressing *pHSe5* in an aminotransferase-deficient strain DL39 *avtA::Tn5*[46] grown with [^{15}N]Glu (Table III, Fig. 4). The transfer of label into alanine appears to be due to the one or more uncharacterized alanine–glutamate aminotransferases. Bacteria with lesions in the glutamine synthetase (glutamate–ammonia lyase) (*glnA*) gene cannot assimilate NH_3 and glutamate to yield glutamine and thus might be used to label glutamate but not glutamine residues with [^{15}N]Glu. Although bacteria lacking glutamate synthase (*gltB*) and glutamate dehydrogenase (*gdhA*) genes are auxotrophic for glutamate in minimal media, these would probably only be useful to label selectively Gln residues with [^{15}N]glutamine. Bacteria, such as DL39 (Table III), which lack the general glutamate-dependent aminotransferases, are useful to prevent transfer of isotopic label from glutamate to Asp, Ile, Leu, Phe, Tyr, and Val.

In a similar fashion, the T4 lysozyme produced using a bacterial strain auxotrophic for aspartate (Table III) grown with [^{15}N]Asp, with or with-

[45] J. H. Miller, "Experiments in Molecular Genetics." Cold Spring Harbor Laboratory, Cold Spring Harbor, New York, 1972.
[46] D. M. LeMaster and F. M. Richards, *Biochemistry,* in press (1988).

out unlabeled Asn, was isotopically labeled equally at the aspartate and asparagine residues and approximately 25% at additional sites. These sites are tentatively assigned to the threonine residues. This undesired isotopic enrichment may be prevented by introduction of threonine auxotrophy into the bacterial strain. To prevent conversion of aspartate to asparagine, bacteria deficient in both the NH_3-dependent (*asnA*) and glutamine-dependent (*asnB*) asparagine synthetases were used. In *E. coli* conversion of asparagine to aspartate by an asparaginase activity appears minor; an aspartate auxotroph such as DL39[46] grows very poorly in media with asparagine substituted for aspartate.

Unfortunately, there are no useful alanine auxotrophs of *E. coli*. Using a bacterial strain deficient in the alanine–valine aminotransferase (Table II)[47–52] grown with DL-[^{15}N]alanine, we selectively labeled the Ala residues of T4 lysozyme (Fig. 4). Misincorporation of the ^{15}N label into glutamate or any of its derivative amino acids was not observed. However, the level of enrichment of the alanine residues was not determined.

Escherichia coli auxotrophic for serine and threonine have been characterized. However, potentially high concentrations of these amino acids required for growth and nonlinear dependence of growth on the amino acid concentrations[38] could make production of protein enriched at these residues prohibitively expensive. Furthermore, both serine and threonine deaminases are found in *E. coli*. Serine is the precursor to glycine, cysteine, and tryptophan, and, perhaps more importantly, is involved in the one-carbon metabolism of the cell.[23,53,54] The interconversion of serine and glycine and the conversion of serine to cysteine and tryptophan can be repressed or prevented with mutations (Table I). Threonine is derived from aspartate and in turn is the precursor of isoleucine. This latter pathway involves deamination of threonine and thus could lead to dilution of [^{15}N] threonine but not to specific transfer of label to isoleucine. Threonine may also be cleaved to yield glycine by an uncharacterized pathway in *E. coli*[55] which involves a constitutive threonine dehydrogenase[56]; this could account for reported scrambling of ^{15}N from supplied threonine (referenced in Ref. 28).

[47] D. Rudman and A. Meister, *J. Biol. Chem.* **200**, 591 (1953).
[48] D. H. Gelfand and R. A. Steinberg, *J. Bacteriol.* **130**, 429 (1977).
[49] F.-C. Lee-Peng, M. A. Hermodson, and G. B. Kohlhaw, *J. Bacteriol.* **139**, 339 (1979).
[50] D. McGilvray and H. E. Umbarger, *J. Bacteriol.* **120**, 715 (1974).
[51] J. O. Falkinham III, *Mol. Gen. Genet.* **176**, 147 (1979).
[52] W. Whalen and C. M. Berg, *J. Bacteriol.* **158**, 571 (1984).
[53] H. E. Umbarger, M. A. Umbarger, and P. M. L. Siu, *J. Bacteriol.* **85**, 1431 (1963).
[54] L. I. Pizer, *J. Bacteriol.* **89**, 1145 (1965).
[55] J. Fraser and E. B. Newmann, *J. Bacteriol.* **122**, 810 (1975).
[56] R. Putter, V. Kapoor, and E. B. Newman, *J. Bacteriol.* **132**, 385 (1977).

The pathways of amino acid biosynthesis are very useful to intentionally direct ^{15}N isotopic label from a precursor to a given amino acid. For example, we have ^{15}N-enriched T4 lysozyme at the glutamate and glutamine residues and the aspartate and asparagine residues using [^{15}N]-glutamate and [^{15}N]aspartate, respectively. We have also enriched the serine and glycine residues of this protein with ^{15}N using a *serA* bacterial strain grown on media lacking serine but containing [^{15}N]glycine. This bacteria is deficient in 3-phosphoglycerate dehydrogenase and can only synthesize serine by the glycine cleavage pathway.[57] A second representative example is to label a protein at the arginine residues using glutamate as the source of the nitrogen isotope. Finally, it should be possible to label amino acids within an amino acid metabolic family or to avoid labeling such amino acids when uniformly labeling the remaining residues with ^{15}NH$_4$Cl by using the appropriate bacterial strains grown on media lacking or containing selected amino acids.

Aminotransferases. Aminotransferases (or transaminases) catalyze the reversible transfer of the amino groups between α-amino acids and α-keto acids.[47,58] Essentially all of the amino acids derive their α-amino groups directly or indirectly from aminotransferase-catalyzed reactions involving glutamate as the nitrogen donor (Table I). Four general aminotransferases with overlapping specificities have been characterized in *E. coli*; the properties of these enzymes are summarized in Table II. In addition, there is at least one glutamate–alanine aminotransferase in *E. coli* for which no mutations are known.[51,52] These aminotransferases can remove the α-amino nitrogen from exogenously supplied Ala, Asp, Glu, Ile, Leu, Phe, Tyr, and Val and thus are a cause of isotopic dilution or of isotope incorporation at undesired sites when labeling with these amino acids. Similar concerns apply for deuterium labeling of proteins in the Cα position, since the aminotransferase-catalyzed reaction removes and replaces the Cα hydrogen.

Two examples from our experience in labeling T4 lysozyme exemplify the potential problems caused by aminotransferases. We observed an approximate 40% loss of ^{15}N when isotopically enriching T4 lysozyme by growing a host bacteria auxotrophic for phenylalanine (*pheA*) on defined media supplemented with [^{15}N]phenylalanine. Presumably this resulted from aminotransferase-catalyzed amino exchange, possibly by the nonrepressible transaminase A (Table II). We have also observed approximately 25% incorporation of ^{15}N label in the alanine residues of T4 lysozyme produced in a bacterial strain grown with [^{15}N]glutamate (Fig. 4).

[57] E. B. Newmann, G. Batist, J. Fraser, S. Isenberg, P. Weyman, and V. Kapoor, *Biochim. Biophys. Acta* **421**, 97 (1976).
[58] R. A. Jensen and D. H. Calhoun, *CRC Crit. Rev. Microbiol.* **8**, 229 (1981).

The strain was deficient in the four general transaminases, including the Ala–Val transaminase C (DL39 $avtA::Tn5$). Therefore, this undesired incorporation of label in the alanine residues is attributed to a Glu–Ala aminotransferase.

Gelfand and Steinberg[48] constructed a polyauxotrophic *E. coli* strain, DG30, with lesions in the $aspC$, $tyrB$, and $ilvE$ genes. The poor growth characteristics of this strain limited its use as a general host for incorporating ^{15}N-labeled amino acids into proteins.[38] Although pyridoxal analogs have been used to inhibit aminotransferases, the importance of pyridoxal enzymes to bacterial metabolism discourage this nonselective method to limit aminotransferase activity. Recently, LeMaster and Richards[46] constructed the *E. coli* strain DL39 carrying mutations only in the three general glutamate-dependent aminotransferase genes, $aspC$, $tyrB$, and $ilvE$. The strain grows robustly and is a good host for expression of T4 lysozyme. We have further introduced a lesion in the $avtA$ gene of this strain by generalized transduction from CBK741 (C. Berg, personal communication). The strain DL39 $avtA::Tn5$ is auxotrophic for Asp, Ile, Leu, Phe, Tyr, and Val by virtue of these aminotransferase deficiencies. These two strains have proved very useful for labeling T4 lysozyme[13] (Table III) and thioredoxin[46] with specific α-[15]N- and α-[2]H-labeled amino acids. The strain DL39 and derivatives thereof should emerge as standard bacterial hosts for selectively labeling proteins with a wide variety of [15]N- and [2]H-labeled amino acids.

Incorporation of ^{15}N into Protein Side Chains. The above discussion primarily emphasized incorporation of ^{15}N into the peptide backbone of a protein. Similar considerations apply to labeling the nitrogen-containing side chains of a protein, although aminotransferase activity is not a problem. We have incorporated [ε_1-^{15}N]tryptophan into T4 lysozyme using a bacterial strain auxotrophic for tryptophan ($trpA$). Several proteins have been labeled with [δ_1-^{15}N] and [ε_2-^{15}N]histidine[5,33] and with [ζ-^{15}N]lysine.[28] We have also utilized the biosynthetic pathways of *E. coli* to enrich the εN and $\eta_{1,2}$N of arginine using ^{15}N]glutamate and ^{15}N]aspartate, respectively, as the source of the isotope. In Table I, a summary of the metabolism of amino acid side chains is also given.

Amino Acid Transport. Bacteria take up all 20 amino acids from the extracellular medium by active transport.[59] With the exception of glycine and glutamine, *E. coli* has multiple transport systems catalyzing the intake of each amino acid, and, with the exception of those serving alanine and serine, these systems are highly stereospecific.[59] Mutations which result in decreased amino acid uptake have been identified in bacteria. A

[59] J. L. Milner, B. Vink, and J. M. Wood, *CRC Crit. Rev. Biotechnol.* **5**, 1 (1987).

potential concern in growing bacteria in a defined medium containing excess amino acids is the competitive inhibition of uptake of a labeled amino acid by unlabeled amino acids which share common transport systems.[60] We have not observed any difficulties attributable to amino acid transport with the media and bacterial strains used to label T4 lysozyme with ^{15}N.

Protocol for Selective ^{15}N Labeling of T4 Lysozyme

Growth Media. Proteins are labeled at specific residues by growing the host bacteria on a defined media enriched with the appropriate ^{15}N-labeled amino acid or precursor. Several such media have been described, including M9 medium supplemented with each unlabeled amino acid at a level of 100 to 200 μg/ml[28,61] and a minimal salt medium containing amino acids in a composition close to that of polypeptone.[36] In general, the defined media contain essentially all amino acids to repress endogenous synthesis by the bacteria. The media must also support bacterial growth to a high cell density for sufficient protein production.

To produce specifically ^{15}N-labeled T4 lysozyme, we use a medium[11,38] containing appropriate unlabeled L-amino acids (0.50 g alanine, 0.40 g arginine, 0.40 g aspartic acid, 0.05 g cystine, 0.40 g glutamine, 0.65 g glutamic acid, 0.55 g glycine, 0.10 g histidine, 0.23 g isoleucine, 0.23 g leucine, 0.42 g lysine hydrochloride, 0.25 g methionine, 0.13 g phenylalanine, 0.10 g proline, 2.10 g serine, 0.23 g threonine, 0.17 g tyrosine, and 0.23 g valine), as well as 0.50 g adenine, 0.65 g guanosine, 0.20 g thymine, 0.50 g uracil, 0.20 g cytosine, 1.50 g sodium acetate, 1.50 g succinic acid, 0.50 g NH$_4$Cl, 0.85 g NaOH, and 10.50 g K$_2$PO$_4$ per 950 ml water. After autoclaving, 50 ml of 40% glucose, 4 ml of 1 M MgSO$_4$, 1.0 ml of 0.01 M FeCl$_3$, and 10 ml of a filter-sterilized solution containing 2 mg CaCl$_2 \cdot$2H$_2$O, 2 mg ZnSO$_4 \cdot$7H$_2$O, 2 mg MnSO$_4 \cdot$H$_2$O, 50 mg L-tryptophan, 50 mg thiamine (B$_1$), 50 mg niacin, 1 mg biotin, and 100 mg ampicillin are added under sterile conditions. The final solution is near pH 7.2. The amino acids corresponding to the residues to be labeled are initially omitted from the media. An inoculating bacterial culture is grown overnight in the defined media supplemented with a limiting amount of unlabeled amino acid. The ^{15}N-labeled amino acids are added along with 20–40 ml of this culture to 1 to 4 liters of the growth media and the remainder of the preparation is as described in Appendix I at the end of this paper.

In Table III a summary of the quantities of labeled amino acids used to

[60] J. R. Piperno and D. L. Oxender, *J. Biol. Chem.* **243**, 5914 (1968).
[61] D. L. LeMaster and F. M. Richards, *Anal. Biochem.* **122**, 238 (1982).

selectively label T4 lysozyme is given. Our standard protocol involves using small-scale test cultures to determine the growth rates, saturating cell densities, and lysozyme production of a particular bacterial host as a function of the concentration of the amino acid to be used in the labeling experiment. The ^{15}N-labeled amino acids are supplied at a minimal level for sufficient cell growth and lysozyme production. This is generally approximately 50% of the concentration normally used in the defined media. The yields of purified T4 lysozyme are 10–60% of those obtained on rich media, varying somewhat with the particular bacterial strain used and the growth conditions.

Availability of ^{15}N-Labeled Amino Acids. Fortunately, many ^{15}N-labeled amino acids are commercially available and are not prohibitively expensive when used conservatively for liter-scale preparations. Several synthetic and biosynthetic methods have been developed to prepare isotopically labeled amino acids or amino acid precursors. LeMaster and Richards[61] have described a procedure for preparative-scale isolation of isotopically labeled amino acids produced by bacteria.

Although we commonly use L-amino acids in the bacterial growth media, we have not observed any problems with racemic amino acids. In general, the nitrogen label from the D-amino acids will not be incorporated into proteins by the bacterial host. The bacterial transport systems for amino acids other than alanine and serine are stereospecific. *Escherichia coli* can grow on a variety of D-amino acids, albeit by deamination of the metabolite.[62] A notable exception is the interconversion of D- and L-alanine by alanine racemase. In the case of serine however, the L-isomer should be used because D-serine is toxic to *E. coli* and is detoxified by D-serine deaminase.[63] After correction for the quantity of the L-isomer, the cost of using commercially available L- or DL-amino acids is often approximately equal.

Determination of Level of Isotopic Enrichment

The most direct assessment of the specificity of the incorporation of ^{15}N label into a protein is to use NMR to determine the number of sites containing the label and the relative intensities of the resonances from these sites. This should agree with that expected from the amino acid sequence of the protein. We have measured the level of isotopic enrichment of specific residues in T4 lysozyme by using NMR to observe multiply labeled sites in the protein. For example, ^{15}N enrichment of the phenylalanine residues in T4 lysozyme was assessed by using ^{13}C NMR to

[62] J. Kuhn and R. L. Somerville, *Proc. Natl. Acad. Sci. U.S.A.* **68**, 2484 (1971).
[63] S. D. Cosloy and E. McFall, *J. Bacteriol.* **114**, 685 (1973).

measure the ratio of $^{13}C-^{15}N$ and $^{13}C-^{14}N$ in a sample of the protein colabeled with [^{15}N]Phe and [1-^{13}C]Leu.[12]

Using mass spectroscopy, the absolute level and specificity of ^{15}N enrichment of a protein can be determined.[36] A convenient protocol involves analysis of the amino acid phenylthiohydantoin derivatives obtained by Edman degradation or from a total protein hydrolysate.[39]

NMR Measurements of ^{15}N-Labeled Proteins

The goal of NMR studies of proteins is to obtain insights into the structure and function of these complex molecules. Enrichment of proteins with ^{15}N, as well as ^{13}C and ^2H, is a powerful aid to these studies. Selective isotope labeling is a means to edit the complex NMR spectra of proteins and thereby address specific questions about the molecules. The applications of ^{15}N labeling of proteins fall into two broad and overlapping categories; the ^{15}N signal is directly or indirectly detected to yield information such as nitrogen chemical shift and relaxation, or the nucleus is employed indirectly as a flag to generate and disperse spectral lines of otherwise complex ^1H NMR spectra.

The ^{15}N nuclei in proteins can be directly observed by ^{15}N NMR in the form of one-dimensional ^{15}N[1,6,7,64] or two-dimensional correlated $^{15}N-^1H$ spectra, including heteronuclear COSY[65,66] and RELAY[67] spectra. Isotopic enrichment and experiments which exploit polarization transfer from geminal protons (such as INEPT, DEPT, and two-dimensional heteronuclear COSY)[6,7,64,68] help alleviate the poor sensitivity associated with ^{15}N NMR. The generally narrower linewidths of ^{15}N NMR peaks compared to those of protons is useful, especially with larger proteins. ^{15}N nuclei in proline residues or in functional groups with protons rapidly exchanging with the solvent may only be observed by direct measurements. Using a variety of ^{15}N NMR techniques, ^{15}N nuclei in proteins can be characterized by chemical shift, ^1H scalar coupling, attached protons, hydrogen exchange, and relaxation including NOE. The heteronuclear $^1H-^{15}N$ Overhauser enhancement ranges from -4.93 to -0.12, with increasing correlation time for the motions modulating the dipolar interaction between the nuclei. Measurements of ^{15}N NOE can be used to distinguish between rigid and mobile sites within a protein.[1,27,28,35] Similarly, ^{15}N

[64] D. T. Pegg, D. M. Doddrell, and M. R. Bendall, J. Chem. Phys. 77, 2745 (1982).
[65] A. A. Maudsley and R. R. Ernst, Chem. Phys. Lett. 50, 368 (1977).
[66] G. Bodenhausen and R. Freeman, J. Magn. Reson. 28, 471 (1977).
[67] S. K. Sarkar and A. Bax, J. Magn. Reson. 63, 512 (1985).
[68] A. Bax, "Two Dimensional Nuclear Magnetic Resonance in Liquids." Delft Univ. Press, Boston, Massachusetts, 1982.

T_1 and T_2 measurements can yield information regarding the environment, motions, and hydrogen exchange of nitrogens within a protein. Structural and dynamic information is also obtained by solid-state ^{15}N NMR experiments.[69,70]

The ^{15}N nuclei can also be detected indirectly by the observation of bound protons to generate two-dimensional $^{15}N-^{1}H$ spectra. Griffey and Redfield[8] have recently reviewed this rapidly developing field. These approaches utilize detection of ^{1}H for the highest sensitivity[71] and indirectly measure ^{15}N shift information by the perturbation of the proton spins through irradiation of the heteronucleus (Fig. 3). A variety of experimental methods utilizing polarization transfer,[65,72] modulation of stimulated echoes,[73] and heteronuclear zero- and double-quantum coherence[40,41,74-76] have been developed to measure $^{15}N-^{1}H$ direct and relayed scalar correlation spectra[77,78] and NOE spectra.[10,79] The sensitivity of ^{1}H NMR is invaluable for studying ^{15}N in proteins, especially when high sample concentrations are not possible because of protein solubility or availability, and when short measurement times are necessary, either due to sample stability or for experimental reasons such as real-time hydrogen exchange. Indirect detection of ^{15}N nuclei with ^{1}H NMR relies on the large[3] ($J^1 \sim 95$ Hz) spin–spin coupling between the nitrogen nucleus and a directly bonded proton. This strong coupling is particularly advantageous for studying larger proteins which have correspondingly larger linewidths. Indirect methods have several additional requirements compared to direct observation of ^{15}N in biomolecules. Resonances from protons not directly bonded to ^{15}N must be efficiently eliminated by appropriate phase cycling, by substraction of spectra, or by additional pulse sequences. These unlabeled proton resonances can be removed in principle, but in practice remain as noise in the spectra because of spectrometer instability. Of course, this is much less of a problem with isotopically enriched samples

[69] S. J. Opella, P. L. Stewart, and K. G. Valentine, *Q. Rev. Biophys.* **19**, 7 (1987).

[70] T. G. Oas, C. J. Hartzell, F. W. Dahlquist, and G. P. Drobny, *J. Am. Chem. Soc.* **109**, 5962 (1987).

[71] D. H. Live, D. G. Davis, W. C. Agosta, and D. Cowburn, *J. Am. Chem. Soc.* **106**, 6104 (1984).

[72] G. Bodenhausen and D. J. Ruben, *Chem. Phys. Lett.* **69**, 185 (1980).

[73] A. G. Redfield, *Chem. Phys. Lett.* **96**, 537 (1983).

[74] A. Bax, R. H. Griffey, and B. L. Hawkins, *J. Magn. Reson.* **55**, 301 (1983).

[75] L. Müller, *J. Am. Chem. Soc.* **101**, 4481 (1979).

[76] P. H. Bolton, *J. Magn. Reson.* **57**, 427 (1984).

[77] P. H. Bolton, *J. Magn. Reson.* **62**, 143 (1985).

[78] D. Brühwiler and G. Wagner, *J. Magn. Reson.* **69**, 546 (1986).

[79] M. Rance, P. E. Wright, B. A. Messerle, and L. D. Field, *J. Am. Chem. Soc.* **109**, 1591 (1987).

Forbidden Echo

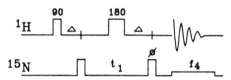

Isotope-Directed NOE

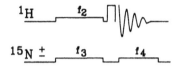

Homo/Hetero Isotope-Directed NOESY

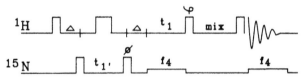

FIG. 3. The pulse sequences used to obtain the spectra discussed in this chapter. The proton 90° pulses are modified jump-return pulses[41,80] to suppress the water resonance and the proton 180° pulses are two sequential jump-return pulses. The nitrogen pulses are hard 90° pulses and the broadband nitrogen decoupling at frequency f_4 is accomplished using a WALTZ-16 sequence. The delay Δ is nominally equal to $1/2\,J$, or 5.5 msec, where J is the nitrogen–proton splitting in hertz, but is usually shortened to 4.5 msec to reduce signal loss due to relaxation. The pulses are phase cycled and the signal is phase shifted within the computer to eliminate signals from unlabeled protons and to obtain quadrature detection and absorption signals in both dimensions. The forbidden-echo experiment (HMQC) is discussed in Refs. 8, 11, 13, and 74. A spin-echo difference spectrum is obtained with $t_1 = 0$ and $\phi = \pm 180°$. The isotope-directed NOE experiment utilizes selective proton preirradiation at f_2 and spectra are recorded as the difference spectra with and without selective ¹⁵N predecoupling at f_3.[9,10] The homo/hetero isotope-directed NOESY experiment is selective for the "source" ¹H (homo) or ¹⁵N (hetero) shift of the ¹⁵N-labeled proton from which the ¹H–¹H NOE originates. The homo-IDNOESY experiment closely resembles a conventional NOESY experiment except that the first ¹H 90° pulse is replaced by a difference echo sequence ($t_{1'} = 0$ and $\phi = \pm 180°$) to select for labeled protons.[16] The hetero-IDNOESY experiment replaces the first proton pulse with a forbidden-echo sequence and $t_1 = 0$ to yield the ¹⁵N shift of the "source" proton. Both t_1 and $t_{1'}$ can be simultaneously incremented to yield a mixture of the ¹H and ¹⁵N chemical shifts of the source-labeled proton (A. G. Redfield, unpublished observations).

than it is for samples with natural abundance ^{15}N. Since all nitrogen-bound protons are exchangeable, labeled proteins are studied predominantly in H_2O buffers and thus suppression of the water resonance is necessary.[80] Finally, proton-detected heteronuclear NMR measurements require instrumental capabilities, such as ^{15}N and ^{13}C broadband decoupling, which have only recently become available on commercial spectrometers.

The ^{15}N nuclei can also be utilized in a more passive manner to filter or edit otherwise complex one- and two-dimensional ^{1}H NMR, yielding difference spectra corresponding to protons labeled or not labeled by the nitrogen or interacting with labeled protons. Obviously, the ^{15}N chemical shift information can be obtained by using selective ^{15}N pulses or ^{15}N decoupling. Difference decoupling (INDOR) and spin-echo difference measurements[8,11,81] produce one-dimensional ^{1}H spectra of ^{15}N-bonded protons. Isotope-directed[9] and -detected[82] NOE measurements yield ^{1}H NOE from or to labeled protons, respectively (Fig. 3). Two-dimensional isotope-filtered ^{1}H-^{1}H scalar (COSY, RELAY, and TOCSY) and NOE experiments have been described[83-87] (Fig. 3). Using these experiments for editing or filtering two-dimensional proton spectra, ^{1}H-^{1}H subspectra corresponding to only labeled or unlabeled protons can be obtained. Clearly selective isotope labeling is a powerful supplement to assigning[10,13,35,88] or testing the assignment[28] of the ^{1}H spectra of proteins obtained by proton only methods.

Uniformly ^{15}N-Labeled T4 Lysozyme

The ^{1}H-^{15}N forbidden-echo spectrum of uniformly ^{15}N-labeled T4 lysozyme in H_2O is shown in Figs. 1 and 5A. A forbidden-echo (or heteronuclear multiple-quantum COSY or heteroCOSY) spectrum shows only peaks from detected protons directly bonded to ^{15}N nuclei[1,40,41] (Fig. 3). The spectrum is detected with the sensitivity of protons but affords the resolution of the ^{15}N dimension. This spectrum serves as a "fingerprint" of T4 lysozyme. On the basis of ^{1}H and ^{15}N chemical shifts, the observed

[80] A. G. Redfield, *NATO Adv. Study Inst. Ser., Ser. A* **107**, 555 (1986).

[81] M. Emshwiller, E. L. Hahn, and D. Kaplan, *Phys. Rev.* **118**, 414 (1960).

[82] M. A. Weiss, A. G. Redfield, and R. H. Griffey, *Proc. Natl. Acad. Sci. U.S.A.* **83**, 1325 (1986).

[83] A. Bax and M. A. Weiss, *J. Magn. Reson.* **71**, 571 (1987).

[84] E. Wörgötter, G. Wagner, and K. Wüthrich, *J. Am. Chem. Soc.* **108**, 6162 (1986).

[85] G. Otting, H. Senn, G. Wagner, and K. Wüthrich, *J. Magn. Reson.* **70**, 500 (1986).

[86] L. Lerner and A. Bax, *J. Magn. Reson.* **69**, 375 (1986).

[87] S. W. Fesik, R. T. Gampe, Jr., and T. W. Rockway, *J. Magn. Reson.* **74**, 366 (1987).

[88] H. Senn, G. Otting, and K. Wüthrich, *J. Am. Chem. Soc.* **109**, 1090 (1987).

peaks can be assigned by functional group. The resonances from ^{15}N–^{1}H pairs in lysine ζ-amine, arginine guanidino (εN and $\eta_{1,2}$N), tryptophan indole ε_1N–H, and glutamine and asparagine primary amide groups and the backbone secondary amides are indicated. The histidine imidazole resonances are not detected due to rapid exchange with the aqueous solvent; the amino and guanidino resonances are broadened for the same reason and are not observed at higher pH.

On the order of 150 peaks are detected in the amide region of the forbidden-echo spectrum of T4 lysozyme (18.7 kDa). Typically these spectra are recorded in a few hours using 250 μl of a 2–4 mM protein sample. However, acceptable spectra can be obtained in as little as 10 min (A. G. Redfield, unpublished observations). This demonstrates the applicability of this technique to studying larger proteins. The forbidden-echo experiment is also well suited for measuring the amide hydrogen exchange kinetics of a protein.[10,13]

T4 Lysozyme Selectively Labeled by Amino Acid Class

The forbidden-echo spectra of T4 lysozyme labeled with ^{15}N at its leucine, alanine, glutamine and glutamate, and serine and glycine residues are presented in Fig. 4 and, at the phenylalanine residues, in Fig. 5. In each case the number of observed peaks was consistent with the amino acid composition of the protein. In general, the resonances are well dispersed in the two dimensions with few degeneracies in both ^{15}N and ^{1}H chemical shifts. The only striking correlation of chemical shift with amino acid type is seen with glycine, having ^{15}N chemical shifts 10–15 ppm upfield from the amino acids with side chains[31,35,38] (Fig. 4).

In Fig. 5D, the forbidden-echo spectrum of [^{15}N]Phe T4 lysozyme near the midpoint of thermal denaturation (pH ~ 2, 48°) is shown. Three additional resonances arise from the phenylalanine residues in the denatured state of the protein. The folded and unfolded forms of the protein are in slow exchange under these conditions. Further studies of native and denatured T4 lysozyme, such as measurements of solvent exchange, are possible by observation of only a limited number of ^{15}N-labeled sites in the protein. We have also selectively measured the relaxation rates of ^{15}N-labeled phenylalanine amide protons in T4 lysozyme using spin-echo difference spectroscopy.[12] Severe overlap of resonances usually prohibits similar studies of native and denatured proteins by ^{1}H NMR alone.

Assignment of ^{15}N–^{1}H Peaks in Forbidden-Echo
   Spectra of T4 Lysozyme

The peaks in the ^{15}N–^{1}H spectrum of uniformly ^{15}N-labeled T4 lysozyme are readily identified by comparison with the spectra of selectively

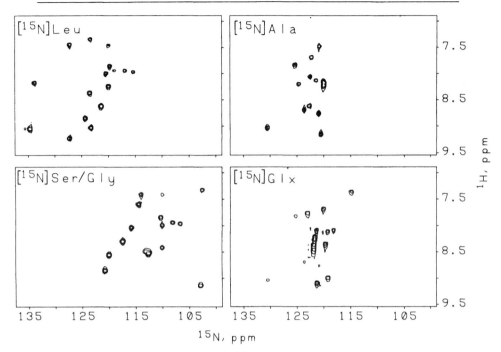

FIG. 4. The forbidden-echo spectra of T4 lysozyme labeled with ¹⁵N at the leucine, alanine, serine plus glycine, and glutamate and glutamine (Glx) residues. The six most downfield peaks in the nitrogen dimension of the spectrum of Ser/Gly-labeled lysozyme arise from the serine amides and are not present in the spectrum of protein labeled only with [¹⁵N]glycine. The weak peaks seen in the spectrum of Glx-labeled lysozyme arise from alanine amides as seen in the spectrum of alanine-labeled protein.

labeled protein (Fig. 5A,B). This approach, using ¹⁵N to enrich T4 lysozyme at one or two amino acid classes (Table I), has enabled us to assign over 120 of 160 total backbone ¹⁵N–¹H resonances by amino acid type.

To further assign resonances to specific residues, several approaches can be taken. For example, numerous variants of T4 lysozyme with single amino acid substitutions have been identified.[89,90] Resonances can be unambiguously assigned to the substituted residues by comparison of the spectra of wild-type and mutant proteins. This approach has proved successful with selectively labeled proteins (Fig. 5C).

A structurally nonperturbing method to assign resonances is through multiple labeling of a protein to uniquely identify a single site. For exam-

[89] T. Alber, S. Dao-pin, J. A. Nye, D. C. Muchmore, and B. W. Matthews, *Biochemistry* **26**, 3754 (1987).
[90] T. Alber and B. W. Matthews, this series, Vol. 154, p. 511.

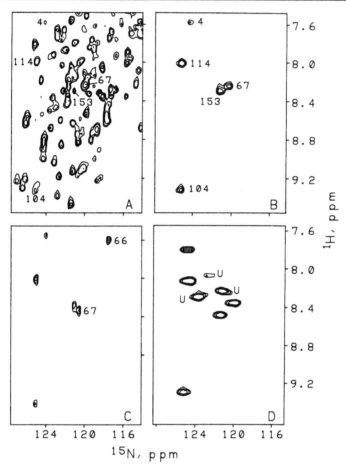

FIG. 5. Forbidden-echo spectra of (A) uniformly ^{15}N-labeled T4 lysozyme and (B) lysozyme labeled with [^{15}N]phenylalanine. The assignments of the peaks to the five Phe residues are indicated.[12] (C) The spectrum of [^{15}N]phenylalanine-labeled T4 lysozyme, containing the substitution Leu-66 to Phe; peaks of Phe-66 and Phe-67 are indicated. (D) The spectrum of [^{15}N]phenylalanine-labeled T4 lysozyme at the midpoint of thermal denaturation (pH 2.0, 45°); the three peaks marked U arise from the denatured form of the protein.

ple, by incorporating a ^{13}C-labeled carbonyl amino acid A and an ^{15}N-labeled amino acid B into a protein, only AB peptides will be ^{13}C–^{15}N labeled.[12,28,36] Using this approach, we have unambiguously assigned three phenylalanine and two tyrosine amide resonances in T4 lysozyme. A disadvantage of using multiple ^{13}C and ^{15}N labeling is the necessity of preparing labeled protein for each new assignment. Also, this approach is limited to proteins with amide ^{1}H or ^{15}N linewidths that are not much greater than the ^{13}C–^{15}N spin–spin coupling constant of approximately 17 Hz.[12,28] A similar assignment strategy is to identify adjacent ^{15}N-labeled residues by isotope-directed NOE measurements[13]; this is not limited by the proton linewidths of T4 lysozyme.

A third approach to assigning the ^{1}H and ^{15}N resonances of a protein involves one- and two-dimensional isotope-directed NOE measurements, as described below.

One of the goals of our research on T4 lysozyme is to investigate the effect of amino acid substitutions on the structure and dynamics of the protein. Unfortunately, in three cases we have observed many changes in the ^{15}N–^{1}H spectra of uniformly ^{15}N-labeled mutant lysozymes relative to the wild-type protein. This limits the extension of assignments of resonances from the wild-type to mutant lysozymes. However, in selectively labeled proteins, changes in peak chemical shifts have not been a problem, thus allowing extension of the assignments (Fig. 5C).

One- and Two-Dimensional Isotope-Directed NOE
Measurements of T4 Lysozyme

One of the central themes of NMR of biomolecules is to obtain structural information in the form of distances from NOE measurements. Griffey and Redfield developed an elegant and simple one-dimensional experiment to measure the NOEs from[8,9] or to[83] isotopically labeled protons. The isotope-directed NOE (IDNOE; Fig. 3) measurement resembles a conventional NOE experiment except that the ^{1}H preirradiation is held at the ^{1}H frequency of a labeled proton, and difference spectra are recorded with and without selective ^{15}N predecoupling at the corresponding ^{15}N frequency of the ^{15}N–^{1}H pair. The method is selective for both ^{1}H and ^{15}N shifts.

The IDNOE experiment is useful for assigning resonances of labeled protons to specific residues. Figure 6 shows a series of IDNOE measurements on three T4 lysozyme samples singly labeled with ^{15}N at the methionine, phenylalanine, and valine residues are presented. The amide ^{15}N–^{1}H resonances of Met-102, Val-102, and Phe-104 in the forbidden-echo spectra of T4 lysozyme were assigned on the basis of the indicated complementary IDNOE peaks. These residues are in an α-helix[91] and display strong amide–amide proton NOEs.

The sensitivity of the two-dimensional forbidden-echo experiment is so high that it can be used as a readout for one-dimensional saturation

[91] L. H. Weaver and B. W. Matthews, *J. Mol. Biol.* **193**, 189 (1987).

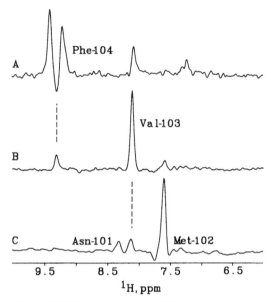

FIG. 6. Isotope-directed NOE spectra of three T4 lysozyme samples labeled separately with [^{15}N]phenylalanine (A), [^{15}N]valine (B), and [^{15}N]methionine (C). The amide–amide NOE connectivity of Asn^{101}-Met^{102}-Val^{103}-Phe^{104} is indicated. The spectra were recorded with preirradiation at the ^{1}H frequency of the labeled amide for 100–150 msec with and without selective ^{15}N predecoupling at the corresponding ^{15}N frequency of the amide. Nitrogen decoupling during acquisition was employed for spectra B and C but not A.

transfer (D. Lowry and A. G. Redfield, unpublished observations). That is, a series of forbidden-echo experiments are performed in which each pulse sequence is preceded by a long, weak presaturation pulse designed to saturate either H$_2$O or a few specific protein resonances, such as αH resonances, as well as by an off-resonance control. These two-dimensional spectra can be subtracted in the same way as one-dimensional NOE spectra. The difference "2-$\frac{1}{2}$ D" spectra can be used to investigate NOE connectivities between the preirradiated protons and specific labeled amide protons or to investigate transfer of saturation from H$_2$O to specific amide protons. Both the ^{1}H and ^{15}N chemical shifts of the labeled amide protons are measured by this experiment. There is no reason why such an experiment could not be done in a true three-dimensional mode, although it is not clear in advance whether the week or more of instrument time and the reduced sensitivity per NOE would be worthwhile.

Selective labeling is also useful in experiments which filter two-dimensional NOESY spectra of proteins. Several methods have been described to obtain ^1H–^1H or ^{15}N–^1H NOESY spectra of labeled proteins.[10,74,79,84,85] In general, either the first or last pulse of a conventional NOESY pulse sequence is replaced by composite ^1H and ^{15}N pulses which differentiate between labeled and unlabeled protons (Fig. 3). An advantage of using the final pulse as the isotope-selective pulse is that hard pulses can be used to develop the NOE interaction and thus NOEs from resonances near water are detectable which otherwise are bleached by water presaturation or eliminated by water suppression techniques. The ^1H or ^{15}N frequency (or a mixture of both) of the proton from which the NOE originates can be measured by appropriate ^1H or ^{15}N pulses during the evolution period.

In Fig. 7, isotope-edited NOESY spectra of uniformly ^{15}N-labeled T4 lysozyme in D_2O are shown. The homo-IDNOESY experiment detects

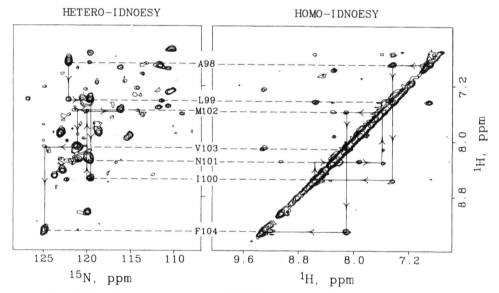

FIG. 7. Homo/hetero isotope-directed NOESY spectra of uniformly ^{15}N-labeled T4 lyso-zyme. The sample was in D_2O for 9 months prior to the measurement and only the 35 slowest exchanging amides remained protonated.[13] The amide–amide NOE connectivity of Ala-98 through Phe-104 is indicated. These residues form an α-helix buried within the C-terminal lobe of the protein.[91] The forbidden-echo peaks of most of these amides were previously assigned by residue type through selective labeling of T4 lysozyme. The horizontal axis is the ^1H shift (homo) or the ^{15}N shift (hetero) of the proton of a ^{15}N–^1H pair from which the NOE emanates. In the hetero-IDNOESY spectrum, peaks arising from NOE interactions between two labeled protons are interdispersed between peaks which would be on the diagonal in the homo-IDNOESY spectrum and which appear at the position they would have in a forbidden-echo spectrum. NOEs from labeled protons to unlabeled protons appear without inverse, symmetry-related peaks in these spectra (for example, the NOE to an aromatic proton at 7.2 ppm is seen from the amide proton of Phe-104).

the proton NOEs *from* ^{15}N-labeled protons in the form of a [1]H–[1]H map (Fig. 3). The NOEs between two labeled protons appear as pairs of cross-peaks symmetrically disposed about the diagonal peaks, whereas the NOEs to unlabeled protons are cross-peaks which lack symmetry-related mates. The heterᴄ-IDNOESY experiment detects the same proton NOEs from ^{15}N-labeled protons but the source (or t_1) axis of the two-dimensional map is the ^{15}N shift of the nitrogen to which the proton is directly bonded (Fig. 3). The NOE cross-peaks are interdispersed between peaks which would be on the diagonal in the homo-IDNOESY spectrum and appear at the position they would have in a forbidden-echo spectrum. By combining the homo- and hetero-IDNOESY measurements, the ^{1}H and ^{15}N shift of a proton from which an NOE emanates can be determined. Using this approach in combination with selective ^{15}N labeling, the forbidden-echo amide peaks corresponding to residues Ala-98 through Phe-104 were assigned. These residues are in a buried α-helix[91] and are characterized by slow hydrogen exchange kinetics[13] and by strong amide–amide NOEs.

Appendix I: Production of T4 Lysozyme from Bacterial
   Hosts Bearing pHSe Plasmids

Several days before the commencement of the protein isolation, freshly grown colonies of bacteria bearing the pHSe plasmids are tested by the lysozyme plate assay.[20] A lysozyme-producing colony is grown at 32° in 5 ml of broth or defined media containing ampicillin and then stored in the cold. On the day preceding the preparation, an aliquot of this culture is diluted 1/50 into the same medium, and the solution is shaken at 32° to a density of 8×10^8 cells/ml. Isopropylthio-β-D-galactoside is then added at a concentration of 1 mM and the cells are shaken for an additional 2 hr. The culture is then cleared by the addition of EDTA at 25 mM and centrifugation, and the yield of T4 lysozyme is determined by a cell wall turbidity assay.[14] Activities equivalent to 20 μg/ml or better indicate that high levels of lysozyme production have occurred and a large-scale preparation is justified.

The original culture is diluted 1/50 into 20–100 ml of the same medium and is grown overnight. On the day of the preparation, this inoculating culture is diluted 1/50 into 1–4 liters of the growth medium, which also contains the ^{15}N-labeled compounds. A 4-liter Erlenmyer flask is used for a 1-liter preparation, and a 5-liter fermenter (Braun Biostat V) is used for a 4-liter preparation. The culture is agitated with good aeration at 32° until the cell density reaches about 8×10^8 cells/ml, at which point lysozyme synthesis is initiated by addition of 1 mM IPTG. The culture is grown for a further time of 2–10 hr, depending on the growth characteristics of the bacterial host. The medium is then centrifuged. The conductivity of the supernatant fraction, which contains T4 lysozyme from cells which had lysed, is reduced to about 5 mmhos by passage of distilled water through a fiber dialysis cassette immersed in the solution. The cell pellet is resuspended in approximately 100 ml of 10 mM EDTA and 10 mM Tris–HCl, pH 7.5 (W. Baase, personal communication). After lysis of the cells is complete, 20 mM MgCl$_2$ and a 0.1 μg/ml of DNaseI are added, and upon reduction in its viscosity, the solution is cleared by centrifugation. The supernatant fractions are combined, adjusted to neutral pH if necessary, and loaded on a CM-Sepharose bed which has been equilibrated against 50 mM Tris–HCl, pH 7.3. The bed volume is approximately 10 ml for each 10^{12} cells in the culture and the column height is three times the diameter. The column is rinsed with one volume of equilibration buffer and then eluted with a linear gradient of NaCl from 0.10 to 0.30 M in the same buffer. The total gradient is 20 bed volumes and the collected fractions are 0.2 bed volumes. The protein is monitored by absorbance at 280 nm and by enzymatic activity.[14] The fractions representing the center 90% of the single lysozyme peak are combined and dialyzed against 50 mM sodium phosphate, pH 5.6. As a final concentration step, the protein is loaded on a small (1–4 ml) bed of SP-Sephadex which has been equilibrated against the same buffer. Each milligram of protein requires 0.01 ml of resin. The lysozyme is eluted from this column with 100 mM Na$_{3/2}$H$_{3/2}$PO$_4$, pH 6.5, and 500 mM NaCl. This is the storage buffer for the protein. The molecular weight homogeneity of the protein is checked by SDS–polyacrylamide gel electrophoresis.

Acknowledgments

We would like to thank Richard Griffey and Alfred Redfield for developing and applying many useful NMR experiments to the study of T4 lysozyme. The NMR spectra presented herein were recorded using Alfred Redfield's homebuilt spectrometer. We also thank Dennis Hare for the FTNMR program which we used to present these data. We are grateful for many fruitful discussions with Terrence Oas, David LeMaster, and Claire Berg and for invaluable help from Joan Wozniak, Christain Neilson, and Amy Roth. This work was funded by NSF Grant DMB8605439. LPM is a recipient of a Canadian N.S.E.R.C. Scholarship and a Alberta Heritage Foundation studentship.

[4] Isotopic Labeling with Hydrogen-2 and Carbon-13 to Compare Conformations of Proteins and Mutants Generated by Site-Directed Mutagenesis, I

By DAVID W. HIBLER, LYNN HARPOLD, MARK DELL'ACQUA, TAYEBEH POURMOTABBED, JOHN A. GERLT, JOYCE A. WILDE, and PHILIP H. BOLTON

Introduction

Site-directed mutagenesis is being used by a wide variety and number of laboratories to create specific amino acid replacements in proteins so that structure–function relationships might be evaluated.[1] Laboratories specializing in mechanistic enzymology are using this approach with the hope that the roles of individual amino acid residues important in catalysis might be dissected and quantitated. However, such analyses require sensitive physical techniques to detect whether conformational changes accompany the amino acid substitutions; if conformational changes do occur, interpretation of the kinetic properties of the mutant enzymes and of their relationship to properties of the wild-type enzyme becomes uncertain. Although X-ray crystallography is traditionally viewed as being suitable for this task, small changes in active site geometry are likely to cause large changes in catalytic efficiency and, at present, it is questionable whether X-ray crystallography can actually detect such changes. For this reason, our laboratories have turned to ^{1}H and ^{13}C NMR spectroscopy to compare the secondary and tertiary structures of wild-type and mutant variants created by site-directed mutagenesis. This chapter will describe the methods we use for isotopic labeling of staphylococcal nuclease (SNase) as well as for one-dimensional NMR characterization of the isotopically labeled samples. Structural inferences from two-dimensional NMR experiments on these samples are described in Chapter [14].

Staphylococcal nuclease is a monomeric protein containing 149 amino acids in a known linear sequence[2,3] and having a molecular weight of approximately 17,000; the secondary and tertiary structures are also known from the X-ray structure determined to 1.5 Å resolution[4] (the

[1] J. A. Gerlt, Chem. Rev. 87, 1079 (1987).

[2] J. L. Cone, C. L. Cusumano, H. Taniuchi, and C. B. Anfinsen, J. Biol. Chem. 246, 3103 (1971).

[3] J. L. Bohnert and H. Taniuchi, J. Biol. Chem. 247, 4557 (1972).

[4] F. A. Cotton, E. E. Hazen, and M. J. Legg, Proc. Natl. Acad. Sci. U.S.A. 76, 2551 (1979).

structure of the wild-type enzyme was recently determined to 1.6 Å resolution in Professor Eaton Lattman's laboratory at Johns Hopkins University, where the structures of mutant versions of the enzyme are also under study). While complete assignment of the ^{1}H NMR resonances is likely to be possible for SNase (and is underway in the laboratories of Dr. Dennis Torchia at the National Institutes of Health and Professor John Markley at the University of Wisconsin), our initial strategy for determining whether and where conformational changes occur in mutant versions of SNase is to focus on the simplest regions of the one-dimensional ^{1}H NMR spectrum, the aromatic and upfield-shifted methyl regions. Because aliphatic protons spatially adjacent to aromatic residues are likely to have resonances in the upfield-shifted methyl region, we can then use two-dimensional nuclear Overhauser effect correlation spectroscopy to ascertain whether the geometric relationships between these pairs of aromatic and aliphatic residues are altered in the mutant proteins. While this initial approach has the disadvantage of studying the relatively limited number of resonances in the best resolved regions of the ^{1}H NMR spectrum, it is the most amenable to unequivocal chemical shift assignments and data analysis. However, the collection of the two-dimensional data used in this investigation also provides the necessary information required for a complete analysis of additional conformational changes that might be apparent only in the more congested regions of the one- and two-dimensional spectra.

We are also using ^{13}C NMR spectroscopy both to assist in unequivocal chemical shift assignments for the protons of selected amino acids of interest (via simultaneous labeling with [α-^{15}N]amino acids[5]) and to compare the environments of the carboxyl carbons of isotopically labeled residues. While the interpretation of carbon chemical shifts is difficult, comparison of the spectra of wild-type and mutant proteins potentially provides another assessment of the impact of active site mutants on secondary and tertiary structures; such use of ^{13}C-labeled proteins requires assignment of the labeled carboxyl carbons by simultaneous ^{15}N labeling of the next amino acid in the sequence, and this is accomplished in the first step of unequivocal chemical shift assignment of the protons on the ^{13}C-labeled amino acid residues.

We are preparing isotopically labeled samples of SNase by the biosynthetic incorporation of ^{2}H- and ^{13}C-labeled amino acids in cells of *Escherichia coli* that are transformed with a plasmid containing the SNase gene that is under the control of a very strong, inducible promoter. Such substitutions allow the more facile observation of subsets of resonances either

[5] M. Kainosho and T. Tsuji, *Biochemistry* **21**, 6273 (1982).

by removing unwanted resonances (^{2}H) or by enriching specific resonances (^{13}C), or by allowing specific observation of the resonances associated through bonds to the labeled site (^{13}C). We believe that the strategy we describe in this chapter should be useful for analogous studies on other proteins and on their site-directed variants.

Placement of Gene for SNase under Control of Strong, Inducible Promoter

The gene for SNase was cloned and sequenced by Shortle.[6] The gene as expressed in *Staphylococcus aureus* contains coding information for a leader peptide to direct secretion of the mature enzyme into the culture medium followed by the 149 amino acids associated with the mature secreted enzyme, which is designated nuclease A by Anfinsen and co-workers.[2,3] These data reveal the presence of a *Sau*3A restriction site six codons upstream of the codon for the N-terminal alanine of nuclease A and a second *Sau*3A site approximately 55 base pairs (bp) downstream of the termination codon. The "sticky ends" produced by *Sau*3A (GATC) are identical to those produced by *Bam*HI-catalyzed restriction of DNA and permit the insertion of the coding sequence for SNase into a variety of expression vectors.

We have placed the *Sau*3A restriction fragment containing the gene for SNase in two expression vectors. The first, pIN-III-*ompA3*,[7] has a unique *Bam*HI site just downstream of the coding sequence for the leader peptide of the *ompA* protein normally synthesized and secreted by *E. coli*; the gene for the hybrid preenzyme made by insertion of the *Sau*3A restriction fragment in this *Bam*HI site is under control of the *lacUV5* promoter and, as such, is inducible by the addition of either lactose or isopropylthio-β-galactoside (IPTG). When the sequence coding for the mature nuclease A was directly fused to the 3' end of the coding sequence for the *ompA* leader peptide (by site-directed deletion mutagenesis) and the plasmid containing this construction, designated pONF1, was transformed into an *E. coli* host, the hybrid preenzyme could be induced by the addition of IPTG, and mature, processed nuclease A accumulated in the periplasmic space of the *E. coli* host.[8] This protein has the same N-terminal sequence as the SNase isolated from the culture medium of *S. aureus* and can be crystallized. However, the level of expression (approximately 10% of the total cellular protein or between 10 and 15 mg of

[6] D. Shortle, *Gene* **22**, 181 (1983).
[7] J. Ghrayeb, H. Kimura, M. Takahara, H. Hsiung, Y. Masui, and M. Inouye, *EMBO J.* **3**, 2437 (1984).
[8] M. Takahara, D. W. Hibler, P. J. Barr, J. A. Gerlt, and M. Inouye, *J. Biol. Chem.* **260**, 2670 (1985).

purified protein per liter of induced bacterial culture) is insufficient for the economically feasible incorporation of most isotopically labeled amino acids with the yields necessary for two-dimensional nuclear Overhauser effect correlation spectroscopy (we normally use 0.6-ml samples that are 2.5 mM in protein and this requires about 25 mg of SNase).

The second expression vector we have used, pCQV2,[9] has a unique *Bam*HI site downstream of the P$_R$ promoter from bacteriophage λ, the *cro* gene ribosome binding site, and an ATG initiation codon (with the G of this codon being the first base in the GGATCC recognition sequence for *Bam*HI); this vector also carries the gene for the *c*I857 temperature-sensitive mutation for the bacteriophage λ repressor. Thus, insertion of the gene into the *Bam*HI site (in reading frame) allows the controlled, high-level expression of SNase by simply (also rapidly) raising the temperature of the growth medium from 30 to 42°, because, at the higher temperature, the temperature-sensitive repressor is denatured and efficient transcription of the gene can occur. The level of expression (approximately 50% of the total cellular protein or between 60 and 80 mg per liter of induced bacterial culture) directed by the plasmid so constructed, designated pNJS, is sufficient for the economically feasible incorporation of isotopically labeled amino acids with yields necessary for two-dimensional NMR experiments. The SNase obtained with this expression system is not identical to that obtained from pONF1, since the former has an N-terminal extension of seven amino acids (methionine and the six amino acids in the leader peptide that are upstream of the processing site) and the latter is identical in sequence to the nuclease A that is produced by *S. aureus;* however, the kinetic properties and the one- and two-dimensional ^{1}H NMR spectra of these enzymes are identical (with the exception of resonances associated with the amino acids in the N-terminal extension in the enzyme encoded by pNJS).

In practice, the chromosome of the host we routinely use for the expression of the gene for SNase contained in pNJS also carries the gene for the *c*I857 mutation of the repressor. This host strain, N4830, is auxotrophic for isoleucine and valine as well as histidine. However, such auxotrophic behavior is not necessary for efficient isotopic labeling.

Incorporation of Isotopically Labeled Amino Acids into
    SNase and Site-Directed Mutants

The strategy we use for labeling is to grow N4830 transformed with either wild-type or mutant versions of pNJS on minimal medium containing glucose as carbon source and all of the L-amino acids except the

[9] C. Queen, *J. Mol. Appl. Genet.* **2**, 1 (1983).

isotopically labeled amino acid(s) to be incorporated; just prior to induction of the gene, the labeled amino acid(s) is (are) added.[10] The presence of the unlabeled amino acids represses and/or inhibits the biosynthesis of at least some of these amino acids[11] as well as dilutes any isotopic scrambling that might occur after the addition of the labeled amino acid(s) to the growth medium. However, we believe that the extremely rapid protein synthesis which follows induction of the gene effectively eliminates these potential problems.

A typical incorporation of isotopically labeled amino acid(s) into SNase is accomplished as follows.[10] A 25-ml overnight culture of N4830 transformed with pNJS containing either the wild-type or a mutant version of the gene for SNase is grown in M9 minimal salt medium with 0.2% glucose as carbon source and supplemented with 0.1 mM CaCl$_2$, 1 mM MgSO$_4$, 50 μg/ml thiamin, 50 μg/ml biotin, and 100 μg/ml concentrations of all the L-amino acids except the isotopically labeled amino acid(s) to be incorporated; this total culture is used as the innoculum for 500 ml of the same medium. When the absorbance at 590 nm of the culture reaches 0.9, 100 mg (200 μg/ml) of the isotopically labeled L-amino acid(s) to be incorporated is (are) added; production of the isotopically labeled SNase is induced 15 min later by shifting the temperature of the culture medium from 30 to 42° by the addition of 150 ml of the M9 medium heated to 68° and containing 200 μg/ml of all L-amino acids except the previously added isotopically labeled amino acid(s). After 2 hr at 42°, the cells are harvested by centrifugation and stored at −20° until isolation of the labeled SNase.

We have also prepared samples of SNase isotopically labeled with L-valine for which our host N4830 is auxotrophic. In the case of readily available samples of isotopically labeled L-valine (e.g., perdeutero or 1-^{13}C), the procedure described in the previous paragraph is followed except that the culture used as the innoculum also contains the labeled valine at 200 μg/ml. We have found that as little as 25 μg/ml of L-valine can be used to grow N4830 without decrease of SNase yield. However, we only optimize the yield of SNase by determining the growth requirements for the labeled amino acid to be incorporated when the amino acid is not readily available or is very expensive to ensure high degrees of incorporation of the labeled amino acid.

The cells containing the labeled SNase are disrupted by passage through a French pressure cell, and the enzyme is purified from the supernatant of the cell extract by cation-exchange chromatography on Bio-Rex

[10] D. W. Hibler, N. J. Stolowich, M. A. Reynolds, J. A. Gerlt, J. A. Wilde, and P. H. Bolton, *Biochemistry* **26,** 6278 (1987).
[11] P. Lu, M. Jarema, K. Mosser, and W. E. Daniel, *Proc. Natl. Acad. Sci. U.S.A.* **73,** 3471 (1976).

70 followed by affinity chromatography on a column of Sepharose 4B containing the competitive inhibitor 3'-[(4-aminophenyl)phosphoryl]-deoxythymidine 5'-phosphate covalently linked by cyanogen bromide activation.[10] Following elution from the affinity column with 0.5 M guanidinium hydrochloride, the fractions containing protein are pooled and dialyzed exhaustively against deionized water. The proteins isolated according to this procedure are homogeneous as judged by SDS–PAGE. Typically, 30 to 35 mg of isotopically labeled SNase is obtained from this procedure. The extinction coefficient of this N-terminal modified SNase is assumed to be $\varepsilon_{280\,nm} = 1.0$ (mg/ml)$^{-1}$ cm^{-1}.

Sample Preparation for NMR Spectroscopy[10]

Samples of 25 mg (1.5 μmol) of protein are dissolved in ^{2}H$_2$O, warmed to 40° for 15 min, and lyophilized; this process is repeated to completely exchange the amide protons with solvent ^{2}H$_2$O. The lyophilized protein is dissolved in 0.6 ml of a ^{2}H$_2$O solution containing 50 mM sodium borate, pH 7.8, 0.1 M NaCl, 1 mM ethylenediaminetetraacetic acid (EDTA), 1 mM ethylene glycol bis(β-aminoethyl ether)-N,N,N',N'-tetraacetic acid (EGTA), 3.75 mM thymidine 3',5'-bisphosphate (pdTp, a competitive inhibitor that binds to wild-type SNase with a binding constant of approximately 1 μM), and 10 mM CaCl$_2$. (These concentrations of pdTp and Ca^{2+} are sufficient to produce greater than 95% saturation of the active site of the wild type and most mutant versions of SNase as judged by measurement of the binding of pdTp monitored by quenching of the tyrosine fluorescence.) The solution is lyophilized, redissolved in 0.6 ml of 100.00% ^{2}H$_2$O, and clarified either by centrifugation in a microfuge or by filtration through a glass fiber filter (Whatman GF/D) directly into a high-precision 5-mm NMR tube, and the pH meter reading of the sample is adjusted to 7.8 with NaO^{2}H or ^{2}HCl. The final protein concentration is 2.5 mM.

One-Dimensional ^{1}H and ^{13}C NMR Spectra of Labeled Proteins

The effectiveness of the isotopic labeling procedure is determined by comparing the ^{1}H and ^{13}C NMR spectra of wild-type SNase and the site-directed mutant E43D in which glutamate-43, the putative general base involved in the hydrolysis reaction, is replaced with aspartate.[10] This mutant has a V_{max} approximately 200-fold less and a K_m approximately 8-fold greater than the values measured for the wild-type enzyme.

SNase contains a single tryptophan, three phenylalanines, four histidines, and eight tyrosines.[2,6] Figure 1 compares the spectra of the aro-

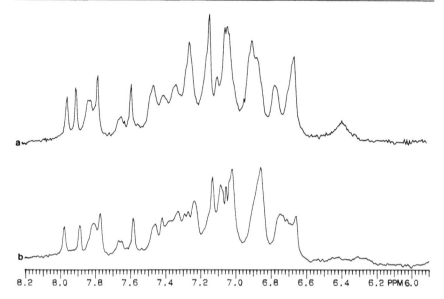

FIG. 1. Comparison of the aromatic region of the 400-MHz ^{1}H NMR spectrum of protiated wild-type SNase (b) and E43D (a).

matic region of wild-type SNase and E43D, which were prepared from perprotiated amino acids. Figure 2 compares the spectra of the same region in which the single tryptophan and the three phenylalanines are labeled with L-[*indole*-^{2}H$_5$]tryptophan and L-[*ring*-^{2}H$_5$]phenylalanine, respectively; the remaining resonances are those associated with the tyrosines and histidines. Figure 3 compares the spectra of the aromatic region in which the single tryptophan and the eight tyrosines are labeled with L-[*indole*-^{2}H$_5$]tryptophan and L-[*ring*-^{2}H$_4$]tyrosine, respectively.

SNase contains 11 leucines and 10 valines.[2,6] Figure 4 compares the spectra of the upfield-shifted methyl region of wild-type SNase and E43D, which were prepared from perprotiated amino acids. Figure 5 compares the spectra of the same region in which the leucines are labeled with L-[^{2}H$_{10}$]leucine. Figure 6 compares the spectra of the same region in which the valines are labeled with L[^{2}H$_8$]valine.

Examination of the spectra reveals that the degree of isotopic enrichment is very high with valine for which the host strain is auxotrophic, as well as for the aromatic amino acids and leucine for which the host strain is prototrophic. Dr. Dennis Torchia at the National Institutes of Health has also used the expression system described in this article to prepare isotopically labeled samples of wild-type SNase and has found similar

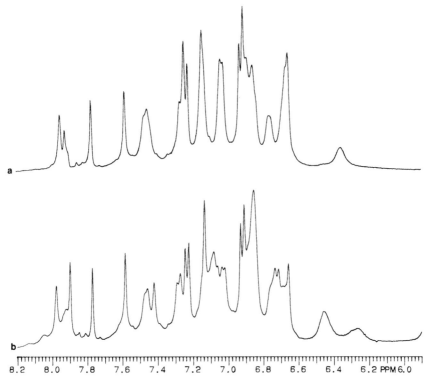

FIG. 2. Comparison of the aromatic region of the 400-MHz ^{1}H NMR spectrum of wild-type SNase (b) and E43D (a) labeled with L-[*indole*-^{2}H$_5$]tryptophan and L-[*ring*-^{2}H$_5$]phenyl-alanine.

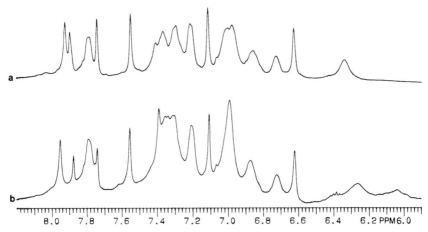

FIG. 3. Comparison of the aromatic region of the 400-MHz ^{1}H NMR spectrum of wild-type SNase (b) and E43D (a) labeled with L-[*indole*-^{2}H$_5$]tryptophan and L-[*ring*-^{2}H$_4$]-tyrosine.

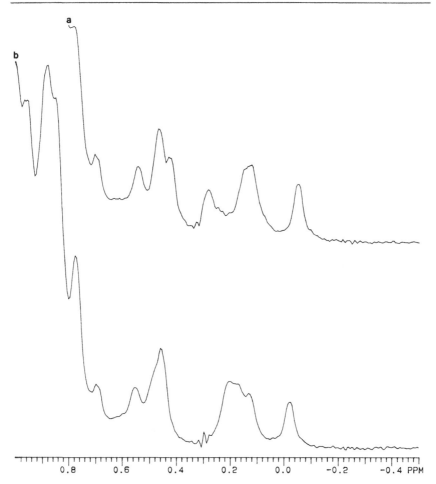

FIG. 4. Comparison of the upfield-shifted methyl group region of the 400-MHz ¹H NMR spectrum of protiated wild-type SNase (b) and E43D (a).

high degrees of enrichment for most amino acids except those that can suffer isotopic dilution from intermediates in the citric acid cycle.

The spectra of the isotopically labeled proteins convincingly demonstrate the power of deuteration of aromatic and aliphatic amino acid residues to accomplish residue-type assignments of resonances in the less congested regions of the ¹H NMR spectrum and to ascertain the types of residues that undergo chemical shift changes when the site-directed substitution is accomplished. These residue-type assignments will be useful

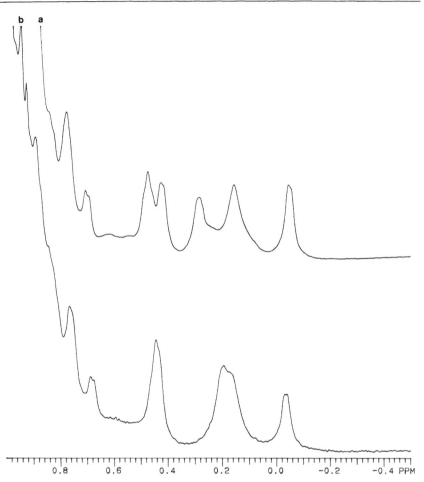

FIG. 5. Comparison of the upfield-shifted methyl group region of the 400-MHz ^{1}H NMR spectrum of wild-type SNase (b) and E43D (a) labeled with L-[^{2}H$_{10}$]leucine.

in our other chapter [14] in this volume which describes the comparison of the two-dimensional nuclear Overhauser effect spectra of the wild-type and E43D mutant.

Figure 7 compares the ^{13}C NMR spectra of wild-type SNase and E43D in which the 10 valines have been isotopically labeled with L-[1-^{13}C]valine. These spectra reveal surprisingly large chemical shift dispersions, suggesting different environments for the valine carboxyl carbons in the wild-type and mutant proteins. Comparison of the spectra reveals that 4 of the

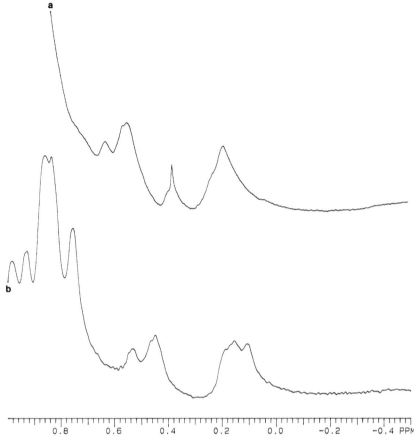

FIG. 6. Comparison of the upfield-shifted methyl group region of the 400-MHz ^{1}H NMR spectrum of wild-type SNase (b) and E43D (a) labeled with L-[^{2}H$_8$]valine.

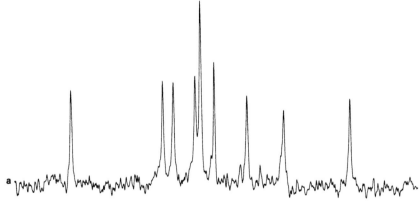

FIG. 7. Comparison of the carbonyl region of the 100-MHz ^{13}C NMR spectrum of wild-type SNase (b) and E43D (a) labeled with [1-^{13}C]valine.

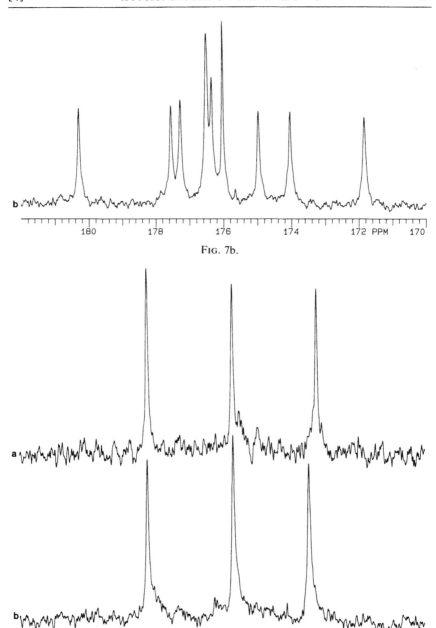

FIG. 7b.

FIG. 8. Comparison of the carbonyl region of the 100-MHz ¹³C NMR spectrum of wild-type SNase (b) and E43D (a) labeled with [1-¹³C]phenylalanine.

10 resonances undergo chemical shift changes in the mutant protein, suggesting a conformational change involving a portion of but not all of the structure of the protein. Figure 8 compares the ^{13}C NMR spectra of wild-type SNase and E43D in which the three phenylalanines have been isotopically labeled with L-[1-^{13}C]phenylalanine. Again, significant chemical shift dispersion is observed, and one of the three resonances experiences a chemical shift change. Successful isotopic incorporation of 1-^{13}C-labeled amino acids is also important, both for unequivocal chemical shift assignments of the carboxyl carbons by additional simultaneous incorporation of α-^{15}N-labeled amino acids so that the one-bond ^{13}C–^{15}N coupling in peptide bonds can be detected[5] and for measurement of the exchange rates of the amide protons in the labeled peptide units.[12]

The use of these 1-^{13}C-labeled samples of SNase to assign the resonances associated with the protons of α-carbons and the side chains of the labeled amino acids will be described in Chapter [14] in this volume.

[12] G. D. Henry, J. D. J. O'Neil, J. H. Weiner, and B. D. Sykes, *Biophys. J.* **49**, 329 (1985).

[5] Spin Labeling of Proteins

By PHYLLIS A. KOSEN

Paramagnetic relaxation probes increase the relaxation rates of the NMR resonances of nearby nuclei in a distance-dependent manner. In principle, distances derived from NMR studies of proteins containing a paramagnetic group can be included as constraints for solution structure calculations.[1] Spin labels usually refer to compounds that contain the paramagnetic nitroxide free radical (stabilized by full alkyl substitution at the α-carbons) and a second group which can covalently bind another molecule. Some spin labels modify the side chains of proteins. These reagents can be subclassified either as affinity labels, with structural complementarity to binding pockets on proteins, or as reagents that covalently modify specific types of side chains, but do not depend on a protein-specific pocket as a prerequisite for labeling.[2] The latter class of spin

[1] I. D. Kuntz, J. F. Thomason, and C. M. Oshiro, this volume [9].
[2] In addition to the class of spin labels discussed in this chapter, and affinity labels, paramagnetic probes of proteins include spin probes, which are nitroxide-containing molecules designed to bind noncovalently at ligand binding sites and paramagnetic ions.

labels provides a general way of incorporating paramagnetic probes. In the best of cases, these reagents bind at a single site. However, absolute reaction specificity is the exception, not the rule. A familiarity with the chemical properties of spin labels and of spin-labeled proteins is needed so that monoderivatized proteins may be prepared for NMR studies.

There are three purposes to this chapter: (1) to outline experimental protocols that have been used in one- and two-dimensional ^{1}H NMR studies of spin-labeled proteins; (2) to discuss spin labels which react with the side chains of lysines (and α-amino groups), tyrosines, cysteines, histidines, and methionines—the utility, limitations, and reaction conditions of these reagents will be evaluated; and (3) to examine purification procedures that have been used to prepare monolabeled proteins. [The protein of interest for my colleagues and myself is bovine pancreatic trypsin inhibitor (BPTI), and studies on spin-labeled derivatives of BPTI will be referred to throughout this chapter.]

Introduction

Since their introduction more than 20 years ago, nitroxide-containing compounds have been exploited by EPR spectroscopists in the study of protein structure, motion, and catalytic activity, with the nitroxide serving as a benign reporter group. The utility of spin labeling in EPR spectroscopy is exemplified by the many reviews and books that have appeared, including previous reports in this series, and the synthesis of hundreds of different nitroxide-containing compounds.[3-11]

The chemistry of affinity labels and spin probes will not be discussed, as the synthesis of each must be individually designed for a unique protein binding pocket.

[3] G. I. Likhtenstein, "Spin Labelling Methods in Molecular Biology." Wiley, New York, 1974.

[4] L. J. Berliner (ed.), "Spin Labeling: Theory and Applications." Academic Press, New York, 1976.

[5] L. J. Berliner, this series, Vol. 49, p. 418.

[6] P. C. Jost and O. H. Griffith, this series, Vol. 49, p. 369.

[7] L. J. Berliner (ed.), "Spin Labeling II: Theory and Applications." Academic Press, New York, 1979.

[8] L. J. Berliner, in "Spectroscopy in Biochemistry" (J. E. Bell, ed.), Vol. 2, p. 1. CRC Press, Boca Raton, Florida, 1981.

[9] G. Benga, in "Biochemical Research Techniques" (J. M. Wrigglesworth, ed.), p. 79. Wiley, New York, 1983.

[10] H. R. Wenzel and H. Tschesche, in "Modern Methods in Protein Chemistry" (H. Tschesche, ed.), p. 385. de Gruyter, Berlin, 1983.

[11] J. L. Holtzman (ed.), "Spin Labeling in Pharmacology." Academic Press, Orlando, Florida, 1984.

Unlike its role in EPR studies, the spin label does not serve as a reporter group in NMR spectroscopy. Instead it is an active agent, specifically altering the magnetic environment of the protein. Ideally, the spin label only perturbs the magnetic environment, and not the conformation of the protein. The structural information that can be obtained via nitroxide-induced relaxation of proton resonances (and, more generally, the resonances of other nuclei) is grouped below into three categories. The emphasis of this chapter is on the last category.

1. Qualitative solvent accessibility and orientation measurements. These involve addition of a chemically inert nitroxide to a protein solution with the caveat that specific noncovalent interactions between the nitroxide and the protein are negligible. The nitroxide must either approach the protein randomly from all directions, or the orientation of the nitroxide in the environment must be known, e.g., for studies of protein embedded in lipid vesicles.[12] The degree to which the relaxation rate of a given resonance is increased is directly proportional to the concentration of the nitroxide and, more importantly, the distance of nearest approach by the nitroxide. Therefore, the relaxation rates of nuclei that are solvent accessible will be increased more than those that are buried within the protein. Because the distances between the nitroxide and the protein's hydrogen atoms are not fixed, quantitative measurements are not obtained. Measurements of the solvent accessibilities of the hydrogens of gramicidin S are typical of this approach.[13] In analogous experiments, the orientation of glucagon in dodecylphosphocholine micelles has been determined.[12]

2. Distance measurements between a protein-bound nitroxide and substrates, cofactors, or inhibitors. The protein-bound nitroxide is incorporated either as a spin label or as a second-site spin probe. This type of experiment provides information on the relative orientation of a ligand with respect to the position of the nitroxide. Fast chemical exchange conditions between the protein-bound and the free ligand are necessary. Typically, relaxation rates of the resonances of the free ligand are measured (in the presence and absence of the nitroxide). Free ligand is usually present in large excess over protein so that the protein resonances do not contribute significantly to the spectra. State-of-the-art instrumentation is not strictly required; the binding sites of proteins, whose size still precludes their direct study by NMR spectroscopy, can be partially mapped. Consequently, this type of distance measurement has dominated the spin-

[12] L. R. Brown, C. Bösch, and K. Wüthrich, *Biochim. Biophys. Acta* **642**, 296 (1981).
[13] N. Niccolai, G. Valensin, C. Rossi, and W. A. Gibbons, *J. Am. Chem. Soc.* **104**, 1534 (1982).

label technique in NMR spectroscopy. Krugh[14] and Likhtenstein[3] have reviewed the field.

3. Distance measurements between a protein-bound nitroxide and protein nuclei. Such measurements were first suggested by Sternlicht and Wheeler[15] and McConnell[16] in 1967 as a means of obtaining solution structural information about proteins. In 1976, Krugh[14] observed that this type of experiment "will, of course, be limited, to the small group of enzymes for which NMR signals may be observed and assigned to particular groups on the enzyme." Since the time of Krugh's review, the sensitivity and resolution capabilities of NMR spectrometers have greatly improved, as have techniques for assigning resonances[17]—at least for proteins with molecular mass of 14,000 Da or less, e.g., lysozyme.[18] In fact, experiments using lysozyme spin-labeled at histidine-15 furnished the first quantitative demonstration that distances could be measured by the spin-label technique.[19] This demonstration occurred nearly 15 years after the method was originally proposed. The feasibility of the lysozyme study was the direct consequence of improvements in instrumentation (cf. Refs. 19–21).

Other improvements in methodology and technology also increase the potential of this type of distance measurement. Spectra can be further simplified and resolution improved through specific or random deuteration of the protein.[22-27] The introduction of many new chromatography matrices, operating at either low or high pressure, increases the likelihood

[14] T. R. Krugh, in "Spin Labeling: Theory and Applications" (L. J. Berliner, ed.), p. 339. Academic Press, New York, 1976.

[15] H. Sternlicht and E. Wheeler, in "Magnetic Resonance in Biological Systems" (A. Ehrenberg, B. G. Malmstrom, and T. Vanngard, eds.), p. 325. Pergamon, Oxford, 1967.

[16] H. M. McConnell, in "Magnetic Resonance in Biological Systems" (A. Ehrenberg, B. G. Malmstrom, and T. Vanngard, eds.), p. 335. Pergamon, Oxford, 1967.

[17] V. J. Basus, this volume [7].

[18] C. Redfield and C. M. Dobson, Biochemistry 27, 122 (1987).

[19] P. G. Schmidt and I. D. Kuntz, Biochemistry 23, 4261 (1984).

[20] J. D. Morrisett, R. W. Wien, and H. M. McConnell, Ann. N.Y. Acad. Sci. 222, 149 (1973).

[21] R. W. Wien, J. D. Morrisett, and H. M. McConnell, Biochemistry 11, 3707 (1972).

[22] J. Anglister, T. Frey, and H. M. McConnell, Biochemistry 23, 1138 (1984).

[23] J. Anglister, T. Frey, and H. M. McConnell, Biochemistry 23, 5372 (1984).

[24] T. Frey, J. Anglister, and H. M. McConnell, Biochemistry 23, 6470 (1984).

[25] J. Anglister, T. Frey, and H. M. McConnell, Nature (London) 315, 65 (1985).

[26] J. Anglister, M. W. Bond, T. Frey, D. Leahy, M. Levitt, H. M. McConnell, G. S. Rule, J. Tomasello, and M. Whittaker, Biochemistry 26, 6058 (1987).

[27] D. M. LeMaster, this volume [2].

that separation of monoderivatized spin-labeled proteins can be accomplished. Finally, gene cloning techniques potentially enhance the ability to introduce new sites at which a protein may be specifically labeled or to produce larger quantities of protein than might be naturally available.

It is however, the preparation and purification of spin-labeled proteins and, to a lesser extent, the ability to measure many relaxation rates that still limit the general usefulness of these experiments. Little emphasis has been placed on the purification of monoderivatized spin-labeled proteins for either EPR or NMR studies. Inactivity in this area may be the result of the following considerations. (1) Much synthetic effort has been directed at the development of spin probes and affinity labels which bind at only one site. (2) In many EPR studies, if more than one site was labeled, there were distinct resonances which could be monitored separately. (3) Until recently, the limitations of NMR spectroscopy have not forced the purification issue. (4) There has been a general belief that existing chromatographic methods would be inadequate for the separation of a mixture of spin-labeled proteins, even in the presence of contrary evidence. (See below for examples.)

Objectively, further development of the chemical and NMR techniques needed for spin-labeling experiments are fundamentally more important than in 1976. The practicality of determining protein structures has been convincingly demonstrated.[1,28-30] Short-range hydrogen–hydrogen distances of about 5 Å or less (derived from NOESY spectra), tortional angles (derived from coupling constants), stereospecific assignments (derived from coupling constants and NOESY spectra), and other constraints, e.g., bond lengths and disulfide bond connectivities, have been used as input for solution structure calculations. Even so, with the currently available sources of input, the solution structures generated are underdetermined and additional distance information would improve the quality of the resulting structures. NMR spectroscopy of spin-labeled proteins is an obvious source of additional distance information.

Only a few attempts to measure distances, employing spin-labeled proteins,[19,31-34] (or, more frequently, interactions between spin probes

[28] A. D. Kline, W. Braun, and K. Wüthrich, J. Mol. Biol. 189, 377 (1986).
[29] G. Wagner, W. Braun, T. F. Havel, T. Schaumann, N. Gō, and K. Wüthrich, J. Mol. Biol. 196, 611 (1987).
[30] G. M. Clore and A. M. Gronenborn, Protein Eng. 1, 275 (1987).
[31] V. T. Ivanov, V. I. Tsetlin, E. Karlsson, A. S. Arseniev, Y. N. Utkin, V. S. Pashkov, A. M. Surin, K. A. Pluzhnikov, and V. F. Bystrov, in "Natural Toxins" (D. Eaker and T. Wadström, eds.), p. 523. Pergamon, Oxford, 1980.
[32] V. F. Bystrov, V. T. Ivanov, V. V. Okanov, A. I. Miroshnikov, A. S. Arseniev, V. I. Tsetlin, and V. S. Pashkov, in "Advances in Solution Chemistry" (I. Bertini, L. Lunazzi, and A. Dei, eds.), p. 231. Pergamon, New York, 1981.

and proteins[20,21,23,24,35-39]), have been reported. Many of these studies employed a protein (with the exceptions of lysozyme,[19] *Naja naj oxiana* neurotoxin II,[31,32] and BPTI[34]) whose NMR spectrum was almost completely unassigned. Rather than compilation of distance information per se, characterization of a binding pocket has been the primary objective of the spin-probe studies.[20,21,23,24,35-39] To date, distances derived from spin-label (or spin-probe) experiments have not been used as distance constraints in *de novo* solution structure calculations. So, the utility of extracting distance information from spin-label studies as input for solution structure calculations has not been fully tested. The final purpose of this chapter is to encourage others to attempt spin-labeling experiments with proteins whose NMR spectra have been assigned and whose solution structures have been calculated. The experimental procedures outlined in the remainder of this chapter examine some of the techniques for preparation of spin-labeled proteins and the measurement of paramagnetic relaxation rates for spin-labeled proteins.

Methods

NMR Spectroscopy of Spin-Labeled Proteins[40]

General Considerations. The following modified Solomon–Bloembergen equations[41] express the relation between the paramagnetic longitudinal or the transverse nuclear relaxation rates, and the distance between the spin label's unpaired electron and a hydrogen nucleus of a protein.

$$\frac{1}{T_{1P}} = \frac{2.46 \times 10^{-32} \text{ cm}^6 \text{ sec}^{-2}}{r^6} \left(\frac{3\tau_C}{1 + \omega_I^2 \tau_C^2} \right) \tag{1}$$

$$\frac{1}{T_{2P}} = \frac{1.23 \times 10^{-32} \text{ cm}^6 \text{ sec}^{-2}}{r^6} \left(4\tau_C + \frac{3\tau_C}{1 + \omega_I^2 \tau_C^2} \right) \tag{2}$$

[33] G. Musci, K. Koga, and L. J. Berliner, *Biochemistry* **27**, 1260 (1988).
[34] I. D. Kuntz, unpublished results.
[35] S. T. Lord and E. Breslow, *Biochem. Biophys. Res. Commun.* **80**, 63 (1978).
[36] S. T. Lord and E. Breslow, *Biochemistry* **19**, 5593 (1980).
[37] P. G. Schmidt, M. S. Bernatowicz, and D. H. Rich, *Biochemistry* **21**, 1830 (1982).
[38] Y.-H. Lee, B. L. Currie, and M. E. Johnson, *Biochemistry* **25**, 5647 (1986).
[39] A. Kuliopulos, E. M. Westbrook, P. Talalay, and A. S. Mildvan, *Biochemistry* **26**, 3927 (1987).
[40] Much of the discussion in this section is appropriate only for spin-labeled proteins. Chemical exchange rates are an important consideration in spin-probe studies. For analysis of spectra of protein and spin-probe mixtures, which is beyond the scope of this chapter, see Refs. 14, 21, and 23.
[41] I. Solomon and N. Bloembergen, *J. Chem. Phys.* **25**, 261 (1956).

Here, $1/T_{1P}$ is the paramagnetic longitudinal relaxation rate, $1/T_{2P}$ is the paramagnetic transverse relaxation rate, ω_I is the nuclear Larmor precession frequency, τ_C is a correlation time for the fluctuation of the electronic–nuclear dipole–dipole interaction, and r is the distance between the unpaired electron (which is localized primarily on the nitrogen of the nitroxide in a polar environment) and the affected hydrogen nucleus. Equations (1) and (2) are appropriate only for relaxation due to the magnetic interaction between a single unpaired electron and a hydrogen nucleus of a macromolecule. Furthermore, assuming τ_C values greater than 10^{-9} sec (which is a conservative lower limit for spin-labeled proteins) and magnetic field strengths corresponding to ^1H resonance frequencies of 400 to 600 MHz, the second term in parentheses in Eq. (2) is negligible. Contained within the numerical constants of Eqs. (1) and (2) is a term for the magnetic moment of the unpaired electron. Given that the magnetic moment of an electron is 657 times that of a proton, the electronic–nuclear dipole–dipole interaction will be effective, i.e., dominant, over a much greater range of distances than are the diamagnetic nuclear interactions that, in the absence of the paramagnetic group, would dominate the relaxation mechanisms. The complete Solomon–Bloembergen equations and discussion of the assumptions inherent in these two equations can be found elsewhere.[14,42,43]

The correlation time for a spin-labeled protein is given by

$$1/\tau_C = 1/\tau_R + 1/\tau_S \tag{3}$$

where τ_R is an isotropic rotational correlation time dependent on the motional characteristics of the protein and τ_S is the electronic longitudinal relaxation time. The value for τ_S of nitroxides is usually greater than 10^{-7} sec.[3,21,38] The rotational correlation time of the protein places an upper limit on τ_R. For proteins whose solution structures can presently be solved by NMR spectroscopy, $\tau_R \leq 0.1\tau_S$, with the qualification that intermolecular electronic spin exchange does not effectively shorten τ_S.[21] So, in practice, τ_C will be determined by τ_R.

The current procedure is to assume a single value of τ_C for all nuclei of the protein, even though the motional characteristics determining τ_R may vary somewhat in different segments of the protein chain. Two methods have been used to estimate τ_C. Either τ_C is taken as the rotational correlation time of the protein or its value is determined from T_{1P} and T_{2P} mea-

[42] J. Kowalewski, L. Nordenskiöld, N. Benetis, and P.-O. Westlund, *Prog. NMR Spectrosc.* **17**, 141 (1985).

[43] T. L. James, "Nuclear Magnetic Resonance in Biochemistry." Academic Press, New York, 1975.

surements. The value of τ_C inserted into Eq. (1) or (2) could vary nearly an order of magnitude—depending on, in large measure, the perspective of those making the calculations. Fortunately, the sixth-root dependence of distance on the functions of τ_C in Eqs. (1) and (2) minimizes the error introduced by an inappropriate, but not grossly incorrect, choice of τ_C.

Most often it has simply been assumed that τ_C is equivalent to the rotational correlation time of the protein estimated experimentally or by the Stokes–Einstein relation for a spherical particle.[19,21,22,33,34,37,38,44] For example, τ_R of lysozyme was measured by depolarized light-scattering experiments and is 10 nsec.[45] For BPTI, τ_R was taken from fluorescence measurements and is about 4 nsec.[46] These values seem to be valid estimate of τ_C (see below). The estimates of τ_R, cited above, are about two-fold larger than those of the Stokes–Einstein value. Often, experimental estimates of τ_R are somewhat larger than predicted by the Stokes–Einstein relation. This discrepancy may be due to water of hydration.

If both T_{1P} and T_{2P} are measured, given that $\omega_I \tau_C \geq 1$, a value for τ_C can be estimated from the ratio of T_{1P}/T_{2P}:

$$\tau_C = \left[\frac{6(T_{1P}/T_{2P}) - 7}{4\omega_I^2} \right]^{1/2} \tag{4}$$

This approach has been used in two spin-probe studies,[36,39] but has not yet been applied to a spin-labeled protein. In both of the spin-probe studies, τ_C was significantly smaller than that expected based on size or other estimates of τ_R.[47] The reasons for these differences are not at all clear. The data base, one set of T_{1P} and T_{2P} values from Ref. 36 and two sets from Ref. 39, is too small for any meaningful conclusions to be drawn. In principle, Eq. (4) should give τ_C. But it would be prudent to average values extracted from many different resonances.

High-frequency segmental motions involving the spin label or portions of the protein chain are reflected as a statistical averaging of r, not as a measure of τ_C. Such motions bias the calculated distances to those of nearest approach between the electron and nucleus. An EPR spectrum of the spin-labeled protein indicates if restricted but rapid motions of the

[44] R. A. Dwek, J. C. A. Knott, D. Marsh, A. C. McLaughlin, E. M. Press, N. C. Price, and A. I. White, Eur. J. Biochem. 53, 25 (1975).
[45] S. B. Dubin, N. A. Clark, and G. B. Benedek, J. Chem. Phys. 54, 5158 (1971).
[46] E. Gratton, personal communication with I. D. Kuntz.
[47] The values, 8.74×10^{-10} sec and 4.73×10^{-10} sec in Ref. 39 refer to $f(\tau_C)$ of Eq. (1) of this chapter, not τ_C. The corresponding values of τ_C are 9.8×10^{-10} sec and 2.4×10^{-9} sec (A. Mildvan, personal communication). The former value of τ_C, 9.8×10^{-10} sec, is derived from a study of the frequency dependency of $1/T_{1P}$ of water bound to the nitroxide in the protein complex. Consequently, it is not an appropriate measure of τ_C for the protein.

label (with respect to the protein) occur or if the mobility of the spin label is no greater than that of the protein.[5,6,8]

The observed relaxation rates of spin-labeled proteins contain terms due to both paramagnetic and diamagnetic relaxation mechanisms. For spin-labeled proteins the observed longitudinal or transverse relaxation rates may be written as

$$1/T_B = 1/T_A + 1/T_P + 1/T_{P*} \tag{5}$$

where $1/T_B$ is the observed relaxation rate for a resonance in the presence of the nitroxide, $1/T_A$ is due to diamagnetic relaxation mechanisms, $1/T_P$ is the paramagnetic longitudinal or transverse relaxation rate as defined in Eqs. (1) and (2), and $1/T_{P*}$ is a relaxation rate due to intermolecular mechanisms (considered to be primarily paramagnetic in nature).[48]

Diamagnetic relaxation rates may be determined by measuring the relaxation rates after reduction of the spin label to the diamagnetic hydroxylamine, although, as discussed below, values for the paramagnetic transverse relaxation rates can be determined without knowledge of the diamagnetic relaxation rates. (Commonly used reducing reagents will also be discussed.) As written, Eq. (5) implies that all contributions to the diamagnetic relaxation rates are the same before and after reduction of the nitroxide. Contributions to the total paramagnetic relaxation rates due to the $1/T_{P*}$ term may be eliminated by extrapolating apparent $1/T_P$ values obtained at various protein concentrations to zero concentration.[34] Such experiments are capable of differentiating between specific self-association of the protein (e.g., monomer–dimer equilibrium in the fast chemical exchange limit) and random collisions between molecules. In the former case, a significant and nonlinear concentration dependence of the apparent $1/T_P$ values will be observed for only a subset of resonances, whereas, in the latter case, the value for $1/T_{P*}$ is linearly dependent on concentration and will most effect those nuclei near or on the surface of the protein.

Reducing Reagents. Ascorbic acid,[49,50] sodium dithionite,[51] and phenylhydrazine[52] have been used as one-electron reducing reagents. None of these compounds is ideal. Phenylhydrazine forms by-products with the hydroxylamine of the reduced label. It also contributes resonances to the aromatic region of the NMR spectrum. Sodium dithionite does not add extraneous signals to the NMR spectrum, but can reduce the nitroxide to the amine. (Reoxidation of the amine to the nitroxide requires

[48] The nomenclature of Ref. 19 is adopted here.
[49] R. D. Kornberg and H. M. McConnell, *Biochemistry* **10**, 1111 (1971).
[50] C. M. Paleos and P. Dais, *J. Chem. Soc., Chem. Commun.*, 345 (1977).
[51] A. J. Ozinskas and A. M. Bobst, *Helv. Chim. Acta* **63**, 1407 (1980).
[52] T. D. Lee and J. F. W. Keana, *J. Org. Chem.* **40**, 3145 (1975).

significantly more vigorous redox conditions than does reoxidation of the hydroxylamine. This may be a consideration if the labeled protein is to be recycled.) Further, dithionite is destroyed at a pH less than 4. The rate of reaction of ascorbic acid and nitroxides is pH dependent, with the acid as the reactive form. Ascorbic acid also contributes resonances to the region upfield of the water resonance.

A stock solution of reducing reagent should be prepared just before reduction, with adjustment of the pH to that of the protein solution. The concentration of the stock solution should be at least 100-fold that of the protein solution so that dilution is minimized. An excess of the reducing reagent is added because the hydroxylamine can be reoxidized by dissolved oxygen if metal ions, even in minute amounts, have not been scrupulously removed. (A 1.5 M excess of ascorbic acid is used in our studies.) It is important to carefully reposition the NMR tube (after having removed it from the magnet to add the reductant) to avoid extensive reshimming.

Spectral Analysis

One-Dimensional Spectra. Relaxation rates have been measured by standard techniques, such as saturation recovery[35,39] or inversion recovery[35,36] (for T_1 measurements) and direct linewidth measurements[35,36,39] or Hahn spin-echo pulse sequences[38] (for T_2 measurements). However, these measurements have only been made for the few resonances that either stand clear of the main spectral envelope, lay sharply above the envelope, or have T_2 relaxation times significantly longer than most of the other hydrogen nuclei of the protein. In comparison to the number of resonances that are perturbed by the nitroxide, direct measurement of relaxation rates distinctly underutilizes the information content of the experiment. Consequently, analysis of a paramagnetic difference spectrum[44,53] is the preferred experimental method, and T_{2P} relaxation times are determined.

Subtracting the spectrum of a spin-labeled protein from that of the reduced protein (which has first been scaled to correct for dilution effects of the added reducing reagent) yields the paramagnetic difference spectrum. The aromatic region of the paramagnetic difference spectrum of lysozyme that is spin-labeled at histidine-15 is shown in Fig. 1. Also shown is the aromatic region of the protein after reduction of the spin label. Fewer resonances appear in the paramagnetic difference spectrum

[53] I. D. Campbell, C. M. Dobson, R. J. P. Williams, and A. V. Xavier, *J. Magn. Reson.* **11,** 172 (1973).

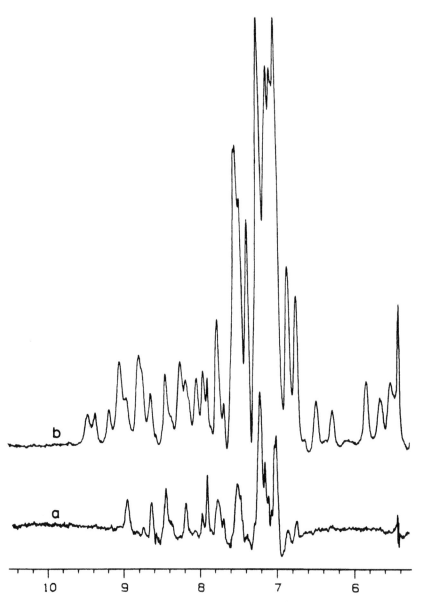

FIG. 1. (a) The aromatic region of the paramagnetic difference spectrum of lysozyme labeled at histidine-15 with (VI), i.e., the spectrum of the hydroxylamine-labeled lysozyme minus the spectrum of the spin-labeled lysozyme. (b) The spectrum of the hydroxylamine-labeled lysozyme at the same gain. Experimental conditions: 1.5 mM protein in 0.1 M oxalate/100% D_2O, pH 5.2, uncorrected for the deuterium isotope effect, 35°; 300-MHz field strength, two blocks of 400 scans each, 2.4-sec repetition delay, 90° pulse = 5.8 μsec, 1-Hz line broadening of the FID, 8192 points after zero filling. [Reprinted with permission from P. G. Schmidt and I. D. Kuntz,[19] *Biochemistry* **23**, 4261 (1984). Copyright 1984 American Chemical Society.]

than in the spectrum of the reduced protein. Only those resonances that are perturbed by the spin label are present. Thus, the resonances which are perturbed by the spin label can be more readily identified. (Resolution is of course decreased for the spectrum of the spin-labeled protein.)

For a homogeneous spin-labeled protein, assuming Lorentzian line shapes, the fractional amplitude of the paramagnetic difference spectrum, ΔAmp,[54] of a singlet is related to the peak intensities, I_A and I_B (at the centers of the resonances and assuming $\omega_A = \omega_B$), or to T_{2A} and T_{2B}, as

$$\Delta Amp = (I_A - I_B)/I_A = 1 - T_{2B}/T_{2A} \tag{6}$$

From Eq. (6) it can be seen that if $T_{2A} = T_{2B}$, $\Delta Amp = 0$, whereas if $T_{2A} \gg T_{2B}$, ΔAmp approaches 1. Consequently, only those nuclei with relaxation rates altered by the spin label appear in the paramagnetic difference spectrum. Figure 2 illustrates the relationship between ΔAmp and the distance separation of the spin label and nuclei of singlet resonances at a ^1H field strength of 500 MHz for two values of τ_C and three values of the linewidth, W_A (with $\pi W_A = 1/T_A$).

Equation (6) and analogous equations, which incorporate the coupling constants and multiplicities of resonances that are not singlets, usually cannot be solved directly because neither T_{2A} nor I_A can be determined directly when resonances overlap. While it is possible, in principle, to carry out a multidimensional iterative fitting procedure in order to extract values for T_{2B} and T_{2A}, direct analysis of the line shapes of the spectra is not necessary. Following the procedures developed by Campbell et al.,[53] Schmidt and Kuntz[19] analyzed the paramagnetic difference spectrum of spin-labeled lysozyme in the following manner. A set of convoluted difference spectra was produced by selectively broadening the reduced protein's spectrum by exponential multiplication of the FID, transforming the FID, correcting for dilution errors, and then subtracting the spin-labeled protein's spectrum from each artificially broadened spectrum. It was also necessary to correct for small chemical shift effects, which arise because of instrumental instabilities or slight differences in the repositioning of the NMR tube after reduction. Chemical shift effects produce distinctive biphasic curves that could usually be recognized, even in the presence of the triphasic peaks of the convolution difference spectrum. Adjustment of the oxidized spectrum by ±2 data points was helpful in removing any biphasic components. Direct peak-by-peak comparison of the lysozyme paramagnetic difference spectrum with each convoluted difference spectrum, with interpolation between spectra, allowed estima-

---

[54] If $1/T_{2P^*}$ contributes significantly to $1/T_{2B}$, ΔAmp will be quite dependent on temperature[19] or concentration.[34]

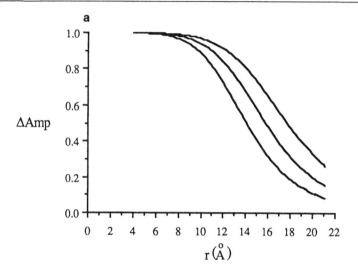

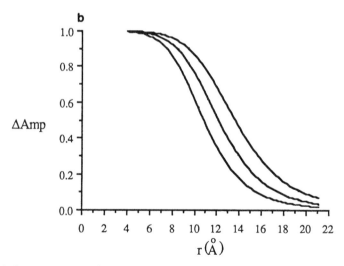

FIG. 2. Calculated ΔAmp for a singlet with a Lorenztian-shaped linewidth as a function of the distance between the hydrogen nucleus and the spin label. The curves are derived from Eq. (6), assuming a field strength of 500 MHz and (a) $\tau_C = 10^{-8}$ sec and (b) $\tau_C = 2 \times 10^{-9}$ sec. In descending order, the three curves in a and b assume a diamagnetic linewidth of 5, 10, and 20 Hz. (Based on similar calculations by Dwek et al.[44])

tion of the paramagnetic contributions to the linewidths with an uncertainty of ±10%.

The paramagnetic contributions to the linewidths were converted into distances using Eq. (2) and the validity of the experimental approach was tested as follows. The experimentally derived distances were used as input along with the heavy atom coordinates for lysozyme[55] to solve for the position of the spin label by searching all of the conformational space. Optimization of the distance geometry error function with various starting positions of the label and various error limits always returned a single, well-defined position with respect to the protein, with uncertainties of ±1 Å in each dimension. Schmidt and Kuntz[19] reported data for 21 proton resonances extracted from spectra recorded between 23 and 45°. Figure 3 compares 40 experimental distances (i.e., solution distances) derived from T_{2P} measurements at 75° to the distances between the heavy atom positions of the associated hydrogen atoms of the protein and the optimized position of the spin label (i.e., crystal distances). The solid line is a linear least-squares fit of the data with a slope of 0.91 and a y intercept of 1.2 Å and serves primarily as a visual aid. Note that a subset of paramagnetic relaxation rates was highly temperature dependent. It was concluded that the temperature dependency was due to a rapid self-association of a specific dimeric form of the protein. At 75° these effects were unimportant.

Musci et al.[33] used a somewhat different method of spectral analysis, whereby T_{2A} (as the linewidth, W_A) and I_A were first indirectly estimated in cases where resonance overlap could be eliminated by resolution enhancement of the reduced protein's spectrum. (Here, the protein was a spin-labeled bovine α-lactalbumin.) Two convoluted difference spectra were obtained by applying exponential multiplication factors of either 5 or 30 Hz to the FID of the reduced protein and by subtracting each of these spectra from the spectrum of the reduced protein to which a 0.5-Hz broadening factor had been applied. As Eq. (6) holds for singlets of these convoluted difference spectra, values for W_A and I_A were estimated when the relative peak intensities of the two convoluted difference spectra, I_5 and I_{30}, could be measured, because

$$W_A = 30(I_{30} - I_5)/(6I_5 - I_{30}) \tag{7}$$
$$I_A = 5I_5[I_{30}/(6I_5 - I_{30})] \tag{8}$$

Having thus indirectly determined I_A and W_A in regions where peak overlap was eliminated by application of convoluted difference resolution enhancement, Musci et al.[33] could then directly apply Eq. (6) to the para-

[55] R. Diamond, J. Mol. Biol. 82, 371 (1974).

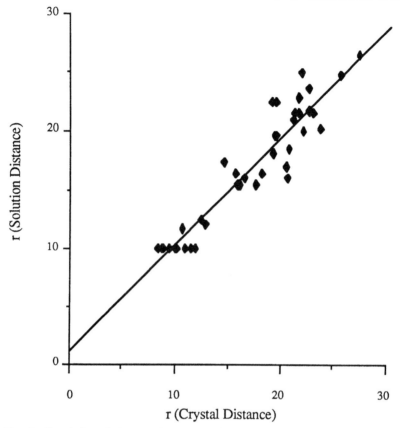

FIG. 3. Correlation of the experimental solution distances and the optimized crystal distances for spin-labeled lysozyme. Distances are given in angstroms. Experimental conditions are similar to those of Fig. 1 except that the temperature was 75°.

magnetic difference spectrum. When peak overlap was significant, similar manipulations of the data were made for doublets and triplets incorporating the appropriate coupling constants. This type of analysis is a less general method than that used by Schmidt and Kuntz.[19]

Two-Dimensional Spectra. Even though a one-dimensional paramagnetic difference spectrum reduces the spectral complexity, there will be peak overlap, introducing additional experimental uncertainties and decreasing the number of distances that can be calculated. Selective or nonselective deuteration of the protein is one means of decreasing spectral overlap.[22–27] Another way of circumventing the overlap problem is to determine the paramagnetic relaxation rates by analyzing the cross-peaks of two-dimensional NMR spectra. An additional advantage of two-dimensional spectroscopy is the smaller degree of overlap between resonances of the protein and resonances of the hydroxylamine substituent and the reducing reagent, which normally obliterate portions of a one-dimensional paramagnetic difference spectrum. While there are many two-dimensional experiments available, each has some advantages and some disadvantages for the spin-label technique (Table I). Presently, the best option is to use the absolute-value COSY experiment.

The two-dimensional experimental protocols for spin-labeled proteins are similar to those described for one-dimensional spectroscopy. Absolute-value COSY spectra of the spin-labeled protein and the protein after reduction are acquired and processed in the usual manner. A set of artificially broadened spectra (starting with the spectrum of the reduced protein) are calculated and the cross-peaks of these spectra are matched, peak-by-peak, to those of the spin-labeled protein's COSY spectrum.[34] Alternatively, a semiautomatic procedure can be used in which each cross-peak of the reduced spectrum is back transformed, broadened by exponential multiplication, forward transformed, and compared iteratively to the oxidized cross-peak until the integrated intensities of the two cross-peaks match.[56] In this procedure, both the real and the imaginary parts of the interferogram are retained during processing. Both approaches give similar results.

Two cautionary notes need be given. First, it is not possible to resolve the two paramagnetic components that modulate the linewidths of the two resonances of a COSY cross-peak. Therefore, the geometric mean average of the paramagnetic contribution to the total linewidth, W_P, is assumed:

$$W_P = (W_{P1} W_{P2})^{1/2} \tag{9}$$

where W_{P1} and W_{P2} are the paramagnetic line-broadening components in each of the two dimensions. Second, the same uncertainty then applies to the location of the point in space which is subject to the paramagnetic field. This point is displaced, as expected due to the $1/r^6$ distance dependence, along the line of centers of the hydrogen nuclei whose resonances constitute the broadened cross-peak, and is biased to the nucleus closest to the spin label.

Analysis of the NH–C_α region of COSY spectra of the BPTI lysine-26 monoderivative gives experimentally derived distances with an average deviation of ±2 Å compared to the crystal structure.[34] Figure 4 correlates the experimentally derived distances (for the geometric mean average of

[56] R. M. Scheek and S. Manogaran, unpublished results.

TABLE I
COMPARISON OF ADVANTAGES AND DISADVANTAGES OF TWO-DIMENSIONAL NMR
SPECTRA FOR PURPOSES OF DISTANCE MEASUREMENTS

Two-dimensional experiment	Advantages	Disadvantages
Absolute-value COSY	High signal-to-noise ratio	W_{P1} and W_{P2} cannot be measured separately
	Spectral simulation is simple Integrated intensity is strongly dependent on linewidth when the change in linewidth is approximately equal to the coupling constant	
Double-quantum filtered COSY	High signal-to-noise ratio Spectral simulation is simple Moderate dependence of peak height on linewidth W_{P1} and W_{P2} can be measured independently	Integrated intensity is zero
Phase-sensitive NOESY	Simulation is not necessary W_{P1} and W_{P2} can be measured independently Many cross-peaks	Low signal-to-noise ratio $T_{1P} < T_{2P}$ Uncertainty in distance measurement increased due to greater separation of nuclei, if W_{P1} and W_{P2} are not evaluated separately
	Direct measurement of T_{1p} is possible	
Phase-sensitive HOHAHA	High signal-to-noise Moderate dependence of peak height on linewidth W_{P1} and W_{P2} can be measured independently Many cross-peaks	Spectral simulation is not possible

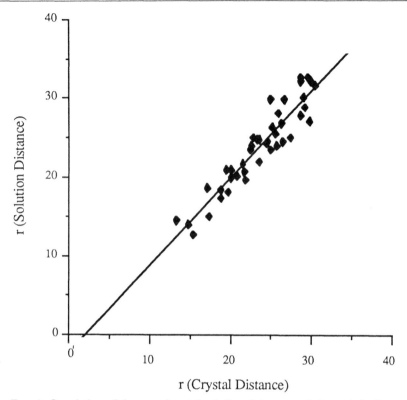

FIG. 4. Correlation of the experimental solution distances and the optimized crystal distances for spin-labeled BPTI. Distances are given in angstroms. Solution distances were derived from COSY spectra as outlined in the text. NMR samples contained 1–5 mM protein at pH 2 in 90% H_2O/10% D_2O. The spectra were recorded at 68°. For each spectrum, 512 free induction decays, each of 1024 complex data points with a ±3512-Hz spectral width, were acquired, with eight scans per FID. The data were weighed with an unshifted sine–bell function and were zero filled once in both domains before Fourier transform.

the NH–C$_\alpha$ hydrogen nuclei of BPTI) with the optimized geometric mean average distances of the same nuclei and the spin label using a neutron diffraction crystal structure of BPTI[57] for the nuclear positions. Again, the solid line is a linear least-squares fit of the data with a slope of 1.1 and a y intercept of −2.4 Å.

As with lysozyme, there seems to be a rapid exchange dimer–monomer equilibrium for spin-labeled BPTI, even at 68°. Therefore, COSY

[57] A. Wlodawer, J. Walter, R. Huber, and L. Sjölin, *J. Mol. Biol.* **180,** 301 (1984).

spectra were recorded with protein concentrations between 1 and 5 mM. The line-broadening contributions due to the electronic–nuclear interactions for the monomeric protein were determined by extrapolation of the apparent paramagnetic line-broadening contributions to zero concentration.

Preparation and Purification of Spin-Labeled Proteins

Chemical Properties of Nitroxides. Commercially available spin labels for protein modification are based on derivatives of 2,2,6,6-tetramethylpiperidine-1-oxyl (I), 2,2,5,5-tetramethylpyrrolidine-1-oxyl (II), or 2,2,5,5-tetramethylpyrroline-1-oxyl (III), with R the reactive portion of the molecule:

(I)   (II)   (III)

Spin labels can be obtained from several commercial sources; Refs. 4, 5, 9, and 10 list suppliers of spin labels. (Commercially available reagents have sometimes been found to be impure and should be routinely recrystallized.[58,59])

The stability of the nitroxide radical per se should not be the limiting factor in the preparation and purification of spin-labeled proteins. Very acidic solutions should be avoided.[60] (Spin-labeled BPTI is stable at pH 2 for at least 72 hr at 68° and for months at 4°.) Of more concern is the presence of reducing reagents, particularly cysteine residues, which can be oxidized to cysteic acid with the concomitant reduction of the nitroxide in the presence of dissolved oxygen and trace metal ions.[61,62] This reaction may be minimized by the elimination of oxygen through deaeration and the presence of an anaerobic atmosphere. It has been shown that catalytic amounts of superoxide dismutase will decrease by 80% the redox reaction between cysteine and 2,2,6,6-tetramethylpiperidine-1-oxyl,[63]

---

[58] G. L. Jones and D. M. Woodbury, *Arch. Biochem. Biophys.* **190**, 611 (1978).
[59] P. A. Kosen, unpublished results.
[60] B. M. Hoffman and T. B. Eames, *J. Am. Chem. Soc.* **91**, 2169 (1969).
[61] J. D. Morrisett and H. R. Drott, *J. Biol. Chem.* **244**, 5083 (1969).
[62] G. J. Giotta and H. H. Wang, *Biochem. Biophys. Res. Commun.* **46**, 1576 (1972).
[63] E. J. Rauckman, G. M. Rosen, and L. K. Griffeth, *in* "Spin Labeling in Pharmacology" (J. L. Holtzman, ed.), p. 175. Academic Press, Orlando, Florida, 1984.

but the generation of hydrogen peroxide by superoxide dismutase may incur other oxidative damage of a protein, e.g., oxidation of methionines. Additionally, metal ions can be sequestered using chelators (ethylene-diaminetetraacetic acid or diethylenetriaminepentaacetic acid) during labeling and purification. Metal ions also can be removed by chromatography of solutions through chelating matrices.[63] Thiols may be blocked by alkylation with iodoacetamide or iodoacetic acid before spin labeling the protein, if no conformational alteration results.[64,65]

Note that derivatives of structure (II) contain an asymmetric carbon at position 3. The racemic mixture does not seem to interfere in the analysis of the spectrum of spin-labeled lysozyme discussed above, but may, for other proteins, complicate the analysis if the structure of one enantiomer fortuitously complements a cleft on the protein. As noted below the modifying group may also introduce sites of asymmetry.

General Considerations of the Purification Procedure. Except for the restrictions noted for the nitroxide, the purification of a spin-labeled protein will depend primarily on the chemical and physical properties of the protein. Knowledge of these properties is of paramount importance for the successful isolation of a spin-labeled protein. It is, of course, not possible to recommend a single purification procedure. As an example, the considerations that led to the final purification design of spin-labeled BPTI derivatives are discussed below. Monographs and reviews on protein purification are widely available (e.g., Refs. 66–70). What follows now are a few general comments that pertain especially to the purification of spin-labeled proteins.

1. HPLC systems which involve direct exposure of the spin-labeled protein and its solvent system to stainless-steel components must be avoided, because, in particular, the paramagnetic Fe^{3+} ion is readily leached into the eluants.[71] (This restriction is valid whenever a protein is purified for NMR spectroscopy.)

2. For a given type of chromatography, e.g., ion exchange, differences in the interfacial interactions due to the chromatographic sorbants and matrices may influence the relative mobilities of the components of a

[64] F. R. N. Gurd, this series, Vol. 11, p. 532.

[65] F. R. N. Gurd, this series, Vol. 25, p. 424.

[66] R. Scopes, "Protein Purification: Principles and Practice." Springer-Verlag, Berlin and New York, 1982.

[67] R. Burgess (ed.), "Protein Purification Micro to Macro." Liss, New York, 1987.

[68] H. Hagestam Freiser and K. M. Goodline, *BioChromatography* **2,** 186 (1987).

[69] F. E. Regnier, *Science* **238,** 319 (1987).

[70] C. H. Suelter, "A Practical Guide to Enzymology." Wiley, New York, 1985.

[71] J. M. Anderson, Jr., R. Steffensen, and W. S. Foster, *BioChromatography* **2,** 190 (1987).

mixture of spin-labeled proteins. Therefore, comparison of the elution patterns using different resins, which have nominally the same function, should be made at an early stage in the design of the purification procedure.

3. Except for proteins spin labeled at tyrosines, modification by the spin labels discussed below leads to a change in the net charge of the protein at pH values where the nitroxide and the covalent bond between the label and the protein are stable. Ion-exchange chromatography should be considered as one or more of the steps in the purification.

4. The size of the spin label may influence the ease of labeling, especially if the reactive side chain is located in a cleft of the protein.[72] When it is necessary, one might compare the relative reactivities of the five- and the six-membered ring nitroxides or reagents, which vary in the number of spacer groups between the nitroxide and the reactive portion of the reagent.

5. Nitroxides are immunogenic. Antibodies that cross-react specifically with nitroxides (and the hydroxylamine derivative) have been isolated.[73-75] This type of chromatography has yet to be attempted, but affinity chromatography with antinitroxide antibodies as the sorbant may be a useful step in purification.

6. Finally, it may be tempting to consider recycling the hydroxylamine-labeled protein by reoxidation after an NMR experiment. However, even very gentle oxidation procedures, e.g., air oxidation in the presence of catalytic amounts of Cu^{2+}, may also oxidize methionines. For BPTI, the facile oxidation of methionine causes many changes in the NMR spectrum.[59]

Monitoring Purification. The purification of the spin-labeled monoderivatives of BPTI was monitored using UV spectroscopy to follow the changes in the maximum (~277 nm) to minimum (~250 nm) ratio of absorbance during the course of purification.[76] The absorbance of nitroxides in the region between 350 to 240 nm is not very large.[77] (The estimated contribution by the nitroxide to the extinction coefficient of the BPTI monoderivatives is 500 M^{-1} cm^{-1} at 280 nm, with a total extinction

[72] J. S. Taylor, J. S. Leigh, Jr., and M. Cohn, *Proc. Natl. Acad. Sci. U.S.A.* **64,** 219 (1969).
[73] P. Rey and H. M. McConnell, *Biochem. Biophys. Res. Commun.* **73,** 248 (1976).
[74] K. Balakrishnan, F. J. Hsu, D. G. Hafeman, and H. M. McConnell, *Biochim. Biophys. Acta* **721,** 30 (1982).
[75] D. C. Eichler, M. J. Barber, and L. P. Solomonson, *Biochemistry* **24,** 1181 (1985).
[76] P. A. Kosen, R. M. Scheek, H. Naderi, V. J. Basus, S. Manogaran, P. G. Schmidt, N. J. Oppenheimer, and I. D. Kuntz, *Biochemistry* **25,** 2356 (1986).
[77] J. D. Morrisett, *in* "Spin Labeling: Theory and Applications" (L. J. Berliner, ed.), p. 273. Academic Press, New York, 1976.

coefficient of 5900 M^{-1} cm^{-1}.) Nonetheless, for BPTI, the absorbance of the nitroxide is sufficiently large to use UV spectroscopy as a simple monitor of purification.

For proteins with an extinction coefficient significantly greater than that of BPTI, the absorbance by the nitroxide will not significantly change the UV spectrum of the protein. Other methods must then be used to monitor the extent of spin-label incorporation. Double integration of an EPR spectrum of a known concentration of the spin-labeled protein can be used in these cases. Alternatively, the spin-labeled protein can be hydrolyzed in 1.0 M NaOH at 60° for 24 hr, followed by acquisition of an EPR spectrum.[78] In either case, the EPR spectrum is compared to a standard solution of the free spin label treated in the same manner. Additional experimental details are given in Refs. 6 and 8.

If cysteines, histidines, or lysines are alkylated with spin-label derivatives of iodo- or bromoacetamide, hydrolysis with 6 M HCl at 110° produces unique carboxymethylated amino acids. These can be quantitated upon amino acid analysis either as a loss of the unmodified amino acid or as the appearance of a new amino acid. For spin-labeled methionines, which are sulfonium salts and are not stable to the conditions of hydrolysis, performic acid oxidation should first be used to convert unlabeled methionines to sulfones.[64,65] Sulfonium salts are resistant to oxidation and sulfones are stable to hydrolysis. Therefore, for acetamido derivatives of cysteine, histidine, lysine, or methionine, amino acid analysis may also be used as a purification monitor. However, because experimental uncertainties in amino acid analysis are at best ±5%, amino acid analysis should not be used as the only proof of homogeneity.

Purification Criteria. The interpretation of a paramagnetic difference spectrum is most straightforward if the spectrum of a single monoderivative is analyzed. Care should be taken to prove both the identity of the labeled site and the extent of homogeneity of the product. It should not be assumed that if the reagent (from which the spin label was derived) reacted at a specific site, that the same reaction will occur with the spin label. In fact, the specificities of the two reagents may differ. For example, in an EPR study,[79] where no product separation was attempted and characterization of the products was limited to amino acid analysis, the reaction of ribonuclease A with a bromoacetamide spin label was concluded to have occurred primarily at histidine-12. This conclusion was based on finding 3-carboxymethylhistidine in the amino acid hydrolysate.

[78] C. T. Cazianis, T. G. Sotiroudis, and A. E. Evangelopoulos, *Biochim. Biophys. Acta* **621**, 117 (1980).
[79] I. C. P. Smith, *Biochemistry* **7**, 745 (1968).

Under similar reaction conditions, iodoacetamide reacts at position 3 of histidine-12.[80] However, a more thorough study,[81] using the identical reaction conditions but including separation and peptide mapping of the products, proved that the spin-labeled position was histidine-105, not histidine-12.

Verification that isolation of monoderivatized spin-labeled proteins has been achieved may be based on as many of the following criteria as possible:

1. All possible monoderivatives have been separated.[59]
2. Identification of the labeled site by protease digestion and peptide mapping.
3. The EPR spectrum of a putative monoderivative does not show multiple components due either to contamination Or to the presence of two populations of slowly interconverting derivatives at the same protein site but with different degrees of motional freedom.[8,81]
4. Characterization by several different types of analytical chromatography showing unique mobilities for each of the putative monoderivatives.

Amino Group Modification

General Considerations. Succinimidyl 2,2,5,5-tetramethyl-3-pyrroline-1-oxyl-3-carboxylate [(IV): Fig. 5] is a preferred reagent when amino groups, including α-amino and ε-amino groups of lysines, are to be spin labeled.[82,83] Other spin labels designed to modify amino groups have been reported, including a water-soluble ketone which forms a Schiff's base with the amino group,[84] dinitrofluorophenyl and dinitrofluorobenzyl derivatives,[78,85] and iodo- and bromoacetamide derivatives.[86] These reagents do not have any inherent advantage over structure (IV). Formation of the Schiff's base requires alkaline conditions. The reduction of the Schiff's base by sodium borohydride, which is needed for stability, must be carefully controlled to avoid concomitant reduction of protein disulfides. The dinitrofluorophenyl and dinitrofluorobenzyl derivatives are not specific

---

[80] R. G. Fruchter and A. M. Crestfield, *J. Biol. Chem.* **242**, 5807 (1967).
[81] W. E. Daniel, Jr., J. D. Morrisett, J. H. Harrison, H. H. Dearman, and R. G. Hiskey, *Biochemistry* **12**, 4918 (1973).
[82] B. M. Hoffman, P. Schofield, and A. Rich, *Proc. Natl. Acad. Sci. U.S.A.* **62**, 1195 (1969).
[83] A. Rousselet, G. Faure, J.-C. Boulain, and A. Ménez, *Eur. J. Biochem.* **140**, 31 (1984).
[84] T. E. Wagner and C.-J. Hsu, *Anal. Biochem.* **36**, 1 (1970).
[85] J. T. Gerig, J. D. Reinheimer, and R. H. Robinson, *Biochim. Biophys. Acta* **579**, 409 (1979).
[86] S. Ogawa and H. M. McConnell, *Proc. Natl. Acad. Sci. U.S.A.* **58**, 19 (1967).

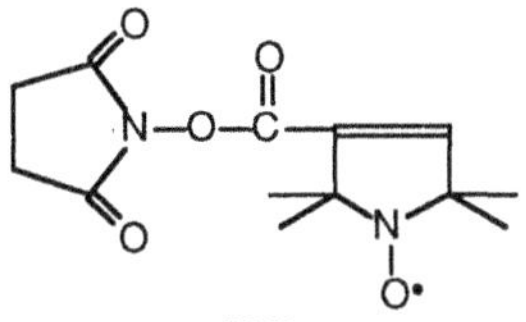

(IV)   LYSINE

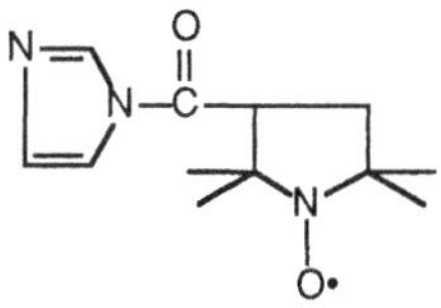

(V)   TYROSINE

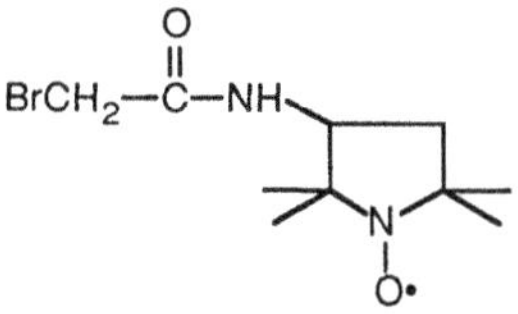

(VI)   CYSTEINE, HISTIDINE, METHIONINE, LYSINE

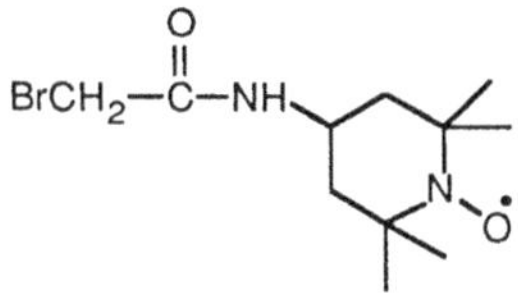

(VII)

FIG. 5. The structures of the spin labels discussed in detail in the text and the amino acid residues these reagents modify.

for amino groups and may require significantly alkaline conditions for reaction. Usually, iodo- and bromoacetamide derivatives react at cysteines, histidines, and methionines, preferentially to amino groups.

The activated ester of structure **(IV)** is resistant to hydrolysis under mild aqueous conditions of neutral or slightly acidic pH. Consequently, concentrations which are stoichiometric with or in slight excess of the protein concentration may be used. Acylation proceeds rapidly. Labeling at sites other than amino groups has not been reported. Spin label [structure (IV)] is readily soluble in organic solvents, e.g., acetonitrile, and can be added as a concentrated solution to a larger volume of aqueous protein solution to start the reaction, without noticeable subsequent precipitation of the spin label. Because an amide bond is formed, the modification is inherently stable.

Unless a single amino group of the protein of interest is fortuitously hyperreactive, the principal experimental challenge is the separation of the mixture of mono-, di-, and multispin-labeled protein products. (In general, the lack of specificity will limit the preparation of mono-derivatized proteins to those proteins which can be obtained in large quantities.) As an example, the preparation and purification of the five spin-labeled derivatives of BPTI are outlined below.[59,76] BPTI has four lysine residues and an α-amino group; all five amino groups react with structure (IV). Obviously, the purification procedure cannot be generalized to all other proteins. Consequently, discussion of the considerations that led to the final purification design is as important, if not more so, than the purification procedure per se. These design considerations will be discussed.

In addition to BPTI, monoderivatives of spin-labeled snake neurotoxins prepared by reaction with structure (IV) have been utilized for both NMR and EPR studies.[31,32,82,87] The purification of these spin-labeled proteins was more straightforward than that for BPTI. One or at most two chromatographic steps were needed to separate the spin-labeled neurotoxins, even though the number of monoderivatives isolated was the same as that of BPTI.

Preparation and Purification of Five Spin-Labeled Derivatives of BPTI. To 10^{-4} mol of BPTI (about 650 mg) in 22 ml of 50 mM sodium phosphate, pH 7.3, a stoichiometric amount of structure (IV) in 3.2 ml acetonitrile is added. The solution is stirred for 5 hr at room temperature, in the absence of light, and is then desalted with exchange of buffer by chromatography using Sephadex G-25 equilibrated with 20 mM imidazole, pH 6.2.

The mixture of BPTI spin-label derivatives is applied to a carboxymethyl-cellulose column (1.5×25 cm) equilibrated with 20 mM imidazole. The protein mixture is fractionated by application of a linear salt gradient between 0 and 0.3 M NaCl in 20 mM imidazole, pH 6.2, of total volume 700 ml. The chromatographic profile is shown in Fig. 6. Peak 3 is

[87] A. S. Arseniev, Y. N. Utkin, V. S. Pashkov, V. I. Tsetlin, V. T. Ivanov, V. F. Bystrov, and Y. A. Ovchinnikov, *FEBS Lett.* **136**, 269 (1981).

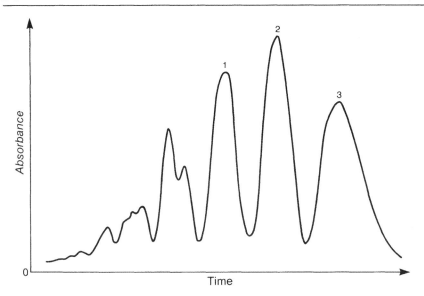

FIG. 6. Carboxymethyl-cellulose chromatography of the reaction mixture of BPTI and structure **(IV)**. Experimental details are given in the text.

unmodified protein. Peak 2 contains the monoderivatives labeled at the α-amino terminus and at the ε-amino group of lysine-26. Peak 1 is a mixture of the three other monoderivatives plus protein that is labeled at more than one position. Protein that elutes earlier than peak 1 is probably labeled at more than one site.

Buffer exchange of peak 2 is accomplished by chromatography using Sephadex G-25 equilibrated in 50 mM ammonium acetate. The two mono-derivatives are separated on an FPLC Mono-S 10/10 column with an isocratic salt gradient that is 0.79 M in ammonium acetate, pH 4. Typically, 20 mg of protein is loaded on the column per round of chromatography. Baseline separation of the two derivatives is not achieved, and two rounds of chromatography are necessary (Fig. 7). Peak 1 of that figure contains the α-amino derivative and peak 2 contains the lysine-26 derivative. The yield of each derivative is about 12% of the starting material.

After a change of buffer to 0.1 M Tris–HCl, pH 8.0 (at 4°), 0.5 M NaCl, and 10 mM CaCl$_2$, again by Sephadex G-25 chromatography, peak 1 of Fig. 6 is applied to chymotrypsin-Sepharose equilibrated in the same Tris–NaCl–CaCl$_2$ buffer at 4°. BPTI binds chymotrypsin via the lysine-15 chain and surrounding residues.[88] The steric bulk of the spin label at

[88] B. Kassell, this series, Vol. 19, p. 844.

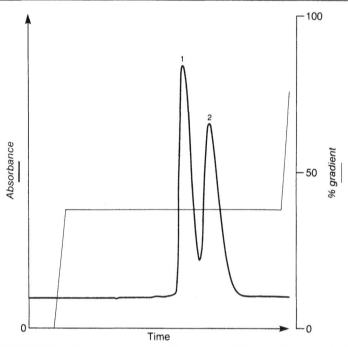

FIG. 7. FPLC Mono-S chromatography of the mixture of the α-amino and lysine-26 BPTI derivatives. Experimental details are given in the text.

lysine-15 precludes binding of the lysine-15 monoderivative to chymotrypsin. That monoderivative is collected upon washing the chymotrypsin-Sepharose with additional Tris–NaCl–CaCl$_2$ buffer. Final yield of the lysine-15 derivative is approximately 9% of the starting material. The proteins bound to chymotrypsin are eluted as a mixture by washing the matrix with 0.01 M HCl. The low pH dissociates the BPTI–chymotrypsin complex.

After lyophilization to remove the acid, the remaining monoderivatives are separated by preparative FPLC reversed-phase chromatography (Pharmacia PEP RPC 16/10 resin) using a linear gradient of 0–50% acetonitrile with 0.1% trifluoroacetic acid. The flow rate is 2 ml/min for a total of 50 min. Again, complete separation is not achieved in one round of chromatography and two or more rounds are used (Fig. 8). Peak 1 of Fig. 8 contains the lysine-41 derivative and peak 2 contains the lysine-46 derivative. The yields of these monoderivatives are each approximately 5%. (Because of the low yields, these monoderivatives have not been used for NMR spectroscopy and, consequently, the separation has not been opti-

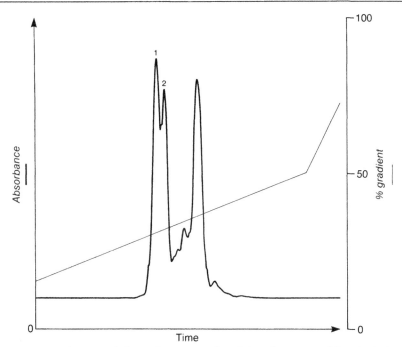

FIG. 8. FPLC reversed-phase chromatography of the mixture containing the lysine-41 and lysine-46 BPTI derivatives. Experimental details are given in the text.

mized.) The remaining fractions contain di- or multispin-labeled derivatives. Mono-S chromatography, which is useful for the separation of the α-amino derivative and the lysine-26 derivative, does not separate the lysine-41 and lysine-46 derivatives.

The identity of each monoderivative is confirmed by peptide mapping of tryptic fragments on thin-layer cellulose plates and amino acid analysis of the peptide fragments. The monoderivatives were also examined by analytical nondenaturing gel electrophoresis. The α-amino derivative and the lysine-26 and lysine-15 derivatives migrate at unique rates upon electrophoresis. The lysine-41 and lysine-46 derivatives migrate as a single band, although at a rate different from the rates of other monoderivatives.

Rationale for the Experimental Design of Purification for the BPTI Monoderivatives. The monoderivatives of BPTI, particularly the lysine-26 derivative, are used in two-dimensional NMR experiments. For each experiment, 3 to 15 mg of protein (about 1 to 5 mM in 0.4 ml) is used. Reoxidation has not been attempted because of oxidation of the methionine. Consequently, considerable amounts of material are required and the primary concern has been to maximize the yield with a minimum time investment.

Because BPTI has a pI of about 10.5,[88] methods that use charge as the fractionation criterion are limited to cationic ion-exchange chromatography. Carboxymethyl-cellulose chromatography has been used to separate BPTI folding intermediates that differ in net charge or in charge distribution.[89] Mono-S FPLC cationic ion-exchange chromatography, which is a "rapid-purification" and metal-free system, was also available when the purification procedure was developed. The two types of cationic ion-exchange chromatography were compared for their ability to fractionate the starting spin-labeled protein mixture. Carboxymethyl-cellulose chromatography was determined to be the appropriate initial step for purification because large amounts of material are partially, although cleanly, separated in a short period of time (and at considerably less cost for the chromatography matrix). The lack of baseline resolution and the much smaller application quantities would have necessitated repeated rounds of Mono-S chromatography, with subsequent pooling of fractions, rechromatography, and continuous loss of material. While the resolution and speed of FPLC (and HPLC) chromatography are often cited as advantages over more traditional chromatographic methods, an impressive difference in resolution was not achieved and the need for repeated rounds of chromatography negated any time advantages.

The well-documented association of BPTI and serine proteases suggested affinity chromatography for isolating the lysine-15 derivative.[88] Obviously, the utility of a chymotrypsin-Sepharose affinity column is maximized if di- and multiderivatized species containing the spin label at lysine-15 are first removed. Once partial fractionation of the monoderivatives was achieved, further purification by either Mono-S chromatography or reversed-phase chromatography became a more realistic possibility. The manipulation of the protein mixtures is much simplified when only a few species must be separated.

The pure lysine-26 monoderivative can be obtained in about 2 weeks. The yield is sufficient for about four to five two-dimensional NMR experiments. While all chromatographic avenues were not explored, the time investment for the quantity of protein isolated is certainly reasonable.

Tyrosine Modification

General Considerations. The syntheses of two spin labels, N-(2,2,5,5-tetramethyl-3-carbonylpyrroline-1-oxyl)imidazole[90] and N-(2,2,5,5-tetra-

---

[89] T. E. Creighton, *J. Mol. Biol.* **95**, 167 (1975).
[90] M. D. Barratt, G. H. Dodd, and D. Chapman, *Biochim. Biophys. Acta* **194**, 600 (1969).

methyl-3-carbonylpyrrolidine-1-oxyl)imidazole **(V)**,[91] both analogous to
N-acetylimidazole[92-94] and both designed to acylate the hydroxyl groups
of tyrosines, have been reported. Neither compound is commercially
available, nor has either reagent been used extensively. The pyrroline
derivative is unstable and cannot be stored, which has undoubtedly discouraged its use. I have been unable to prepare label **(V)** by the recommended published procedure, but a simple modification of that procedure
given below produces label **(V)** in good yields.[95]

Spin label **(V)** is readily hydrolyzed and the labeling reaction should
be performed with an excess of the reagent. Reasonable concentrations of
reactants are 10 : 1 to 50 : 1 label **(V)** to protein, at protein concentrations
of 10^{-3} to 10^{-4} M. The best conditions can be defined only through trial
and error. The rate of hydrolysis of label **(V)** is minimal near pH 6 to 7.[91]
Nucleophilic buffers, e.g., Tris and imidazole, should be avoided since
they catalyze the hydrolysis of label **(V)** and may also hydrolyze the spin-
labeled protein. Instead, borate, phosphate, or veronal buffers should be
used. To start the reaction, label **(V)** may first be dissolved in a small
volume of toluene and added to the reaction tube. The toluene is removed
under a stream of dry nitrogen and the protein solution is then added with
stirring.

Like *N*-acetylimidazole, label **(V)** may not be specific for tyrosines
and may label lysines, cysteines, and to a much smaller extent the hy-
droxyls of threonines and serines. *N*-Acetylimidazole reacts less readily
at lysines than do other acylating reagents, which is why it is a preferred
reagent for tyrosine modification. The same may be true of label **(V)**. In
an EPR study on spin-labeled nucleosome core particles,[96] it was claimed
that more than 90% of the total label was incorporated at tyrosines. But,
since the experimental protocol used to determine the extent of incorporation included hydroxylamine, which reduces nitroxides, the relative
specificity for tyrosines may have been overestimated. For BPTI, preliminary studies, based only on relative migration rates on carboxymethyl-
cellulose and the total number of fractions separated, suggest that lysines
are modified extensively by label **(V)**. Here, the reaction conditions involve a 30 to 50 M excess of label **(V)** in the presence of about 5 mg/ml
BPTI. It may be possible to improve the specificity of label **(V)** for tyro-

[91] M. Adackaparayil and J. H. Smith, *J. Org. Chem.* **42**, 1655 (1977).
[92] J. F. Riordan, W. E. C. Wacker, and B. L. Vallee, *Biochemistry* **9**, 1758 (1965).
[93] J. F. Riordan and B. L. Vallee, this series, Vol. 11, p. 565.
[94] J. F. Riordan and B. L. Vallee, this series, Vol. 25, p. 500.
[95] The modified procedure was suggested by Dr. D. C. F. Chan of the University of Hawaii,
Honolulu.
[96] D. C. F. Chan and L. H. Piette, *Biochemistry* **21**, 3028 (1982).

sines by first irreversibly blocking the lysines by guanidination[97,98] or reversibly blocking with citraconic anhydride or maleic anhydride[99] or blocking cysteines by alkylation.[64,65]

Preparation of N-(2,2,5,5-Tetramethyl-3-carbonylpyrrolidine-1-oxyl)imidazole. Carbodiimidazole (2 mmol) is added to a suspension of *N*-(2,2,5,5-tetramethylpyrrolidine-1-oxyl)-3-carboxylic acid (2 mmol) in 15 ml of dry toluene. The mixture, which is contained in a three-neck flask equipped with a drying tube, is stirred under a nitrogen atmosphere at room temperature for 2 hr. The solvent is decanted from the flask, leaving behind a white solid, presumed to be imidazole. Most of the toluene is removed using a rotoevaporator until only a thick slurry remains. Spin-label (**V**) is crystallized from dry ether at −20°. The final yield is about 60% and the melting point is 128–130° (128–129°, literature value[91]). Spin-label (**V**) can be stored desiccated at −20°.

Cysteine, Histidine, and Methionine Modification

General Considerations. Two classes of spin-labels, maleimide derivatives[100,101] and iodo- or bromoacetamide derivatives,[86] have been used most often to label cysteines. Neither class of reagent is absolutely specific for cysteine. There are spin-labels which should have a complete specificity for cysteine, notably (1-oxyl-2,2,5,5-tetramethylpyrroline-3-methyl)methane thiosulfonate,[102] which forms a mixed disulfide with cysteine. This reagent does not appear to have been used since the initial report of its synthesis. Derivatives of *p*-chloromercuribenzoate,[103,104] also specific for cysteines, have been synthesized, but are not extensively used.

Although commonly used, maleimide derivatives probably should be avoided. The limitations of maleimide reagents have been noted previously.[58,77] Some of the maleimide derivatives are rapidly hydrolyzed. Further, the number of products per modified cysteine may be greater than one. The thiol of cysteine adds across the double bond of the maleimide moiety and this reaction potentially yields a pair of stereoisomers (Fig. 9a;

[97] J. R. Kimmel, this series, Vol. 11, p. 584.
[98] A. F. S. A. Habeeb, this series, Vol. 25, p. 558.
[99] P. J. G. Butler and B. S. Hartley, this series, Vol. 25, p. 191.
[100] O. H. Griffith and H. M. McConnell, *Proc. Natl. Acad. Sci. U.S.A.* **55**, 8 (1966).
[101] S. Ohnishi, J. C. A. Boeyens, and H. M. McConnell, *Proc. Natl. Acad. Sci. U.S.A.* **56**, 809 (1966).
[102] L. J. Berliner, J. Grunwald, H. O. Hankovszky, and K. Hideg, *Anal. Biochem.* **119**, 450 (1982).
[103] J. C. A. Boeyens and H. M. McConnell, *Proc. Natl. Acad. Sci. U.S.A.* **56**, 22 (1966).
[104] A. Zantema, H. J. Vogel, and G. T. Robillard, *Eur. J. Biochem.* **96**, 453 (1979).

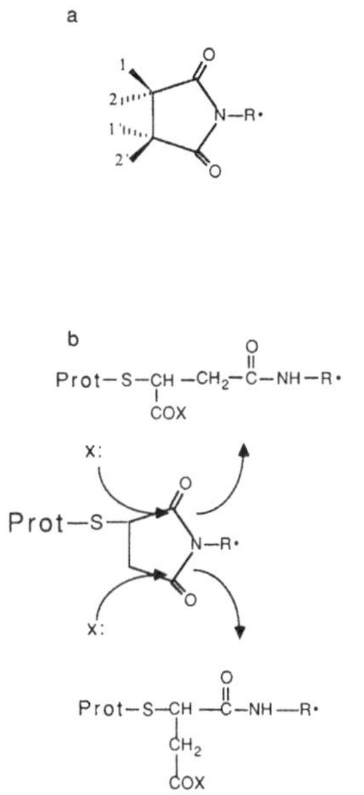

FIG. 9. (a) The stereochemistry of the reaction products of a thiol and a maleimide spin label. The two pairs of stereoisomers are labeled 1 and 1' and 2 and 2'. R represents the spin label. (b) Reaction of a nucleophile (X:) at the carbonyl carbons of the succinimidyl ring of a spin-labeled protein and the expected products. Prot represents the protein.

positions 1 and 1' and 2 and 2' are equivalent in the absence of asymmetry in the nitroxide substituent).[105] In the absence of steric restrictions, or preferential interactions between the protein and the maleimide reagent before covalent addition, both stereoisomers should be found. Further, hydrolytic or nucleophilic attack at either carbonyl carbon of the succinimide ring can lead to an additional spectrum of products (Fig. 9b). Both histidine and cysteine may participate as catalysts of hydrolysis. Amino groups may acylate one of the carbonyl carbons of the maleimide moiety or of the succinimidyl product of cysteine addition.

Iodo- and bromoacetamide spin-label derivatives do not suffer from the limitations of the maleimide derivatives. But, on the other hand, these reagents are less specific and will label cysteines, histidines, methionines, and lysines. The extent and sites of the reaction will depend largely on the pH of the reaction solution and the pK_a values of the accessible side chains. The reactive form of cysteine is the thiolate anion, and the reactive forms of histidine and lysine are the uncharged species. Optimized

[105] J. K. Moffat, S. R. Simon, and W. H. Konigsberg, *J. Mol. Biol.* **58**, 89 (1971).

reactions usually require certain conditions: histidines, a slightly acidic pH (pH ~5 to 6); cysteines, a neutral to slightly basic pH (pH ~7 to 8); and lysines, an alkaline pH (pH >8) (assuming the pK_a values of the residues are not significantly perturbed from their usual values). The thiol ether of methionine reacts independently of pH. Therefore, specific labeling of methionine can be achieved at acidic pH. Only for cysteines can the reaction be expected to be complete in a matter of hours; for the other sites, days may be required and 100% labeling may still not be achieved. This difference in reactivity is useful when cysteines are to be labeled. Iodo- and bromoacetamide derivatives of both the pyrrolidinyl (VI) and piperidinyl (VII) type are commercially available, as are analogs with additional spacer groups in the linker region between the halide and the nitroxide. Usually the reaction is initiated by adding the solid spin label to the protein solution. Alternatively, the spin label may be first dissolved in an organic solvent, e.g., ethanol or methanol. The reaction should be conducted in the absence of light because any iodine or bromine generated can cause oxidation at various sites on the protein.

Preparation and Purification of Proteins Spin Labeled by Bromoacetamide Reagents. There seem to be no examples of spin labeling at cysteines followed by purification of the reaction products, even though cysteines are probably the most commonly targeted site for EPR studies. There are, however, a few examples of labeling at methionine and histidine with subsequent attempts at purification. Three examples will be outlined in detail.

For an NMR study, Berliner and co-workers spin labeled the single methionine of α-lactalbumin with 2,2,6,6-tetramethyl-3-(bromoacetamido)piperidine-1-oxyl (VII).[33] The reaction conditions were 0.2 M acetate, pH 3.6, at 4° for 6 days at a 1 : 10 ratio of approximately 1 mM protein to spin label. Approximately 54% labeling was attained. The specificity of the reaction was shown by lack of incorporation of the spin label if the methionine was first oxidized to the sulfoxide. The authors reported an inability to separate the spin-labeled protein from unlabeled protein by DEAE–Trisacryl chromatography. Without knowing the chromatography conditions, which were not reported, it is difficult to rationalize the poor separation of the labeled and unlabeled α-lactalbumin, because the sulfonium salt is positively charged. For BPTI, after labeling at the methionine with either (VI) or (VII), the protein migrates significantly more slowly than does the unmodified protein upon cationic ion-exchange chromatography.[59]

If methionines are to be labeled, two possible pitfalls must be avoided. In the absence of steric restrictions limiting the accessibility of the methionine's sulfur to the alkylating reagent, the reaction produces dias-

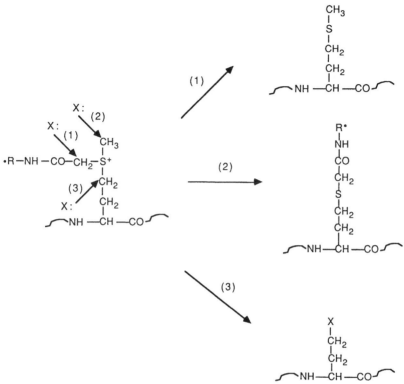

FIG. 10. Reaction of a nucleophile (X·) at three possible sites of a spin-labeled methionine residue and the expected products.

tereomers of the sulfonium salt. Because of the lack of experimental data, it is impossible to comment on what effect the presence of diastereomers will have on the uncertainties in the distance measurements. Second, nucleophiles can attack sulfonium salts in one of three possible reactions (Fig. 10).[106–108] These reactions may also occur slowly at alkaline pH, even in the absence of an extrinsic nucleophile.[108] Usually the type 1 reaction predominates, regenerating the unmodified protein. For example, incubation of tetramethylammonium iodide and BPTI (labeled at the methionine) removes the label.[59] Thiols readily promote the type 1 reaction.

As mentioned above, Daniel and co-workers isolated a derivative of ribonuclease A spin labeled at histidine-105 for EPR studies.[81] The reac-

[106] G. Toennies and J. J. Kolb, *J. Am. Chem. Soc.* **67,** 1141 (1945).
[107] A. Schejter and I. Aviram, *FEBS Lett.* **21,** 293 (1972).
[108] F. Naider and Z. Bohak, *Biochemistry* **11,** 3208 (1972).

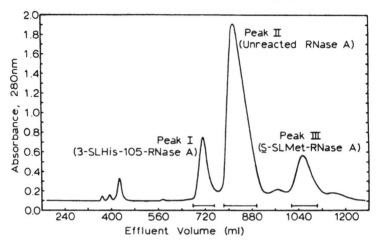

FIG. 11. Bio-Rex 70 chromatography of the reaction mixture of ribonuclease A and structure **(VI)**. Peak I is the derivative labeled at histidine-105. Peak II is unmodified ribonuclease A. Peak III contains protein labeled at methionine. The other minor peaks were not identified. [Reprinted with permission from W. E. Daniel, Jr., J. D. Morrisett, J. H. Harrison, H. H. Dearman, and R. G. Hiskey,[81] *Biochemistry* **12,** 4918 (1973). Copyright 1973 American Chemical Society.]

tion conditions were 450 mg of ribonuclease in 45 ml of 0.1 N sodium acetate, pH 5.5, to which a 16.4-fold excess of 2,2,5,5-tetramethyl-3-(bromoacetamido)pyrrolidine-1-oxyl **(VI)** was added. The reaction proceeded with stirring at room temperature in the dark for 72 hr. The reaction mixture was separated by chromatography on Bio-Rex 70 in 0.2 M sodium phosphate, pH 6.47 (Fig. 11). Two major reaction products were isolated. The identify of the histidine-labeled derivative was confirmed by isolation of a tryptic peptide containing spin-labeled histidine-105 and Edman degradation of that peptide. That a methionine was derivatized was confirmed by amino acid analysis, but the identity of the methionine was not pursued further.

    McConnell and co-workers prepared a spin-labeled derivative of lysozyme labeled at histidine-15, the single histidine of the protein.[20,21] The reaction conditions were 70 mg of lysozyme in 1.0 ml of 0.1 M sodium acetate, pH 5.1, plus 13.9 mg of structure **(VI)**. The mixture was stirred at 40° for 36 hr and the products were separated by chromatography of Bio-Rex 70 (Fig. 12). Schmidt and Kuntz[19] prepared the same spin-labeled derivative of lysozyme and also the 2,2,6,6-tetramethylpiperidine-1-oxyl derivative by reaction with structure **(VII)**, but purified the products by HPLC chromatography on carboxymethylated silica. In both reports the

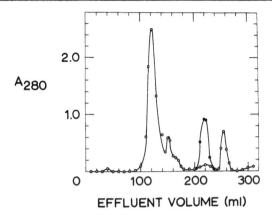

EFFLUENT VOLUME (ml)

FIG. 12. Bio-Rex 70 chromatography in 0.2 M sodium phosphate, pH 7.18, of the reaction mixture of lysozyme and structure (**VI**). The first peak to elute contains the histidine-15 derivative. Elution at 220 ml is characteristic of native lysozyme. The compositions of peaks other than that containing the histidine-15 derivative were not reported. (Reproduced with permission from Morrisett et al.[20])

spin-labeled lysozyme was characterized only by amino acid analysis. It is therefore not possible to strictly rule out the presence of a contaminant, which would bias the distance measurements between the spin-labeled histidine and substrates in the studies by McConnell and co-workers. However, the distances derived in that study are consistent with the crystal structure of lysozyme. The solution structure calculations of Schmidt and Kuntz are consistent with partial labeling of <10% at sites other than the histidine.

Acknowledgments

It is a pleasure to thank Dr. I. D. Kuntz for his continuing support and for the data presented in Figs. 3 and 4. I also thank Drs. V. J. Basus, R. Cooke, C. D. Eads, I. D. Kuntz, N. J. Oppenheimer, and Mr. V. Powers for critical readings of this manuscript in various stages of its preparation. The research reported herein was supported by grants from the National Institutes of Health (GM-19267) to I. D. Kuntz. The UCSF Magnetic Resonance Laboratory was partially funded by grants from the National Science Foundation (DMB 8604081) and the National Institutes of Health (RR-01695 and Shared Instrumentation Grants).

Section II

Protein Structure

[6] Determination of Three-Dimensional Protein Structures in Solution by Nuclear Magnetic Resonance: An Overview

By KURT WÜTHRICH

Introduction

During the last decade a method for the determination of the complete three-dimensional structure of proteins was developed, which uses (1) NMR for data collection and (2) distance geometry, or possibly other mathematical techniques, for the structural interpretation of the NMR data.[1,2] At present, NMR is the only approach, besides diffraction techniques with single crystals,[3] that is available for protein structure determination. Its introduction is of fundamental interest, because NMR can provide data that are in many ways complementary to those obtained from X-ray crystallography and thus promises to widen our view of protein molecules for a better grasp of the relationships between structure and function.

The method of protein structure determination by NMR involves several steps: data acquisition, spectral analysis, and structural interpretation of the experimental data.[2] Chapters [7]–[10] of this volume present separate treatments of these different steps and Chapters [2]–[5] discuss the preparation of proteins for NMR studies. Various aspects of NMR spectroscopy are discussed in Chapters [1]–[11] in Vol. 176 of this series. It is the purpose of this chapter to present a survey of the method, and place the forementioned more specialized contributions into perspective. Furthermore, the range of potential applications for the NMR method of protein structure determination, and the type of results anticipated from it, are briefly discussed.

Information from Protein Structure Determinations by NMR

The complementarity of the structural information obtained from diffraction techniques or from NMR results from the fact that the time scales of the two types of measurements are widely different[2,3] and that, in contrast to the need for single crystals in diffraction studies, the NMR

[1] K. Wüthrich, G. Wider, G. Wagner, and W. Braun, *J. Mol. Biol.* **155,** 311 (1982).
[2] K. Wüthrich, "NMR of Proteins and Nucleic Acids." Wiley, New York, 1986.
[3] T. L. Blundell and L. N. Johnson, "Protein Crystallography." Academic Press, New York, 1976.

measurements use proteins in solution or in other noncrystalline states. The most obvious consequence is that NMR can be applied to proteins for which no single crystals are available, thus yielding new structural data that cannot be determined by any other method. Conversely, the NMR method has so far been applied successfully only to small proteins having molecular weights up to 20,000, and in the future this size limit will, at best, be increased twofold. Furthermore, the solutions used for the NMR measurements must contain protein concentrations of 1 to 5 mM. Therefore, if the size of the protein exceeds a molecular weight of 20,000 (in the future this limit may perhaps be raised to about 30,000), or if it is not possible to prepare stable concentrated solutions, X-ray diffraction remains the only means to determine molecular structure, provided that suitable single crystals are available.

The NMR method is of prime interest for those working with polypeptides and proteins having molecular weights up to approximately 20,000. For molecules in this class the method provides stringent tests for the purity of the protein preparations,[4,5] the amino acid composition, and the amino acid sequence,[6,7] and can yield a three-dimensional structure in solution at high resolution.[8] Comparisons of corresponding structures in single crystals and in noncrystalline states are then highly relevant, because the solution conditions for NMR studies can often be chosen so as to coincide closely with the natural, physiological environment of the protein.[2] Extensive similarities between crystal and solution structures, as well as major conformational rearrangements between the two states, have already been observed. Significant differences between crystal and solution are clearly evident for the molecular surface, which is usually most directly related to the functional properties of the molecule and often includes the active site.

The use of NMR for structure determination with small proteins does not preclude studies of interest with larger systems. For example, individual domains excised by enzymatic or chemical cleavage from large proteins may be accessible for the NMR method.[9] Or, in a more general vein, conclusions from systematic comparisons of the molecular surface of

[4] G. Otting, P. Marchot, P. E. Bougis, H. Rochat, and K. Wüthrich, *Eur. J. Biochem.* **168**, 603 (1987).

[5] H. Widmer, G. Wagner, H. Schweitz, M. Lazdunski, and K. Wüthrich, *Eur. J. Biochem.* **171**, 177 (1988).

[6] P. Štrop, D. Čechová, and K. Wüthrich, *J. Mol. Biol.* **166**, 669 (1983).

[7] G. Wagner, D. Neuhaus, E. Wörgötter, M. Vašák, J. H. R. Kägi, and K. Wüthrich, *Eur. J. Biochem.* **157**, 275 (1986).

[8] A. D. Kline, W. Braun, and K. Wüthrich, *J. Mol. Biol.* **204**, 675 (1988).

[9] E. R. P. Zuiderweg, R. Kaptein, and K. Wüthrich, *Proc. Natl. Acad. Sci. U.S.A.* **80**, 5837 (1983).

small proteins in crystals and in solution may be projected to include larger molecules for which only a crystal structure is available.

Survey of the NMR Method for Protein Structure Determination

The data for protein structure determination are collected using one-, two-, or three-dimensional NMR experiments.[2,10] Two-dimensional Overhauser enhancement spectroscopy (NOESY) is the pivotal experiment in studies of protein structures. In a NOESY spectrum each cross-peak establishes a correlation between two diagonal peaks, as indicated in Fig. 1A for the peaks labeled i, j, and k, and indicates that the protons corresponding to the two correlated diagonal peaks are separated only by a short distance, say less than 4.0 Å. This information is used in the following way. First, sequence-specific resonance assignments must be obtained, i.e., for all the protons in the polypeptide chain the corresponding diagonal peaks must be identified. This is indicated in Fig. 1A and B by the two dashed arrows linking the protons i' and j' with their diagonal positions i and j in the NOESY spectrum. Each NOESY cross-peak shows that two protons in known locations along the polypeptide chain are separated by a distance of less than approximately 4.0 Å in the three-dimensional protein structure. Since the overall length of an extended polypeptide chain with n residues is 3.5 n, the NOEs may thus impose stringent constraints on the polypeptide conformation. In Fig. 1C this is schematically indicated by the formation of a circular structure through the near approach of the two protons i' and j', which are correlated via the NOESY cross-peak k. Typically, the NOESY spectra of proteins contain hundreds of cross-peaks (Fig. 1A), indicating that the three-dimensional fold of the polypeptide chain contains a large number of circular structures of the type shown in Fig. 1C. For the structural interpretation of these data, mathematical techniques are available to identify those three-dimensional arrangements of the linear polypeptide chain which satisfy all the experimental constraints. Figures 2 and 3 show the solution structure of bull seminal protease inhibitor IIA (BUSI IIA), which was determined with this method.

Sequence-specific resonance assignments have a key role in protein structure determinations by NMR.[1,2] This is readily appreciated from an inspection of Fig. 1. Without the resonance assignments the size of the circular structure in Fig. 1C and its location in the amino acid sequence would not be defined. The same would be true for the circular structures manifested by all other NOESY cross-peaks, and hence without the reso-

[10] R. R. Ernst, G. Bodenhausen, and A. Wokaun, "Principles of Nuclear Magnetic Resonance in One and Two Dimensions." Oxford Univ. Press (Clarendon), London and New York, 1987.

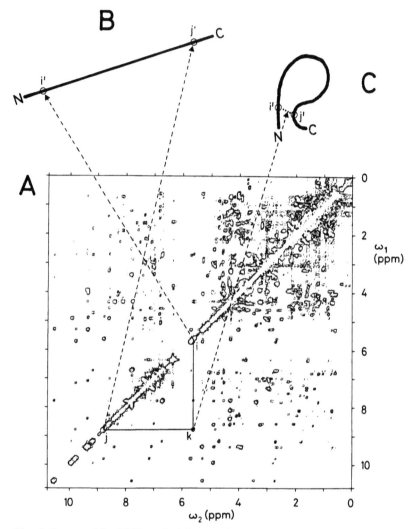

FIG. 1. Survey of the NMR method for protein structure determination in solution. (A) Contour plot of a 500-MHz ^{1}H NOESY spectrum of the protein basic pancreatic trypsin inhibitor (BPTI). A cross-peak k is connected by horizontal and vertical lines with the diagonal peaks i and j. (B) Straight line representing a polypeptide chain. Two protons are identified by circles and the letters i' and j'. Two broken arrows connect the corresponding diagonal peaks i and j in the NOESY spectrum with these protons. (C) Schematic presentation of the circular structure imposed on the polypeptide chain by the fact that the protons i' and j' are separated by a distance of less than 4.0 Å, as is evidenced by the appearance of the NOESY cross-peak k.

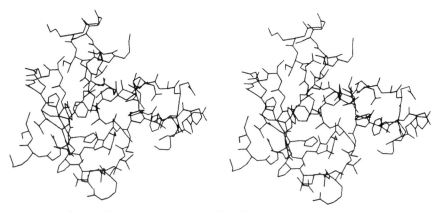

FIG. 2. Stereo view of the structure of BUSI IIA, which was determined with the distance geometry program DISGEO[12] from NMR data recorded in aqueous solution; the structure was subsequently refined by energy minimization.[11] The disulfide bonds are not shown.

nance assignments there would be no way to determine a molecular structure from the information that can be derived from NOEs. The sequential assignment technique for obtaining the desired resonance assignments is described in Chapter [7] in this volume and individual two-dimensional NMR experiments needed for this procedure are presented primarily in Chapters [2], [5], [6], and [8] in Vol. 176 of this series. Obtaining ¹H NMR

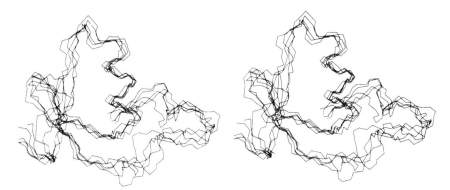

FIG. 3. Stereo view of five structures of BUSI IIA calculated with the distance geometry program DISGEO[12] from the same NMR distance constraints (see the text). The structures were superimposed so as to minimize the pairwise root-mean-square distance with respect to the structure shown in Fig. 2. All main-chain backbone atoms (C^α, C', and N) are shown. For clarity, residues 1 to 4, which are essentially unconstrained by the NMR data, have been omitted, and the disulfide bonds are not shown.[11]

assignments for proteins is quite laborious, and at least partial automation to improve the efficiency of the procedure is of considerable interest. Initial work on the use of computers for this purpose is presented in Chapter [8] in this volume.

The initial protein structure calculations from NMR data used metric matrix distance geometry algorithms,[11-14] and presently most work is done based on the principles of distance geometry. Each individual structure calculation may find a protein conformation that is compatible with all the experimental constraints collected from NOESY and possibly supplementary experiments (Fig. 2).[2] To investigate further whether a given set of NMR input data defines a unique conformation type, a group of structures is obtained by repeating the calculation with the same NMR data but with different, randomly generated starting conditions. Since the experimental data represent only an incomplete set of intramolecular interatomic connectivities, and each constraint describes an allowed distance range rather than a precise value for the distance,[2] the individual structures are similar but not identical. The result of a structure determination from NMR data is therefore commonly represented by a group of conformers, each of which represents a solution to the geometric problem of fitting the polypeptide chain to the ensemble of all experimental constraints. The spread among the different structures, usually expressed by the average of the pairwise RMSDs,[2,11,14] indicates the precision of the structure determination. As an illustration, Fig. 3 shows a superposition of the BUSI IIA structure of Fig. 2 with four additional structures calculated from the same input data. For the sake of clarity only the backbone atoms have been drawn. The figure demonstrates that the apparent precision of the structure determination varies significantly for different segments of the polypeptide chain.

A detailed account of distance geometry and of its use for protein structure determinations from NMR data is presented in Chapter [9] in this volume. The use of molecular dynamics calculations either as an alternative to distance geometry or as a supplementary technique for refinement of the distance geometry conformations is discussed in Chapter [10] in this volume. Other techniques for structure calculations have been reviewed elsewhere.[15]

[11] M. P. Williamson, T. F. Havel, and K. Wüthrich, *J. Mol. Biol.* **182,** 295 (1985).
[12] T. F. Havel and K. Wüthrich, *Bull. Math. Biol.* **46,** 673 (1984).
[13] W. Braun, C. Bösch, L. R. Brown, N. Gō, and K. Wüthrich, *Biochim. Biophys. Acta* **667,** 377 (1981).
[14] W. Braun, G. Wider, K. H. Lee, and K. Wüthrich, *J. Mol. Biol.* **169,** 921 (1983).
[15] W. Braun, *Q. Rev. Biophys.* **19,** 115 (1987).

A structural interpretation of NOE distance constraints and supplementary NMR data, in particular spin–spin coupling constants and information on individual amide proton exchange rates, may also be achieved by empirical model considerations and is briefly described here. An empirical pattern recognition approach is particularly attractive for secondary structure identification,[2,16,17] because it requires very little work in addition to that needed for obtaining sequence-specific resonance assignments by the sequential method. This is best explained by way of an example: in an extended polypeptide chain, the distances from the hydrogen atoms in the first amino acid residue to those in the fourth residue are approximately 12 Å, and thus are far too long to be manifested by NOEs. In contrast, in an α helix the proton–proton distances between residues 1 and 4 are of the order 3.5–4.5 Å, and thus are within the range accessible to NOE observation. Therefore, the presence of $^1H-^1H$ NOEs between residues in the relative sequence positions i and $i + 3$ over a certain polypeptide segment shows that this part of the chain forms an α helix. In practice, all of these medium-range NOEs may not be observed, but there are additional data patterns which are characteristic of an α helix, including short sequential distances d_{NN},[16] small vicinal spin–spin couplings $^3J_{HN\alpha}$,[17] and slowed exchange of the hydrogen-bonded amide protons in all but the first three helical residues.[2] Similarly, characteristic patterns of NOE data, spin–spin couplings, and amide proton exchange rates have been found for the other common regular polypeptide secondary structures.[2,16,17] Once the secondary structure elements have been identified, and additional NOEs are identified between protons located in different secondary structure elements, a low-resolution model of the three-dimensional polypeptide fold may be derived interactively using, for example, mechanical molecular models or an interactive computer graphics program such as CONFOR.[18]

Acknowledgments

The author's research was supported by the Schweizerischer Nationalfonds and by special grants of the ETH Zürich. I thank Mrs. M. Schütz for the careful processing of the manuscript.

[16] K. Wüthrich, M. Billeter, and W. Braun, *J. Mol. Biol.* **180,** 715 (1984).
[17] A. Pardi, M. Billeter, and K. Wüthrich, *J. Mol. Biol.* **180,** 741 (1984).
[18] M. Billeter, M. Engeli, and K. Wüthrich, *J. Mol. Graphics* **3,** 79 and 97 (1985).

[7] Proton Nuclear Magnetic Resonance Assignments

By VLADIMIR J. BASUS

Introduction

Sequence-specific resonance assignments are essential for the study of the solution structure of proteins by NMR, as discussed in Chapter [6] in this volume. The most successful and rigorous procedure for sequential assignments is that proposed by Wüthrich and co-workers,[1-5] based on the observation of sequential NOESY cross-peaks. This procedure was derived from a statistical analysis of 19 proteins whose structures were known from X-ray crystallography, including a total of 3224 residues. The reliability of this procedure is quite evident from the fact that for several proteins that have now been assigned, the original sequence, arrived at by chemical means, was proved to be in error as the NMR assignments neared completion.[6-9] These errors were subsequently verified by resequencing the protein. The sequential NOESY cross-peaks, which are of importance in this procedure, are shown in Fig. 1. The overall strategy can be divided as follows: (1) identification of spin systems using mainly $^1H-^1H$ spin-spin connectivities and (2) sequential assignments of these spin systems based on the observed NOEs between adjacent residues, starting with unique residues and unique pairs of residues for which the NOEs give the highest probability of being sequential.

Before going into the details of steps 1 and 2, the following experimental considerations should be noted. Acquisition of spectra in H_2O is essential because sequential assignments are based completely on the NOEs to the labile backbone NH protons. Spectra in D_2O are essential for stereospecific assignments; they are also very useful in the identification of spin

[1] K. Wüthrich, G. Wider, G. Wagner, and W. Braun, *J. Mol. Biol.* **155**, 319 (1982).
[2] M. Billeter, W. Braun, and K. Wüthrich, *J. Mol. Biol.* **155**, 321 (1982).
[3] G. Wagner and K. Wüthrich, *J. Mol. Biol.* **155**, 347 (1982).
[4] G. Wider, K. H. Lee, and K. Wüthrich, *J. Mol. Biol.* **155**, 367 (1982).
[5] K. Wüthrich, "NMR of Proteins and Nucleic Acids." Wiley, New York, 1986.
[6] P. Štrop, G. Wider, and K. Wüthrich, *J. Mol. Biol.* **166**, 641 (1983).
[7] G. Wagner, D. Neuhaus, E. Wörgötter, M. Vašák, J. H. R. Kägi, and K. Wüthrich, *Eur. J. Biochem.* **157**, 275 (1986).
[8] D. E. Wemmer, N. V. Kumar, R. M. Metrione, M. Lazdunski, G. Drobny, and N. R. Kallenbach, *Biochemistry* **25**, 6842 (1986).
[9] P. A. Kosen, J. Finer-Moore, P. McCarthy, and V. J. Basus, *Biochemistry* **27**, 2775 (1988).

METHODS IN ENZYMOLOGY, VOL. 177

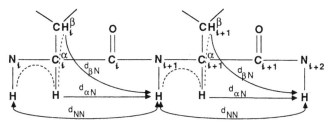

FIG. 1. Polypeptide segment showing the sequential NOE connectivities (arrows) d_{NN}, $d_{\alpha N}$, and $d_{\beta N}$ used for sequential proton assignments. Scalar coupling connectivities are shown with dashed lines and connect only protons within the same residue. (Redrawn with permission from Billeter *et al.*[2])

systems, although they are not essential due to the availability of relayed-COSY[10,11] and TOCSY experiments.[12,13] NOESY spectra in D_2O simplify the observation of cross-peaks to the aromatic protons for the identification of the spin systems of the aromatic residues. Furthermore, spectra in D_2O may be necessary for the identification of proline spin systems which have no backbone NH proton, and for other spin systems under the rare situation of a near-zero coupling constant between the NH and α proton, or when the α proton chemical shift is near that of the water solvent resonance. COSY spectra in H_2O require saturation of the solvent resonance, which may eliminate some of the protein resonances as well, due to overlap or saturation transfer. This problem can be easily overcome by taking advantage of the temperature dependence of the chemical shift of the H_2O resonance (0.012 ppm/°C), with the acquisition of spectra obtained at a minimum of two different temperatures. Recently, a new technique called pre-TOCSY has been developed to observe cross-peaks near the H_2O resonance for both NOESY and COSY spectra.[14]

As shown elsewhere in this volume [8], the task of following the logical steps for sequential assignments, which may be particularly difficult with larger proteins, can be made easier with the aid of computer programs.

Identification of Spin Systems

The resonances of each amino acid residue can be identified through experiments correlating the proton resonances via spin–spin coupling.

[10] G. Eich, G. Bodenhausen, and R. R. Ernst, *J. Am. Chem. Soc.* **104**, 3731 (1982).
[11] G. Wagner, *J. Magn. Reson.* **55**, 151 (1983).
[12] L. Braunschweiler and R. R. Ernst, *J. Magn. Reson.* **53**, 521 (1983).
[13] A. Bax and D. G. Davis, *J. Magn. Reson.* **63**, 207 (1985).
[14] G. Otting and K. Wüthrich, *J. Magn. Reson.* **75**, 546 (1987).

Because no coupling between protons can be observed through a peptide bond, amino acid residues can be thus identified as individual spin-coupling networks, or spin systems. Often these spin systems appear identical for several different amino acid residues, and such spin systems may be one of several possible residue types. The two-dimensional NMR experiments that should be performed for this purpose are COSY, relayed-COSY, and TOCSY (or HOHAHA) (see Bax [8], Vol. 176, this series). It may also be helpful to run double-relayed COSY and double-quantum spectra.[15-17] Spin systems can be classified in terms of the pattern of the coupling network, as shown in Fig. 2. Included in Fig. 2 are the typical chemical shifts. These will vary, due to the local environment in the protein or peptide, typically by ±0.5 ppm. Aromatic rings, however, can cause much larger shifts, as in the case of Pro-9 of BPTI[18] with its β protons, at 0.14 and 0.27 ppm, due to their proximity to the aromatic rings of Phe-22 and Phe-33. Another example is the Val-39 of α-bungarotoxin, whose β proton, at 0.34 ppm, lies in front of the ring of Trp-28.[19] Therefore, chemical shifts can only be used as a guide in the identification of spin systems. The spin systems can be subdivided into three classes as described below.

Methyl-Containing Residues

Valines and Leucines. These residues contain a network of two methyl doublets coupled to one common proton. In valines this is the β proton, and in leucines this is the γ proton. Relayed-COSY and TOCSY (or HOHAHA) experiments may be necessary in order to identify the complete spin system for leucines, because the β and γ protons often appear in a very crowded region, making it impossible to identify direct COSY connectivities between them, in addition to having similar chemical shifts placing these COSY cross-peaks near the diagonal.

Isoleucines. A unique kind of spin system is involved here, because, of the two methyl groups, one is a doublet and the other is a triplet. The connectivity pattern is consequently unique, as shown in Fig. 2. Care must be taken when interpreting the fine structure in terms of a triplet,

[15] L. Braunschweiler, G. Bodenhausen, and R. R. Ernst, *Mol. Phys.* **48,** 535 (1983).
[16] G. Wagner and E. R. P. Zuiderweg, *Biochem. Biophys. Res. Commun.* **113,** 854 (1983).
[17] M. Rance, O. W. Sorensen, W. Leupin, H. Kogler, K. Wüthrich, and R. R. Ernst, *J. Magn. Reson.* **61,** 67 (1985).
[18] G. Wagner, W. Braun, T. F. Havel, T. Schaumann, N. Gō, and K. Wüthrich, *J. Mol. Biol.* **196,** 611 (1987).
[19] V. J. Basus, M. Billeter, R. A. Love, R. M. Stroud, and I. D. Kuntz, *Biochemistry* **27,** 2763 (1988).

however, due to the antiphase nature of the COSY cross-peaks causing the central peak of the triplet to have zero intensity.

Alanines and Threonines. In both of these spin systems the methyl is coupled to a single proton. In threonine, however, this proton is a β proton with a chemical shift usually very close to that of its α proton. Occasionally there may be overlap between the α and β protons. Inspection of the fine structure of the COSY cross-peak between the methyl and its adjacent proton in phase-sensitive spectra may reveal a pattern that is more complex than the quartet which is expected for the α proton of an alanine. This should be sufficient in identifying this as a threonine. It may occur that the α-to-β-proton coupling constant is close to zero and no cross-peak is observed between α and β protons of a threonine, as is found for Thr-6 of α-bungarotoxin.[19] In such a case there would be no double-relay COSY or TOCSY cross-peak between the NH and the methyl proton in H_2O, and there may be no way to distinguish between an alanine or threonine until the sequential assignments have progressed far enough to eliminate one of the possibilities.

Short Side-Chain Residues

Glycines. This is the simplest possible spin system. Often the α-proton chemical shifts are not identical, and the large geminal coupling of 16 Hz is easily observed in the fine structure of the COSY cross-peak between two α protons. This may only be visible in D_2O spectra because the chemical shifts of the α protons are often too close to that of the water solvent resonance to be observable in H_2O. There should be, however, two NH-to-α-proton COSY cross-peaks in H_2O, which is a unique characteristic of glycines. Occasionally both of these cross-peaks are not observable because either one or both NH-to-α-proton coupling constants may be very small. Acquisition of HOHAHA spectra with a mixing time of about 30 msec is quite useful in this case, because in these spectra the COSY-type cross-peaks are more intense due to the efficiency of the magnetization transfer and the in-phase nature of the HOHAHA spectra. Furthermore, in the case where just one of the cross-peaks is not visible, efficient relay through the large geminal coupling constant will give rise to an intense relay cross-peak. In the case of overlap of the two α protons, close inspection of the fine structure of the COSY cross-peak between the NH and the two α protons will reveal the triplet nature of the NH component, that is, relative intensities of 1, 0, and -1, along the amide proton direction and 1 and -1 in the α-proton direction. With small coupling constants this may be difficult to see, and other experiments may be useful, such as a double-quantum spectrum in H_2O^{16} (see Rance *et al.* [6],

A

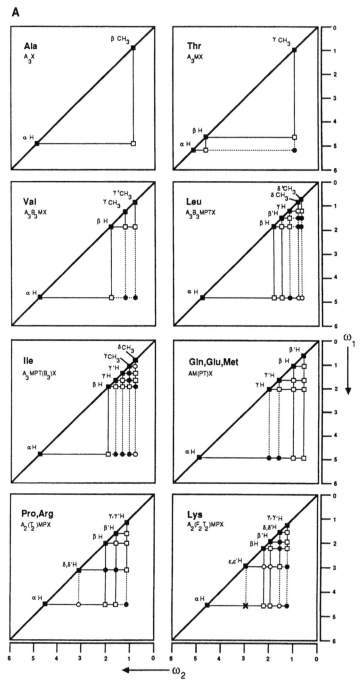

FIG. 2. Connectivity diagrams (A and B) for all the nonexchangeable protons of the common amino acid residues. Shown are diagonal peaks (■), COSY connectivities (□),

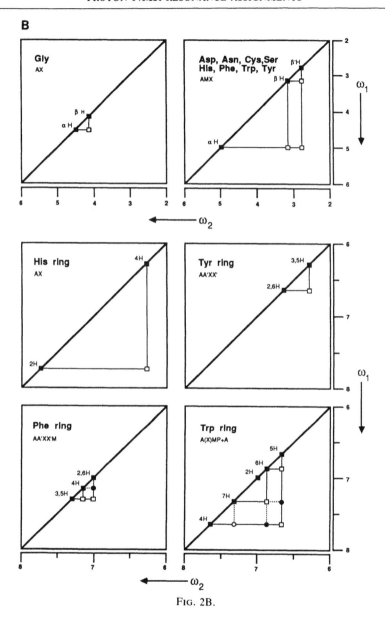

FIG. 2B.

relayed-COSY connectivities (●), and double-relayed-COSY connectivities (○). TOCSY or HOHAHA experiments will show all of these cross-peaks plus the ones indicated with crosses. Only direct COSY connectivities are joined with solid lines. The most common chemical shifts are indicated. (Adapted from Wider *et al.*[4])

Vol. 176, this series). In such a spectrum, glycine resonances in the amide region will show remote peaks corresponding, in the double-quantum dimension, to the sum of the frequencies of the α protons. Under favorable conditions of pH and temperature, lysine side-chain NH_2 protons and the arginine side-chain NH proton show similar peaks at the sum of the frequencies of the side-chain methylene protons to which they are coupled. These side-chain methylene protons usually appear near 3.5 ppm, so that glycine residues may still be distinguished by the chemical shift of this remote peak, as their protons are expected to be at a chemical shift larger than 4 ppm.

Phenylalanines, Tyrosines, Histidines, and Tryptophans. For these residues there is no observable coupling between the β protons and the protons in the aromatic ring. Thus, using COSY, relayed-COSY, or TOCSY spectra, no cross-peak is visible linking the β protons to the aromatic protons. Without these connectivities, these spin systems look identical to those described in the following section. However, due to the necessarily close proximity of the β protons to the closest ring protons demonstrated by Billeter *et al.,*[2] these spin systems can be identified by large NOESY cross-peaks between the β protons and the closest ring protons. In the case of histidines, this distance is somewhat larger, and the corresponding NOESY cross-peaks may be difficult to identify from other long-range NOESY connectivities, making it indistinguishable from the other short-chain residues described in the following section. As shown in Fig. 2, the ring protons of these residues are easily distinguished from each other.

Aspartic Acid, Asparagines, Cysteines, and Serines. The only side-chain protons that can be connected via spin coupling in these residues are α and β protons. Therefore, all of these residues are not distinguishable from each other based on their coupling networks. They are, however, distinguishable from the long-chain residues described next on the basis of the lack of relay cross-peaks farther than the β protons. Also, inspection of the fine structure of phase-sensitive COSY spectra will reveal the absence of further coupling necessary in the presence of two γ protons, as required for the residues with longer side chains. In the case of serines, the analysis of the fine structure of phase-sensitive COSY or E.COSY[20] spectra will reveal a geminal coupling constant of about 12 Hz versus the 14-Hz values between β protons of the other residues. The exchangeable side-chain NH protons of asparagines may be visible provided that their exchange rates with water are slow enough. There is no observable coupling between either of these two protons and the β protons, although a COSY cross-peak between the side-chain NH protons

[20] C. Griesinger, O. W. Sorensen, and R. R. Ernst, *J. Am. Chem. Soc.* **107,** 6394 (1985).

will be observed if their chemical shifts are different enough. However, these NH protons are close enough to the β protons to show large NOESY cross-peaks, thus allowing identification of the spin system of asparagine residues.

Long Side-Chain Residues

All of these residues contain side-chain protons farther than the β protons. The β protons of these residues usually come in the range of 1.5 to 2.5 ppm. The last methylene protons of the chain often appear in the 2.5- to 4-ppm region. All of the intermediate methylene protons normally appear in the 1.5- to 2.5-ppm region, thus overlapping each other as well as the β protons, making this a very crowded region particularly difficult to analyze in COSY spectra. In addition the coupling patterns are very complex for these protons. Therefore, the use of relayed-COSY and especially HOHAHA experiments is quite valuable in order to identify the spin systems of the long side-chain residues.

Glutamic Acids, Glutamines, and Methionines. All of these residues contain α, β, and γ protons. Methionines have a methyl group that is not coupled to the γ protons. In principle, the methyl protons should be close enough to create a strong NOESY cross-peak to the γ protons. However, the chemical shift range where these protons appear is usually a very crowded region, so that these NOESY cross-peaks cannot be used as a diagnostic for methionines. Analogous to asparagines, the side-chain NH protons of glutamines, which may be observable under favorable conditions of pH and temperature, will be close enough to the γ protons to show NOEs, thus allowing identification of glutamine spin systems.

Prolines and Arginines. Both of these residues contain side-chain protons up to the δ position. Arginines can be easily distinguished from prolines because they have a backbone NH proton, whereas prolines do not. Under favorable conditions of pH and temperature, the side-chain ε-NH proton is visible and couples to the δ protons. Figure 3 shows the HOHAHA cross-peaks from the backbone NH all the way to the δ protons, and cross-peaks from the ε-NH all the way to the α protons for Arg-71 of α-bungarotoxin. For Arg-36, the cross-peak from the backbone NH to the δ protons has almost disappeared, and the weak cross-peak from the ε-NH protons to the α proton is lost in the overlap with another cross-peak. Nevertheless, this spin system can be easily identified as an arginine without these cross-peaks. When the exchange rate for the ε-NH protons is fast enough to make these protons or their cross-peaks unobservable, although it still is possible to distinguish arginines from prolines

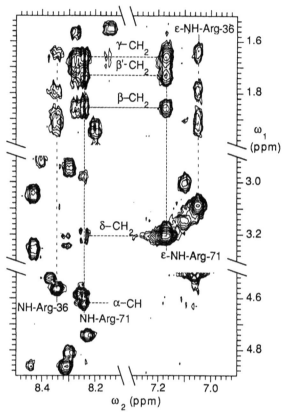

FIG. 3. Portions of the HOHAHA spectrum of α-bungarotoxin at pH 4.8 and 35° with a 70-msec mixing time. Relayed connectivities to the same side-chain protons can be observed for Arg-71 from the backbone NH proton and from the ε-NH proton, clearly identifying this spin system as an arginine. Also shown are connectivities from each of the NH protons of Arg-36. [Reprinted with permission from V. J. Basus, M. Billeter, R. A. Love, R. M. Stroud, and I. D. Kuntz,[19] *Biochemistry* **27**, 2763 (1988). Copyright 1988 American Chemical Society.]

due to the presence of the backbone NH, it may not be possible to distinguish from glutamic acid, glutamine, or methionine residues due to the common overlap between any of the β, γ, or δ protons.

Lysines. These spin systems can be distinguished from the other long side-chain spin systems if there are enough chemical shifts associated with them in relayed-COSY or HOHAHA spectra, i.e., greater than the number of possible chemical shifts for the other long residues. Because there is usually significant overlap of the side-chain protons due to the complex coupling patterns, it is not normally possible to distinguish lysines from an arginine whose ε-NH proton is not observable. Lysines are, however, usually discernible from glutamic acid, glutamine, or methionine residues, because the latter cannot have more than five different side-chain chemical shifts in their spin systems. Occasionally there is even too much overlap for this type of argument.

Sequential Resonance Assignments

Having identified the spin systems as much as possible, we can now proceed to obtain sequential assignments by detailed analysis of NOESY spectra in H_2O. The short sequential distances that are important for sequence-specific assignments are the distance between the α proton of one residue and the NH proton of the next residue $d_{\alpha N}(i, i + 1)$, the distance between the β proton of one residue and the NH proton of the next residue $d_{\beta N}(i, i + 1)$, and the distance between sequential NH protons $d_{NN}(i, i + 1)$ (see Fig. 1).

NOESY cross-peaks can be classified as strong, medium, and weak for the purposes of assignments corresponding to distances approximately less than 2.4, 3.0, and 3.6 Å, respectively. It has been shown[2] that for strong $d_{\alpha N}$ connectivities (i.e., ≤ 2.4 Å), there is a 98% probability for this connectivity to be sequential. For d_{NN} this probability drops to 94%, and for $d_{\beta N}$ it is only 79%. Thus, while a strong $d_{\alpha N}$ or d_{NN} connectivity can be considered sufficient to assign the involved spin systems as sequential, this cannot be done with very high reliability for $d_{\beta N}$ alone. Furthermore, the NOESY cross-peaks are typically weaker, so that these probabilities are then considerably lower. The statistics for pairs of short distances give a much greater reliability for sequential assignments. In particular, for $d_{\alpha N} \leq 3.6$ Å and $d_{NN} \leq 3.0$ Å, there is a 99% probability that the involved spin systems are from sequential residues. For $d_{\alpha N} \leq 3.6$ Å and $d_{\beta N} \leq 3.4$ Å, this probability drops to 95%, and for $d_{NN} \leq 3.0$ Å and $d_{\beta N} \leq 3.0$ Å, this probability drops to 90%.

For small peptides the search for sequential connectivities can be easily done graphically. Figure 4 exemplifies the assignment based on sequential $d_{\alpha N}$ connectivities and shows the assignment of residues 23–28 of α-bungarotoxin, a segment later identified as belonging to the central strand of a triple-stranded β sheet. This conformation shows typically large sequential $d_{\alpha N}$ connectivities. The position of the COSY cross-peaks for each spin system has been indicated in Fig. 4. There is normally a NOESY cross-peak at each of these positions because the intraresidue distance is always short enough. In this molecule, Trp-28 is unique and the assignment can be started from this point and proceeding, in this case, backward to the α proton of residue 27 along the vertical line to the large

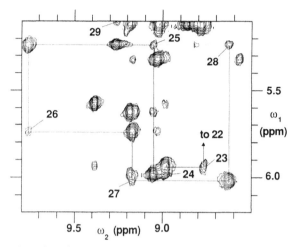

Fig. 4. A small portion of the NOESY spectrum of α-bungarotoxin at pH 4.0 and 35° with a 150-msec mixing time, showing the sequential assignment of residues 23 to 29. COSY peak positions are indicated, each having a NOESY cross-peak at these positions between the α proton and the NH proton of the same residue.

NOESY cross-peak at this α-proton chemical shift. A horizontal line connects this NOESY cross-peak to the position of the COSY peak of 27, whose spin system has been identified as M, Q, or E, thus completing one assignment cycle. Then one proceeds again along the NH chemical shift (vertical line) to the NOESY cross-peak at the chemical shift of the α-proton of residue 26. After several cycles, the assignments then become (D,N,C,S, or H)-Y-R-(M,Q,E, or K)-(M,Q, or E)-W, which unambiguously fits the sequence C-Y-R-K-M-W for this segment. Notice that the spin system of residue 26 could not be identified as a lysine before sequential assignment. This was due to severe overlap between its side-chain resonances. Also, it was very helpful to have all of the arginines identified based on their ε-NH HOHAHA connectivities. Residues 25 and 28 have nearly the same α-proton chemical shifts. Thus the sequential connectivity from residue 25 to residue 26 could have been erroneously selected as the large NOESY cross-peak from residue 28 to residue 29. However, residue 29 is a short side-chain spin system (i.e., containing coupled protons up to the β position only) and residue 26 is a lysine, which would be inconsistent with the assignment of the spin system of residue 29 to residue 26. Assignments for this segment were actually extended to Leu-22 and to Cys-29 before overlap between α protons of different residues made it difficult to continue.

In this case, Trp-28 was a unique spin system for this molecule. The next step would be to look for unique pairs of residues, which can be

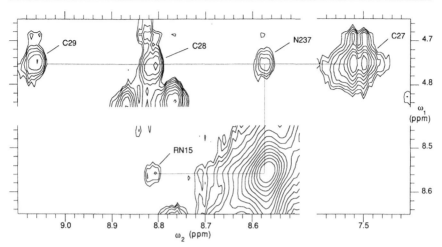

FIG. 5. Portion of the HOHAHA–relayed-NOESY spectrum of α-bungarotoxin showing the assignment of the α-proton chemical shift of the NOESY cross-peak N237 to the spin system of C28. The positions of the COSY cross-peaks C29, C28, and C27, whose α-proton chemical shifts are unresolvable, are indicated. The relayed NOE cross-peak RN15 with the same chemical shift in ω_1 as N237 has the same shift in ω_2 as C28, identifying the α proton of C28 as the one involved in the NOESY cross-peak N237. [Reprinted with permission from V. J. Basus and R. M. Scheek,[22] *Biochemistry* **27**, 2772 (1988). Copyright 1988 American Chemical Society.]

identified as sequential with high probability, and then unique tripeptide segments, proceeding as far as possible with assignments from each of these unique segments. As assignments progress, more of the remaining unassigned residues form unique short segments, giving us new starting points. Finally, some spin systems may only be assigned by the process of elimination.

Overlap of the NH chemical shifts can usually be removed by changing the temperature, because these shifts have temperature dependencies which are different for each NH. The α-proton chemical shifts, however, are not usually sensitive to temperature, so that such overlap cannot be resolved by varying the temperature. However, where the NOESY cross-peaks are large enough, we can overcome the overlap problem between different α-proton chemical shifts by the use of the relayed-NOESY experiment,[21] provided that the relevant α-proton-to-NH coupling is also large enough. The example shown in Fig. 5 is from work on α-bungaro-toxin,[19,22] where the position of the COSY peaks labeled C27, C28, and C29 have the same α-proton chemical shift, and the relayed-NOESY

[21] G. Wagner, *J. Magn. Reson.* **57**, 497 (1984).
[22] V. J. Basus and R. M. Scheek, *Biochemistry* **27**, 2772 (1988).

cross-peak RN15 shows that the NOESY cross-peak N237 is in fact between the α proton of the spin system of C28 and the NH of the spin system of C7, which is the only spin system assignable to the NH chemical shift of N237. Furthermore, for α-bungarotoxin the completely buried NOESY cross-peak between the α proton and the NH of His-4 was demonstrated to exist with the use of HOHAHA–relayed-NOESY,[22] an improved version of the relayed-NOESY sequence. If there is still an overlap problem, changing the pH may also be helpful.

Stereospecific Resonance Assignments

In the study of solution structures of proteins by NMR, the lack of stereospecific assignments of diastereotopic protons in CH_2 groups and of methyl groups in leucine and valine residues creates large uncertainties in the distance constraints. Such stereospecific assignments would be valuable, and are indeed possible, in many cases, through a detailed analysis of the fine structure in phase-sensitive COSY and E.COSY spectra,[20,23] along with the observation of certain NOESY cross-peaks.

Stereospecific Assignment of β Protons. For these protons the assignment is centered around the definition of the rotation angle χ^1 around the C^α–C^β bond. The energetically most favorable conformations are those with the atoms around this bond in a staggered conformation. Therefore, it is useful to talk about the β protons in terms of being gauche or trans with respect to the α proton. In these terms, there are three possible side-chain rotamers: g^2g^3, g^2t^3, and t^2g^3 where the superscripts indicate the β^2 and β^3 protons defined according to the IUB–IUPAC convention[24] as shown in Fig. 6. The case of g^2g^3 can be easily identified from the coupling constants alone, since it would require both α-proton-to-β-proton coupling constants to be small. It is, however, not sufficient for the stereospecific assignment of β^2 and β^3, because both couplings are the same. Analysis of the α-proton-to-β-proton NOEs will not assist in this assignment, because both β protons are approximately equidistant from the α proton. Analysis of the NOEs between the backbone NH proton and the β protons will, however, be useful in distinguishing the two β protons. Rotating the ϕ torsional angle through its complete allowed range[2] gives a range of distances from the NH proton of 2.4 to 3.4 Å to the β^3 proton and 3.5 to 4.0 Å to the β^2 proton. Thus one of these shows a strong to medium NOE while the other must show a weak NOE, thus assigning both protons. For the other two possibilities, g^2t^3 and t^2g^3, although the two cou-

[23] S. G. Hyberts, W. Märki, and G. Wagner, *Eur. J. Biochem.* **164**, 625 (1987).
[24] IUB–IUPAC Commission of Biochemical Nomenclature, *J. Mol. Biol.* **52**, 1 (1970).

Conformation	g^2g^3	g^2t^3	t^2g^3
χ^1	$60°$	$180°$	$-60°$
	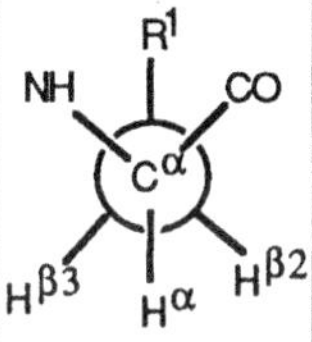	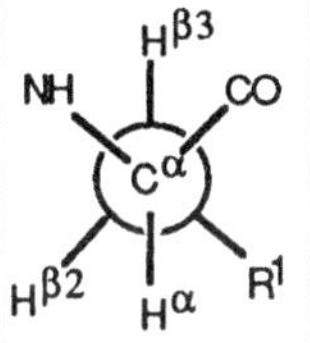	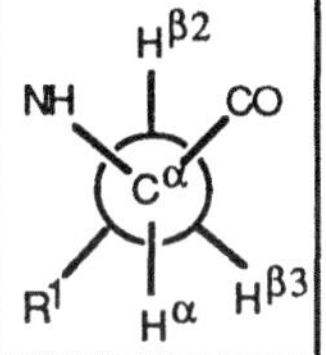
$^3J_{\alpha\beta^2}$ (Hz)	2.6-5.1	2.6-5.1	11.8-14.0
$^3J_{\alpha\beta^3}$ (Hz)	2.6-5.1	11.8-14.0	2.6-5.1
NOE(α,β^2)	Strong	Strong	Weak
NOE(α,β^3)	Strong	Weak	Strong
NOE(NH, β^2)	Weak	Strong-Medium	Strong
d(NH, β^2)(Å)	3.5-4.0	2.5-3.4	2.2-3.1
NOE(NH, β^3)	Strong-Medium	Strong	Weak
d(NH, β^3)(Å)	2.5-3.4	2.2-3.1	3.5-4.0

FIG. 6. Characterization of the three possible staggered conformations around the C^α–C^β bond. Gauche and trans orientations are indicated as g and t, respectively. (Adapted from Wagner et al.[18]) The β^2 and β^3 positions are defined according to the IUB–IUPAC convention.[24] A Newman projection is shown for each of the conformations and the χ^1 angle is indicated. The ranges of distances indicated are based on a rotation of the ϕ tortional angle through all of the allowed range in residues other than glycines. The coupling constants are based on the empirical equation[25] $^3J(\theta) = (9.5 \pm 0.3) - (1.6 \pm 0.2) \cos\theta + (1.8 \pm 0.6)$ Hz.

pling constants are quite different, they cannot be distinguished from each other prior to stereospecific assignment. Observation of α-proton-to-β-proton NOEs will not assist in distinguishing these two cases. The NH-proton-to-β-proton NOEs can be used, however, for this distinction, because in the first case the NOEs will be strong or strong to medium to both protons, and in the second case will be strong to the β^2 proton and weak to the β^3 proton. After distinguishing between these two possible confor-

Conformation	g⁻	t	g⁺
	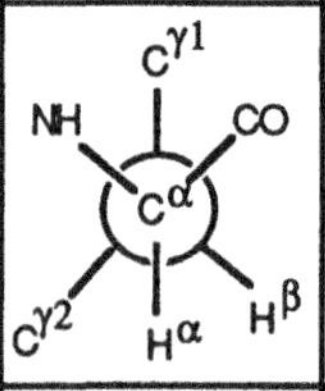	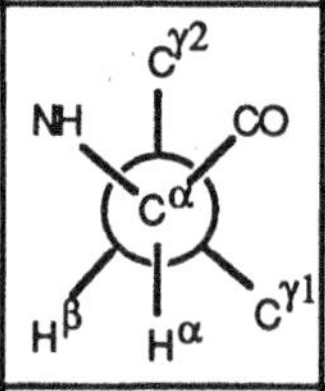	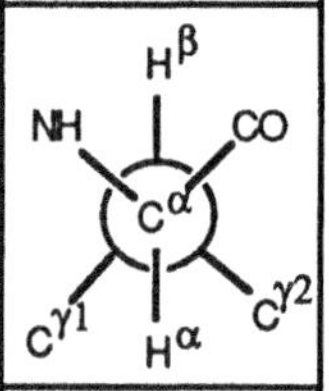
$^3J_{\alpha\beta}$ (Hz)	2.6 -5.1	2.6 -5.1	11.8-14.0
NOE(α,γ^1)	Weak	Strong	Strong
NOE(α,γ^2)	Strong	Weak	Strong
NOE(NH, γ^1)	Strong	Weak	Strong-Medium
NOE(NH, γ^2)	Strong-Medium	Strong	Weak

FIG. 7. Characterization of the three possible rotamers around the C^α–C^β bond for valines. A Newman projection is shown for each conformation. The coupling constants were calculated as in Fig. 6. The γ^1 and γ^2 methyl protons are defined in accordance with the IUB–IUPAC convention. (Adapted from Zuiderweg et al.[26])

mations, the stereospecific assignment can be easily obtained from the coupling constants.[25]

Stereospecific Assignments of Valine Methyl Protons. It has been demonstrated that stereospecific assignments of the valine methyl protons can be obtained in some cases.[26] The possible rotamers around the α-to-β bond are shown in Fig. 7. Of these, the g^+ conformation is the most predominant, as is shown in a statistical analysis of valine side-chain conformations of proteins with known X-ray crystal structures.[27] In this case the α-proton-to-β-proton coupling constant is about 12 Hz and is quite distinct from the other two rotamers. Stereospecific assignments can be accomplished by observation of the NOEs from the valine NH proton to the methyl protons showing a much larger NOE, to the C^{γ^1} methyl than to the C^{γ^2} methyl protons.

For the other two rotamer conformations, the coupling constant between the α proton and the β proton should be less than 5 Hz, and it

[25] A. Marco, M. Llinás, and K. Wüthrich, *Biopolymers* **17**, 617 (1978).
[26] E. R. P. Zuiderweg, R. Boelens, and R. Kaptein, *Biopolymers* **24**, 601 (1985).
[27] J. Janin, S. Wodak, M. Levitt, and B. Maigrait, *J. Mol. Biol.* **125**, 357 (1978).

cannot usually be reliably measured to distinguish these two cases from each other. The only way to distinguish between these two cases would be through the NOEs from the NH to the β proton, which should be large for the t conformation and small for g^-.

Coupling constants of about 7–8 Hz between the α proton and β proton are frequently observed. In the free amino acid, the side chain is certainly free to rotate rapidly, thus averaging the coupling constant, which is observed to be 7 Hz. Measurement of a 7- to 8-Hz coupling constant for a valine residue is then indicative of free rotation, thus precluding stereospecific assignments, although an energetically unfavorable eclipsed conformation would also show such coupling constants.

Stereospecific Assignments in Other Residues. Stereospecific assignments for the glycine α protons have not yet been possible without assuming a structural model. The assignment of the δ protons for leucines should be possible after its β protons have been stereospecifically assigned, although no report of such assignment has yet appeared. In general, however, stereospecific assignments may be possible based on long-range NOEs.[28]

Summary

The procedures outlined here have been used successfully for more than 30 proteins to date,[3,4,6–8,19,29–56] and are nearly routine for molecules

[28] H. Senn, M. Billeter, and K. Wüthrich, *Eur. Biophys. J.* **11**, 3 (1984).

[29] M. Ikura, O. Minowa, and K. Hikichi, *Biochemistry* **24**, 4264 (1985).

[30] P. L. Weber, D. E. Wemmer, and B. R. Reid, *Biochemistry* **24**, 4553 (1985).

[31] T. A. Holak and J. H. Prestegard, *Biochemistry* **25**, 5766 (1986).

[32] R. E. Klevit, G. P. Drobny, and E. B. Waygood, *Biochemistry* **25**, 7760 (1986).

[33] A. C. Bach, M. E. Selsted, and A. Pardi, *Biochemistry* **26**, 4389 (1987).

[34] O. Lichtarge, O. Jardetzky, and C. H. Li, *Biochemistry* **26**, 5916 (1987).

[35] P. L. Weber, S. C. Brown, and L. Mueller, *Biochemistry* **26**, 7282 (1987).

[36] C. Redfield and C. M. Dobson, *Biochemistry* **27**, 122 (1988).

[37] D. M. LeMaster and F. M. Richards, *Biochemistry* **27**, 142 (1988).

[38] G. T. Montelione, K. Wüthrich, and H. A. Sheraga, *Biochemistry* **27**, 2235 (1988).

[39] K. H. Lee, J. E. Fitton, and K. Wüthrich, *Biochim. Biophys. Acta* **911**, 144 (1987).

[40] V. I. Kondakov, A. S. Arseniev, K. A. Pluzhnikov, V. I. Tsetlin, and V. F. Bystrov, *Bioorg. Khim.* **10**, 1606 (1984).

[41] E. R. P. Zuiderweg, R. M. Scheek, and R. Kaptein, *Biopolymers* **24**, 2257 (1985).

[42] P. R. Gooley and R. S. Norton, *Biopolymers* **25**, 489 (1986).

[43] E. R. P. Zuiderweg, R. Kaptein, and K. Wüthrich, *Eur. J. Biochem.* **137**, 279 (1983).

[44] C. I. Stassinopoulou, G. Wagner, and K. Wüthrich, *Eur. J. Biochem.* **145**, 423 (1984).

[45] P. R. Gooley and R. S. Norton, *Eur. J. Biochem.* **153**, 529 (1985).

[46] P. C. Driscoll, H. A. Hill, and C. Redfield, *Eur. J. Biochem.* **170**, 279 (1987).

[47] E. Wörgötter, G. Wagner, M. Vašák, J. H. R. Kägi, and K. Wüthrich, *Eur. J. Biochem.* **167**, 457 (1987).

up to a molecular weight of 10,000. Some of the proteins assigned have a molecular weight greater than 10,000.[36,37,46] For these larger proteins, relayed-COSY and TOCSY experiments have been essential for the identification of spin systems, although for thioredoxin[37] these experiments could not be used. In this case, assignments were accomplished using nonspecific deuteration to the level of 75% and specific, nearly complete, deuteration of certain kinds of residues (see LeMaster [2], this volume). Nonspecific deuteration reduces the cross-relaxation rates of each proton to the rest of the molecule, thus reducing the linewidths. The cross-peak patterns were also narrowed due to simplification of the coupling patterns. Such a laborious procedure of nonspecific deuteration may not be necessary for complete proton assignments of proteins in this size range, as evidenced by the fact that this method was not used for the other two molecules mentioned above. It may prove, however, to be quite valuable in the study of larger molecules, where linewidths are expected to increase due to longer rotational correlation times.

Overlap problems in the NH chemical shifts can be dealt with by making use of the differential temperature dependence of these shifts. Another technique is to take advantage of the wide range of exchange rates between these protons and the solvent. Spectra containing only the slowly exchanging NH protons can be obtained by acquiring spectra of the protein soon after dilution in D_2O, and spectra of only the rapidly exchanging protons can be obtained by obtaining spectra in a freshly prepared H_2O solution of the protein after having completely exchanged all the NH protons with deuterium. Variation of the pH will resolve problems of overlap in all regions of the spectrum, although many chemical shifts may be unaffected by pH. In some cases, pH variation may change the conformation of the molecule.[40] This may, in fact, assist in the sequential assignment if the chemical shifts can be followed with pH. Finally, the relayed-NOESY experiments can resolve overlap problems

[48] H. Widmer, G. Wagner, H. Schweitz, M. Lazdunski, and K. Wüthrich, *Eur. J. Biochem.* **171,** 177 (1988).

[49] A. S. Arseniev, G. Wider, F. J. Joubert, and K. Wüthrich, *J. Mol. Biol.* **159,** 323 (1982).

[50] R. M. Keller, R. Baumann, E. H. Hunziker-Kwik, F. J. Joubert, and K. Wüthrich, *J. Mol. Biol.* **163,** 623 (1983).

[51] A. D. Kline and K. Wüthrich, *J. Mol. Biol.* **192,** 869 (1985).

[52] F. J. M. vandeVen and C. W. Hilbers, *J. Mol. Biol.* **192,** 389 (1986).

[53] G. M. Clore, D. K. Sukumaran, A. M. Gonenborn, M. M. Teeter, M. Whitlow, and B. L. Jones, *J. Mol. Biol.* **193,** 571 (1987).

[54] J. Zarbock, G. M. Clore, and A. M. Gronenborn, *Proc. Natl. Acad. Sci. U.S.A.* **83,** 7628 (1986).

[55] A. D. Robertson, W. M. Westler, and J. L. Markley, *Biochemistry* **27,** 2519 (1988).

[56] G. I. Rhyu and J. L. Markley, *Biochemistry* **27,** 2529 (1988).

with the α-proton chemical shifts. Thus, it is very likely that the assignment methods outlined here will be successful for the assignment of the proton spectra of even larger molecules if there is significant secondary structure and significant variety of residues to provide enough dispersion of the chemical shifts.

Another problem is the increase in the rotational correlation times causing increased linewidths. Thus, complete proton assignments for larger proteins may only be achievable if the overall structure is not greatly elongated in any direction, as this would increase the rotational correlation time compared to a more spherical and compact structure. Aggregation is also a concern, as this would increase the correlation time, thus broadening the lines as well as increasing the complexity of the spectra by either showing a second set of chemical shifts or extra NOESY cross-peaks, or both. In this case, lower concentrations may provide better spectra. So far, the lowest concentration for which there is sufficient intensity in the cross-peaks to allow for a complete or nearly complete assignment is 4 mM at 500 MHz.[48] Slightly lower concentrations should be acceptable, especially with probes that use higher sample volumes, or with the use of more spectrometer time. Higher fields will also provide better sensitivity.

Structural studies will require the assignment of as many of the NOESY cross-peaks as possible involving nonsequential residues. It is very important then to have assigned all of the protons in the spectrum in order to minimize the possibility of erroneous assignments for these NOESY cross-peaks. Overlap of the assigned resonances may cause ambiguities in the assignment of these cross-peaks, which would not allow their use in structure determination. A set of low-resolution structures based on the available unambiguous data will assist in removing these assignment ambiguities, as some of the possibilities may involve protons that cannot be close enough to give an NOE based on any of these low-resolution structures.

Acknowledgments

I thank Dr. K. Wüthrich, Dr. R. M. Scheek, and Dr. P. A. Kosen for a critical reading of this manuscript. Financial support by Grants GM-19267 and RR-01695 from the National Institutes of Health is gratefully acknowledged.

[8] Computer-Assisted Resonance Assignments

By Martin Billeter

Introduction

During the last few years the determination of complete spatial structures of small proteins (less than 100 amino acid residues) by high-
resolution ¹H NMR has evolved to a well-established procedure,[1] although the effort in terms of man-years is still significant. Today both the
data acquisition, including initial steps of data preparation, such as
Fourier transforms,[2] and the calculation of the three-dimensional structures[3] are based on automatic computer programs. The intermediate steps
of locating the individual peaks in the NMR spectra and of assigning them
to uniquely identified protons in the protein were described several years
ago,[4] but they are, to date, still accomplished on a manual basis. This is
one of the reasons why the NMR investigation of a protein is a time-
consuming process.

Because the characterization and assignment of peaks in NMR spectra
are embedded among other steps, a definition follows of what is here
considered as "computer-assisted resonance assignments." The starting
data are (frequency domain) ¹H NMR spectra of various types, e.g.,
COSY and related spectra such as relay-COSY and TOCSY, multiple-
quantum filtered spectra, NOESY, and possibly other spectra (e.g., relayed-NOESY).[1,2] Changing measuring conditions (temperature, pH,
etc.) may result in different sets of such spectra. All these spectra are
assumed to have been well prepared using routines to reduce noise and to
enhance the resolution. The long-term goal is the completely automated
inspection and explanation of the measured data, resulting in the assignment of all resonances to individual protons of the protein. On a shorter
term, it would be desirable to have semiautomatic programs available that
take care of the tedious and error-prone bookkeeping as well as solve
independently simple subproblems. More complex situations would be

[1] K. Wüthrich, "NMR of Proteins and Nucleic Acids." Wiley, New York, 1986.
[2] R. R. Ernst, G. Bodenhausen, and A. Wokaun, "Principles of Nuclear Magnetic Resonance in One and Two Dimensions." Oxford Univ. Press (Clarendon), London and New
York, 1987.
[3] A recent review on this topic is that by W. Braun, Q. Rev. Biophys. 19, 115 (1987).
[4] K. Wüthrich, G. Wider, G. Wagner, and W. Braun, J. Mol. Biol. 155, 311 (1982); M.
Billeter, W. Braun, and K. Wüthrich, J. Mol. Biol. 155, 321 (1982); G. Wagner and K.
Wüthrich, J. Mol. Biol. 155, 347 (1982); G. Wider, K. H. Lee, and K. Wüthrich, J. Mol.
Biol. 155, 367 (1982).

METHODS IN ENZYMOLOGY, VOL. 177

left to the user in an interactive environment where the computer should assist based both on its hardware facilities (computer graphics) as well as on the ease with which a program can scan large data sets and extract essential information. Later improvements of such a software package would enable the program to handle more and more complex situations. The computer-assisted resonance assignments are usually followed by the determination of the secondary[5,6] and eventually tertiary[3] structures of the protein.

The inherent problems of the resonance assignment process are well known: overlap of peaks resulting in the loss of a complete peak or of its fine structure information, random and nonrandom noise, missing data (e.g., bleached areas due to water suppression), and finally the interdependence of structural and dynamic effects on the measured data.[1,2] On the other hand, it should be noted that ^{1}H NMR yields a wealth of structural data, including to a significant extent redundant information. Various approaches of having the computer assist in assigning ^{1}H NMR resonances are described in the following sections; emphasis is put on automatic or semiautomatic procedures.

Use of Computer Graphics Tools

The first and today still most practical use of computers for the interpretation of NMR spectra consists of adding the power of a high-resolution computer graphics system to human experience. This offers the following advantages. Handling a spectrum on an interactive graphics device is more flexible compared to plotted data (provided that the resolution of the screen is sufficient). Simple operations, such as calculating peak coordinates and comparing them, are performed instantly. An important aspect consists of the fact that the process of inspecting NMR spectra may be incorporated into a software package, including also procedures of spectrum preparation. This allows easy switching between steps that serve to improve the spectra (e.g., flattening of baselines) and commands aiming at the characterization of peaks. These software packages are, however, designed to merely assist the human in making decisions rather than to follow a given logic.

Several programs are in use; many of them are adapted to the local hardware and therefore do not accommodate other laboratories. Distributed to a wider public are FTCGI and NMR1/NMR2[7] (for one- and two-

---

[5] K. Wüthrich, M. Billeter, and W. Braun, *J. Mol. Biol.* **180,** 715 (1984).
[6] A. Pardi, M. Billeter, and K. Wüthrich, *J. Mol. Biol.* **180,** 741 (1984).
[7] See Appendix at the end of this volume for technical details on the programs.

dimensional NMR data, respectively). The scope of these programs covers all basic operations needed for analyzing the spectra: Fourier transforms, phasing, baseline corrections, linewidth determination and integration of peaks, curve fitting, etc. (For details, consult the corresponding user's manuals.) Especially the NMR1/NMR2 package is becoming a well-supported and complete software tool, including, for example, maximum entropy and linear prediction methods as well as simulation programs. Even in view of automating the assignment procedure (see below), these packages represent an extremely useful tool, because it is still necessary to inspect the input data, to check intermediate and final results of the automated programs, and to manually treat all the "special" cases where the automatic procedures fail.

Automation of the Resonance Assignments

Efforts are being made in several laboratories to speed up the resonance assignment process by designing programs capable of automatically interpreting large portions of the data contained in the NMR spectra. The underlying principles are based on the existence of characteristic patterns both for spin systems and for the fine structure of individual cross-peaks. The basic COSY experiment exhibits three kinds of symmetries.[8] A *global* symmetry about the main diagonal of the spectrum duplicates all peaks. Each cross-peak reflects a *local* symmetry bearing coupling information in its fine structure. In addition one could consider the internal symmetry of every single resonance given by its shape. Original attempts aimed at locating patterns for complete spin systems in the raw spectral data, i.e., without prior extraction of information individual to cross-peaks.[9] A library of all occurring spin systems was needed and special care had to be taken to handle situations where peaks were missing in the spectra or peaks from other spin systems distorted the pattern being considered. Furthermore, the shifts of the protons were not known *a priori*. As a consequence it was necessary to systematically vary a set of parameters for each spin system in order to describe the relative positions of the cross-peaks. The method was applied to a 10-residue-long neuropeptide, where it successfully identified all spin systems except for two prolines.[9] The method failed for the prolines because of overlapping peaks, a feature that will occur more frequently as the polypeptides grow in size.

[8] J. C. Hoch, S. Hengyi, M. Kjaer, S. Ludvigsen, and F. M. Poulsen, *Carlsberg Res. Commun.* **52**, 111 (1987).
[9] K. P. Neidig, H. Bodenmüller, and H. R. Kalbitzer, *Biochem. Biophys. Res. Commun.* **125**, 1143 (1984).

More recent methods follow the approach used in a manually accomplished resonance assignment,[4] i.e., the problem is split into subtasks. Initially, spectra are scanned to locate individual cross-peaks, and their multiplet structure is analyzed. In a next step the peaks are grouped into networks or spin systems, and finally these are sequentially assigned to individual amino acid residues of the protein. The following three subsections give an overview of these approaches.

Peak Analysis

Currently a number of laboratories are putting considerable efforts into the design of algorithms that locate cross-peaks and analyze their fine structure. The basic pattern of a COSY cross-peak for a two-spin system AX consists of four peaks with alternating signs at positions ($\Omega_A \pm J_{AX}^{act}$, $\Omega_X \pm J_{AX}^{act}$), where Ω_A and Ω_X are the shifts of the two protons and J_{AX}^{act} is their (active) coupling constant.[10] Another set of four peaks is found at the mirror position with respect to the main diagonal. More complex spin systems include further splitting by the passive couplings. In general, a cross-peak multiplet consists of a superposition of squares or rectangles.[11]

A direct approach is to search the spectrum for known cross-peak patterns by moving a corresponding mask over the entire spectrum.[11] This mask has to be defined for each basic pattern and for each possible coupling constant J^{act}. In order to speed up the process and to reduce the memory requirements, one can replace the spectral matrix by two Boolean matrices[12] using the following criteria: The first (second) Boolean matrix has an entry of "1" whenever the corresponding value in the original matrix is positive (or negative) and its absolute value exceeds a given noise level. In addition, elements occurring only on one side of the main diagonal are eliminated (requirement of global symmetry). These Boolean matrices are then jointly searched for the basic patterns by varying J^{act} and the shape of the pattern from squares to rectangles as needed. (Note that positive peaks of the basic pattern are found in one Boolean matrix and negative peaks are found in the other.) This yields for each pattern shape a new three-dimensional matrix, where the coordinates define the centers of the basic patterns, i.e., the chemical shifts of the two protons involved and the active coupling constant. Using spectra from two complementary experiments (e.g., a z-filtered and an anti-z-filtered COSY[13]) that are both processed as described above, one can obtain

[10] B. U. Meier, G. Bodenhausen, and R. R. Ernst, *J. Magn. Reson.* **60**, 161 (1984).
[11] P. Pfändler, G. Bodenhausen, B. U. Meier, and R. R. Ernst, *Anal. Chem.* **57**, 2510 (1985).
[12] P. Pfändler and G. Bodenhausen, *J. Magn. Reson.* **70**, 71 (1986).
[13] H. Oschkinat, A. Pastora, P. Pfändler, and G. Bodenhausen, *J. Magn. Reson.* **69**, 559 (1986).

values for the apparent passive coupling constants[12]: a pair of complementary spectra exhibit, if processed as described above, patterns similar to the two Boolean matrices with the positive and negative peaks of an original spectrum. Therefore, the same step of recognizing joint patterns can be applied again, but this time the resulting coupling constants are functions of the passive couplings. (A similar procedure can be designed for z-filtered double-quantum spectra.[14])

Another approach locates the cross-peaks by systematically searching for regions in the spectrum with high local symmetry.[8,15–17] Because this step is time-consuming, one may first determine small areas covering only one or a small number of peaks. The extent of each region is defined by the requirement that every peak inside is separated from all peaks outside by more than the largest expected coupling.[8,15,16] An alternative method to locate cross-peaks consists of decrementing the constant phase correction by 90° in both dimensions; this results in large peaks at the centers of the multiplets, and regions of interest are then defined by squares around these peaks.[8] Next a measure for local symmetry is defined and this symmetry is evaluated for each point of a region. Various measures[8,15–17] have been introduced that are all based on the same idea. To evaluate the symmetry measure of point A, one looks at all point pairs (B, B') inside the region, where B' is the reflection of point B with respect to A. The difference of the spectral values at points B and B' is computed, and the sum over all point pairs of the absolute differences corresponds inversely to the amount of local symmetry around point A. (An alternative procedure consists of the two-dimensional convolution of the spectral region.[17]) This general idea of defining a symmetry measure can be adapted to various types of symmetry. For example, antisymmetry can be detected by taking the sum of spectral values rather than their difference. One may also consider point pairs that are reflected at an axis rather than at a single point.[16] Those points A that exhibit a larger value of this symmetry measure than their neighbors (and a given threshold) are called symmetry centers. Assuming that the region under consideration contains a single cross-peak, one can define several classes of symmetry centers.[17] With respect to the highest symmetry center all multiplet peaks have a symmetric partner. The coordinates of this center correspond to the chemical shifts of the two protons. Symmetry centers on a next level are reflection points for only one-half the multiplet peaks to other peaks; their positions relative to the highest center determine the passive couplings. A further

[14] M. Novic, H. Oschkinat, P. Pfändler, and G. Bodenhausen, *J. Magn. Reson.* **73**, 493 (1987).
[15] Z. Mádi, B. U. Meier, and R. R. Ernst, *J. Magn. Reson.* **72**, 584 (1987).
[16] S. Glaser and H. R. Kalbitzer, *J. Magn. Reson.* **74**, 450 (1987).
[17] B. U. Meier, Z. L. Mádi, and R. R. Ernst, *J. Magn. Reson.* **74**, 565 (1987).

level would then describe the active coupling. The latter information is, however, preferably extracted from a search of basic patterns in the original spectral data. At this point the proton shifts are known and the pattern search is fast; global symmetry may be considered as well. The list of cross-peaks obtained by the above procedures may then be further reduced by testing versus characteristic features of t_1 noise and other artifacts.[16]

All variations of the above method have been demonstrated on molecules consisting of a single or a small number of spin systems.[10–12,14,15] Two publications[8,17] report applications to polypeptides, each consisting of 10 residues. Meier et al.[17] describe the identification of sufficient peaks to characterize eight spin systems of the cyclic decapeptide antanamide (see also next subsection). Two groups applied their programs to spectra of proteins with an approximate molecular weight of 9000: barley serine protease inhibitor,[8] and the heat-stable protein (HPr) from Streptococcus faecalis.[16] However, the descriptions end after the step of locating peaks (about 400 in Ref. 16). Therefore, one cannot judge the correctness and completeness of these intermediate results. The computation time looks rather attractive. For the HPr protein it is of the order of 15 min (on a NORD-500 computer).

Spin System Identification

At this point one has a more or less complete list of peaks describing the scalar proton–proton couplings in the entire protein. For each cross-peak this list may include, besides the chemical shifts of the two protons involved, the active and passive coupling constants; many of the latter entries will, however, not be detectable. Error limits should be determined for all these data. In addition, the symmetry measure provides for each peak a number indicating the confidence level with which the peak was identified. The buildup of cross-peak networks that correspond to amino acid spin systems or fragments thereof is based primarily on a comparison of the chemical shifts. The coupling constants can help to resolve situations where the chemical shifts of two or more protons cannot be distinguished. Programs based on these ideas have been described,[11,12,17] however, only one of these publications includes results for a polypeptide: Meier et al.[17] report the purely automatic identification of eight spin systems of the cyclic decapeptide antamanide; two strongly coupled proline spin systems could not be fully derived.

Sequential Assignment of Spin Systems

Once the cross-peaks in ^1H NMR spectra arising from scalar proton–proton couplings are gathered into groups that identify individual spin systems, one may proceed to assign sequentially these spin systems to unique residue positions in the protein sequence. The data used in this step are the protein sequence, a description of the spin systems, and a list of NOE distance measurements that allow the determination of sequential neighborhoods between spin systems. The list of spin systems gives for each detected spin system a list of amino acid residue types that are compatible with the spin system. For example, some spin systems will uniquely determine residue types such as glycine, whereas others will only carry the information that they should be assigned to a residue with a long side chain. The problems encountered at this stage are, first, that the spin system list is incomplete both with respect to the number of spin systems and to the uniqueness of the spin system description, and, second, that the list of NOEs may lack some sequential as well as include some long-range distances.

SEQASSIGN[7] is a program written specifically to automate this task.[18] It consists of an automatic and an interactive part. The basic ideas of this program are as follows: It continuously keeps track of all assignments consistent with the information currently available. The automatic part performs logical decisions and takes care of the bookkeeping. These steps are guaranteed never to loose the correct solution. They always run to completion, i.e., until no further changes can be obtained. At this point the program provides, through an interactive dialogue, detailed information on the current situation indicating what types of new data would be most promising. The user may enter such new data, if available, and switch back to the automatic part. The interactive dialogue and the frequent use of the automatic part should ensure that a maximum of assignments is obtained with minimal use of ambiguous NMR data. A further advantage of this approach is that the individual steps are documented and thus reproducible. This allows testing of several interpretations of a piece of ambiguous information by jumping back to the situation reached before introducing the uncertain data.

The program has been applied to a number of test cases based on simulated input data, and to α-bungarotoxin, a 74-residue neurotoxin.[18,19] Complete sequential assignments were obtained. The most remarkable result is that it led to the identification of two errors involving four residues of the proposed sequence; these were later confirmed by resequencing.[20] Earlier NMR studies of small proteins have yielded similar sequence errors in the chemically determined primary structures (see

[18] M. Billeter, V. J. Basus, and I. D. Kuntz, *J. Magn. Reson.* **76**, 400 (1988).
[19] V. J. Basus, M. Billeter, R. A. Love, R. M. Stroud, and I. D. Kuntz, *Biochemistry* **27**, 2763 (1988).
[20] P. A. Kosen, J. Finer-Moore, M. P. McCarthy, and V. J. Basus, *Biochemistry* **27**, 2775 (1988).

examples in Ref. 1). The needs of memory and CPU time are low, so that the program can well be used in a truly interactive manner even on mini-computers (e.g., microVAX). At present, this program is not part of a larger software package and thus requires manual preparation of the input files. It is designed to run on any terminal, and therefore makes no use of graphics capabilities.

Summary

Investigation of NMR spectra by automatic procedures began only a few years ago. Many projects are still in progress and results have not been published; completed projects, with published results, have not yet found their final form. Other approaches might be incorporated. One example is the simulation of cross-peaks.[21–23] These calculations explicitly take into account spectral parameters set for data acquisition and processing. They thus allow realistic simulations of expected peak patterns. Comparison of experimentally obtained peaks with a collection of simulated peaks helps characterize the spin system and determine coupling constants, especially in strongly coupled systems. Whereas automatic assignment procedures will require improvement for quite some time, it seems that useful and time-saving software tools complementing pure computer graphics will be available in the very near future.

Appendix: List of Referred Software

This list is not meant to be complete; if so, it would include all software useful for resonance assignments in proteins. One reason this is not possible is that many programs are currently in development and therefore have not been released at the time of submission of this manuscript (December 1987).

Computer Graphics Tools

NMR1, NMR2

Language: Fortran 77
Implementation: VAX and μVAX (VMS and ULTRIX), Sun (UNIX), and Tektronix (and similar color graphics) terminals
Inquire: New Methods Research, Inc., 719 East Genesee Street, Syracuse, New York 13210
Description: user's manual

[21] H. Widmer and K. Wüthrich, J. Magn. Reson. 70, 270 (1986).
[22] H. Widmer and K. Wüthrich, J. Magn. Reson. 74, 316 (1987).
[23] I. Bock and P. Rösch, J. Magn. Reson. 74, 177 (1987).

FTCGI

Language: Fortran
Implementation: various systems, including μVAX and SUN
Inquire: Infinity Systems, 14810 2156 Avenue NW, Woodinville, Washington 98072
Description: user's manual

Automatic Assignment

Software for locating cross-peaks, analyzing their multiplet structure, and identifying spin systems: inquire with the authors of Refs. 8–12 and 14–17.

SEQASSIGN

Language: Pascal
Implementation: VAX (VMS and Berkeley UNIX) and SUN (UNIX)
Inquire: M. Billeter, Institut für Molekularbiologie und Biophysik, ETH-Hönggerberg, Ch-8093 Zürich, Switzerland
Description: see Ref. 18

Simulations

SPHINX, LINSHA

Language: Fortran
Implementation: DEC10, VAX (UNIX), and SUN (UNIX)
Inquire: H. Widmer and K. Wüthrich, Institut für Molekularbiologie und Biophysik, ETH-Hönggerberg, Ch-8093 Zürich, Switzerland
Description: see Refs. 21 and 22

MARS

Language: Pascal
Implementation: NORSK DATA ND500 (soon: UNIX, MS.DOS)
Inquire: authors of Ref. 23
Description: Ref. 23

Acknowledgments

Financial support by the Komission zur Förderung der wissenschaftlichen Forschung, Project 1615, is gratefully acknowledged. Thanks go to Professors Wüthrich and Ernst for fruitful discussions, as well as to the authors of Refs. 16, 17, and 20, for providing preprints of their publications.

[9] Distance Geometry

By I. D. KUNTZ, J. F. THOMASON, and C. M. OSHIRO

Introduction

Distance geometry, a branch of mathematics developed by L. M. Blumenthal, is concerned with building structures from internal distances.[1] Blumenthal's ideas were adapted for use in chemical structure problems by Crippen and co-workers.[2-4] The name distance geometry has come to refer to the computer programs that convert geometric constraints into molecular coordinates. This chapter will explain the method and its application to the interpretation of nuclear magnetic resonance (NMR) data. The approach is quite general and can be used to search conformation space subject to a wide variety of constraints beyond those obtainable from NMR.[5] The same mathematics has also been used as a tool for statistical analysis of data that is known as "multidimensional scaling."[6] Our purpose in this chapter is to provide a descriptive introduction to the mathematics and a user's guide to the distance geometry computer programs. We will also briefly discuss other approaches to the problem of determining molecular structures in noncrystalline environments. Recent reviews in this area[7,8] and more advanced treatments of distance geometry are available.[3,9,10]

The General Problem

In brief, the goal of the distance geometry program is to produce one or more molecular structures that meet a set of arbitrary constraints. We use the word "arbitrary" to suggest that the constraints can apply to any

[1] L. M. Blumenthal, "Theory and Applications of Distance Geometry." Chelsea, Bronx, New York, 1970.

[2] G. M. Crippen, *J. Comput. Phys.* **24**, 96 (1977).

[3] T. F. Havel, I. D. Kuntz, and G. M. Crippen, *Bull. Math. Biol.* **45**, 665 (1983).

[4] I. D. Kuntz, G. M. Crippen, and P. A. Kollman, *Biopolymers* **18**, 939 (1979).

[5] G. M. Crippen, *J. Med. Chem.* **22**, 988 (1979).

[6] J. D. Carroll and P. Arabie, *Annu. Rev. Psychol.* **31**, 607 (1980).

[7] W. Braun, *Q. Rev. Biophys.* **19**, 115 (1987).

[8] G. M. Clore and A. M. Gronenborn, *Protein Eng.* **1**, 275 (1987).

[9] G. M. Crippen, "Distance Geometry and Conformational Calculations." Wiley, Chichester, England, 1981.

[10] G. M. Crippen and T. F. Havel, "Distance Geometry and Molecular Conformation." Research Studies Press Ltd., Taunton, England, and Wiley, Chichester, England, 1988.

part of the molecule and come from a wide variety of experimental or theoretical sources. From its earliest use for molecular problems, distance geometry was considered only as a first stage to be followed by molecular mechanics or molecular dynamics calculations to obtain structures of suitable energy. The idea was to use the relatively high speed of the distance geometry algorithm and its ability to explore conformation space to supply starting structures as input for more sophisticated refinement methods.

General Features of Conformational Searches

The general problem of searching conformation space is challenging because of the very rapid increase in computer time required as the molecular system gains in complexity. Only for small molecules is there any hope of undertaking a comprehensive search. The computation effort is exponential in the number of rotatable bonds. Systematic grid search methods have been developed.[11,12] These are limited to systems containing about 10 dihedral angles. Thus, most systems require significant assumptions to limit or direct the search effort. Other methods forego exhaustive searches and explore either randomly or selectively. For example, molecular dynamics programs search the thermally accessible vicinity of a particular starting conformation.[13,14] Monte Carlo procedures employing the Metropolis algorithm[15] have similar characteristics because of the small displacement steps used in both methods. More sophisticated approaches include thermal annealing[16,17] and force biasing.[18] These methods replace random searches and find a higher percentage of interesting conformations. Distance geometry is one such method.

Another fundamental issue is selecting the geometric variables to describe the molecular conformations. One common choice is to use internal coordinates. Typically, bond lengths and bond angles are held con-

[11] H. A. Scheraga, *Pure Appl. Chem.* **36**, 1 (1973).

[12] G. R. Marshall, C. D. Barry, H. E. Bosshard, R. A. Dammkoehler, and D. A. Dunn, *in* "The Conformation Parameter in Drug Design: The Active Analog Approach" (E. C. Olson and R. E. Christofferson, eds.), pp. 205–226. Amer. Chem. Soc., Washington, D.C., 1979.

[13] E. R. P. Zuiderweg, R. M. Scheek, R. Boelens, W. F. van Gunsteren, and R. Kaptein, *Biochimie* **67**, 707 (1985).

[14] G. M. Clore, A. T. Brünger, M. Karplus, and A. M. Gronenborn, *J. Mol. Biol.* **191**, 523 (1986).

[15] N. Metropolis, A. W. Rosenbluth, A. H. Teller, and E. Teller, *J. Chem. Phys.* **21**, 1087 (1953).

[16] S. Kirkpatrick, C. D. Gelatt, Jr., and M. P. Vecchi, *Science* **220**, 671 (1983).

[17] M. Nilges, A. M. Gronenborn, A. T. Brünger, and G. M. Clore, *Protein Eng.* **2**, 27 (1988).

[18] P. K. Mehrotra, M. Mezei, and D. L. Beveridge, *Proc. NRCC Workshop (1981)*, p. 63.

stant and each rotatable bond is characterized by a dihedral angle. Alternatively, Cartesian coordinates can be used as the basic variables. There are advantages to both representations. Use of internal angles requires approximately one-third as many variables and hence less computer memory. On the other hand, optimization often proceeds more smoothly in Cartesian space.[9]

Constraints

In general terms, our problem is to identify the parts of conformation space that are available to a molecule when certain average properties are known, either from experimental measurements or theory. For example, it is desirable to look for the "low-energy" conformations of a molecule, because it is these that will be the most significantly populated at the temperatures of biological systems. The problem we focus on in this chapter is to produce a set of conformers that meet the distances derived from experiments such as nuclear Overhauser effect (NOE) measurements used in NMR. It is important to realize that structural information can be obtained from many types of experiments. For instance, in addition to distances from NOE measurements, NMR experiments can yield data on dihedral angles from coupling constants, and distances from spin labels (see Chapter [5] by Kosen in this volume). Often the positions of hydrogens near aromatic rings can be determined from the effect of ring currents on the proton chemical shifts. Chemical cross-linking experiments yield distances between specific atoms or residues; fluorescence energy transfer measurements are sources of distance and angular information. CDNIP and chemical reactivity can identify the parts of a molecule that are near the molecule–solvent interface. Information about chirality is also valuable. From solution scattering data one can obtain radial distribution functions and their moments. Theory is also a rich source of constraints. However, it is important to realize that it is not always possible to construct a single algorithm that can simultaneously enforce all these types of constraints. Present programs can deal easily only with distance, angle, and chiral constraints.

Constraints are often expressed in terms of an objective (or error) function. There are two common ways to do this. The first method assigns a target value for the parameter of interest. The objective function measures deviations from the optimum value. Alternatively, one can specify upper and lower bounds on a particular parameter. No penalty is assessed when a distance or an angle falls within its bounds, but a term is added to the objective function when the boundary conditions are violated. Most versions of distance geometry use this second method for handling constraint violations. In the literature, "objective function" and "error function" are often used interchangeably. Strictly, an objective function can contain other terms beyond those dealing with violations of constraints.

Another general issue worth mentioning at this point is whether it is better to fit original data or to process the data to yield constraints. The brief answer is that it is generally better to fit to the data themselves rather than to quantities derived from the data. However, as we shall see, there are some practical advantages to compromising on this issue. Certainly in the NMR field it is customary to convert NOE intensities into semiquantitative distance constraints (see below).

General Features of Generating Structures from NMR Experiments

There are two general methods used to build structures from NMR-derived constraints. The first developed was distance geometry. It uses a matrix method, called embedding, for converting distances between points into coordinates. These coordinates are a best-fit multidimensional solution to all the distances taken simultaneously. However, to enforce a three-dimensional solution, the coordinates must be optimized, and to get reasonable molecular energies, energy minimization is required. The other approach begins with random or model-built structures and applies the constraints directly, without the embedding step. Such programs can be modifications of conventional molecular mechanics or molecular dynamics routines,[19-21] or more complex mathematical search procedures.[22,23] All these programs share certain basic features. First, one must have some form of data input procedure. Second, there must be some method of generating an initial structure or family of structures. Third, all procedures use some type of optimization to bring the constraints and the structure(s) into as close a correspondence as possible. Optimization itself requires a scoring formula that measures the goodness-of-fit to the constraints and contains, if desired, other terms, such as an interatomic potential function. We briefly describe the choices that have been made for each of these areas.

Input routines normally allow descriptions of standard geometry taken from small-molecule crystallography, electron diffraction, or other structural measurements. This is particularly pertinent to bond lengths and

[19] R. Kaptein, E. R. P. Zuiderweg, R. M. Scheek, R. Boelens, and W. F. van Gunsteren, *J. Mol. Biol.* **182,** 179 (1985).
[20] G. M. Clore, A. M. Gronenborn, A. T. Brünger, and M. Karplus, *J. Mol. Biol.* **186,** 435 (1985).
[21] T. A. Holak, J. H. Prestegard, and J. D. Forman, *Biochemistry* **26,** 4652 (1987).
[22] W. Braun and N. Gō, *J. Mol. Biol.* **186,** 611 (1985).
[23] B. Duncan, B. C. Buchanan, B. Hayes-Roth, O. Lichtarge, R. Altman, J. Brinkley, M. Hewett, C. Cornelius, and O. Jardetzky, *Bull. Magn. Reson.* **8,** 111 (1986).

bond angles. We call such data "holonomic," because they are known prior to the NMR experiment. Input routines should allow the user to enter constraints in the most natural way (i.e., distances as distances, angles as angles), letting the computer do the conversions as needed internally.

We also point out that NMR studies on proteins and nucleic acids normally involve materials of known peptide or nucleotide sequence. While it is possible to carry out the sequencing directly by NMR, in practice this would make the assignment efforts much more difficult.

Generation of a starting structure can be relatively simple or quite time-consuming. The choices include random starting conformations, model-built starting conformations, or starting conformations calculated from distance geometry. The procedures of Braun and Gō[22] and early efforts from Karplus' group[20] used a random start, either in dihedral or Cartesian variables, the former being preferable because the angular variables are sensibly bounded. Others have used model-built structures. These often come about naturally from qualitative interpretation of the NOE data both in terms of secondary structure and the packing of the secondary elements. The *lac* repressor headpiece, the acyl carrier protein, and ubiquitin all began as "hand-built" structures that were then refined by various optimization procedures,[19,21,24] Distance geometry combines some aspects of the "model" and the "random-start" approaches. The mathematical algorithm guarantees that the initial coordinates are the best multidimensional fit to the trial distances. The randomization is associated with the trial distances, which are chosen as randomly distributed between upper and lower boundary conditions. This procedure provides a useful way to sample the allowed conformation space.

After generating a starting conformation, one generally finds that it does not meet all the constraints—either some of the distances are not those desired or some other function (energy, angles, fit to NOE intensities, etc.) has not been adjusted to its best value. An optimization step is used to reduce the value of the objective function using an iterative process. There are many possible optimization routines. The best (most efficient) involve storing the second derivatives of the objective function with respect to the variables. These approaches (i.e., Newton–Raphson) are guaranteed to find minima on harmonic surfaces in the fewest possible steps.[25] However, they require storage in proportion to the square of the number of variables. For most applications, the best compromise between

[24] P. L. Weber, S. C. Brown, and L. Mueller, *Biochemistry* **26**, 7282 (1987).
[25] R. H. Pennington, "Introductory Computer Methods and Numerical Analysis." Macmillan, New York, 1965.

storage and efficiency is the conjugate gradient algorithm.[26,27] It requires retention of the first derivatives and, in practice, it is probably not the limiting aspect of convergence to a minimum. Another procedure, called the ellipsoid method,[28] has been used, but its behavior when the number of degrees of freedom becomes large is not well understood.

In our experience, with constraints from NMR data sets, it is not possible to produce structures in which there are no violated distances. However, it is generally possible to make any remaining violations very small, e.g., 0.1 Å. We will discuss the implications of nonzero error values later. At this point, it is important to note that there are many conformations of a complex molecule that will give equally good agreement to distance constraints. How should these be distinguished? It is possible that some of the conformers will be of sufficiently high energy that they can be discarded directly. However, it is quite common to still be left with many conformers that are essentially equally good fits to the data. If the structures span too wide a range to answer the questions originally posed at the beginning of the study, one must then return to experiment to try to gather more constraint information. Alternatively, the user may decide that the range of conformers is small enough that the original question has been answered. In any case the set of conformers lies within an "envelope" that is bounded by the experimental uncertainties plus procedural uncertainties.

Comparison of Methods

As documented in a later section (see Results), all the major approaches have been used successfully at the level of simulated data and to solve at least one structure from NMR experimental data. But there has been little opportunity for a set of comparative tests. The main points of comparison in such tests would be (1) computer time required per successful structure, (2) number of acceptable structures divided by the total number of trials, (3) efficiency of sampling conformation space, and (4) sampling bias. A beginning has been made,[29-31] but a full comparison is clearly of interest. We must leave such efforts for a future study. In

[26] R. Fletcher and C. M. Reeves, *Comput. J.* **7,** 149 (1964).

[27] A. Perry, *Int. Comput. Math.* **6,** 327 (1978).

[28] M. Billeter, T. F. Havel, and K. Wüthrich, *J. Comput. Chem.* **8,** 132 (1987).

[29] G. M. Clore, M. Nilges, A. T. Brünger, M. Karplus, and A. M. Gronenborn, *FEBS Lett.* **213,** 269 (1987).

[30] G. Wagner, W. Braun, T. F. Havel, T. Schauman, N. Gō, and K. Wüthrich, *J. Mol. Biol.* **196,** 611 (1987).

[31] A. E. Torda, "The Solution Structure and Dynamics of a Polypeptide Cardiostimulant Studied by ¹H NMR Spectroscopy," Ph.D. thesis. Univ. New South Wales, New South Wales, Australia, 1988. (Available from *Diss. Abstr. Int.*)

TABLE I
PROGRAMS FOR DISTANCE GEOMETRY

Program name	Language	Authors	Contact and location[a]
EMBED	Fortran	Crippen, Kuntz, and Havel	Kuntz (UCSF)
VEMBED	Fortran	Thomason	Kuntz (UCSF)
DISGEO	Pascal	Havel	Havel (UM)/QCPE
DGEOM	Fortran	Blaney	Blaney (DuPont)
DSPACE	Fortran	Hare	Hare (Infinity Systems)

[a] Contacts: I. D. Kuntz, Dept. of Pharmaceutical Chemistry, University of California, San Francisco, CA 94143; T. F. Havel, Div. of Biophysics, University of Michigan, Ann Arbor, MI 48109; D. Hare, Infinity Systems, 14810 216th Avenue NE, Woodinville, WA 98072; J. Blaney, DuPont Experimental Station, Building E328, Wilmington, DE 19898.

this chapter we will concentrate on how to use the distance geometry program.

Distance Geometry

We now turn to the detailed procedures involved in the distance geometry method. There are a number of programs that differ somewhat from each other, but essentially all were developed from the work of Crippen.[2-4] A list is given in Table I. Each of these programs has four major groupings of subroutines: (1) input preparations, (2) bounds generation and bounds smoothing, (3) embedding, and (4) optimization. The logical flow diagram is given in Fig. 1. We have not included the DISMAN program in Table I because it does not use the metric matrix method of structure generation. Instead, it takes an initial random structure which is then optimized by a variable target function.[22] It has proved to be a useful alternative approach (for comparisons, see Torda[31]).

Atom Representations

The mathematical routines place no limitations on the way in which molecules are described in terms of atoms. Distance geometry only deals with distances between points. These points can be "real" atoms or united atoms or even united residues. A common feature in NMR studies is the use of virtual atoms, or pseudoatoms. This representation places atoms, for convenience, where no real atom exists. Such NMR studies frequently use real atom descriptions for all the heavy atoms and for those hydrogens that can be resolved. Unresolved hydrogens associated with β-methylenes, the α-hydrogens of glycine, or various ring hydrogens can be treated in three different ways. The first approach makes the fewest as-

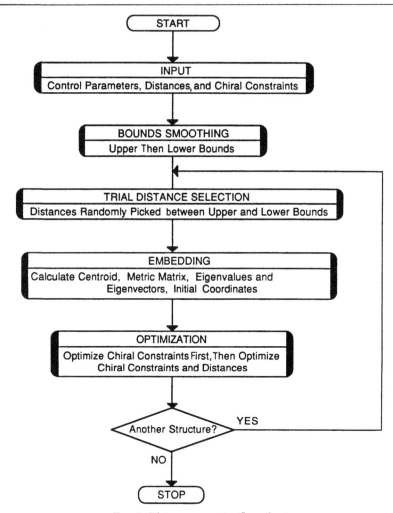

FIG. 1. Distance geometry flow chart.

sumptions. It uses a virtual atom that is positioned at the geometric center of the unresolved group of hydrogen atoms. The user must simply assign the distance constraints for the set of atoms to a single center. Because any member of the set can be responsible for a particular NOE or coupling constant, the boundary conditions are loosened to permit the virtual atom to meet the constraints properly. For example, if the two protons of a β-carbon methylene (labeled LB in Fig. 2) are unresolvable, then a virtual atom can be used. In such a case the pseudoatom replaces the two methylene protons. It is "bonded" to the β-carbon at a distance of 0.63 Å. This

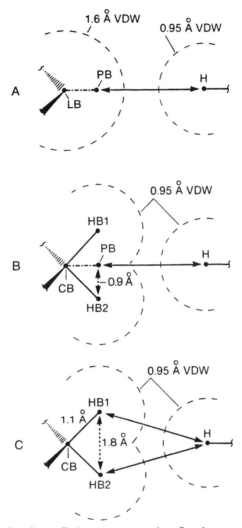

FIG. 2. (A–C) Atom representation. See the text.

distance is then used as a holonomic constraint (see later, Holonomic Distances and Bounds). The pseudoatom, PB, is treated as a point and has no van der Waals radius. In order to compensate, the van der Waals radius for a methylene group, 1.6 Å, is given to the LB atom (see Fig. 2A). Because of the positional averaging, virtual atoms cannot, in general, be placed to satisfy the experimental NOE data with high accuracy. Thus the constraints must be loosened by the maximum distance from the hydrogen atoms to the pseudoatom. Virtual atom approximations have been developed by the Wüthrich group for all the common situations in peptides and proteins.[32] One of the benefits of this method is that it reduces the number of atoms needed to describe the molecule. Because the computational time depends on the number of points (atoms), the use of virtual atoms reduces the time it takes to generate structures.

A second method of atom representation is very similar to the previously described method. It makes use of the virtual atoms, but it does not replace the hydrogen atoms. This method allows for a more accurate placement of the pseudoatom because the distances from the pseudoatom to each hydrogen atom are considered to be holonomic distances. The real atom van der Waals radii are used. The experimental distances are loosened as described above. The benefits of this approach include improved placement of the pseudo atom and positioning of all atoms for use with the energy refinement methods. Its major drawback is that the number of points used to describe a molecule is greater than the number of real atoms (see Fig. 2B).

The third method makes use of the pseudoatom idea but does not actually use pseudoatoms. Instead the experimental constraint distance is added to the maximal distance among the set of protons in the unresolvable group. Though this method creates less restrictive bounds than either of the other two methods, it has the benefit of not requiring explicit pseudoatoms. Often, but not invariably, the user can inspect the results of a preliminary calculation and assign specific constraints to specific atoms of an unresolved (degenerate) set. The reason for doing this is that the constraints are more restrictive when assignable to specific atoms than when attributed to virtual atoms. This method makes use of the full atom representation and in general makes management of the data processing much easier (see Fig. 2C).

Of much current interest is the attempt to fix stereospecific assignments for all resolvable prochiral pairs of hydrogens: the α-hydrogens of glycine, the β-hydrogens of all residues except Ile and Thr, and the appropriate β, δ, and ε pairs. Two sources of information can be used: coupling constants[33] and NOE data.[34] A clever approach suggested by Blaney and Weber[35] is to remove any chiral constraints (see below for a discussion of chiral constraints) for prochiral hydrogens, run the distance geometry program to produce a number of structures, and see if the NOE informa-

[32] K. Wüthrich, M. Billeter, and W. Braun, *J. Mol. Biol.* **169,** 949 (1983).

[33] H. Widmer, G. Wagner, H. Schweitz, M. Lazdunski, and K. W. Zurich-Hoenggerberg, *Eur. J. Biochem.* **171,** 177 (1988).

[34] K. Wüthrich, "NMR of Proteins and Nucleic Acids." Wiley, New York, 1986.

[35] J. Blaney and P. Weber, personal communication (1987).

tion is sufficient to select one hydrogen assignment over the other. If the procedure always favors one assignment, it can be assumed to be correct. Otherwise, the current practice is to use one of the pseudoatom representations described above, leaving the stereospecific assignments for a later stage.

Two general comments should be made about the use of virtual atoms. First, distance geometry structures will normally be refined using molecular mechanics or molecular dynamics (see below). These methods are best done with full atom representations. The constraints can then be defined in terms of real atoms or pseudoatoms as one prefers, but the program should always maintain a full set of real atom coordinates. Second, there can be no doubt that there is more structural information contained in stereospecific assignments compared with pseudoatom descriptions. Hence stereospecific constraints should be used whenever possible.

Parameter Input File

Table II is a list of the fundamental parameters required by all of the distance geometry programs. Specific values for these parameters are discussed in a later section (see Optimization). There are many more parameters dealing with control of embedding, optimization, and output not shown in Table II. A discussion of these more specialized items is beyond the scope of the present paper. Interested readers should consult the documentation for the program they choose.

Generation of Data Input File

After choosing the molecular representation, the next step is to generate the input data file. The input procedures described here are used for proteins and nucleic acids. However, conformational searching of small

TABLE II
PARAMETERS REQUIRED FOR DISTANCE GEOMETRY PROGRAMS

Parameter name	Purpose
NATOMS	Defines number of atoms in molecule
NUMBER_STRUC	Defines number of structures to be generated
ISEED	Seeds random number generator
ITER_LIMIT	Number of cycles of optimization
FUNC_MIN	Target function value for optimization
GRAD_MIN	Target gradient value for optimization
SAVE_BOUNDS	Binary flag for saving smoothed bounds
READ_BOUNDS	Binary flag for reading in already smoothed bounds

molecules requires a much more elaborate entry system. Blaney has developed such an input routine for DGEOM. There are two basic types of input data: distances and chiral volumes. The distances are assigned upper and lower bounds whereas the chiral volumes are described by a signed volume, where the sign indicates the chirality of the asymmetric center. These distance bounds and chiral volumes then form the actual input data file. The distances can be divided into three categories: (1) holonomic, (2) experimental, and (3) secondary structure distances, and these are discussed below.

Holonomic Distances and Bounds. The holonomic distances are those that can be determined by knowledge of the sequence of the molecule under investigations. We use a templating procedure which calculates bond lengths, geminal distances, fixed dihedral distances such as those in the peptide bond, and all distances involved in rigid structures such as the aromatic ring of phenylalanine, tyrosine, and tryptophan. Also calculated are the cis and trans distances for rotatable dihedrals. The templates of each amino acid are stored in a library. They are taken from the three-dimensional coordinates of amino acids developed from the ECEPP parameters given by Némethy et al.[36] Our practice is to increase the bond lengths and geminal distances by 2% for the upper bounds and to decrease them by 2% for the lower bounds. This difference improves the behavior of the molecule during the embedding and optimization procedures without having any detrimental effects on local geometry. We presently do not extend the holonomic information beyond a two-residue segment, except, possibly, where secondary structure has been identified (see below).

Experimental Distances and Bounds. Currently, experimental information is essential for the accurate determination of the structures of biomacromolecules. NMR is the most productive experiment for obtaining structural information in noncrystalline environments.[37] Most of this information comes from NOE data. However, there is some concern about how to interpret NOE data in terms of distance constraints. A complete discussion of the NOE experiment is beyond the scope of this chapter. A proper treatment would take into account the full range of molecular motions and the full interaction of all the spin systems using the complete relaxation matrix (see Borgias and James [9], Vol. 176, this series). So far such an approach has only been used for a few systems. What is commonly done, instead, is to adopt simplifying assumptions. Typically, NOE data are analyzed in the limit that the molecule undergoes

[36] G. Némethy, M. S. Pottle, and H. A. Scheraga, *J. Phys. Chem.* **87**, 1883 (1983).
[37] I. D. Kuntz, *Protein Eng.* **1**, 147 (1987).

TABLE III
BOUNDARY CONDITIONS FROM NOEs

Bound	← Stronger intensity			Weaker intensity →	Reference[a]
			Distances calculated from NOE intensity (Å)		
Upper	2.5	—	3.0	—	4.0	1
Lower	—	—	2.5	—	3.0	
Upper	2.8	—	3.3	—	5.0	2
Lower	1.8	—	1.8	—	1.8	
Upper	2.8	—	3.3	—	5.0	3
Lower	1.8	—	1.8	—	2.5	
Upper	3.0	—	3.5	—	5.0	4
Lower	2.0	—	2.0	—	3.0	
Upper	2.5	3.0	—	3.5	5.0 + pseudo	5
Lower	1.8	1.8	—	1.8	—	
Upper	—	4.0	—	5.0	—	6
Lower	—	2.0	—	2.0	—	

[a] References: (1) T. Havel and K. Wüthrich, *Bull. Math. Biol.* **46,** 673 (1984); (2) G. M. Clore, D. K. Sukumaran, M. Nilges, J. Zarbock, and A. M. Gronenborn, *EMBO J.* **6,** 529 (1987); (3) G. M. Clore, D. K. Sukumaran, M. Nilges, and A. M. Gronenborn, *Biochemistry* **26,** 1732 (1987); (4) G. M. Clore, A. T. Brünger, M. Karplus, and A. M. Gronenborn, *J. Mol. Biol.* **191,** 523 (1986); (5) G. Wagner, W. Braun, T. F. Havel, T. Schauman, N. Gō, and K. Wüthrich, *J. Mol. Biol.* **196,** 611 (1987); (6) M. P. Williamson, T. F. Havel, and K. Wüthrich, *J. Mol. Biol.* **182,** 295 (1985).

no motion except for a rigid isotropic rotation characterized by a single correlation time. Spin–spin interactions are limited to consideration of individual pairs of protons (isolated spin-pair approximation). Further, the NOE intensity is dependent on the mixing time and the normal analysis is limited to the initial buildup rates. With all these assumptions, it is not straightforward to convert a NOE intensity directly to a distance. Instead, most NOE data are converted into simple proximity measurements or quite loose distance constraints. Early work in our laboratory[38] showed that such constraints are productive, and a conservative choice helps to avoid inconsistent constraints. Assigning a 5-Å upper bound to any NOE (not arising from spin diffusion) is normally a safe assumption for proteins and nucleic acids. Some workers divide NOEs into classes such as "strong and weak" or "strong, medium, and weak," thus allowing more restrictive upper and lower bounds (Table III). These choices

[38] T. F. Havel, G. M. Crippen, and I. D. Kuntz, *Biopolymers* **18,** 73 (1979).

can certainly lead to better defined structures, since they contain more information, but they can lead to systematic errors. The best procedure, at present, is to try more than one of these schemes and to check carefully for model-dependent biases. It is important not to overinterpret the NOE information. For example, too restrictive a set of upper bounds would result in structures that explore an incorrect part of conformation space. It is even possible to develop sets of bounds that are not consistent with any three-dimensional structures.

Normally, all lower bounds are initially set to the sum of the appropriate van der Waals radii. However, it is interesting to ask whether one can usefully raise the lower bounds for a hydrogen pair if a NOE is *not* observed. Rigorously, the answer is "no," because there can always be some mechanism, e.g., motional averaging, that can drive NOE intensities to zero even when the hydrogens are close in space. However, if both hydrogens of a pair show strong NOEs to at least one other hydrogen not rigidly connected, and there is no question of saturation transfer or spectral overlap, then it is reasonable to put a 3.5- to 4.0-Å lower bound on that hydrogen pair.[39]

Secondary Structure Distances and Bounds. Another source of input constraints comes from predefined secondary structure. Because the NMR data directly yield assignments of helical and β-sheet conformations, some workers have used various features implicit in such structures. Approaches vary from just replacing the observed NOEs with crystallographic distances, to using hydrogen bond constraints on hydrogen–oxygen or hydrogen–nitrogen distances, to assuming full secondary units with idealized geometries. Typically used hydrogen bond distances in amides are 2.0 and 1.8 Å for the H–O upper and lower bounds, and 3.0 and 2.7 Å for the N–O upper and lower bounds, respectively. We use upper bounds of 110% and lower bounds of 90% of the ideal distances to allow for small distortions. As long as these assumptions are clearly identified and are consistent with the total data set, they represent a reasonable method of reducing the allowed conformation space. The major concern is whether the experimental data are sufficiently accurate to detect irregularities in the idealized structures.

Chiral Constraints. Other holonomic inputs are the chiral constraints. This information is not used for the generation of the trial structure because the embedding procedure is dependent only on distances. Instead, these constraints are used during optimization to ensure that the handed-

[39] J. D. Vlieg, R. Boelens, R. M. Scheek, R. Kaptein, and W. F. van Gunsteren, *Isr. J. Chem.* **27**, 181 (1986).

ness of the chiral centers are proper. For the four ligands about the central atom, a signed volume can be calculated by the following equation:

$$f_{ch} \equiv (v_1 - v_4)[(v_2 - v_4)(v_3 - v_4)] \qquad (1)$$

Where v_1, v_2, v_3, and v_4 are the ligands ordered in the Cahn–Inglod–Prelog system of nomenclature.[40] As an example, the ordering of the ligands about the C_α of an amino acid would be N, C, C_β, H_α. This ordering is independent of which stereoisomer is being examined. The stereoisomer is distinguished by the sign of the volume as given by the above calculation. This sign is positive for an L-isomer. The input routine library or the user must supply a list of chiral centers as part of the initial constraints. This list can include any mixture of D and L centers, but the default should be conventional L-amino acids unless there is specific information to the contrary. It is sometimes forgotten that Ile and Thr side chains have chiral β-carbons. These should normally be chosen as R for Thr and S for Ile.[41] Another use of this type of constraint is to force planarity. This is simply accomplished by equating the volume to zero. It is a beneficial procedure for aromatic rings and peptide bonds. The order of the atoms is irrelevant for planarity constraints.

Ordering of Input Data. The various distance geometry programs assimilate the geometric input in different ways. Some programs have included the holonomic constraints in the program. Others use a preprocessing method. We favor the second approach as it keeps the distance geometry program completely independent of the specific system being studied. This means that the EMBED and VEMBED programs are not restricted to any one type of molecule or atomic representation. To accomplish this separation of input procedures from the actual data processing, each atom in the input data file is given a unique number. Within this file the distance and chiral constraints must be grouped separately, but within each group the entries can be in any order. The record format for the two types of constraints is different. The distance constraints are ordered as atom number i, atom number j, upper bound, lower bound, atom name i, and atom name j. For the chiral constraints there are two types of formatting. The first deals with the chiral constraints that have nonzero volumes. For the C_α of an amino acid the ordering of the atoms is fixed as N, C, C_β, H_α, C_α. The atom numbers are followed by the (signed) chiral volume. Only the first four atoms are used for the chiral volume calculation. The fifth atom is used during distance optimization of the

[40] R. S. Cahn, S. C. Ingold, and V. Prelog, *Angew. Chem., Int. Ed. Engl.* **5**, 385 (1966).
[41] T. E. Creighton, "Proteins," pp. 8–9. Freeman, San Francisco, California, 1984.

chiral center and is optional though recommended. For planar constraints a specific ordering of the atoms is not necessary and only four atoms are used to define the constraint. Table IV shows a sample of the input to the program VEMBED for the dipeptide Ala-Gly.

The user should be aware that there are opportunities for repeats of the same distance in the input file. This can occur in several ways. First, an experimental distance can be determined for a distance that also has a holonomic distance. In this case it is usually the holonomic distance that gives the most restrictive bounds, so it is this distance that should be used. Second, the two-dimensional (2D) NOESY data often occur in the spectrum on both sides of the diagonal. If multiple entries occur in the input file, the last distance entered into the input program is the one used currently. The user is encouraged to examine carefully the input file for multiple entries and to decide on the appropriate distance to be used. Our procedure will be modified to alert the user whenever two entries exist for the same atom-pair distance.

Bounds

Organization of Boundary Matrices. Once the input has been prepared, the holonomic, experimental, and secondary structure bounds are used to form the bounds matrix. The distance matrix is a symmetric matrix with $(N^*N - 1)/2$ distances for an object with N points (atoms). The bounds matrix we use is not symmetric. We store the upper bounds in the upper triangular portion of the matrix and the lower bounds in the lower triangular portion of the matrix. The matrix is initialized prior to entry of the input distances so that upper bounds are set to 999 Å, and the lower bounds are fixed as the sum of the van der Waals radii of the atoms or virtual atoms (Table V). The large value for the upper bounds is necessary to ensure that the bounds-smoothing procedure will process correctly.

Bounds Smoothing. Normally, only a small fraction of the approximately $N^*N/2$ interatomic distances is available from experimental and holonomic considerations. However, there are powerful constraints on the undetermined distances. The bounds-smoothing routine applies such geometric constraints to the bounds matrix to ensure that the input data are geometrically self-consistent. This step also reduces the allowed conformation space by lowering the upper bounds and raising the lower bounds where possible. The constraint that is applied is the triangle inequality. It has a somewhat different form for the upper and lower bounds. For the upper bounds matrix, the triangle inequality is defined: for all atoms taken three (i, j, k) at a time, the distance between the atoms

TABLE IV
INPUT TO PROGRAM VEMBED FOR DIPEPTIDE Ala-Gly[a]

Atom i	Atom j	Upper bound (Å)	Lower bound (Å)	Atom name i			Atom name j		
1	2	1.005	0.995	N	Ala	1	HN	Ala	1
1	3	1.458	1.448	N	Ala	1	CA	Ala	1
1	4	2.070	2.060	N	Ala	1	HA	Ala	1
1	5	2.441	2.431	N	Ala	1	CB	Ala	1
1	6	3.377	2.403	N	Ala	1	HB1	Ala	1
1	7	3.377	2.403	N	Ala	1	HB2	Ala	1
1	8	3.377	2.403	N	Ala	1	HB3	Ala	1
1	9	2.438	2.428	N	Ala	1	C	Ala	1
1	10	3.585	2.653	N	Ala	1	O	Ala	1
1	11	3.636	2.600	N	Ala	1	N	Gly	2
1	12	3.998	2.167	N	Ala	1	HN	Gly	2
1	13	4.852	4.015	N	Ala	1	CA	Gly	2
2	3	2.088	2.078	HN	Ala	1	CA	Ala	1
2	4	2.942	2.211	HN	Ala	1	HA	Ala	1
2	5	3.348	2.446	HN	Ala	1	CB	Ala	1
2	9	3.346	2.441	HN	Ala	1	C	Ala	1
3	4	1.095	1.085	CA	Ala	1	HA	Ala	1
3	5	1.535	1.525	CA	Ala	1	CB	Ala	1
3	6	2.159	2.149	CA	Ala	1	HB1	Ala	1
3	7	2.159	2.149	CA	Ala	1	HB2	Ala	1
3	8	2.159	2.149	CA	Ala	1	HB3	Ala	1
3	9	1.535	1.525	CA	Ala	1	C	Ala	1
3	10	2.406	2.396	CA	Ala	1	O	Ala	1
3	11	2.415	2.405	CA	Ala	1	N	Gly	2
3	12	2.596	2.586	CA	Ala	1	HN	Gly	2
3	13	3.790	3.780	CA	Ala	1	CA	Gly	2
3	14	4.495	4.025	CA	Ala	1	HA1	Gly	2
3	15	4.495	4.025	CA	Ala	1	HA2	Gly	2
3	16	4.844	4.230	CA	Ala	1	C	Gly	2
4	5	2.151	2.141	HA	Ala	1	CB	Ala	1
4	6	3.046	2.244	HA	Ala	1	HB1	Ala	1
4	7	3.046	2.244	HA	Ala	1	HB2	Ala	1
4	8	3.046	2.244	HA	Ala	1	HB3	Ala	1
4	9	2.153	2.143	HA	Ala	1	C	Ala	1
4	10	3.266	2.510	HA	Ala	1	O	Ala	1
4	11	3.310	2.451	HA	Ala	1	N	Gly	2
4	12	3.642	2.186	HA	Ala	1	HN	Gly	2
4	13	4.562	3.903	HA	Ala	1	CA	Gly	2
5	6	1.095	1.085	CB	Ala	1	HB1	Ala	1
5	7	1.095	1.085	CB	Ala	1	HB2	Ala	1
5	8	1.095	1.085	CB	Ala	1	HB3	Ala	1
5	9	2.548	2.538	CB	Ala	1	C	Ala	1
5	10	3.690	2.760	CB	Ala	1	O	Ala	1
5	11	3.740	2.730	CB	Ala	1	N	Gly	2
5	12	4.088	2.330	CB	Ala	1	HN	Gly	2
5	13	4.961	4.118	CB	Ala	1	CA	Gly	2
6	7	1.785	1.775	HB1	Ala	1	HB2	Ala	1

(*continued*)

TABLE IV (*continued*)

Atom i	Atom j	Upper bound (Å)	Lower bound (Å)	Atom name i			Atom name j		
6	8	1.785	1.775	HB1	Ala	1	HB3	Ala	1
6	9	3.478	2.506	HB1	Ala	1	C	Ala	1
7	8	1.785	1.775	HB2	Ala	1	HB3	Ala	1
7	9	3.478	2.506	HB2	Ala	1	C	Ala	1
8	9	3.478	2.506	HB3	Ala	1	C	Ala	1
9	10	1.235	1.225	C	Ala	1	O	Ala	1
9	11	1.330	1.320	C	Ala	1	N	Gly	2
9	12	2.063	2.053	C	Ala	1	HN	Gly	2
9	13	2.424	2.414	C	Ala	1	CA	Gly	2
9	14	3.307	2.505	C	Ala	1	HA1	Gly	2
9	15	3.307	2.505	C	Ala	1	HA2	Gly	2
9	16	3.712	2.860	C	Ala	1	C	Gly	2
10	11	2.267	2.257	O	Ala	1	N	Gly	2
10	12	3.176	3.166	O	Ala	1	HN	Gly	2
10	13	2.785	2.775	O	Ala	1	CA	Gly	2
10	14	3.839	2.341	O	Ala	1	HA1	Gly	2
10	15	3.839	2.341	O	Ala	1	HA2	Gly	2
10	16	4.277	2.680	O	Ala	1	C	Gly	2
11	12	1.005	0.995	N	Gly	2	HN	Gly	2
11	13	1.458	1.448	N	Gly	2	CA	Gly	2
11	14	2.095	2.085	N	Gly	2	HA1	Gly	2
11	15	2.095	2.085	N	Gly	2	HA2	Gly	2
11	16	2.464	2.454	N	Gly	2	C	Gly	2
11	17	3.605	2.691	N	Gly	2	O	Gly	2
12	13	2.088	2.078	HN	Gly	2	CA	Gly	2
12	14	2.961	2.246	HN	Gly	2	HA1	Gly	2
12	15	2.961	2.246	HN	Gly	2	HA2	Gly	2
12	16	3.365	2.479	HN	Gly	2	C	Gly	2
13	14	1.095	1.085	CA	Gly	2	HA1	Gly	2
13	15	1.095	1.085	CA	Gly	2	HA2	Gly	2
13	16	1.535	1.525	CA	Gly	2	C	Gly	2
13	17	2.406	2.396	CA	Gly	2	O	Gly	2
14	15	1.757	1.747	HA1	Gly	2	HA2	Gly	2
14	16	2.162	2.152	HA1	Gly	2	C	Gly	2
14	17	3.273	2.522	HA1	Gly	2	O	Gly	2
15	16	2.162	2.152	HA2	Gly	2	C	Gly	2
15	17	3.273	2.522	HA2	Gly	2	O	Gly	2
16	17	1.235	1.225	C	Gly	2	O	Gly	2
1	1	0.000	0.000	N	Ala	1	N	Ala	1
1	4	5	9	3	8.28				
3	9	11	12	0.00					
3	9	11	13	0.00					
10	9	11	12	0.00					
10	9	11	13	0.00					

[a] Sample input to the VEMBED program for a dipeptide Ala-Gly. The distance information is terminated by the last line in the set of distances. The chiral information shows input for the chiral center of the Ala C_α and the peptide bond between the two residues. This example shows only the holonomic input. CA, C_α; CB, C_β.

TABLE V
VAN DER WAALS RADII

Atom	Radius (Å)
Real	
C	1.43
H	0.95
N	1.30
O	1.25
S	1.50
Pseudo[a]	
L	1.6
M	1.8
K	1.5
P	0.0
Q	0.0

[a] L, Methylene pseudoatom; M, methyl pseudoatom; K, CH fragment; P and Q are dimensionless pseudoatoms.

i and j must be less than the sum of the distances from i to k and from k to j.

$$D_{ij}^{U} = \text{MIN}(D_{ij}^{U}, D_{ik}^{U} + D_{kj}^{U}) \qquad (2)$$

for $k = 1, N; k \neq i, j$. If the distance between i and j is greater than $D_{ik}^{U} + D_{kj}^{U}$, then D_{ij}^{U} is reduced to the sum of the other two distances. This process is repeated until no further changes occur. If, at any time, the sum of the two distances is less than the lower bound for D_{ij}, then a geometric inconsistency, or triangle violation, has occurred. Once the upper bounds can be reduced no further, the triangle inequality can be applied to the lower bounds. This form of the inequality (often called the inverse triangle inequality) is defined: the upper bound on the distance from k to j must be greater than or equal to the sum of the upper bound from i to j and the lower bound from i to k for all points i, j, k. From this definition, we obtain

$$D_{ij}^{L} = \text{MAX}(D_{ij}^{L}, D_{ik}^{L} - D_{kj}^{U}, D_{jk}^{L} - D_{ki}^{U}) \qquad (3)$$

for $k = 1, N; k \neq i, j$. If any one of the distances in the above set is greater than D_{ij}^{L} then that value is substituted for D_{ij}^{L}. However, if the maximum distance is greater than the D_{ij}^{U} then an inverse triangle inequality violation has occurred. These constraints apply to any object that can be built using Euclidean geometry, regardless of how many dimensions it spans. The triangle inequality can be implemented as a "shortest-path" procedure.[42] A more complicated constraint limits distances among four atoms simultaneously. Four centers are described by six interpoint distances. Any one of the distances can be written in terms of the other five and will have lower and upper bounds given (in chemical nomenclature) by the cis and trans geometries associated with the four atoms. This tetrangle inequality is more difficult to implement. It is very computationally intensive for problems involving large numbers of atoms ($n > 100$). Subroutines have been written by Havel and Blaney to use this procedure,[10] providing considerably tighter bounds than the triangle inequality.

Our program flags all violations of the triangle inequalities. The user is encouraged to resolve these inconsistencies before proceeding with embedding. Whereas the embedding and optimization algorithms will function even if the boundary conditions contain triangle violations, there is no Euclidean structure that can be found to satisfy the constraints exactly.

Initial Structure Generation

Trial Distances. After smoothing the boundary matrices with the triangle and, possibly, the tetrangle procedures, one must generate a set of distances among all the pairs of atoms. These distances are chosen to lie between the upper and lower bounds. They may be picked randomly or to fit some predetermined distribution function. All of the programs allow the simplest option: random choice of all distances. It would be straightforward to bias the choice with distribution functions of various types. For example, bond lengths are distributed harmonically about an average value. The relation between the NMR NOE intensity and the distance follows a well-known dependence of r^{-6}. Other distances, derived from the triangle limits, would be expected to have quite complex distribution qualities. The use of Cartesian variables introduces complications, because there are many more distances than there are degrees of freedom. Thus, as noted above, some implicit correlations must exist among the distances. Correlation subroutines exist in the EMBED, VEMBED, and DISGEO programs to avoid inconsistent choices among the distances. These routines reduce such problems but do introduce biased sampling. Our experience so far is that a full NOE set (i.e., several hundred long-range NOEs for a protein of 50–100 amino acids) provides enough information so that the use of the correlation routines can be eliminated.

Embedding. The operation that converts the trial distances into coordinates is called embedding. The process is quite general and can be

---

[42] A. W. N. Dress and T. F. Havel, Conference on Computer-Aided Geometric Reasoning, INRIA. Sophia Antipolis, France, 1987.

applied to any number of Euclidean dimensions. The first step of the process is to define a reference point to which the molecule can be related. Havel showed that the geometric center, or centroid, of the molecule can be calculated from the distance matrix.[43] Distances from each point (atom) to the origin 0 are entered as the diagonal elements of the distance matrix. The next step is to apply the law of cosines to obtain the metric matrix, $\mathbf{G}$, the elements of which are the dot products of the vectors from atom i to 0 and atom j to 0. This matrix is central to the process of embedding. Its unique property is that the square roots of the eigenvalues of this matrix are the principal moments of the molecule with the origin of the coordinate system at the molecular centroid, and the eigenvectors are the distributions of the atoms along the axes. The coordinates of the molecule can be calculated from the eigenvalues and the eigenvectors by

$$c_m = \lambda_m^{1/2} \mathbf{W}_m \qquad (4)$$

Where the vectors c, λ, and $\mathbf{W}$ are the coordinates, eigenvalues, and eigenvectors, respectively, for each dimension. Thus the problem of finding coordinates has been transformed into a straightforward problem of finding the nonzero eigenvalues of a square matrix.

There are many algorithms one could use to determine the eigenvalues and eigenvectors. We use an eigenvalue routine that determines the largest eigenvalues and eigenvectors in decreasing order.[44] Alternatively, there are procedures to find all $n - 1$ values (many of which should be zero or very small relative to the largest eigenvalues). Though the coordinates of the molecule can be directly determined from the metric matrix, there are generally more than three nonzero eigenvalues, so that the molecule fits into more than three dimensions. To derive the coordinates of the three-dimensional object, only the largest eigenvalues and their associated eigenvectors are used. This process of truncation "shrinks" the molecule. This is one of the reasons that optimization is applied subsequent to the embedding procedure. Other projection methods are available.[45,46] The dimensionality of the object merits further discussion. First, it is important to note that the looser the boundary conditions and the more limited the measurements on a system, the more likely one is to find higher dimensional forms. Thus looking at the magnitude of the first four or five eigenvalues can be instructive. If they are much smaller than the

[43] G. M. Crippen and T. F. Havel, *Acta Crystallogr. Sect. A* **34**, 282 (1978).
[44] D. K. Faddeev and V. N. Faddeeva, "Computational Methods of Linear Algebra," pp. 307 and 328. Freeman, San Francisco, California, 1963.
[45] G. M. Crippen, *J. Comput. Chem.* **5**, 548 (1984).
[46] E. Purisima and H. A. Scheraga, *Proc. Natl. Acad. Sci. U.S.A.* **83**, 2782 (1986).

first three, the structure is reasonably well defined. If, on the other hand, there is not much gradation, the boundary conditions are not strong. Negative eigenvalues are more troublesome. They imply that the molecule built from that particular set of trial distances is not a Euclidean object. In fact, such a molecule must incorporate imaginary distances, an unsettling idea to most physical scientists. To develop proper molecular structures, some method must be found to guarantee that the final coordinates are three-dimensional. Normally, structures with negative eigenvalues are discarded. If they must be used, take the absolute value of the eigenvalue and continue through the program. However, such structures are relatively poor fits to the initial conditions.

In actual practice, for NMR problems dealing with small proteins or nucleic acids, one can measure several hundred NOEs, and similar numbers of bond lengths and bond angles are available. Normally, such a data set provides satisfactory embedding. However, minimally determined systems can cause some problems. The use of correlation routines for the generation of the trial distances can sometimes help. The better approach is to provide more data to correct the underdetermined conditions.

A common concern is whether it is possible to weight the embedding process to emphasize the fit to certain distances. There is no elegant way to introduce weights directly into the metric matrix. It is possible to increase the importance of one part of the structure by inserting extra (dummy) atoms that coincide with other atoms. This stratagem will produce extra rows and columns in the metric matrix that will have the effect of improving the embedding of those particular atoms. Alternatively, extra constraints can be added to emphasize particular features. For example, the input routine we use provides a long list of holonomic distances which are derived from ring planarity constraints and redundant descriptions of internal angles. These additional terms provide a stronger weighting of local geometry, but they do not constitute new information about the structure. Weighting schemes can be directly introduced during the optimization procedures that are described in the next section.

Optimization

As mentioned earlier, it is not expected that the structures immediately after embedding will satisfactorily meet all the boundary conditions. The problem is that the random choice of trial distances can create some quite unusual conformations. Furthermore, the truncation of higher eigenvalues will always produce some errors. Thus the distance geometry programs use an optimization routine that measures the error between the distances and the desired boundary conditions and alters the coordinates in such a way as to minimize this error. The actual optimization proceeds as follows: an objective function is evaluated, a gradient is calculated, and a set of changes to the coordinates is determined. The objective function must then be evaluated with the new coordinates. If the new value is less than that in the previous step, the new coordinate set is accepted and the whole procedure begins anew. If the objective function value is larger than in the previous step, the coordinates are rejected and a new set is found by taking a different step: either along the gradient to a different extent or in a new direction in conformation space. If no satisfactory step can be executed, the procedure terminates. The error function we use has two parts, one related to chirality, the other to the distances themselves.

Distance Function. The objective function that we use for all distances is

$$\mathbf{E}(\text{distances}) = \sum_{\text{violated } ij \text{ distances}} k'(B_{ij}^2 - D_{ij}^2)^2 \tag{5}$$

where B_{ij} is the bound being violated, D_{ij} is the associated structure distance, and k' is the constant 1 Å^{-4}. It should be noted that $\mathbf{E}(D_{ij}) = 0$ when $B_{ij}^L \le D_{ij} \le B_{ij}^U$. The form of the function is chosen to be easily computed, with analytical and continuous first derivatives with respect to the Cartesian coordinates. An alternative objective function normalizes each term by dividing by the square of the bound.[47]

$$\mathbf{E}(\text{distances}) = \sum_{\text{violated } ij \text{ distances}} \left(\frac{B_{ij}^2 - D_{ij}^2}{B_{ij}^2} \right)^2 \tag{6}$$

The form in Eq. (5) weights the longer distances more than the shorter distances, forcing the NOE constraints to a closer tolerance than the bond lengths and bond angles. The latter function tends to weight all distances equally, giving better local geometry at the expense of satisfying the NOE constraints.

Chiral Function. We must consider both local chiral centers (e.g., α-carbon atoms) and the global handedness of the entire molecule. The chiral constraints are used only during optimization. The procedure for the local centers is to calculate a signed volume using Eq. (1) for each center identified in the input data and to compare the calculated value with the target value:

$$\mathbf{E}(\text{chiral}) = \sum_{\text{chiral centers}} k''(f_{\text{ch}} - f_{\text{ch}}^*)^2 + \sum_{\substack{\text{chiral center} \\ \text{distances}}} k'(B_{ij}^2 - D_{ij}^2)^2 \tag{7}$$

[47] T. F. Havel, "The Combinatorial Distance Geometry Approach to the Calculation of Molecular Conformation," Ph.D. thesis. Univ. California, 1982. [Available from *Diss. Abstr. Int. B* **43**(8), 2425 (1983), Order No. DA8300523.]

where f^*_{ch} is the target volume, f_{ch} is the corresponding volume calculated from the structure, and k'' is the constant 1 Å^{-6}. This objective function also includes local distance information. A similar function can be used to hold sets of four atoms in a planar configuration. The target volume for such a set is zero.

Practice of Optimization. Before beginning the optimization procedure on a new system it is good practice to examine the global handedness of the embedded structure and determine which global isomer has the lowest initial chiral error. This is done by comparing the magnitude of the chiral error for the initial structure to the error for the same structure passed through a mirror. This approach works well if the two errors differ appreciably.

Once the global handedness of the structure has been chosen, the program begins with optimization of the chiral error function. It proceeds rapidly and always should achieve very low error values if the data are self-consistent. Failure to reach chiral error values less than 1.0 must be looked at carefully. In our hands, it has always signaled a problem in data entry. The optimization of the chiral centers proceeds until one of the criteria for termination has been met.

In the second stage of the optimization procedure, the error function includes the sum of the chiral error function and the full distance error function:

$$E(\text{total}) = E(\text{chiral}) + E(\text{distances}) \qquad (8)$$

There are several features of the optimization routine that merit further discussion. As noted earlier, we make use of the conjugate gradient procedure. It is a variant of the steepest descent method that incorporates a memory function. This function biases the search in the direction of previously successful steps. While it is a pure minimization algorithm, it does allow for automatic restarting if a particular route is unproductive.

There are a number of parameters that control the termination of optimization algorithms. Termination is achieved by one of three conditions: the value of the objective function is reduced below a threshold (FUNC_MIN); second, the magnitude of the gradient is reduced below a threshold (GRAD_MIN); third, a fixed number of iterations has occurred (ITER_LIMIT).

It is often difficult to set these parameters in advance. For NMR problems, the value of the function given in Eq. (8) can, in principle, fall to zero if all the constraints are satisfied. This is rarely achieved with real experimental data, but very low values, for example, <100.0, are possible. These correspond to average errors of less than 0.01 Å and no errors greater than 0.1 Å in the NOE constraints for a protein the size of bovine pancreatic trypsin inhibitor (BPTI), which is about 900 atoms (see Fig. 3). As a rough rule, NOE violations of ~0.1 Å contribute about 1 unit to the value of the error function per constraint violation.

The gradient will behave in a similar fashion, rarely going to zero with a conjugate gradient procedure, but reaching quite low values. If a particular application (such as finding the normal modes) requires reaching a true minimum in the objective function, the problem should be cast in terms of internal angle variables, and more complex optimization procedures (such as the Newton–Raphson method) should be used.[25]

It is a common experience in all minimization that the objective function drops rapidly at first and then proceeds quite slowly and unpredictably toward lower values. Thus it is better to do an initial survey with a limited number of iterations (say 100–200) to make sure that the data set is well behaved. This can be followed with more extensive runs using 500–1000 iterations.

What are the tactics to be used when the optimization proceeds slowly (or not at all) to unsatisfactory "final" values of the objective function? First, if the error values actually increase steadily with iteration number, a significant problem must exist within the program. The only part of the EMBED or VEMBED program that can change simply on being used on a new computer system is the random number generator, which has no connection with the optimization routine. Unless the program has been modified at the local site, there are no input parameters or data that can cause the error function to increase steadily using EMBED.

If the error function decreases but ends at an unsatisfactory value, the first concern should be the data set. It is quite common to have mistakes on data entry, either in the distances themselves or the way in which the bounds are established. Sometimes the atom descriptors are mistyped. These kinds of errors often lead to violations of the triangle inequality. Any such violations should be scrutinized very carefully and corrected. The only circumstances under which a proper holonomic and experimental data set for molecules can lead to violations of the triangle inequality are extreme conformational averaging (see discussion below).

It is rather common to have large values of the error function before optimization, even though the embedded coordinates obey the triangle inequalities. The triangle constraints are relatively mild restrictions. Furthermore, real data can contain a number of inconsistencies that do not cause violations of the triangle checks. These include improper chirality, inconsistent treatment of prochiral centers, incorrect assignments, overly strong interpretation of the NOE data, and impurities in the sample.

Even though the chiral centers are optimized successfully early in the procedure, they can be inverted during later stages. Thus, if the complete

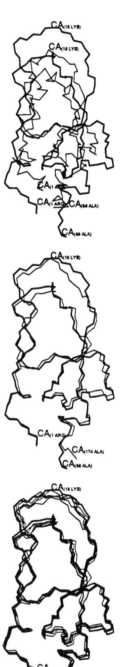

A

B

C

optimization is proceeding poorly, it is useful to check the local handedness of the amino acids and helical features. If any inversions have occurred, it is possible to include another stage of optimization against only the chiral constraints. We note in passing that similar inversions can occur in crystallographic refinement.

It is always possible that some NOE peaks have been misassigned. This can seriously impact the optimization if inconsistent constraints are introduced. Sometimes careful inspection of the residual errors, after optimization, can help identify such problems. It is always helpful to see if there are NOE peaks for all the close distances in the optimized structures, but there are a number of valid reasons why all such peaks need not be seen (see below).

Another reason for poor optimization is misinterpretation of the NOE data. Fundamentally, there are three concerns with the conventional practice of using the NOEs to provide interatomic distances. First, we should be calculating the NMR spectrum directly, that is, we should utilize optimization against the NOE intensities, not some quantity derived from them. There are indications that the isolated spin pair approximation introduces systematic compression that may not be relieved by any of the structure-building techniques now in use. The magnitude of this problem is hard to ascertain at this stage, but it might approach 1 Å, on average. The CORMA program of James and Keepers (see Borgias and James [9], Vol. 176, this series) offers a way to approach this problem.

The neglect of internal motions produces an error of opposite sign. The NOE intensities are reduced by reorientations that are faster than the rigid tumbling time of the protein. It is common, for example, to have no detectable NOEs for residues at either end of the peptide chain or for residues in flexible loops. Lack of NOEs in such cases cannot be reliably interpreted as long proton–proton distances because of the motions.

A third problem is conformational diversity. It is very likely that the amino acid side chains and even the peptide backbone exist in diverse conformations in solution (see, for example, the calculations of Elber and Karplus[48]). There is direct evidence of such diversity in the crystalline

[48] R. Elber and M. Karplus, *Science* **235**, 318 (1987).

FIG. 3. Backbone atom (N, CA, C) stereoimages of distance geometry-generated structures superimposed onto the crystal structure of BPTI (thick line). The distance geometry structures were generated from simulated data. (A) Distance geometry structure after embedding. (B) Same structure as in A after optimization. (C) Five distance geometry structures superimposed onto the crystal structure.

state.[49] To date, there have been only a few examples for which NMR data were inconsistent with a single conformational minimum.[50,51] Jardetzky has consistently warned against overinterpretation of NOE data in this regard and is developing methods that offer an alternative representation of conformation space.[23]

These are all important issues. They cannot be resolved in a very satisfactory manner at this time. The safest course, at present, is to use the NOE data to derive proximity constraints without attempting to convert NOE intensities into high-precision distances. For proteins, the basic structures seem well reproduced simply using a 5-Å upper bound for all hydrogen pairs involved in NOEs or by dividing the NOEs into two or three classes (strong, weak; strong, medium, weak) with the upper bounds given in Table III. Simulations indicate that, if no mistakes are made, these more restrictive sets do improve the structural accuracy. Each investigator must decide how tightly to interpret the experimental data. It is clearly desirable to try alternative assumptions and methods to see which structural features are robust.

Even if the data set is correct and the underlying motional model is self-consistent, any optimization procedure has some chance of finding local minima that have higher residual errors than true global minima.

This situation is a consequence of optimization algorithms that only proceed in a downhill direction. They can get trapped at local minima far from the best solutions. The magnitude of this problem depends in a critical way on the formulation of the constraints and the details of the search procedures. There is no general solution to the local minimum problem now available, although progress is being made.[45,46] The distance geometry error function was selected to be relatively free of such difficulties, and this has proved to be true in practice.

There are four steps the user can take to see if local minima are a serious impediment. First, it is always advisable to repeat the embedding procedure many times (5–50) with the same boundary conditions. Given the random initial distances, each attempt is sensibly independent. If all starts yield approximately the same final error, this error probably reflects the small inconsistencies in the constraints. This observation is based on the high dimensionality of the conformation space, and the resultant unlikelihood of the same function values being achieved simply by chance.

[49] J. L. Smith, W. A. Hendrickson, R. B. Honzatko, and S. Sheriff, *Biochemistry* **25**, 5018 (1986).

[50] J. Lautz, H. Kessler, J. M. Blaney, R. M. Scheek, and W. F. van Gunsteren, in press (1988).

[51] H. Pepermans, D. Tourwe, G. Van Binst, R. Boelens, R. M. Scheek, and W. F. van Gunsteren, *Biopolymers* **27**, 323 (1988).

Second, the use of four-dimensional (4D) optimization significantly speeds convergence, especially for multiple fused rings in cyclic peptides.[52] While the idea of a four-dimensional object is foreign to molecular considerations, it is easily achieved mathematically. One simply retains the fourth eigenvalue after embedding and forms a fourth spatial coordinate (w). Distances in four dimensions are calculated in analogy with three-dimensional distances:

$$E(\text{distance}) = [(\Delta x)^2 + (\Delta y)^2 + (\Delta z)^2 + (\Delta w)^2]^{1/2} \qquad (9)$$

The optimization errors and gradients are readily determined from this formula. The only decision required is how to project the results of such a step back into three dimensions. The simplest way is just to set the 4D coordinates to zero at some point in the optimization and continue optimizing in three dimensions. Alternately, one can "squeeze out" the 4D component over many steps.

A third approach to exploring better solutions is to turn off a subset of the constraints for some period of time and then reinstate them. This is frequently done in optimizing the chiral constraints separately from the NOEs. Other possibilities are to use the upper bounds alone, the lower bounds alone, or the local (close in sequence) versus the long-range constraints. Conversion from Cartesian to angular coordinates, or vice versa, can be helpful. The goal in each case is to get over a barrier in the error function by changing the potential surface.

The last method used to get out of a local minimum is to inspect the coordinates on a graphics system and make manual rearrangements interactively. For example, occasionally it is clear that a loop is entangled and progress cannot be made without atoms interpenetrating. Displacing a few atoms can often solve such problems.

While all these procedures work by redefining the potential surface or by altering the optimization pathway, as long as one is careful to restore the full constraint set for the final stages, there is no mathematical inconsistency. The effort is to achieve the lowest value of the error function. The path by which this is accomplished is irrelevant and has no physical significance.

Parameters for Structure Generation

The procedure for the generation of structures using distance geometry is given in Fig. 1. The decisions the user must make are (1) the number of structures to be generated, (2) saving the bounds, (3) use of the distance correlation routine, and (4) control of optimization.

---

[52] J. Blaney, personal communication (1987).

Number of Structures. Deciding on the number of structures to be generated for any one data set is fairly straightforward. For now we will assume that computer time is not a factor in this decision (see Computational Performance and Requirements). When structures are first generated it is rare that the data set will be entirely correct. The input data set must usually be carefully inspected in light of the initial results. During this process one needs only a limited number of structures (3–5). We do not recommend generating only one structure per run, because it is possible to get atypical structures in individual runs that will become obvious when multiple structures are available. Once the data set is satisfactory, it is up to the user to decide how many structures are necessary to answer the scientific questions. The number of structures required to yield meaningful thermodynamic sampling is impracticably large. In practice, the available computer time limits the number of generated structures, with 10–100 a reasonable range. Most of the published efforts are based on 10 or fewer independent structures.

Saving Bounds. We recommend saving the smoothed bounds if the disk space is available. The bounds-smoothing procedure is computationally intensive and need only be done once per data set. Frequently, there are reasons to continue an optimization after preliminary inspection of the coordinates. If the bounds have been saved, restarting is a simple process of resubmitting the coordinate set with the other input and output files. VEMBED recognizes the presence of a coordinate set on input as a flag to proceed directly to the optimization routine.

Distance Correlation. As noted above, use of the distance correlation routine is not generally advisable since it causes biased sampling. However, if the molecule is not well defined by the data set, and negative eigenvalues are being produced, then this routine is helpful. The structures generated will have improved behavior during embedding but will not sample the conformation space as well. In short, this procedure should not be used if structures can be embedded without it.

Termination of Optimization. The primary parameters that control the termination of the optimization are FUNC_MIN, GRAD_MIN, and ITER_LIM. It is difficult to estimate what values the first two parameters will achieve during optimization. In order to avoid this difficulty, we usually set the FUNC_MIN and GRAD_MIN values to be so small (0.01 and 0.005, respectively) that the iteration limit will determine how long the optimizer will run.

Computational Requirements and Performance

The computational requirements of distance geometry are important to understand prior to serious work with the program. We describe the

TABLE VI
TIMING FOR VEMBED PROGRAM ON COMPUTERS

	Timing (sec) on type of computer		
Procedure	CRAY XMP-48	Convex C1[a]	VAX 8650
Bounds smoothing	200	2400	~32,000
Embedding	4	40	450
Optimization	230	2200	~29,000

[a] Not fully optimized for the Convex.

performance of the VEMBED program, which includes extensive vectorization. The bounds-smoothing routine is the most computationally complex of the main procedures, nominally scaling as N^3, where N is the number of points. Fortunately, this procedure vectorizes easily. Furthermore, it only needs to be run once for each new data set. Because of the cubic dependence of the bounds smoothing, various attempts have been made to alter the algorithm.[42] The time required for the embedding and optimization routines depend quadratically on the number of points and linearly on the number of structures generated. Thus the total time for optimization is on the order of $k \times m \times N^2$, where k is the number of iterations of the optimizer and m is the number of structures. Table VI shows the amount of time the VEMBED program takes on several different machines for a molecule of ~900 atoms. The timing for the optimization is based on 750 iterations of optimization of a single structure of ~900 atoms.

The computational requirements have resulted in several techniques to reduce the size of a particular problem. The simplest idea is the use of united atoms. For example, the pseudoatom representations reduce the number of hydrogens by about one-half. This means that the number of total points it takes to define a molecule is about 75% of all the atom representation. Another method to help reduce the computational complexity of embedding is the use of substructure embedding.[53] This method selects a subset of atoms, usually the backbone atoms and one atom of each side chain. This subset is then embedded. The resultant structures are then analyzed and a few of the best structures (i.e., structures with the lowest errors) are picked. These are used to calculate triangle-smoothed bounds for a full atom set which is reembedded. This technique allows for examination of the conformation space in a reasonably efficient manner.

[53] T. Havel and K. Wüthrich, *Bull. Math. Biol.* **46**, 673 (1984).

However, with the increasing availability of supercomputers, use of the full atom representation is becoming the accepted procedure.

Another issue is the memory requirements of the distance geometry programs. Since the bounds matrix and the distance matrix are both square matrices, the memory scales as N^2. This means that one has to consider the amount and type of memory when running large systems. For systems with large physical memory (50–100 Mbytes), this will not be a problem. However, some machines have hard memory limits, which will then limit the size of the problem that can be done efficiently.

Refinement

A structure produced by distance geometry methods meets all the upper and lower bound conditions deduced from holonomic bond lengths, bond angles, and the NOE constraints rather well. However, such a structure has three major undesirable characteristics that require further calculations. First, the structure is likely to have a very high (unfavorable) potential energy because of the large number of small distortions in geometry and close nonbonded contacts. These are acceptable errors in the distance geometry objective function, but they make large repulsive contributions to the atom pair potential functions. A few dozen iterations of molecular mechanics programs such as AMBER,[54,55] CHARMM,[56] or ECCEP[36] dramatically reduce the energy without much change in the fit to the distance constraints. These programs allow minimization of the energy while restraining the atomic positions close to their initial values. The rule of thumb is that the internal energy in the refined structures should be negative by approximately -5 kcal/residue. Further, there should be very small contributions from the constraint terms, approximately 0.1 to 1.0 kcal/residue when using harmonic constraints of 5 kcal/$\mathring{A}^2$.[31] More sophisticated treatments would include solvent representation and even molecular dynamics simulations. These latter improvements are quite time-consuming calculations that are best done in collaboration with experts in the field.

Second, the distance geometry structures, as well as structures produced by other current methods, meet all the constraints simultaneously. While this may seem to be a desirable trait, recall that the experimental measurements report the average behavior for an ensemble of molecules. Thus, in general, structures which meet all constraints simultaneously

[54] P. K. Weiner and P. A. Kollman, *J. Comput. Chem.* **2**, 287 (1981).
[55] S. J. Weiner, P. A. Kollman, D. A. Case, U. C. Singh, C. Ghio, G. Alagona, S. Profeta, and P. Weiner, *J. Am. Chem. Soc.* **106**, 765 (1984).
[56] B. R. Brooks, R. E. Bruccoleri, B. O. Olafson, D. J. States, S. Swaninathan, and M. Karplus, *J. Comput. Chem.* **4**, 187 (1983).

must sample too narrow a portion of conformation space, because there can always be physically reasonable *distributions* of conformers that have the same average value of a particular property. There is presently no method of refinement that addresses this problem. What is needed is the generation of sets of conformers, properly distributed in energy, whose average behavior is directly constrained. Molecular dynamics would seem to be the most promising technique for this purpose.[57]

Third, neither the individual distance geometry structures nor the average structure have been fit to the actual experimental data. The bulk of the raw data are the intensities of the cross-peaks in the NOE spectra. These intensities are, in turn, complicated functions of the time-averaged distances *and* motions of an ensemble of molecules. A formal treatment of this problem has been presented by Keepers and James. This approach uses the complete relaxation matrix for the nuclear spins and is called CORMA (see Borgias and James [9], Vol. 176, this series).

CORMA serves two possible functions. First, it permits direct calculation of NOE intensities from a coordinate set and a motional model. Thus one can generate a calculated NOE spectrum for any dynamic structure. Carrying this a step further, it is possible to compare the calculated and observed intensities and to use their differences to obtain figures of merit similar to the R factor used in crystallographic analysis. To date this calculation has only been done for a few test cases and an oligonucleotide structure. It offers the most demanding test of quality of the NMR-derived structures now available.

The second use of CORMA is to actually refine a structure. To do this one must calculate the derivatives of the relaxation matrix elements with respect to the structural coordinates. These derivatives can be used to form a gradient that allows a search for an improved set of cross-peak intensities. There are three hurdles that must be overcome before this is a practicable approach. First, the elements of the relaxation matrix emphasize the short distances and are dominated by noise for the long distances. Second, one must develop an explicit motional model. Third, CORMA involves numerically intensive calculations that currently require large amounts of computer time per cycle. The combination of distance geometry and CORMA is being explored as an efficient method producing structures that fit the NOE intensities. Some of these issues are also addressed in a related approach called IRMA, under development by Kaptein and co-workers.[58] This method combines the relaxation terms from a structural model and the experimental data to calculate a new relaxation matrix and a new set of distances. This leads to a new structure and the whole

---

[57] W. van Gunsteren and R. M. Scheek, personal communication (1988).
[58] R. Boelens, T. M. G. Koning, and R. Kaptein, *J. Mol. Struct.* **173**, 299 (1988).

process can be repeated until the structures converge. Initial tests are quite promising.

Analysis of Structures

A serious question is how to assess the quality of the generated structures. The first concern is how well the structures fit the initial constraints. This is easily determined. It is measured directly as a part of the objective function. Current experience is that it should be possible to meet all the NOE constraints with an average violation of 0.1 Å or better and a worse-case violation of less than 1.0 Å (assuming the NOE constraints are simply upper bounds of 5 Å). As noted earlier, each such violation contributes approximately 1 unit to the objective function value of Eq. (8). To help analyze the results, violations can be summed by atom or by residue so that "hot spots" can be identified for further study. It is very helpful to have an interactive real-time graphics display system[59] to inspect the structures and violations in detail.

A second issue is the amount of conformational space allowed by the constraints. Two procedures are available for comparing the average root-mean-square (rms) displacement for the set of structures. Interatomic distances can be directly compared. Coordinates require removing any rigid displacements. A simple numerical factor relates the RMSDs from these two methods.[60] The average displacement among equally "good" structures represents the minimum diversity of conformers that meet the constraint equations. One useful parameter is the standard deviation of the structures at each atom. Typical results for globular proteins with 5–10 NOEs per residue are 1-Å rms deviations for all backbone hydrogens (excluding ill-defined residues at the N and C termini) and 2-Å rms deviations for side-chain hydrogens. Inspection of Table VII suggests that the results do not depend strongly on which computer program is used. Nor do the details of the interpretation of NOE data seem crucial.

At a deeper level, it would be interesting to know what is the conformational space actually permitted by the constraints and how well this space is sampled by the structures obtained. These points are not easy to address. One problem, noted earlier, is that all current methods impose all constraints simultaneously, whereas, in fact, it is only the ensemble average that is actually measured. The real conformation space consistent with the experimental constraints must therefore be substantially larger than that currently being explored. Some steps to examine this situation

[59] R. Langridge, T. E. Ferrin, I. D. Kuntz, and M. L. Connolly, *Science* **211,** 661 (1981).
[60] F. E. Cohen and M. J. E. Sternberg, *J. Mol. Biol.* **138,** 321 (1980).

TABLE VII
REPORTED SOLUTION STRUCTURES DETERMINED BY COMPUTATIONAL METHODS

Molecule	Number of experimental constraints[b]	Program[c]	Number of structures generated	rms displacement[a] (Å)	Energy (kcal/mol)	Reference[d]
α_1-Purothionin	341	DISGEO/RMD	9	2.1 ± 0.3[c,f] 2.3 ± 0.3[c,g]	8040 ± 1200[f] 47 ± 68[g]	1
Barley serine protein-ase inhibitor 2	463	DISGEO/RMD	11	1.6 ± 0.1[e,f] 2.1 ± 0.3[e,g]	7104 ± 755[f] -167 ± 116[g]	2
Bleomycin	~50	EMBED	5	—		3
Carboxypeptidase inhibitor	309	DISGEO/RMD	11	1.7 ± 0.2[c,f] 2.1 ± 0.4[c,g]	3707 ± 1461[f] -107 ± 139[g]	4
Cyclosporin A	58	DGEOM	27	2.4 ± 0.8[h] 2.3 ± 1.0[h]	264 ± 33[i] 48.3 ± 19[i]	5
DNA binding helix F for cAMP receptor protein	87	RMD	3	1.7 ± 0.2[e]	-132 ± 17[j]	6
Enterotoxin	11	DISMAN	10	3.6[e]	—	7
Epidermal growth factor	333	DISMAN	5	1.3	—	8
Glucagon	123	EMBED	10	—		9, 9a
Hirudin	359	DISGEO/RMD	7	1.7 ± 0.2[c,k] 2.6 ± 0.5[c,g]	5507 ± 779[k] -90 ± 96[f,l]	10
Histone H5	317	DISGEO/RMD	4	2.1 ± 0.4[c,k] 4.3 ± 0.7[e,l]	-365 ± 150[l]	11
lac repressor headpiece	186	VEMBED/RMD	10	1.1 ± 0.1[f,h] 1.9 ± 0.6[f,g]	2980 ± 194[f] -671 ± 21[g]	12
Metallothionein 2	54 76	DISMAN	1 1	— —	— —	13

(continued)

TABLE VII (continued)

Molecule	Number of experimental constraints[b]	Program[c]	Number of structures generated	rms displacement[a] (Å)	Energy (kcal/mol)	Reference[d]
Pancreatic trypsin inhibitor	200	DISGEO	5	$1.6 \pm 0.4^{c,f}$ $2.3 \pm 0.3^{c,m}$	—	14
		DISMAN	5	$2.9 \pm 0.7^{c,n}$ $2.4 \pm 0.5^{c,o}$	—	
Phoratoxin	331	DISGEO/RMD	8	$2.1 \pm 0.3^{c,f}$ $2.6 \pm 0.7^{c,g}$	5911 ± 2429^{f} -116 ± 108^{g}	15
Plastocyanin	490	DISGEO	41	$1.2 \pm 0.8^{c,f}$	—	16
Proteinase inhibitor IIA	>202	DISGEO	10	$2.2 \pm 0.3^{f,h}$	—	17
Scorpin insectotoxin I₅A	>100	EMBED	15	2.1 ± 0.3^{f}	—	18
Secretin	52	RMD	12	$1.9 \pm 0.6^{c,p}$ $1.3 \pm 0.4^{c,p}$ $2.6 \pm 0.5^{c,p}$ $1.6 \pm 0.1^{c,p}$	—	19
Somatostatin	29	VEMBED/RMD	92	—	—	20
Tendamistat	619	DISMAN	4	$1.6^{c,n}$	—	21
Adenosine stacks	96	DSPACE	11	—	—	22
DNA decamer 5'd(CTGGATCCAG)₂	160	RMD	2	0.9	—	23
DNA hairpin	240	DSPACE	26	—	—	24
Wobble dG·dT	92	DSPACE	~5	—	—	25
RNA hexamer 5'r(GCAUGC)₂	110	RMD	8	0.8 ± 0.2^{j}	—	26

[a] The number of structures used to determine the root-mean-square displacement (rmsD) is not necessarily equal to the number of structures generated.

[b] Sum of NOE, hydrogen bonds, and torsion angles used in the calculations.

[c] RMD, Restrained molecular dynamics.

[d] References: (1) G. M. Clore, M. Nilges, D. K. Sukumaran, A. T. Brünger, M. Karplus, and A. M. Gronenborn, *EMBO J.* **5,** 2729 (1986); (2) G. M. Clore, A. M. Gronenborn, M. Kjaer, and F. M. Poulsen, *Protein Eng.* 1, 305 (1987); (3) G. M. Crippen, N. J. Oppenheimer, and M. L. Connolly, *Int. J. Pept. Protein Res.* **17,** 156 (1981); (5) J. Lautz, H. Kessler, J. M. Blaney, R. M. Scheek, and W. F. van Gunsteren, in press (1989); (6) G. M. Clore, A. M. Gronenborn, A. T. Brünger, and M. Karplus, *J. Mol. Biol.* **186,** 435 (1985); (7) T. Ohkubo, Y. Kobayashi, Y. Shimonishi, Y. Kyogoku, W. Braun, and N. Gō, *Biopolymers* **25,** S123 (1986); (8) G. T. Montelione, K. Wüthrich, E. C. Nice, A. W. Burgess, and H. A. Scheraga, *Proc. Natl. Acad. Sci. U.S.A.* **84,** 5226 (1987); (9) W. Braun, G. Wider, K. H. Lee, and K. Wüthrich, *J. Mol. Biol.* **169,** 921 (1983); (9a) W. Braun, C. Bösch, L. R. Brown, N. Gō, and K. Wüthrich, *Biochem. Biophys. Acta* **667,** 377 (1981); (10) G. M. Clore, D. K. Sukumaran, M. Nilges, J. Zarbock, and A. M. Gronenborn, *EMBO J.* **6,** 529 (1987); (11) G. M. Clore, A. M. Gronenborn, M. Nilges, D. K. Sukumaran, and J. Zarbock, *EMBO J.* **6,** 1833 (1987); (12) J. DeVlieg, R. M. Scheek, W. F. van Gunsteren, H. J. C. Berendsen, J. Thomason, and R. Kaptein, in press (1988); (13) W. Braun, G. Wagner, E. Worgotter, M. Vasak, J. H. R. Kagi, and K. Wüthrich, *J. Mol. Biol.* **187,** 125 (1986); (14) G. Wagner, W. Braun, T. F. Havel, T. Schauman, N. Gō, and K. Wüthrich, *J. Mol. Biol.* **196,** 611 (1987); (15) G. M. Clore, D. K. Sukumaran, M. Nilges, and A. M. Gronenborn, *Biochemistry* **26,** 1732 (1987); (16) J. M. Moore, O. A. Case, W. J. Chazin, G. P. Gippert, T. F. Havel, R. Powls, and P. E. Wright, *Science* **240,** 314 (1988); (17) M. P. Williamson, T. F. Havel, and K. Wüthrich, *J. Mol. Biol.* **182,** 295 (1985); (18) A. S. Arseniev, V. I. Kondakov, V. N. Maiorov, and V. F. Bystrov, *FEBS Lett.* **165,** 57 (1984); (19) G. M. Clore, M. Nilges, A. Brünger, and A. M. Gronenborn, *Eur. J. Biochem.* **171,** 479 (1988); (20) H. Pepermans *et al.,* *Biopolymers* **27,** 323 (1988); (21) A. D. Kline, W. Braun, and K. Wüthrich, *J. Mol. Biol.* **189,** 377 (1986); (22) D. Hare, L. Shapiro, and D. J. Patel, *Biochemistry* **25,** 7456 (1986); (23) M. Nilges, G. M. Clore, A. M. Gronenborn, N. Piel, and L. W. McLaughlin, *Biochemistry* **26,** 3734 (1987); (24) D. R. Hare and B. R. Reid, *Biochemistry* **25,** 5341 (1986); (25) D. Hare, L. Shapiro, and D. J. Patel, *Biochemistry* **25,** 7445 (1986); (26) C. S. Happ, E. Happ, M. Nilges, A. M. Gronenborn, and G. M. Clore, *Biochemistry* **27,** 1735 (1988).

[e] The rmsD of the backbone atoms N, C, C$_\alpha$.

[f] The rmsD and energy values from the distance geometry structures versus each other.

[g] The rmsD and energy values from the RMD distance geometry structures versus each other.

[h] The rmsD determined from the C$_\alpha$ atoms.

[i] The rmsD and energy values from the RMD structures versus the average RMD of the crystal structure.

[j] The rmsD and energy values from the RMD structures.

[k] The rmsD and energy values from the distance geometry structures versus the average distance geometry structure.

[l] The rmsD and energy values from the RMD distance geometry structures versus the average structure.

[m] The rmsD value from the distance geometry structures versus the X-ray crystal structure.

[n] The rmsD value from the DISMAN structures.

[o] The rmsD value from the DISMAN structures versus the X-ray crystal structure.

[p] The rmsD value from the RMD structures versus the average RMD structure.

are underway.[61] Various methods have been compared for simulated data. At this writing, procedures with truly random (uncorrelated) starting configurations do the best. There does seem to be a small systematic difference between the use of internal angles and internal distances as the fundamental variables. In a test with simple dipeptides the ellipsoid algorithm gave the best sampling of the methods we tried.[62] The distance geometry algorithms sample poorly when the system is poorly constrained.[63] The trade-off is that the total computation time and the attrition rates are worse for methods using angular variables so that some compromise method is worth developing.

More sophisticated statistical analyses are possible to explore correlations between the constraints and the final structures. Intense study probably should await resolution of some of the other issues.

Accuracy of Generated Structures

It is not possible to make quantitative assertions about the accuracy of the structures derived from NMR at this time. We cannot use direct comparison with crystallographic results to prove that the NMR structures are correct because there are possibly genuine differences in structure for the same protein in different environments. Furthermore, there is still concern that details of the motions, assignments, and conformational averaging could, in specific cases, lead to systematic errors in the interpretation of NMR data.

What is possible is to demonstrate that the mathematical procedures reproduce known structures to an accuracy commensurate with the information content of the constraints. These simulation studies[14,64] produce an average structure that fits the known coordinates with errors commensurate with the uncertainty in the constraints. Thus, it appears that the basic mathematical procedures do not introduce significant errors.

It would be interesting to estimate how many constraints are required to provide a structure of a given quality. Although there is no rigorous way to do this, a few simple ideas can be helpful. The most reliable guidelines to relate constraints and quality come from crystallography. The concept of resolution is especially useful. This idea is implicit in all discussions of protein structure. Atoms are 1-Å resolution descriptions of collections of subatomic particles. The "united atom" constructs of molecular mechanics have characteristic radii of 1–2 Å. Amino acid residues

[61] R. M. Scheek, W. F. van Gunsteren, and R. Kaptein, this volume [10].
[62] J. F. Thomason, M. Billeter, and I. D. Kuntz, unpublished data (1986).
[63] A. Pardi and D. Hare, manuscript in preparation (1989).
[64] T. F. Havel and K. Wüthrich, *J. Mol. Biol.* **182**, 281 (1985).

have been represented by single spheres, providing a 3- to 4-Å resolution picture. Finally, secondary features such as helices or β-sheets can be recognized at 5-Å resolution. The number of independent constraints needed to fix a structure at a given resolution increases as R^{-3}, where R is the nominal resolution. The number of constraints is also linear in the size of the object. For crystallographic systems, the size is related to the molecular weight *and* the number of molecules in the unit cell. For NMR, the molecular weight suffices as long as the molecule is monomeric in solution. Based on crystallographic experiments, approximately 1000 measurements (reflections) are needed to solve the structure of a small protein (molecular weight ~5000) to a resolution of 3 Å. Approximately 30,000 reflections would be needed for the same protein at 1 Å.[37]

Another way to estimate the number of constraints required is to count the number of degrees of freedom as a function of the resolution of the object. A protein composed of N atoms has nearly $3N$ degrees of freedom, which can be reduced to about N degrees of freedom if bond lengths and bond angles are taken as fixed. This description fixes the position of each atom and is equivalent to the best protein crystallographic efforts (~1-Å resolution). Since a small protein contains about 1000 atoms, including the hydrogens, the typical X-ray experiment clearly contains a very large excess of experimental data points over degrees of freedom (i.e., 30,000 reflections for 3,000 degrees of freedom). Some of the extra data go to fixing additional parameters such as the N isotropic thermal factors or even the $6N$ anisotropic thermal factors. A rule of thumb is that six measurements are required for each degree of freedom.

Structures at lower resolution have, effectively, fewer degrees of freedom. A residue-level description of the same protein might only require 50×3 parameters. If the only need is to locate the secondary features, perhaps only a few dozen parameters are needed.

Throughout the discussion we have made two assumptions. First, we have assumed that the precision of the measurement was not an important variable. Second, we have tacitly assumed that the measurements are independent. To a first approximation, highly precise measurements do not seem to be essential. We know this both from the protein crystallographic experience where the final fits to the data are never as good as 10% and rarely as good as 20% and from the NMR analyses in which a very qualitative treatment of the NOE constraints still yields respectable results. This counterintuitive result probably comes from the highly correlated nature of high-precision data. Such results do not have as much impact as the same number of bits of information spread throughout the structure. The second issue is the independence of the basic measurements. The crystallographic reflections are genuinely orthogonal. So are coupling constants. However, internal distances derived from NOE experiments have a degree of correlation that is hard to establish in advance of knowing the structure.

To complete the argument, these ideas suggest that the best of the NMR structures obtained on small proteins from ~1000 constraints will be roughly equivalent in quality to 3-Å resolution X-ray structures. If only ~100 constraints are available, the structures would be comparable to 5- to 6-Å resolution.

On a practical level, it is tempting to compare the rms deviations of the NMR structures with the errors in crystallography. The relationship is not a simple one because the spread in coordinates from the NMR studies is not directly related to a physical quantity the way disorder is in crystals. Nevertheless, it is a useful analogy.[37] Of course, one cannot interpret the distance geometry structures in the sense of a statistical sampling of conformation space. They are not distributed in energy as required by the Boltzmann equation and do not necessarily reflect the best fit to the underlying physical phenomena. Thus it is not true that the NMR structures span the allowed thermal or conformational motion. Qualitatively, the constraints must give more defined structures for the central features of proteins than for surface side chains or loops, so it is reasonable that there is some empirical relation between atomic rms displacements and the thermal factors from crystallography.

Results

Solution structures for a variety of proteins and nucleic acids are arranged in alphabetical order in Table VII. The most important aspect of Table VII is that the NMR experiments clearly contain enough information to provide useful structures with random errors (rms displacement), typically 1–2 Å for backbone atoms and 2–3 Å for side-chain atoms. The quality of the final structures in fitting the NMR constraints does not seem to depend strongly on which mathematical procedure was used to generate the structures, although this point cannot be explored in depth until several methods are compared for a specific data set. It is also encouraging that the methods seem to handle a wide variety of protein tertiary structure types equally well. The only obvious features that are reported with large uncertainties are the N and C termini of the polypeptide chains and occasionally individual loops.

If we turn our attention to the internal energies of the structures, the distance geometry methods and the DISMAN procedure do not yield low-energy conformations. This result was anticipated. In most cases, the energy is dramatically reduced when these conformers are subjected to restrained molecular dynamics and/or energy minimization. It is usually possible to achieve a negative potential energy for conformers that meet essentially all the NMR constraints.

To date, none of the protein structures has been refined to fit the NOE intensities. In fact, no group has yet reported any direct comparison of calculated versus observed NOE intensities except that of James *et al.*[65] Such comparisons are expected to be common in the near future.

Another issue is how well the structure-generating programs sample the allowed conformation space. While a definitive test has not yet been performed, there are some indications in the entry for pancreatic trypsin inhibitor that the DISMAN program samples more widely (larger rms displacement) than does DISGEO. One word of caution is that the comparison must be made at the same value of the objective function. It is not clear that this condition has been met for any studies done to date. Interestingly, the deviations of the DISMAN and DISGEO structures from that of the crystal structure are not much different. There is some evidence from studies of the *lac* repressor headpiece and cyclosporin studies that restrained molecular dynamics samples the conformational vicinity of the distance geometry structures in an efficient manner.

A few trends are clear from the data of Table VII. First, the number of constraints reported per structure seems to be increasing. Second, while the early work often stopped at the distance geometry level, it is now customary to carry out some level of energy refinement. These trends reflect the improvements in NMR experiments—especially the increased sensitivity of new probes and the ability to collect coupling-constant data.[33] Also apparent is the improved access to computational tools, particularly the increasing use of the regional supercomputing facilities.

The primary focus of this chapter has been toward the modeling of proteins, however, we do show information pertaining to the generation of DNA structures. The use of distance geometry to generate DNA structures is of particular interest because modeling such structures is more difficult than it appears to be for proteins. There are several structural motifs, which are implicit to a DNA structure, that the data set needs to describe. Such constraints include planarity of the bases, proper orientation of the base-pairing partners, stacking relationships of the bases, and information pertaining to the pitch of the helix. The first three can be described through the use of holonomic distance and chirality/planarity constraints. However, it is the pitch of the helix that is the most difficult to describe within the data set. Information about the pitch of the helix

---

[65] T. L. James *et al.*, *in* "NMR Spectroscopy and Drug Development" (J. W. Jaroszewski, K. Schaumburg, and H. Kofod, eds.). Munksgaard, Copenhagen, 1987.

cannot be determined from the NMR information, and the holonomic information requires specific assumptions, which may then improperly bias the structure. In many cases the DNA structures generated by distance geometry methods appear to be underwound.

Reporting Structural Data

The NMR community has not yet decided upon a standard presentation of their results. One possibility is to follow the format used by the crystallographers, reporting the coordinates averaged over the replica runs, with the uncertainty given as a separate error reminiscent of the B factor in crystallographic results. It would also be extremely valuable for the NMR spectroscoptists to report and archive the NOE and other data that went into the structural constraints. Only in this way could we easily adhere to the scientific dictum that results should be reproducible by other laboratories.

Another problem is atom-naming conventions. There is presently no convention adhered to within (or outside) the NMR community for the naming of hydrogen atoms in proteins and nucleic acids. The pseudoatom representations only complicate the situation further.

Some form of standard nomenclature must be put into practice for archival purposes. The naming convention that we presently use is given in Table VIII. It allows stereospecific assignments to be made because each atom is uniquely named. It also allows for easy mapping of the names into the naming conventions used for empirical energy methods.

Conclusions

In spite of the complicated issues raised above, it is quite clear that the mathematical procedures for developing coordinates are soundly based. Real NMR data, interpreted conservatively, provide plausible structures with considerable detail. The best guess is that the average random errors in these structures are roughly measured by their rms deviations among the structures found in replicate runs. While it is difficult to determine systematic errors, the rather close correspondence of X-ray and NMR structures of rigid proteins such as BPTI and tendamistat (Table VII) suggests that the systematic errors for proteins are not much worse than the random errors. The same may not be true for nucleic acid structures. These molecules have lower hydrogen densities than do proteins, leading to fewer NOEs. They are frequently more elongated, and the long distances in such systems may be significantly distorted.

TABLE VIII
ATOM NAMES USED AT UCSF

Backbone atoms (except Gly, Pro)	Residue: Asp (continued)
01: C	10: CG
02: O	11: OD1
03: N	12: OD2
04: HN	13: HD2
05: CA	Residue: Cys (oxidized)
06: HA	07: CB
Residue: N terminus	08: HB1
01: H	09: HB2
02: N	10: SG
03: HN	Residue: Cys (reduced)
Residue: Ala	07: CB
07: CB	08: HB1
08: HB1	09: HB2
09: HB2	10: SG
10: HB3	11: HG
Residue: Arg	Residue: Gln
07: CB	07: CB
08: HB1	08: HB1
09: HB2	09: HB2
10: CG	10: CG
11: HG1	11: HG1
12: HG2	12: HG2
13: CD	13: CD
14: HD1	14: OE1
15: HD2	15: NE2
16: NE	16: HE1
17: HE	17: HE2
18: CZ	Residue: Glu
19: NH1	07: CB
20: HH1	08: HB1
21: NH2	09: HB2
22: HH2	10: CG
23: HH3	11: HG1
Residue: Asn	12: HG2
07: CB	13: CD
08: HB1	14: OE1
09: HB2	15: OE2
10: CG	16: HE2
11: OD1	Residue: Gly
12: ND2	01: C
13: HD1	02: O
14: HD2	03: N
Residue: Asp	04: HN
07: CB	05: CA
08: HB1	06: HA1
09: HB2	07: HA2

(continued)

TABLE VIII (*continued*)

Residue: His	Residue: Lys (*continued*)
07: CB	14: HD1
08: HB1	15: HD2
09: HB2	16: CE
10: CG	17: HE1
11: ND1	18: HE2
12: CD2	19: NZ
13: HD1	20: HZ1
14: CE1	21: HZ2
15: NE2	Residue: Met
16: HD2	07: CB
17: HE1	08: HB1
Residue: Ile	09: HB2
07: CB	10: CG
08: HB	11: HG1
09: CG2	12: HG2
10: HG1	13: SD
11: HG2	14: CE
12: HG3	15: HE1
13: CG1	16: HE2
14: HG4	17: HE3
15: HG5	Residue: Phe
16: CD1	07: CB
17: HD1	08: HB1
18: HD2	09: HB2
19: HD3	10: CG
Residue: Leu	11: CD1
07: CB	12: CD2
08: HB1	13: HD1
09: HB2	14: CE1
10: CG	15: CE2
11: HG	16: HD2
12: CD1	17: HE1
13: HD1	18: CZ
14: HD2	19: HE2
15: HD3	20: HZ
16: CD2	Residue: Pro
17: HD4	01: C
18: HD5	02: O
19: HD6	03: N
Residue: Lys	04: CD
07: CB	05: CA
08: HB1	06: HA
09: HB2	07: CB
10: CG	08: HB1
11: HG1	09: HB2
12: HG2	10: CG
13: CD	11: HG1

TABLE VIII (*continued*)

Residue: Pro (*continued*)	Residue: Trp (*continued*)
12: HG2	23: HZ2
13: HD1	24: HH2
14: HD2	Residue: Tyr
Residue: Ser	07: CB
07: CB	08: HB1
08: HB1	09: HB2
09: HB2	10: CG
10: OG	11: CD1
11: HG	12: CD2
Residue: Thr	13: HD1
07: CB	14: CE1
08: HB1	15: CE2
09: OG1	16: HD2
10: HG1	17: HE1
11: CG2	18: CZ
12: HG2	19: HE2
13: HG3	20: OH
14: HG4	21: HH
Residue: Trp	Residue: Val
07: CB	07: CB
08: HB1	08: HB1
09: HB2	09: CG1
10: CG	10: HG1
11: CD1	11: HG2
12: CD2	12: HG3
13: CE3	13: CG2
14: CE2	14: HG4
15: NE1	15: HG5
16: HD1	16: HG6
17: HE3	Residue: C terminal
18: CZ3	01: OCOH
19: CZ2	01: C
20: HE1	02: O
21: HZ3	03: O
22: CH2	04: H

How far the NMR experiment can be extended to include finer structural details and motional models remains to be seen. It will be most interesting to learn if subtle structural modifications occur as a function of temperature, pH, or solvent. In those cases where some differences have been proposed for crystal and solution structures, decisive tests must be devised. These may well involve biological, chemical, and spectroscopic assays to probe the putative rearrangements.

It is an exciting time for structural biology to have NMR and crystal methods available. Two powerful approaches, both with sensitivity to the atomic level, will certainly provide important new insights into the structure and function of biomacromolecules.

Acknowledgment

We are grateful to our friends and colleagues who read this manuscript. Special thanks go to Dr. Charles Eads and George Seibel for their comments. This work was supported through research grants from the NIH and NSF. Funds for the UCSF Magnetic Resonance Center came from the Research Resources Division of the NIH; DARPA and the NSF and NIH shared-instrument grants provided the NMR facilities and some of the computational facilities used in this project. The UCSF Computer Graphics Laboratory (R. Langridge, PI) was also most helpful. Finally, a joint study with IBM is gratefully acknowledged.

[10] Molecular Dynamics Simulation Techniques for Determination of Molecular Structures from Nuclear Magnetic Resonance Data

By R. M. SCHEEK, W. F. VAN GUNSTEREN, and R. KAPTEIN

Introduction

With the introduction of two-dimensional NMR techniques,[1] nuclear Overhauser enhancements (NOEs) became measurable for many proton pairs in a single experiment.[2] This NOE effect can be observed when two protons are close together (<0.5 nm) and it provides by far the richest available source of structural information for dissolved molecules.[3] Today the measurements are practical for molecules with molecular weights up to 15,000. For a protein such as the bovine pancreatic trypsin inhibitor, several hundreds of NOEs can be measured and, under certain assumptions, each NOE can be translated into an upper bound on the distance between the protons involved. The interpretation of this type of information in terms of three-dimensional structures is not a trivial problem. Much uncertainty and even scepticism remained for several years about

[1] R. R. Ernst, G. Bodenhausen, and A. Wokaun, "Principles of Magnetic Resonance in One and Two Dimensions." Oxford Univ. Press (Clarendon), London, 1987.
[2] S. Macura and R. R. Ernst, *Mol. Phys.* **41,** 95 (1980); see also A. Kumar, R. R. Ernst, and K. Wüthrich, *Biochem. Biophys. Res. Commun.* **95,** 1 (1980).
[3] K. Wüthrich, "NMR of Proteins and Nucleic Acids." Wiley, New York, 1986.

the quality of the molecular structures that could be derived from NMR data.[4] This situation is now changing, because several techniques have been tested and among these the technique of molecular dynamics (MD) was chosen by several groups, either alone or in combination with other approaches, such as distance geometry (DG).

We shall start with a short description of the MD technique and explain how the force field can be modified in order to bring or keep protons close together when a NOE between them has been measured. Next we shall describe some protocols that lead from a set of NOEs to a structure in which not only all these NOEs are accounted for but which also has a favorable energy in the force field chosen. We shall illustrate the weak and strong points of the MD technique when used for this purpose. Finally, we shall show how a simplified MD algorithm can be used to estimate the accuracy of a structure, as determined by NMR.

Molecular Dynamics: The Technique

Each MD simulation starts with some starting set of coordinates. There are several ways to arrive at such a starting structure, as we shall discuss later. What we need next is a force field. Here the available programs (GROMOS,[5] CHARMM,[6] and AMBER[7] are the most popular ones) differ in subtle details. Usually the force field has the following contributions:

$$V = V_{\text{van der Waals}} + V_{\text{coulomb}} + V_{\text{dihedral}} + V_{\text{torsion}} + V_{\text{bond}} + V_{\text{angle}}$$

which in GROMOS is as follows:

$$V_{\text{van der Waals}} = \sum_{\text{pairs}(ij)} [C_{12}/r_{ij}^{12} - C_6/r_{ij}^6]$$

$$V_{\text{coulomb}} = \sum_{\text{pairs}(ij)} q_i q_j / 4\pi\varepsilon_0 \varepsilon_r r_{ij}$$

[4] O. Jardetzky, *Biochim. Biophys. Acta* **621,** 227 (1980).

[5] The GROMOS (Groningen Molecular Simulation) software package (by W. F. van Gunsteren and H. J. C. Berendsen) is available from BIOMOS b.v., Nijenborgh 16, 9747 AG Groningen, The Netherlands.

[6] B. R. Brooks and M. Karplus, *J. Chem. Phys.* **79,** 6312 (1983); see also B. R. Brooks, R. E. Bruccoleri, B. D. Olafson, D. J. States, S. Swaminathan, and M. Karplus, *J. Comput. Chem.* **4,** 187 (1983).

[7] U. C. Singh, P. K. Weiner, J. W. Caldwell, and P. A. Kollman, AMBER (UCSF) version 3.0, Department of Pharmaceutical Chemistry, University of California, San Francisco, 1985.

$$V_{\text{dihedral}} = \sum_{\text{dihedrals}} K_\varphi[1 + \cos(n\varphi - \delta)]$$

$$V_{\text{torsion}} = \sum_{\text{torsions}} \tfrac{1}{2}K_\xi(\xi - \xi_0)^2$$

$$V_{\text{bond}} = \sum_{\text{bonds}} \tfrac{1}{2}K_b(b - b_0)^2$$

$$V_{\text{angle}} = \sum_{\text{angles}} \tfrac{1}{2}K_\vartheta(\vartheta - \vartheta_0)^2$$

The last three terms are harmonic potentials that constrain bond angles (ϑ), bond lengths (b), and certain torsion angles (ξ) to their ideal values, indicated with the subscript 0. In the first term we recognize the Lennard–Jones potential, with its repulsive and attractive terms. The second term describes the coulombic interactions between two charged particles (i, j) with partial charges q that are a distance r_{ij} apart in a dielectricum described by $\varepsilon_0\varepsilon_r$. The term V_{dihedral} models the energy differences between staggered and eclipsed conformations of atoms with respect to a rotatable bond.

The derivatives of this potential with respect to the coordinates **r** yield the forces

$$\mathbf{F}_i = -\partial/\partial\mathbf{r}_i V(\mathbf{r}_i, \mathbf{r}_j, \ldots, \mathbf{r}_N)$$

Initial velocities $\mathbf{v}_i$ for all atoms are taken from a Maxwellian distribution corresponding to the desired temperature. Rotation around and translation of the center of mass are removed from the system. During the simulation, the temperature T is held constant by scaling the velocities **v** after each step Δt (coupling to a bath of constant temperature T_0)[8]:

$$\mathbf{v}_i \leftarrow \lambda\mathbf{v}_i$$
$$\lambda = [1 + (T_0/T - 1)\Delta t/\tau_T]^{1/2}$$

where τ_T is the time constant for the coupling.

$$T = 2E_{\text{kin}}/(3N - 6)k$$

where k is the Boltzmann constant and N is the total number of atoms. E_{kin} is the total kinetic energy given by

[8] H. J. C. Berendsen, J. P. M. Postma, W. F. van Gunsteren, A. DiNola, and J. R. Haak, *J. Chem. Phys.* **81**, 3684 (1984).

$$E_{kin} = \tfrac{1}{2} \sum_{i=1}^{N} m_i v_i^2$$

with m_i standing for the atomic masses.

Newton's equations of motion are solved by integration over very small time steps Δt (in which the forces can be regarded constant) using a leap-frog algorithm[9]:

$$v_i(t + \Delta t/2) = v_i(t - \Delta t/2) + m_i^{-1} F_i(t) \Delta t$$
$$r_i(t + \Delta t) = r_i(t) + v_i(t + \Delta t/2) \Delta t$$

These are all the equations needed to perform a MD simulation *in vacuo*, using parameters that are available from the literature.

To save computer time the time step Δt must be much smaller than the inverse of the highest frequency that occurs in a molecular system, i.e., bond stretching. Fortunately, bond lengths can be constrained to their mean values (e.g., using an algorithm called SHAKE[10]) without significantly affecting the overall dynamics. This allows us to integrate the equations in time steps of 0.002 psec.

Instead of evaluating the pairwise interactions for all atom pairs, one can define a radius around each atom and only take into account atoms within this radius. Pair lists for each atom need to be updated only after every 10 steps.

Molecular Dynamics and NMR: Restrained Molecular Dynamics[11]

The detection of a NOE between two protons is usually assumed to define an upper bound on the distance between them. The value of this upper bound is determined from the intensity of the NOE cross-peak (or, more precisely, the initial buildup rate of the NOE) relative to that of a reference NOE between two protons that are a known distance apart. This procedure implies several assumptions (cf. Borgias and James [9], Vol. 176, this series) and usually the upper bound is set to a somewhat higher value to account for uncertainties, both in the intensity measurement and in the assumed model for the cross-relaxation. Now suppose we wish to modify a given starting structure in such a way that all protons (i, j) between which a NOE has been measured are brought within an

[9] R. W. Hockney and J. W. Eastwood, "Computer Simulation Using Particles." McGraw-Hill, New York, 1981.

[10] J. P. Ryckaert, G. Ciccotti, and H. J. C. Berendsen, *J. Comput. Chem.* **23**, 327 (1977).

[11] W. F. van Gunsteren, R. Kaptein, and E. R. P. Zuiderweg, *in* "Nucleic Acid Conformation and Dynamics" (W. K. Olson, ed.). Report of the NATO/CECAM Workshop, Orsay, 1983.

upper bound u_{ij}, but not closer than a lower bound l_{ij}, which in many cases will equal the sum of the atomic radii of the two protons. One way this can be achieved is by defining a potential as follows:

$$V^{dc} = K^{dc}(d_{ij} - u_{ij})^2 \quad \text{for } d_{ij} > u_{ij}$$
$$= K^{dc}(l_{ij} - d_{ij})^2 \quad \text{for } d_{ij} < l_{ij}$$
$$= 0 \text{ otherwise}$$

where d_{ij} stands for the distance $|\mathbf{r}_i - \mathbf{r}_j|$; now simply add this term to the other terms of the force field. When large violations of the bounds are to be expected (this depends on the quality of the starting structure), it is advisable to limit the constraining forces to a maximum value to avoid them dominating the force field. The important question arises what weight we shall give to the constraining potentials relative to the other terms in the force field. If we choose $K^{dc} = 125$ kJ mol^{-1} nm^{-2}, a violation of 0.1 nm will add $kT/2$ to the total potential energy at 300 K. Because the average kinetic energy of an atom per degree of freedom is $kT/2$, this implies that during the simulation violations of the order of 0.1 nm are tolerated. An advantage of such a low value for the force constant is that the molecule will have more freedom to find low-energy conformations during the simulation.

Some MD programs do not simulate explicitly all protons but use the concept of united atoms. Thus, in GROMOS, all carbon-bound protons are omitted from the calculations and a slightly larger carbon atom is used instead. This has consequences for the constraining potential described above, which needs to be referenced to just these protons. Another problem arises when it can not be deduced from the NMR spectra exactly to which proton the constraining force must be referenced. Such situations occur when two protons (or two groups of equivalent protons) cannot be individually assigned in the spectra (this often occurs for methylene protons and for the methyl groups of Leu and Val) or when they undergo rapid dynamic processes (methyl rotation, ring flips). This technical problem is solved by introducing pseudopositions on which the constraining forces act.[12,13] If such a pseudoposition does not coincide with a real proton position, a correction factor is added to the upper bound and subtracted from the lower bound, which equals the maximum possible distance between the actual proton position and the pseudoposition on

[12] K. Wüthrich, M. Billeter, and W. Braun, J. Mol. Biol. **169**, 949 (1983).
[13] W. F. van Gunsteren, R. Boelens, R. Kaptein, R. M. Scheek, and E. R. P. Zuiderweg, in "Molecular Dynamics and Protein Structure" (J. Hermans, ed.), p. 92. Polycrystal Book Service, Western Springs, Illinois, 1986.

which the constraining force acts. Thus, such a pseudoposition can be defined between two methyl groups (one site at the averaged position of the six methyl protons; correction factor, 0.22 nm). Their positions r_{ps} are expressed as a function of the coordinates of the "real" atoms:

$$r_{ps} = f(r_i, r_j, \ldots)$$

so that the effect of the constraining forces on those real atoms can be found by applying the chain rule as follows:

$$F_i^{dc} = -(\partial V^{dc}/\partial r_{ps})(\partial r_{ps}/\partial r_i)$$

NOEs involving the δ and ε protons of rapidly flipping tyrosine or phenylalanine rings are referenced to the Cγ and Cζ protons, respectively (correction factor, 0.21 nm).

The J coupling data may yield a unique value or an allowed range for a dihedral angle and provide an additional type of restraint.[14] This may take the form of a harmonic potential similar to $V_{torsion}$ (see above).

The high-frequency motions that are simulated during a MD simulation, with inverse frequencies less than 1 nsec, cause the vector linking two protons to vary, both in amplitude and in orientation. To evaluate NOE intensities from such a MD trajectory, it is necessary to average the proton–proton distances in the proper way.[15] As demonstrated by Kessler et al.,[16] the resulting NOE is proportional to r_{eff}^{-6}, where $r_{eff} = \langle r^{-3} \rangle^{-1/3}$.

The ^3J coupling constants are evaluated from a MD trajectory using a Karplus relation as follows:

$$^3J_{KLMN} = \langle A \cos^2 \vartheta + B \cos \vartheta + C \rangle$$

where ϑ is the dihedral angle between atoms K and N.

Protocols

Now that we have introduced the machinery, we shall critically examine some of the ways it was used for the purpose of finding structures that are in agreement with experimental constraints (NOEs, J couplings) and at the same time have low energies in a given force field.

[14] J. de Vlieg, R. Boelens, R. M. Scheek, R. Kaptein, and W. F. van Gunsteren, *Isr. J. Chem.* **27**, 181 (1986).
[15] J. Tropp, *J. Chem. Phys.* **72**, 6035 (1980).
[16] H. Kessler, C. Griesinger, J. Lautz, A. Müller, W. F. van Gunsteren, and H. J. C. Berendsen, *J. Am. Chem. Soc.* **110**, 3393 (1988).

Restrained MD (Protocol 1)

Kaptein *et al.*[17] started to build a plausible model of the *lac* repressor headpiece in which the majority of a set of 169 NOEs[18] was already accounted for. A restrained MD run was started subsequently to remove the remaining violations, but only after several "manual" interventions (on a graphics screen), after the structure got trapped in local minima, could a good solution be obtained. A slightly different approach was chosen by Clore *et al.* They showed, using both synthetic data calculated from the crystal coordinates of crambin[19] and experimental data,[20] that in favorable cases one can start even with a stretched polypeptide chain and use distance constraints to fold it into a plausible structure. A drawback of these approaches, however, apart from the obvious problem of local minima in the energy function, is that most of the computer time is spent in the evaluation of the molecule's energy to calculate the interatomic forces for structures that are still far from a good solution that satisfies the boundary conditions. It was felt that a more efficient method was needed to generate starting coordinates that already satisfy these constraints and that restrained MD should be used only to search for low-energy conformations within the allowed parts of conformation space (see below).

Nevertheless, the approach was definitely successful in several cases. Cyclosporin A, an 11-amino-acid cyclic peptide that has been extensively studied by Kessler's group, was "dynamically modeled" into a perfectly plausible conformation.[21] No violations of the constraints were introduced during a subsequent unrestrained MD simulation, in which the solvent molecules were simulated as well. Similarly, antamanide,[16] another cyclic peptide studied in the same group, could be modeled such as to satisfy all the NOEs. Here, however, a problem was met. Although the NOEs could be accounted for satisfactorily in one "dynamic" structure, a conflict arose with the *J* coupling data that were available. The case could be resolved only by introducing an equilibrium between different conformations, and a good fit with all available data was obtained when a structure similar to the crystal structure was postulated to participate in

[17] R. Kaptein, E. R. P. Zuiderweg, R. M. Scheek, R. Boelens, and W. F. van Gunsteren, *J. Mol. Biol.* **182,** 179 (1985).

[18] E. R. P. Zuiderweg, R. M. Scheek, and R. Kaptein, *Biopolymers* **24,** 2257 (1985).

[19] M. G. Clore, A. T. Brünger, M. Karplus, and A. M. Gronenborn, *J. Mol. Biol.* **191,** 523 (1986); see also A. T. Brünger, M. G. Clore, A. M. Gronenborn, and M. Karplus, *Proc. Natl. Acad. Sci. U.S.A.* **83,** 3801 (1986).

[20] M. G. Clore, A. M. Gronenborn, A. T. Brünger, and M. Karplus, *J. Mol. Biol.* **186,** 435 (1985).

[21] J. Lautz, H. Kessler, R. Kaptein, and W. F. van Gunsteren, *J. Comput. Aided Mol. Design* **1,** 219 (1987).

the equilibrium. Similarly, Pepermans *et al.*[22] found evidence for the occurrence of four families of structures in the case of a somatostatin analog, but their procedure includes an initial distance geometry step, as in the next protocol.

Criticism. Once a "good" structure is found following this protocol, the question is legitimate whether that solution is unique. In other words, to what degree does it depend on the starting structure that was chosen? Clearly the only way to answer this question is by repeating the procedure a number of times with different starting structures, preferably random (random in the sense that the covalent structure is intact, but with random values taken for the dihedral angles). Here the main drawback of the protocol is felt: restrained MD is not the most efficient method to search a large portion of conformation space for different structures that satisfy experimental constraints.

Another disadvantage of this approach is that structural information from experiments is mixed up in an early stage with structural information that is implicit in the force field parameters used. Many workers prefer a clear-cut separation between experiment and theory, even more so since there is no consensus about which force field is best. The next protocols will solve, or at least alleviate, these drawbacks.

First Distance Geometry, Then Restrained MD (Protocol 2)

This protocol implies two steps. In the first step, one of the presently available distance geometry (DG) algorithms is used to generate a large number of structures consistent with the observed NOEs (and possibly other experimental information on interatomic distances). In the second step restrained MD is used to minimize the internal energies of the resulting structures. Here MD is superior to other minimization techniques. It generates structures far lower in energy than do techniques which simply follow gradients (steepest descents, conjugate gradients, and Newton-Raphson, to mention a few). This is due to the simulation of kinetic energy in the system, which enables atoms to overcome barriers of the order of kT so that the algorithm searches a larger part of conformation space than do the energy minimizers.

The terminology "distance geometry" used to be reserved for the metric matrix embedding algorithm developed by Havel, Kuntz, and Crippen[23] (the EMBED program) and recently implemented in the

[22] H. Pepermans, D. Tourwe, G. van Binst, R. Boelens, R. M. Scheek, W. F. van Gunsteren, and R. Kaptein, *Biopolymers* **27**, 323 (1988).
[23] T. F. Havel, I. D. Kuntz, and G. M. Crippen, *Bull. Math. Biol.* **45**, 665 (1983).

program DISGEO.[24] Now the term is used for several other algorithms that serve the same goal: to generate coordinates for a set of points (molecular conformations) using only information on distances between those points (atoms) (cf. Kuntz *et al.* [9], this volume). These other algorithms have in common that an error function is set up (comparable to the constraining potential used in restrained MD). Starting with a random conformation this error function is subsequently minimized. The procedure is repeated a number of times with different starting structures to result in a cluster of structures that satisfy the distance constraints.

The program DISMAN[25] tackles the problem of local minima in the error function hypersurface by adapting the range of the error function in a clever way: initially only constraints that involve protons on nearby positions in the covalent chain are worked on, and subsequently the range is extended until all constraints are active. For reasons of efficiency, dihedrals instead of the Cartesian coordinates are taken as variables. A conventional conjugate gradient method is used for the actual minimization.

The ellipsoid program[26] follows a somewhat different strategy: only one constraint is worked on at each step and a very efficient global optimizer (the ellipsoid algorithm) is used for the optimization. Again, dihedrals are chosen as the variables. A comparison of these and other DG algorithms, however, is beyond the scope of this chapter (but see Kuntz *et al.* [9], this volume).

de Vlieg *et al.*[27] tested this protocol with the set of 169 NOEs measured for the *lac* repressor headpiece, augmented with 17 "hydrogen bond constraints" to model the three helices. The location of these helices was inferred from a qualitative interpretation of short- and medium-range NOEs and amide–proton exchange with water. Ten structures were generated using a distance geometry program that was developed from Crippen's original EMBED program. Each of these 10 structures was then used as a starting structure for a restrained MD simulation. A short and simple protocol was used: first, 300 steps of restrained energy minimization to reduce the strain in the molecules, then 5 psec (2500 steps) of MD, including the restraining potential described above with $K^{dc} = 250$ kJ mol^{-1} nm^{-2}. During the next 5 psec the force constant was gradually increased to a final value of 4000 kJ mol^{-1} nm^{-2} after 10 psec of MD. Finally, the kinetic energy was removed from the system by another 300

[24] T. F. Havel and K. Wüthrich, *Bull. Math. Biol.* **46**, 673 (1984).

[25] W. Braun and N. Gō, *J. Mol. Biol.* **186**, 611 (1985).

[26] M. Billeter, T. F. Havel, and K. Wüthrich, *J. Comput. Chem.* **8**, 132 (1987).

[27] J. de Vlieg, R. M. Scheek, W. F. van Gunsteren, H. J. C. Berendsen, R. Kaptein, and J. Thomason, *Proteins* **3**, 209 (1988).

TABLE I
AVERAGE rms POSITIONAL DIFFERENCES
BETWEEN EIGHT *lac* REPRESSOR
HEADPIECE STRUCTURES

Cluster[a]	rms (nm) ($C\alpha$ 1–51)	rms (nm) ($C\alpha$ in helices)
After DG	0.14	0.10
After MD	0.26	0.11

[a] Eight structures are compared after superposition of the $C\alpha$ atoms. Cluster DG was generated by a DG algorithm. It was converted in cluster MD by a restrained MD protocol (two times 5 psec following protocol 2; see the text).

steps of restrained energy minimization, with the same high value of the force constant. Of the 10 starting structures, 8 reached low energies after the refinement step without significant violations (<0.02 nm) of the distance constraints. Somewhat surprising, however, was the observation that the range of conformations after MD refinement is significantly larger than before (see Table I). This is most pronounced in the termini and in the loop region between the second and third helices, which are not well defined by the available NOEs. Similar observations were made by Clore *et al.*,[28] who applied a very similar protocol to a variety of small proteins; by Cooke *et al.*,[29] who studied the epidermal growth factor (EGF); and by Fesik *et al.*,[30] who modeled a cyclic peptide. In all these cases poorly defined parts of the molecules were seen to adopt a wider range of conformations after the MD refinement step than was generated by the DG algorithm. In all cases, however, the final structures were acceptable by the two criteria: no significant violations of the distance constraints, and low energies in the force field that was chosen.

Criticism. The spread among structures that are generated by a DG algorithm is usually seen to increase significantly during a restrained MD

[28] M. G. Clore, D. K. Sukumaran, M. Nilges, and A. M. Gronenborn, *Biochemistry* **26**, 1732 (1987); M. G. Clore, D. K. Sukumaran, M. Nilges, J. Zarbock, and A. M. Gronenborn, *EMBO J.* **6**, 529 (1987); M. G. Clore, A. M. Gronenborn, M. Nilges, D. K. Sukumaran, and J. Zarbock, *EMBO J.* **6**, 1833 (1987); and M. G. Clore, M. Nilges, D. K. Sukumaran, A. T. Brünger, M. Karplus, and A. M. Gronenborn, *EMBO J.* **5**, 2729 (1986).

[29] R. M. Cooke, A. J. Wilkinson, M. Baron, A. Pastora, M. J. Tappin, I. D. Campbell, H. Gregory, and B. Sheard, *Nature (London)* **327**, 339 (1987).

[30] S. W. Fesik, G. Bolis, H. L. Sham, and E. T. Olejniczak, *Biochemistry* **26**, 1851 (1987).

refinement step. Therefore, we must conclude that the DG algorithms that were used do not sample the conformation space very exhaustively. The next protocol was designed to remedy this problem.

Furthermore, when the MD simulations are carried out without including the effects of solvent (*in vacuo*), as is usually the case, this may cause amino acid side chains at the surface, and in the case of the *lac* repressor headpiece even the complete C-terminal region, to collapse onto the core of the protein, thus biasing the sampling behavior of the protocol. Other artifacts of *in vacuo* simulations are expected, affecting, e.g., hydrogen bond formation and breaking and frequencies and amplitudes of motions involving solvent-accessible parts of the molecule. This problem is felt most seriously with MD simulations of solvent-accessible molecules such as small peptides and nucleic acids. Presently no consensus exists on how solvation effects are best simulated.

First DG, Then DDD, Followed by Restrained MD (Protocol 3)

We recently tested a third protocol, which is really an extension of the previous one. Because it appears to be the most promising protocol that we used thus far, we shall describe it in detail. We reasoned as follows: the tendency of MD to increase the spread among DG-generated structures is due to the presence of kinetic energy, which will always tend to increase the entropy of a system, if there is room for it. Hence we can improve the performance of the DG algorithm by including a MD step, using a simplified force field that is determined only by the experimental (and the holonomic) distance constraints that were also used during the DG step: distance bounds-driven dynamics (DDD). We chose the following force field:

$$V^{dc} = K^{dc}\left[\sum_{d>u} (d_{ij}^2 - u_{ij}^2)^2 + \sum_{l>d} (d_{ij}^2 - l_{ij}^2)^2\right]$$

where u and l are the upper and lower bounds, respectively, on the interatomic distances. The sum runs over all pairs of atoms for which one of the bounds is violated in the actual structure. K^{dc} was taken as 10000 kJ mol^{-1} nm^{-4} and all atoms were given a mass of 1 a.u.

Again we used the 169 NOEs and the 17 hydrogen-bond constraints that were measured in the *lac* repressor headpiece.[27] Twenty-one headpiece structures were generated with a DG algorithm that was developed from the original EMBED program and adapted to run on a Convex C1-XP minisupercomputer. After the embedding step, the remaining violations were removed by minimizing an error function that consists of the V^{dc} term introduced above, plus a term that ensures the proper chiralities

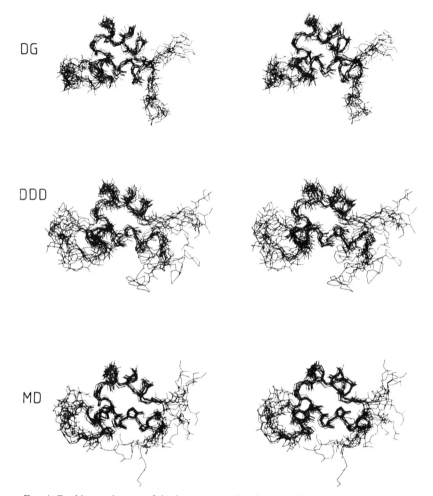

FIG. 1. Backbone pictures of the *lac* repressor headpiece at the three stages of protocol 3 (see the footnote to Table II for a description of these clusters). The structures were superimposed such as to minimize the positional rms differences of the helical Cα atoms. The N termini are at the upper right-hand side; the loop is at the left-hand side of the drawings.

around the asymmetric centers in the molecule. Distance bounds-driven dynamics was switched on in stages using exactly the same bounds as were used for the DG step. The result is shown in Fig. 1 and Table II. After the first 1000 steps (formally corresponding to about 2 psec) of DDD, not much further divergence occurs among the structures during another 1000 steps at 300 K and during a third run of 1000 steps at 1000 K. After each stage the structures were frozen to 1 K and optimized as

TABLE II
AVERAGE rms POSITIONAL DIFFERENCES BETWEEN
21 *lac* REPRESSOR HEADPIECE STRUCTURES

Cluster[a]	rms (nm) ($C\alpha$ 1–51)	rms (nm) ($C\alpha$ in helices)
DG	0.13	0.09
DDD (1)	0.24	0.12
DDD (2)	0.29	0.13
DDD (3)	0.30	0.13
MD	0.30	0.08

[a] Comparisons are of 21 structures after superposition of the $C\alpha$ atoms. Cluster DG resulted from a distance geometry calculation. It was converted into cluster DDD (1) by 1000 steps (2 psec) of DDD at 300 K and into cluster DDD (2) by another 1000 steps of DDD at 300 K. Cluster DDD (3) was generated by a final 1000 steps of DDD at 1000 K. After each stage of DDD the kinetic energy was removed (by cooling to 1 K) and the structures were further optimized with respect to the V^{dc} term plus a chiral term (same procedure as after the DG step; see the text). Of the structures of cluster DDD (3), 10 were energy minimized by a restrained MD protocol as described above (two times 5 psec as in protocol 2; see the text), resulting in cluster MD.

described above, to make sure the chiral centers were not distorted during the DDD step. Clearly this simplified MD technique succeeds in improving significantly the sampling properties of the DG step in the sense that a wider variety of structures is found satisfying the same set of constraints.

Note that, apart from the holonomic constraints, which follow directly from the known bond lengths, bond angles, and atomic radii, no other input than the experimental distance constraints was used up to this point. Hence the root-mean-square (rms) variation in the atomic positions that can be calculated for each atom from the cluster of superimposed structures after DDD is a better measure of the accuracy with which the structure was determined by NMR. Clearly, if this rms variation were calculated from the cluster of structures immediately after the DG step, the obtained accuracy of the structure would have been seriously overestimated (see Fig. 2).

The next step is to start restrained MD, as in step 2 of the previous protocol, in order to minimize the high internal energies of the structures.

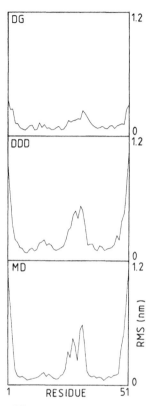

FIG. 2. The positional rms differences between the Cα positions in the three clusters of *lac* repressor headpiece structures shown in Fig. 1.

The result for 10 of the 21 structures is shown in Table II.[31] The picture that emerges is now consistent in the sense that the input of new structural information during the refinement step (via the force field parameters) results in some convergence among the structures, most notably in the helical regions of the protein.

Criticism. As discussed above, care must be taken in the interpretation of the solvent-accessible parts of the molecule after the MD refinement step, when the effects of solvent were not properly included. Furthermore, one can never be absolutely sure that no (families of) conformations were missed by the combined DG/DDD protocols (nor by

[31] R. Kaptein, R. Boelens, R. M. Scheek, and W. F. van Gunsteren, *Biochemistry* **27**, 5389 (1988).

any other protocol). Hence one is forced to generate a large number of structures by DG techniques to minimize this possibility.

Concluding Remarks

The combined use of distance geometry calculations and distance bounds driven dynamics gives insight into the accuracy of structure determinations by NMR. Refinement by restrained molecular dynamics improves the quality of the structures in terms of energetics.

It is important in this respect to discriminate between uncertainties in the structure determination and thermal fluctuations. Increased flexibility in some parts of the molecule often results in diminished NOE intensities. In such situations the experimental uncertainty, caused by the absence of NOEs, will result in large rms variations in the corresponding atomic positions after the DG/DDD steps, which happen to parallel the molecule's flexibility. This situation is reminiscent of the interpretation of B factors in X-ray crystallography. Positive evidence of increased local flexibility may come from relaxation measurements (e.g., the sharpening of resonances in flexible regions causes more intense cross-peaks in two-dimensional J-correlated spectroscopy, such as COSY).

Computer simulation of at least the high-frequency modes of a molecule's thermal motions by MD and calculation of the experimentally observed parameters by proper averaging over computed trajectories can be used to support a dynamic model, in which more than one "dynamic conformation" may have to be included to explain all experimental results.

[11] Heuristic Refinement Method for Determination of Solution Structure of Proteins from Nuclear Magnetic Resonance Data

By Russ B. Altman and Oleg Jardetzky

Principles

Spatial Distribution Function as the Most General Description of Protein Solution Structure

There is a critical difference between the study of protein structure in solution and in crystalline form. For the most part, all individual protein molecules in a crystal assume the same conformation.[1] It is therefore appropriate to model the structure as a rigid conformer, with some measure of uncertainty (typically the "temperature factor") associated with each atom.

It may not be reasonable to expect such well-defined and rigid structures from solution data. First, in solution the molecules may take on different conformations and the structural data must be interpreted as an average over the population.[2] Second, there is no single source of abundant high-resolution data comparable to X-ray crystallography, from which the structure can be simply calculated. NMR is to date the most abundant source of distance information, but even the best sets of data are insufficient to completely determine the structure even *with* prior knowledge of the sequence and of all covalent distances, van der Waals radii, and, insofar as possible, bond angles.

A unique single set of atomic coordinates from solution data, especially for larger proteins, is generally impossible to obtain without a large number of assumptions, and a single structure may in fact be misleading. The most general and accurate description of the solution structure of a protein is a spatial distribution function for each atom. A spatial distribution function describes the probability that a certain atom is at a certain location in space. In general, a spatial distribution function is a $3N$-dimensional joint probability density function (where N is the number of atoms and $3N$ is the number of Cartesian coordinates to be determined).

In the case of crystals, the distribution function for most atoms is a relatively sharp and approximately normal curve independent of other atoms. The variance in position is proportional to the temperature factor. Solution studies, on the other hand, may imply a nonnormal and broader distribution for many atoms. It is desirable to characterize these distributions in some detail in order to understand the extent of disorder and motion implied by the solution data and the degree to which the structure is underdetermined.

Methods for Interpretation of NMR: Adjustment or Exclusion

Methods within the adjustment paradigm generate starting structures, usually at random, and then search the neighboring conformational space until the mismatch between the data predicted from the adjusted structure and the experimental data is minimized, in terms of the chosen target function. The specific target functions and the criteria for ending the

[1] T. L. Blundell and L. N. Johnson, "Protein Crystallography." Academic Press, New York, 1976.

[2] M. Madrid and O. Jardetzky, *Biochim. Biophys. Acta* **953**, 61 (1988).

search may vary, but always depend on the discrepancy between the predicted and the observed data being at a minimum.

By contrast, methods within the exclusion paradigm generate starting structures *systematically* and test them for agreement with the given set of data. All structures compatible with the data are retained as possible solutions, and all structures that are incompatible are excluded from further consideration. The two elements of an exclusion method are (1) a structure generator which systematically produces candidate structures and (2) a structure evaluator which calculates the predicted values of the data and compares them to the observed data set.

It might be noted that adjustment methods for the interpretation of solution data are modeled after the methods of crystallographic refinement, which perform iterative refinement on rough X-ray structures. Such methods are entirely appropriate for the analysis of X-ray data: coarse starting structures can be used to iteratively increase the match between the structural hypothesis and the copious and complete data set. Adjustment methods have been reasonably successful in NMR structural studies on small proteins, and the two principal examples—distance geometry and restrained molecular dynamics—are discussed elsewhere.[3-6] In general, however, it must be borne in mind that, unlike X-ray data, solution data are often not of sufficient information content to uniquely determine the structure.[7] Indeed, for solution studies it is not self-evident that single

---

[3] G. Clore, A. Brünger, M. Karplus, and A. Gronenborn, *J. Mol. Biol.* **191,** 523 (1986).

[4] M. P. Williamson, T. F. Havel, and K. Wüthrich, *J. Mol. Biol.* **182,** 295 (1985).

[5] W. Braun and N. Gō, *J. Mol. Biol.* **186,** 611 (1985).

[6] T. Havel and K. Wüthrich, *Bull. Math. Biol.* **46,** 665 (1984).

[7] In general, a minimum of $4N - 10$ exact distances are required to position N points in space (where N is four or greater). This figure is derived from the following argument. Given four fixed points to describe a coordinate system, any additional point can be unambiguously positioned by providing the distances to the four fixed points (the distance to the first point provides a shell of possible locations, the distance to the second point provides a two-dimensional circle of possible locations, the distance to the third point selects two points on the circle, and the fourth distance disambiguates between these two points). Additional points can then be unambiguously positioned with four distances to any of the previously fixed ones. Thus, for all points after the first four, we require four distances or $4(N - 4) = 4N - 16$. In order to position the first four points, we need only six distances: one point can be placed arbitrarily at the origin (no distances required); its distance to the second point enables us to place the second point on the x axis (one distance). The distances of the first two points to the third allow us to place the third point on the positive xy plane (two more distances). Finally, the fourth point can be positioned in the positive z hemisphere using the distances to the first three points (three more distances). Thus, the total number of distances required is, minimally, $4N - 16 + 6 = 4N - 10$. This figure represents a minimum and is only achieved if the distances are distributed correctly throughout the structure. In experimental situations, the number of distances required may approach N^2, because the distribution is arbitrary.

structures are the desirable goal. An estimate of positional uncertainty may be more important for many purposes.

Heuristic Refinement as an Exclusion Method

The method described in this chapter is based on the exclusion paradigm. Simply stated, its purpose is to sample the conformational space systematically and to determine the entire set of positions for each atom that is compatible with the given set of constraints. In its design it falls into the general class of constraint satisfaction methods, which are the subject of a sizeable literature.[8,9] The method is coded in a program called PROTEAN.[10]

With the aim of accurately defining the spatial distribution of atomic positions allowed by the given data, the value of using a systematic search of conformational space and the exclusion paradigm is almost self-evident. Attempts to characterize the distribution of nonunique structures derived from solution data using an adjustment paradigm are necessarily limited to multiple trials with multiple starting conformations. This type of study can not detect biases inherent in the method. It is entirely possible that multiple trials will all converge in a neighborhood which is actually much smaller than the neighborhood implied by the data. This neighborhood may completely, partially, or not at all overlap the neighborhood of the "correct" solution. The neighborhood defined by an exclusion method may be much larger than is typically seen in crystallographic studies, but it will contain the "correct" solution to the problem (Fig. 1).

The principal difficulty in applying exclusion methods to problems of the size of a protein structure determination (with the positions of hundreds, if not thousands, of atoms to be specified) is their computational cost. With a small number of objects, coarse sampling, and narrow distributions (tight constraints), the coordinate tables are small and readily manageable. With fine sampling, broad distributions, and an increasing number of objects, they become cumbersome and ultimately unmanageable. To maintain computational feasibility, the method outlined here solves the protein structure problem in a hierarchy of steps: starting with the prior knowledge of the protein sequence it determines the secondary structure from NMR data first, then defines the coarse topology of the folded structure by coarse systematic sampling, and finally specifies the

---

[8] A. K. Mackworth, *Artif. Intell.* **8,** 99 (1977).

[9] A. K. Mackworth, *Artif. Intell.* **25,** 65 (1985).

[10] R. B. Altman, "Exclusion Methods for the Determination of Protein Structure from Experimental Data," Ph.D. thesis. Department of Medical Information Science, Stanford University, Stanford, California, 1989.

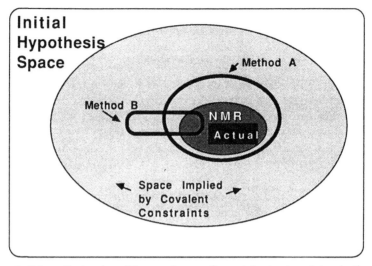

FIG. 1. Schematic drawing of the search space for protein structure determination from experimental solution data. NMR and other data imply a set of solutions that contain the actual structures. Structure determination methods should produce a tight upper bound on the experimental solution space (Method A) and should avoid bias, which introduces extra solutions or excludes legal solutions (Method B).

spatial distribution of atomic positions using a parametric description of accessible volumes. From these, the original data are predicted by solving the relaxation matrix (generalized Bloch equations) to verify the correctness of the resulting family of structures.[11] We might note that this type of method, allowing solution of the problem in stages, is more likely to remain tractable when applied to larger proteins—where adjustment methods may become prohibitively expensive.

PROTEAN Overview

The modules of the PROTEAN system and their flow of control, with brief descriptions, are shown in Fig. 2. These modules perform all the tasks from data preparation to final analysis. A complete documentation and validation of the system has been published.[12–14]

[11] A. Kalk and H. J. Berendsen, *J. Magn. Reson.* **24,** 343 (1976).
[12] R. B. Altman and O. Jardetzky, *J. Biochem.* **100,** 1403 (1986).
[13] J. F. Brinkley, R. B. Altman, B. S. Duncan, B. G. Buchanan, and O. Jardetzky, *J. Chem. Inf. Comput. Sci.* **28**(4), 194 (1988).
[14] O. Lichtarge, C. W. Cornelius, B. G. Buchanan, and O. Jardetzky, *Proteins: Struct. Funct. Genet.* **2,** 340 (1986).

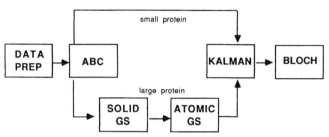

FIG. 2. The modules of the PROTEAN system. DATA PREP prepares the raw data for use by the other modules. ABC uses symbolic knowledge to identify the location of secondary structures within the protein. SOLID GS models secondary structures as rigid objects for rough determination of topology using discrete sampling. ATOMIC GS refines the secondary structures and coils using discrete sampling for atoms. KALMAN employs a probabilistic refinement method for determination of uncertainty in each atom. BLOCH calculates the expected time course for NMR data and evaluates the match between observed and predicted data. In this chapter we stress the functionality of SOLID GS and KALMAN.

Determination of Secondary Structure

Subset of NMR Data Which Defines Secondary Structure versus the Subset Defining Tertiary Folding. The information needed to define the secondary structure is for the most part contained in the section of the NMR spectrum containing backbone resonances (involving amide and α hydrogens), whereas the information for defining the tertiary structure is found in the spectra of the side chains. The PROTEAN module ABC uses as input the following data.[15]

1. Hydrogen exchange rates. These provide evidence of hydrogen-bonded structures, but do not offer evidence supporting any structure type in particular.

2. Short- and medium-range NOEs. These are principally from the backbone–backbone sequential distances $\Delta\alpha\alpha$ (α to α), $\Delta N\alpha$ (amide to α), and ΔNN (amide to amide), and the medium-range connectivities $\Delta\alpha N_{i,i+2}$, $\Delta\alpha N_{i,i+3}$, $\Delta\alpha N_{i,i+4}$, $\Delta NN_{i,i+2}$, and $\Delta\alpha\alpha_{i,i+3}$. These distances are potentially short enough to manifest an NOE (generally less than 4 Å) and are typically found within secondary structures.

3. Spin–spin coupling constants ($^3J_{HN\alpha}$). These can provide supporting evidence for structures identified with NOE information, or can be used as an independent criterion for locating helical segments.

4. Long-range NOEs. These are principally backbone–backbone connections $\Delta\alpha_{i,j}$, $\Delta\alpha N_{i,j}$, and $\Delta NN_{i,j}$, where i and j are separated in the

[15] J. A. Brugge, B. G. Buchanan, and O. Jardetzky, *J. Comput. Chem.* **9**, 662 (1988).

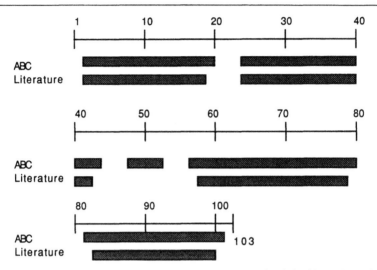

FIG. 3. The performance of ABC on test data sets shows that it is able to determine the location of secondary structure with an accuracy similar to manual methods. The locations of helices in cytochrome b-562 as determined by ABC using artificial NOE data and as reported in literature are shown aligned with the primary sequence. They match substantially, except for an extra short helical portion predicted by ABC.

sequence by more than six residues. Such connections often indicate a bonding between secondary structures such as β strands.

Secondary Structure Inferred from NMR Data Using Symbolic Reasoning Only. The diverse nature of NMR data used to define secondary structure makes symbolic reasoning a more appropriate method of analysis than numerical computation, which necessarily involves information loss. This is quite evident from the early successes of the "visual" method for secondary structure determination.[16,17] ABC formalizes the logical rules inherent in this method and is able to identify regions of helices and β sheets with a high degree of accuracy.[15] An example of its performance is shown in Fig. 3.

Constraint Satisfaction in Sampled Space

The second step in the structure determination by PROTEAN involves systematic sampling of the conformational space available for pro-

[16] E. Zuiderweg, R. Kaptein, and K. Wüthrich, *Proc. Natl. Acad. Sci. U.S.A.* **80,** 5837 (1983).
[17] A. A. Ribeiro, D. Wemmer, R. P. Bray, and O. Jardetzky, *Biochem. Biophys. Res. Commun.* **99,** 668 (1981).

tein folding, given the units of secondary structure and a set of experimental constraints. The input to the SOLID GS module, which generates candidate structures and checks them against the given set of constraints, consists of primary structure (from protein or nucleic acid sequencing), secondary structure output of ABC, distances from NMR experiments, other distance information, and surface and volume information, if available. The output is the accessible volume for the units of secondary structure, defined by the given set of constraints.

Critical to maintaining computational feasibility of the exclusion paradigm is the initial stage of coarse sampling and the use of abstract representations to reduce the number of objects whose positions need to be sampled systematically. For example, both helices and β strands are represented by cylinders enclosing all atoms proximal to the β carbons for initial placement.

Refinement of Accessible Volume. The basic operations that we have defined for the reduction of accessible volumes are general to all geometric constraint satisfaction problems. Each of these operations has many possible implementations. In fact, we have implemented each of them for three different representations (three-dimensional sampled accessible volumes, six-dimensional sampled accessible volumes, and three-dimensional parametrically described accessible volumes). Despite detailed differences in implementation, the basic definitions remain constant. In the description below, we include a generic definition of the operation, as well as a definition for geometric constraint satisfaction. In the language of constraint satisfaction used in our descriptions, a node corresponds to a three-dimensional object, a label is the set of possible locations for that object, and an arc is the constraint between two nodes. The computational complexities refer to cases in which location lists (accessible volumes) are discretely sampled. These operations provide a logical vocabulary for any exclusion method.

CREATE: Creates a new constraint satisfaction network. Initially the network is empty. For geometric problems, this corresponds to creating a fixed coordinate system in which objects are to be positioned. Computational complexity: constant time.

INCLUDE: Adds a new node to the network and associates all possible labels with the node. This corresponds to adding an object to the coordinate system and associating all possible locations with it. Computational complexity: constant time.

ORIENT: Chooses a single variable binding for a node in the network. By fixing this value, it defines the value for that node for all possible solutions. Fixing one value in a geometric problem corresponds to picking a single object to fix in the coordinate system and defining the frame of reference relative to all other objects. We call the fixed objects the "anchor" of the coordinate system. Solutions which are related by simple translations and rotations are thereby excluded. Fixing more than one object biases the solution toward only those structures which have the fixed objects arranged in the same way. Computational complexity: constant time.

ANCHOR: Uses the constraints between fixed-valued nodes and a nonfixed node to generate the list of variable values that are consistent with the constraints. This corresponds to considering constraints between the fixed "anchor" and a movable "anchoree" and determining which positions in space satisfy the constraints between them. Computational complexity: linear in the initial search space for legal values of the movable object.

APPEND: Uses the constraints between (1) fixed valued object A and variable valued object B and (2) object B and variable valued object C to define the composition of values C relative to A. Geometrically, this corresponds to finding positions of B relative to A and C relative to B and then composing them (using the cross-product of transformations) to position C relative to A. Computational complexity: proportional to product of label lists A–B × B–C.

PRUNE: Similar to ANCHOR, PRUNE acts on a previously defined variable value list (such as produced by APPEND) and uses constraints between the fixed-value node and a nonfixed node to reduce a list of variable values by removing those that are not consistent with the constraints. Corresponds to considering constraints between the fixed coordinate system and an object introduced into the coordinate system only by consideration of its constraints to another movable object. Computational complexity: linear in size of the location table.

YOKE: Uses the constraints between two non-fixed-valued nodes to prune each of their variable value lists. A value for one node is pruned if it is incompatible with every value in the value list for the second node. This corresponds to checking constraints between two "anchorees" and throwing out positions within each accessible volume. Computational complexity: proportional to the product of the length of each.

NYOKE: A generalization of yoke, uses the constraints between N non-fixed-valued nodes to prune each of their value lists. A value for one node is pruned if there is not a set of variable bindings for the other nodes with which it is compatible. This corresponds to checking constraints that involve N or less objects and throwing out positions that are incompatible. Computational complexity: (worst case) proportional to the product of variable value lists and (best case) proportional to the length of the longest variable value list.

INSTANTIATE: Uses all constraints within the network to conduct a backtrack search for all possible sets of variable bindings that satisfy the constraints. This corresponds to picking one location from each accessible volume and checking for mutual consistency. Computational complexity: proportional to the product of the lengths of the variable value lists.

The sequence of steps and their characteristics can be illustrated by the assembly of the first domain of T4 lysozyme (EC 3.2.1.17)—a protein fragment of 80 amino acids with a known X-ray structure.[18] A more detailed description of the SOLID GS has also been published.[13]

Details of assembly of T4 lysozyme. The primary and secondary sequences of T4 lysozyme as well as NMR, covalent tether, hydrogen-bond, surface, and volume constraints used are shown in Fig. 4. The construction of the protein at the solid level of abstraction illustrates the full range of decisions that must be made in the construction of a protein

---

[18] S. J. Remington, L. F. Ten Eyck, and B. W. Matthews, *Biochem. Biophys. Res. Commun.* **75,** 265 (1977).

PRIMARY STRUCTURE:

MET ASN ILE PHE GLU MET LEU ARG ILE ASP GLU GLY LEU ARG LEU LYS ILE TYR LYS
ASP THR GLU GLY TYR TYR THR ILE GLY ILE GLY HIS LEU LEU THR LYS SER PRO SER
LEU ASN ALA ALA LYS SER GLU LEU ASP LYS ALA ILE GLY ARG ASN CYS ASN GLY VAL
GLY ILE LEU ARG ASN ALA LYS LEU LYS PRO VAL TYR ASP SER

SECONDARY STRUCTURE:

[intervening coils not shown]
Helix-1   3 - 11   Beta-subunit-3A   31 - 32
Beta-subunit-1A   14 - 17   Beta-subunit-3B   33 - 34
Beta-subunit-1B   18 - 20   Helix-2   39 - 50
Beta-subunit-2A   23 - 25   Helix-3   60 - 80
Beta-subunit-2B   26 - 28

NOES:

3-71   17-43   27-46   33-49   07-71   25-33   33-42   49-66
4-68   17-46   31-49   34-41   16-28   25-39   33-45   50-62
4-71   18-26   31-66   34-42   17-39   27-31   33-46   50-66
7-67   20-24   31-70   34-45   17-42

SURFACE AMINO ACIDS: 5,48,61,79,80
ENCLOSING ELLIPSOIDS: 22.5 x 15.0 x 13.0
SURFACE TOLERANCE: 0.58
VOLUME TOLERANCE: 0.35

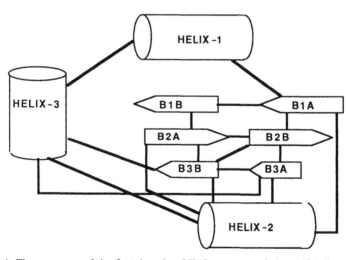

FIG. 4. The sequence of the first domain of T4 lysozyme and the artificially generated data used by PROTEAN. The solid-level secondary structures are shown in a constraint network. The arcs denote distance constraints between structures. The β strands have been divided into subunits to decrease the error in this approximation to their structure.

with PROTEAN. Figure 4 also shows schematically the solid level objects in T4 lysozyme and the NMR constraints between them. Surface and volume constraints are also artificially generated and are used to test coherent instances. The construction of the protein is outlined here.

1. Definition of Secondary Structure: The protein has three β strands. Because these are likely to be curved, they are each broken into two β subunits. For reasons of energetic stability, β strands exhibit less curving behavior as they get larger, and so the subunits are smaller for short β strands then they are for long ones.

2. Selection of Anchor: The distribution of artificially generated NOEs indicates that β-subunit 2B is the most evenly constrained object. Thus, its short length and nonhelicity is outweighed by its constraining power, and so it is chosen as the anchor.

3. Construction of β Sheet: Each of the β subunits with direct covalent connections or hydrogen bonds to β-subunit 2B are anchored to it. These are chosen first because they have the most constraints to the anchor and the strongest constraints (hydrogen bonds are more constraining and less variable than are NOE data). The average length of a location list resulting from the anchor operation is 400.

4. Refinement of β-Sheet Accessible Volumes: The accessible volumes for each β subunit is reduced by successive application of the YOKE operation between them. The average location list for the β subunits goes from length 400 to length 20.

5. Completion of β Sheet: β-subunit 1B has no direct constraints to the anchor. It must be anchored to β-subunit 1A in a different coordinate system. It is then appended to 1A in the partial arrangement of β-subunit 2B. Because 1B has 51 locations relative to 1A and 1A has 438 locations relative to 2B, 1B has $51 \times 438 = 22,338$ locations relative to 2B. Another set of successive yokes is applied to reduce the size of this location list to 1500. Because the locations for β-subunit 1A are more tightly sampled than are the other nonappended β-subunits, 10% (150) are taken to normalize the effective sample interval.

6. Introduction of Helix-2: With the β sheet constructed and yoked, we can bring in the helix elements. Helix-2 has the most constraints to the anchor, β-subunit 2B, and so it is chosen for introduction into the partial arrangement. It is anchored with 4991 positions.

7. Introduction of Helix-3: Helix-3 is then anchored to the anchor using its constraints to the anchor. It is anchored with 2349 positions.

8. Refinement of Helices and β Sheet: Helix-2 and helix-3 are yoked with each other and with the other β subunits with which they have constraints to reduce their locations lists to length 347 and 953, respectively.

9. *Introduction of Helix-1:* Helix-1 has no direct constraints to the anchor, so must be appended in fashion similar to β-subunit 1A in step 5. This time, it is first anchored in the space of helix-3 and then appended to the β subunit. It has a large number of locations for reasons outlined in step 5. It can also be sampled, however, to normalize its effective sample interval. After the sampling, it has 1211 locations.

10. *Refinement of All Secondary Structures:* Helix-1 is then yoked repeatedly with the other helices and β subunits. At this point, all NOE constraints have been introduced and applied to the secondary structure elements. The β subunits were already substantially yoked, and the average number of locations for them remains 20. The location tables for the helices, however, are dramatically reduced by multiple yoking with each other and with the β sheet. The location lists for helix-2 goes from length to 45; for helix-3, from 953 to 17; and for helix-1, from 1211 to 17.

11. *Generation of Coherent Instances:* The generation of coherent instances follow. The total number of coherent instances possible from the location lists of these nine objects is the product of their lengths. Because there are too many combinations to check, the system randomly selects 330,000 potential instances to test for coherency (a number that can be checked in less than 1 hr). The coherent instance generator only finds 1581 coherent instances that satisfy all the constraints.

12. *Test of Volume and Surface Constraints:* For each of the sampled coherent instances, we calculate an ellipsoid enclosing 95% of the atoms. If the ellipsoid axes are not within some tolerance factor of the known axes, then the structures are excluded. Similarly, if surface atoms are found within a core region of the ellipsoid (again using a tolerance factor), they are excluded. These operations reduce the number of coherent instances from 1581 to 244.

Analysis of intermediate T4 lysozyme accessible volumes. The topology of T4 lysozyme from PROTEAN produced by the sampled system can be compared with the crystal structure from which the artificial data set was generated. The accessible volume of each secondary structure contains the crystal structure in all cases. The variance around these structures shows that the average atomic root-mean-square (rms) deviation from crystallographic position is 5 Å. These errors are similar to the errors of starting structures for many optimization techniques,[19] but

[19] A. Kline, W. Braun, and K. Wüthrich, *J. Mol. Biol.* **189,** 377 (1986).

have been derived solely from experimental data, without any theoretical assumptions.

The variance is greatest in the N-terminal end of helix-3. There were no artificial constraints generated for this end of the helix, since it is closely packed with elements of the second domain of the protein, which were not considered. Lacking any constraints, the N-terminus has a large accessible volume. The result is shown in Fig. 5 (see color insert).

The structures calculated with this technique are sufficient to provide a clear indication of the topology of the molecule. It is necessary, however, to increase the precision of these structures. A finer sampling of the solution space becomes intractable if we consider all the positions for each atom within the structure. We have therefore developed a parametric representation of the accessible volume, which allows more precise structural calculations, still relying on the exclusion paradigm. In this representation it is possible to define both the average position of each atom and the uncertainty of this position implied by the data.

Parametric Description of Accessible Volumes

Probabilistic Representation and Semantics. The parametric system that we have implemented is based on work in probabilistic signal-processing data literature.[20] The basic premise is that any system can be described by a vector of parameters, called a state vector. At each moment in the calculation, this vector contains the best estimate of the value for each parameter (given the current information). In the case of PROTEAN, the state vector is a vector of (xyz) values for each atom. (Alternatively, we could use a six-dimensional representation of position and orientation or a dihedral space representation.) With the Cartesian system, a protein with 1000 atoms is represented as a state vector of 3000 elements. As an example, consider a molecule with four atoms. Its state vector would look like this:

$$[x_1 y_1 z_1 x_2 y_2 z_2 x_3 y_3 z_3 x_4 y_4 z_4]^T \qquad (1)$$

Associated with a state vector is a covariance matrix which summarizes the covariance between each pair of elements of the state vector. The covariance between two random variables is defined as

$$\sigma_{xy} = E(xy) - E(x)E(y) \qquad (2)$$

It is related to the correlation coefficient in the following way:

$$\rho_{xy} = \sigma_{xy}/\sqrt{\sigma_{xx}\sigma_{yy}} \qquad (3)$$

[20] A. Gelb, "Applied Optimal Estimation." MIT Press, Cambridge, Massachusetts, 1984.

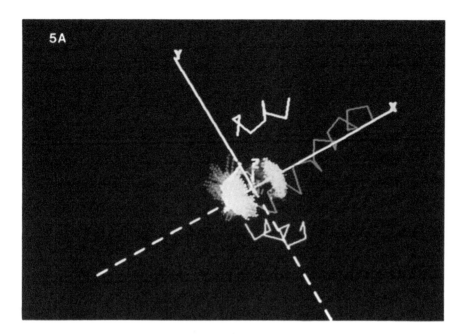

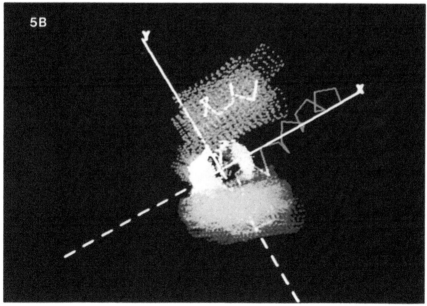

FIG. 5. Looking down the long axis of the β sheet of T4 lysozyme, are the three crystallographic locations of helix-1, helix-2, and helix-3, along with the accessible volumes of the β subunits (A). The accessible volumes computed by PROTEAN for helix-1 and helix-2 are also shown (B). These accessible volumes enclose the crystallographic positions of all the α carbons in the helices and β strands.

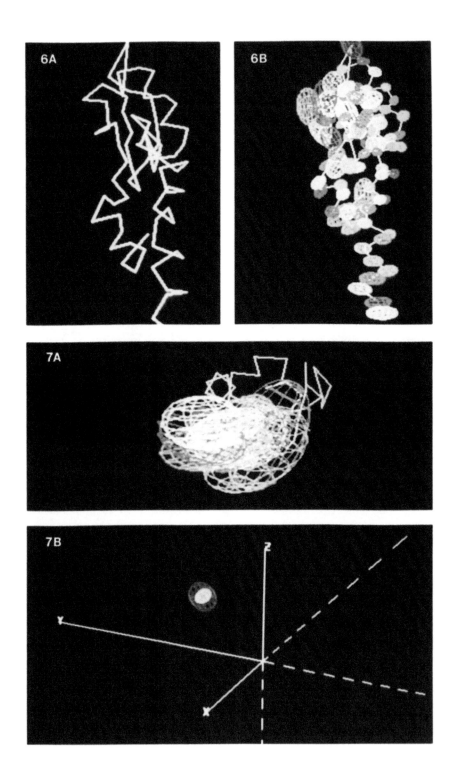

Fig. 6. The mean positions of the α carbons of the first domain of T4 lysozyme calculated by PROTEAN are shown (A). The associated uncertainty for each carbon (drawn as a contour at two standard deviations) is shown as well (B).

Fig. 7. The *lac* repressor headpiece. (A) The initial uncertainty ellipsoids are shown (at two standard deviations) for the α carbons of helix-3, obtained from the sampled cloud (not shown). (B) Uncertainty ellipsoid is shown for a single atom before and after one cycle of the full atomic Kalman filter iteration. (C and D) Uncertainty ellipsoids are shown for helix-1 and helix-2 with the side chain attached before (C) and after (D) one full atomic Kalman iteration.

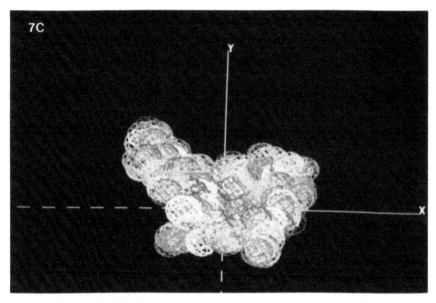

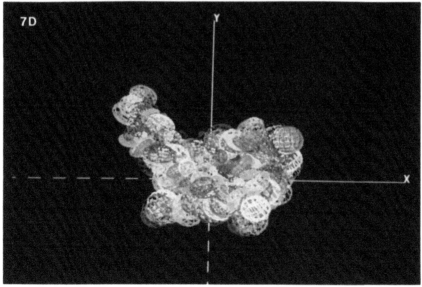

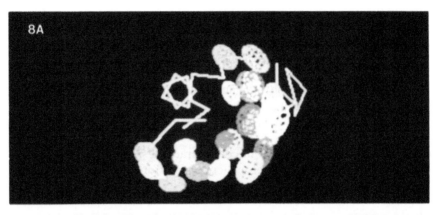

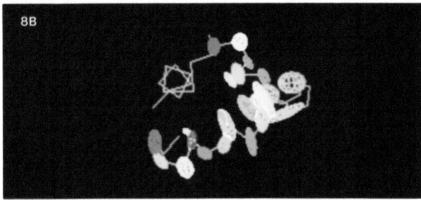

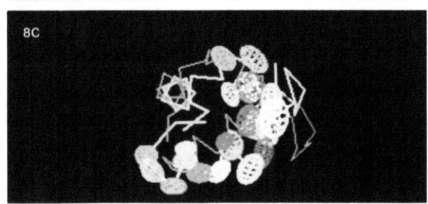

FIG. 8. The mean α carbon positions of the *lac* repressor (as calculated by PROTEAN) are shown superimposed with the uncertainty ellipsoids for helix-2 and helix-3 (A). The mean α carbon positions of the 10 *lac* repressor structures calculated by the Groningen group are shown with the associated uncertainties (B, same orientation as A). The structure calculated by Frayman is shown superimposed on the mean structure calculated by the Kalman filter method with the uncertainty ellipsoids for helix-2 and helix-3 (C).

Thus, the covariance matrix for our three-atom molecule would look like this:

$$C(x) = \begin{bmatrix} C(x_{1,1}) & C(x_{1,2}) & C(x_{1,3}) & C(x_{1,4}) \\ C(x_{2,1}) & C(x_{2,2}) & C(x_{2,3}) & C(x_{2,4}) \\ C(x_{3,1}) & C(x_{3,2}) & C(x_{3,3}) & C(x_{3,4}) \\ C(x_{4,1}) & C(x_{4,2}) & C(x_{4,3}) & C(x_{4,4}) \end{bmatrix} \tag{4}$$

where each submatrix in $C(x)$ is given by

$$C(x_{i,j}) = \begin{bmatrix} \sigma_{x_i x_j} & \sigma_{x_i y_j} & \sigma_{x_i z_j} \\ \sigma_{y_i x_j} & \sigma_{y_i y_j} & \sigma_{y_i z_j} \\ \sigma_{z_i x_j} & \sigma_{z_i y_j} & \sigma_{z_i z_j} \end{bmatrix} \tag{5}$$

The covariance of a random variable with itself is just its variance. Thus, the diagonal elements of the matrix represent the variance (or uncertainty) in each of the parameters (Cartesian coordinates in this case). A variance of zero indicates complete certainty about the value of the parameter in the state vector. A large variance indicates the degree of uncertainty in the value of the variable. The covariance between two random variables is a linear estimate of the relationship between the two variables. A positive value indicates that one variable increases as the other does. A negative value indicates that the variable decreases as the other increases. A value of zero indicates that the two variables are independent. The magnitude of the covariance indicates the strength of the coupling.

The 3×3 variance–covariance matrix on the diagonal of the covariance matrix for each coordinate of an atom [Eq. (5)] indicates the extent of three-dimensional uncertainty in the position of the atom. The mean position of the atom is given in the state vector, and the parameters of the distribution around that mean is indicated by the 3×3 diagonal matrix.[21] In principle, there is an infinite number of three-dimensional distributions that can be characterized by the same mean and variance (but which will

---

[21] In order to graphically display the variance of an atom as an ellipsoid, we use an eigenanalysis of the covariance matrix. Jacobi's method is used to find the eigenvalues and eigenvectors of the covariance matrix, which is guaranteed to be symmetric. The eigenvalues are the variances in each of the principal axes of the atom, while the eigenvectors are the columns of a rotation matrix that provides the proper orientation of the axes. We can therefore construct an ellipsoid based on calculating two standard deviations along each principal axis, and then rotating into the proper orientation. The two standard deviations are meaningful only if the actual atomic distribution is roughly normal, and this is one of the primary drawbacks of our method.

differ in higher order moments). Thus, in order to draw a distribution in space, it is necessary to assume some shape. One reasonable possibility is an assumption of uniform distribution. The mean position is given by the state vector, and the 3×3 covariance matrix indicates the shape of a uniform ellipsoidal distribution. Another reasonable possibility is a three-dimensional normal distribution in which the mean is the most likely position and the density trails off in each of the three directions as a normal curve with variance given by the 3×3 matrix. It has been shown that as more and more data are introduced in the pruning of these accessible volumes, the assumption of normality introduces the least statistical bias.[20] However, in the case of protein structure determination from solution data, it has been shown that the distributions for some atoms are quite eccentric, and not well described by the first two moments of a probability distribution.[22] If there is independent information about the final shape of the distribution, it is possible to use this in the choice of shape. Otherwise, it is possible to keep track of more degrees of freedom for the distributions (third moments and higher) in order to get a more precise description of the eccentricities.

Finally, the 3×3 matrices on the off-diagonal of the covariance matrix indicate the correlation in position between the coordinates of two different atoms. Constraints between two objects result in nonzero entries into this off-diagonal position that indicate how the constraints constrain the points to move in a dependent way. The off-diagonal position is therefore a summary of the dependencies each atom has on the position of other atoms.

Conversion between Sampled Representation and Probabilistic Representation. The conversion from a sampled representation to the probabilistic requires calculation of the mean position and its variance in three dimensions of each object within the sampled volumes. The mean position is placed in the state vector in the appropriate place, and the corresponding 3×3 matrix is placed on the diagonal. Similarly, the covariance between two different sets of clouds can be computed and inserted into the off-diagonal elements to represent the dependencies of the two accessible volumes.

Mechanism for Update. Given the data structures and their semantics, we require a mechanism to estimate and update the covariance matrix using new data. The mechanism for updating our estimates is called the double-iterated Kalman filter. We have extended the iterated Kalman filter for use in highly nonlinear problems such as protein structure determination. The filter is a conditional probability-based solution to the prob-

[22] B. S. Duncan, "Computing Protein Structures from Experimental Data," Ph.D. thesis. Biophysics Program, Stanford University, Stanford, California, 1989.

lem of updating.[20] In order to incorporate a new piece of data, we need to specify a model (or strong theory) that relates the new data to the parameters in the state vector. In general, this model has the following form:

$$z = h(x) + v \qquad (6)$$

The measured value z is a function of the state vector $\mathbf{x}$ and a source of noise v (v is assumed to be normally distributed with mean zero). Thus, a measurement of distance between two points has the following form:

$$z = |\mathbf{x}_i - \mathbf{x}_j| + v \qquad (7)$$

where $\mathbf{x}_i$ and $\mathbf{x}_j$ are the coordinates of the two points within the state vector. The measurement model has no other dependencies on other elements of the state vector.

If the measurements are known to be precise, then v is very small. If the error is known, then it can be summarized in the v parameter. The Kalman filter works by calculating the expected value of z from the current state vector ($\mathbf{x}$) and comparing it with the actual value of z. If the state vector predicts a value of z that is in agreement with the measured value, then there is no update of the dependent variables in the state vector. If the state vector predicts a value of z that does not agree with the measured value, then the update depends on the ratio of the uncertainties in the measurement with the uncertainty in the dependent state variables. If the certainty in the state vector is low (as measured in the covariance matrix of the state), and the certainty in the measurement is high (as measured by the v parameter), then the state and covariance are updated to maximally "correct" the error in the current estimate. If the certainty in the state vector is high, and the certainty in the measurement is low, then the state estimate is only corrected a negligible amount. If the certainty of both is comparable, then there is a correction in the state commensurate with the ratio of certainties.

A simple one-dimensional case shows the behavior of Bayes' law in combining independent pieces of data. If m_1 and m_2 are two independent scalar estimates of a parameter's mean value, and v_1 and v_2 are the estimates of their variances, then the new estimate of the parameter's mean value and variance is taken as the weighted average of the evidence:

$$m_3 = v_1 m_2 / v_1 v_2 + v_2 m_1 / v_1 v_2 \qquad (8)$$

and the variance is given as

$$v_3 = v_1 v_2 / (v_1 + v_2) \qquad (9)$$

In the case of a multiple-parameter problem, the general equation for updating the state estimate and covariance given the model of the data is given by the general Kalman filter.[20] The new state vector is the sum of the old state vector plus a weighted difference between the measured data z and the predicted data $h(x)$.

$$\mathbf{x}_{\text{new}} = \mathbf{x}_{\text{old}} + K[z - h(\mathbf{x}_{\text{old}})] \tag{10}$$

The new covariance matrix is the difference between the old covariance matrix and a linear, weighted product of the gain matrix K, derivative of the data model H, and the old covariance matrix.

$$C(\mathbf{x}_{\text{new}}) = C(\mathbf{x}_{\text{old}}) - KHC(\mathbf{x}_{\text{old}}) \tag{11}$$

where K is given by

$$K = C(x_{\text{old}})H^{T}[HC(x_{\text{old}})H^{T} + C(v)]^{-1} \tag{12}$$

and H is given by

$$H = \frac{\partial h(\mathbf{x})}{\partial \mathbf{x}}\bigg|_{\mathbf{x}_{\text{old}}} \tag{13}$$

K is the Kalman gain matrix and is simply the ratio between the certainty in the estimate of the data by the old model and the certainty in the measurement itself. H^{T} is the transpose of matrix H. Note that the term within the inverse is the first-order Taylor approximation to the variance of the measurement:

$$C(z) = C[h(x)] + C(v) \tag{14}$$

or, taking the first term of the expansion of $C[h(x)]$:

$$C(z) = HC(x)H^{T} + C(v) \tag{15}$$

The derivative of the data model H is used to calculate the direction of updating when moving the state vector values. It shows how the changes in the values of state variables affect the state estimate of the data. If $h(x)$ is not linear in the state vector, then an iterated Kalman filter can be used to minimize errors in the update[20]:

$$x_{\text{new}_i} = x_{\text{old}} + K_i\{z - [h(x_{\text{new}_{i-1}}) + H(x_{\text{old}} - x_{\text{new}_{i-1}})]\} \tag{16}$$
$$C(x_{\text{new}}) = C(x_{\text{old}}) - K_iHC(x_{\text{old}}) \tag{17}$$

where

$$K_i = C(x_{\text{old}})H^{T}[HC(x_{\text{old}})H^{T} + C(v)]^{-1} \tag{18}$$

Even with this correction, we have found that the filter can have residual inaccuracies. When a piece of data is introduced, the optimal new estimate must be reached from the old estimate by traveling along a vector of appropriate length and direction. The length is corrected using the iterated Kalman filter, but the direction may be inaccurate. We have developed a method for iterative correction of the direction of update of the mean so that it is optimal in nonlinear cases.

Our extension uses iteration to increase the accuracy of the filter solution. After sequential introduction of all new constraints, we have an estimate of the mean values which is an improvement over the starting means, but which may not yet be the best estimate. The mean positions can, however, be used as starting locations for another round of updating. In order to allow the atoms freedom to move in response to the constraints, all the covariances are reset to their initial (large) values and all constraints are reintroduced into the system. By repeating this operation, we perform an operation that has some similarity to simulated annealing[23] in the case of first-order optimization methods. Unlike simulated annealing, however, the "heating up" phase of our method is not based on random statistics, but on the second-order covariance information indicating the uncertainty in the position of an atom. Resetting the covariance after each iteration to its initially large size is equivalent to heating up the system. As more constraints are introduced, the mean position is refined and the covariance matrix settles down to a "cool" low value. If the mean position does not satisfy the constraints to within a tolerable (arbitrarily set) number of standard deviations, then the cycle is repeated until all constraints are satisfied to within some threshold. This is typically set to be one standard deviation. The iteration is an essential step in our method. The nonlinear nature of the protein structure problem requires that a system based on linear estimate be iterated in order to reduce residual errors.

Mapping Basic Constraint Satisfaction Operators to This Method. The probabilistic approach outlined here can be considered as another formulation of a constraint satisfaction problem. Most of the terms and operations defined in the preceding sections have equivalents in this representation.

The construction of the empty state vector and state covariance matrix corresponds to the CREATE action for constraint satisfaction. The addition of each object into the state vector and covariance matrix corresponds to the INCLUDE action. If an object is fixed in the state vector (covariance = 0), it corresponds to the ORIENT action. Finding the prior mean position and variance of the accessible volume by considering constraints to the fixed object is the ANCHOR procedure. Introduction of constraints and update as discussed above corresponds to YOKE and

---

[23] P. J. M. van Laarhoven and E. H. L. Aarts, "Simulated Annealing: Theory and Applications." Reidel, Dordrecht, The Netherlands, 1987.

NYOKE, because all dependencies (to first order) are updated simultaneously. Finally, the INSTANTIATE action corresponds to selecting a value for a variable from its accessible volume, setting its variance to 0 and then propagating this information through the network to get decreased accessible volume for other objects. If this operation is repeated successfully for consecutive objects, a single sampled structure can be generated.

The double-iterated Kalman filter remains within the exclusion paradigm: (1) the starting variance for each atom is always set so that all positions consistent with the size of the protein are initially represented; (2) the exclusion of hypotheses comes by the updating of mean and variance estimates, which makes the probability of certain hypotheses negligible; and (3) the mechanism for exclusion (probabilistic update) is based solely on the information contained in each constraint individually: constraints are introduced sequentially and not as a combined global optimization criterion. The significant difference of this parametric approach from the sampled approach is that the generator of hypotheses is implicit in the probabilistic representation of accessible volume; there is no need to generate specific hypotheses because the space of hypotheses is represented parametrically. In addition, instead of absolutely excluding hypotheses, this approach determines that the hypotheses have negligible probability. Finally, the accessible volumes are approximate, in the sense that they are represented with only two moments of the probability distributions.

Construction of Models. In order to illustrate the use of the Kalman filter, it is necessary to construct models for the kinds of data that are to be applied in the filter. In this section, we describe the construction of data models for distance data in detail and describe how analogous models for bond-angle and torsion-angle data are constructed. For each of these models, we formulate a model and variance and calculate the derivative of the model. In cases for which the derivative is a very difficult analytic calculation, we can rely on a method of finite differences to estimate the derivative based on multiple evaluations of the function.

Distance data model. Distance information is the most important structural information that we typically receive. In the parametric system, distance information is interpreted as coming from a sensor with the following form:

$$z = \sqrt{(x_i - x_j)^2 + (y_i - y_j)^2 + (z_i - z_j)^2} + v \qquad (19)$$

where x and y are the two points involved in the constraint, z is the measured value (from the data), and v is the noise associated with the measurement. In the normal Kalman filter, v is assumed to be normal with mean of zero, but the double-iterated filter can correct for inaccuracies in this assumption. To the first approximation, the variance of z is simply the variance of the term under the square-root symbol.

$$=C[h(x)] = HC(x)Ht + \text{higher order terms} \qquad (20)$$

Since z is nonlinear in the variables x and y, the H matrix (defined previously) is not a constant matrix. The derivative of z with respect to the state vector $\mathbf{x}$ is taken in order to update H in the iterated Kalman filter and to calculate the variance of z.

The formulas for z and $C(z)$ allow us to introduce distance information into the model. Consider, for example, information that an atom is 3 Å away from another atom with a variance of 2 Å. This tells us that the observed values for z, $C(z)$ are $z = 3$ and $v = 2$. However, we also have formulas for calculating z and $C(z)$ from our current state vector $\mathbf{x}$. If the state vector and distance measurement do not agree in value or variance, then both are updated according to Eqs. (16)–(18) to bring them into agreement.

If we are given information that the distance between two atoms is in the interval $[a, b]$ and is uniformly distributed in that interval, then we can apply the same method by setting

$$z = (a + b)/2 \qquad (21)$$
$$C(z) = (a - b)^2/12 \qquad (22)$$

where Eq. (22) is the variance of a uniform distribution. In general, however, it is extremely unlikely that experimental data come in a truly uniform distribution. Usually, there is a most likely central value with a dropoff on either side. If a distribution is known to be bimodal, the update formulas will still work for two moments, but the detailed form of the distribution that results will be uncharacterized.

Defining a coordinate system using distances. The task of orienting the state vector requires that a coordinate system be defined with the location of four points. If the relative positions of four (or more) points is already known or assumed (such as the relationship of atoms within the peptide bond or the knowledge that some set of atoms is part of a helix), then they can be positioned in a coordinate system with mean positions and associated variances and covariances. A coordinate system can also be calculated if a set of distances between four points is known (outlined earlier[7]).

Bond-angle model. For many atoms, there are strict constraints on bond angle. We have implemented models, analogous to that for distance, for testing and updating structures based on these data. The bond-angle model is interesting, because it is best expressed as a vector of measurements:

$$z = \begin{bmatrix} \cos(\theta) \\ \sin(\theta) \end{bmatrix} + \begin{bmatrix} v_{\cos} \\ v_{\sin} \end{bmatrix} \tag{23}$$

where θ is a function of x, y, and z (three points involved in the angle). The derivatives are not given here, but are calculated similarly as outlined in the distance model example. A finite-difference method can be used for numerically calculating derivatives of vector-valued functions. This model works best for highly constrained angles which allow the linear approximation of $C(z)$ to converge most quickly.

Secondary structure model. A second class of models that can be used with the probabilistic system are those which represent standard geometric relationships between atoms. Knowledge that atoms are arranged in an α helix, for example, allows us to incorporate the known dihedral angles for helices and their associated covariance matrix onto the current state estimation. For a helix with five amino acids, this model would look like this:

$$z = \begin{bmatrix} \psi_1 \\ \phi_1 \\ \psi_{1+1} \\ \phi_{1+1} \\ \psi_{1+2} \\ \phi_{1+2} \\ \psi_{1+3} \\ \phi_{1+3} \\ \psi_{1+4} \\ \phi_{1+4} \end{bmatrix} + \begin{bmatrix} v_\psi \\ v_\phi \\ v_\psi \\ v_\phi \\ v_\psi \\ v_\phi \\ v_\psi \\ v_\phi \\ v_\psi \\ v_\phi \end{bmatrix} \tag{24}$$

where ψ and ϕ are mean values for dihedral angles and v_ψ and v_ϕ are their associated variances. The proper values for each of these parameters and their associated variances can be obtained directly from the empirical analysis of secondary structure. If the user is suspicious of the model gained from empirical studies, the variance can be increased arbitrarily in order to weaken its effect on the final solution.

If an α-carbon abstraction is used, the distances between each of the α carbons in an α helix can be calculated, along with appropriate variance estimates. These distance constraints can be included in a list of constraints to the double iterated filter. They have the effect of imposing a model of an α helix onto the solution. This technique was used in the calculation of the structure of both T4 lysozyme and the *lac* repressor headpiece discussed below.

Methods for Solving Big Problems. In tackling large, combinatorially expensive problems with the double-iterated Kalman filter, we have used the same methods that the sampled PROTEAN uses for reduction of computational complexity: ideal arrangements, partial arrangements, and refinement of detail. The general rule in the case of the parametric system is that tractability comes with the decrease in the size of the global covariance matrix.

Refinement of detail. In order to include all the atoms in the backbone and amino acid side chains we would require an extremely large state vector. By using approximate representations of a protein, we can drastically reduce the size of the state vector. In the case of T4 lysozyme, we initially calculated the positions of the α carbons. This representation is sufficient to define the general topology of each of these molecules. We can use this solution as a starting point for more refined representations. Partial arrangements and ideal arrangements are then used to define the structure in more detail.

Ideal arrangements of atoms. When atoms are known to be in a low-variance secondary structure, such as a helix or β strand, they can be positioned in the coordinate system with fixed positions and no variance. In the case of T4 lysozyme, we chose a small stretch of β strand as the anchor. The atoms that are part of an anchor do not appear in the covariance matrix, and so the size of the problem is decreased. As individual atoms are positioned by the filter with low variance, they can be removed from the filter and added to the list of relatively fixed points.

Partial arrangements. When subproblems within a larger problem can be identified as highly constrained (many constraints between atoms are known) and relatively independent (no strong constraints to outside atoms), then they can be solved separately and combined based on the constraints between them. For instance, solution of the entire T4 lysozyme protein can be facilitated by separately solving each domain first, and then combining them. The existence of the domains can be inferred from coarse solutions which indicate general topology.

Solution of the First Domain of T4 Lysozyme. The coarse solution of the first domain of T4 lysozyme described in the preceding section can be used as input to the parametric implementation of PROTEAN.

We took each of the nine secondary structures in the T4 sampled solution (three helices and six β subunits) and summarized the positions of each of the α carbons within these clouds as a mean and variance. The resulting variances were multiplied by 1.5 to account for possible missed locations at the edge of the sampled space. The α carbons belonging to coils were placed at the origin of the anchor coordinate system.

Given this initial state vector and covariance matrix, the constraints were prepared for serial introduction into the state estimates. There were three kinds of constraints used to calculate this structure. (1) The covalent distance between neighboring α carbons is 3.8 Å with a variance of 0.1 Å. (2) The distances between all of the α carbons in helices 2 and 3 are introduced with a variance of 1.0 Å (as determined empirically).[10] These distances provide strong constraints that force the α carbons to assume roughly helical conformations. (3) Each of the NMR constraints was abstracted to an equivalent distance distribution between α carbons. Lichtarge[24] has calculated the range of distances between α carbons, given that there is an NOE between different types of hydrogens (α, β, γ, etc.). We took each of these ranges and calculated the variance so that the minimum and maximum values were one standard deviation from the mean. We then multiplied this variance by 1.5 in order to partially account for the limited sample size.

Finally, a partial van der Waals check was added. This is an adjustable constraint that is introduced during processing only if the system determines that two atoms violate the van der Waals distance. Since we are dealing with an abstraction of the actual protein, we cannot use actual van der Waals distances. Instead, we made an empirical study of the protein data bank and determined that virtually no α carbons approach each other closer than 4.0 Å. Therefore, at the end of each cycle of refinement, we checked for α carbons that were closer than 4.0 Å. For any pair that violated this constraint, we introduce a new constraint between the atoms with a mean of 8.0 Å, and a variance of 16 Å. The net effect of this constraint is simply to push away atoms that are too close. It is not an ideal approximation to van der Waals constraints because they are not well represented as a simple mean and variance. Although there are ways to represent threshold functions such as van der Waals radius of closest approach, these are expensive to implement and use, and we have been able to use this approximation successfully.

All these constraints were used in the double-iterated Kalman filter starting with the means and covariances extracted from the sampled clouds. The anchor of the system remained β-subunit 2B, and so the α

[24] O. Lichtarge, "Structure Determination of Proteins in Solution by NMR," Ph.D. thesis. Biophysics Program, Stanford University, Stanford, California, 1986.

carbons of this secondary structure were fixed. There were thus $82 - 4 = 78$ movable α carbons in the state vector.

The double-iterated Kalman filter ran for 200 cycles (168 hr) on a 68020-based Macintosh II microcomputer. We compared the expected value of each constraint with the value calculated from the state estimate, and scaled them by the variance of the constraint. The results showed that the average error of all experimental and secondary structural constraints was 1.3 standard deviations (less than 1% of these constraints was violated by more than 3 standard deviations). The van der Waals constraints were satisfied by an average of 0.4 standard deviations, with no constraints violated by more than 1.1 standard deviations. The mean positions of the result match the crystal structure to within 4.2 Å. The variances for most atoms is on the order of 4 Å, so the α carbon representation of the protein is sufficient to clearly establish topology and moderate detail in the structure. Figure 6 (see color insert) shows the backbone of the structure along with the uncertainties in each of the α carbon positions (drawn at two standard deviations).

Calculation in Full Atomic Detail and Comparison of Structures Obtained by Different Methods

Given the uncertainties and limitations of the existing methods for the structural interpretation of NMR data, it is instructive to make comparisons between structures obtained by different methods using the same or a similar data set. We have carried out such a comparison for the structure of the *lac* repressor headpiece obtained by the methods described here with structures determined by restrained molecular dynamics and distance geometry. We have also used this structure to test the performance of the method at full atomic representation. The coordinates for the comparisons were kindly provided to us by Dr. Ruud Scheek, University of Groningen, and Dr. Felix Frayman of Xerox, Inc., Palo Alto, California.[25,26] The full atomic level calculations were carried out by Dr. Ruth Pachter. The original NMR distance constraints have been published elsewhere.[24]

Full Atomic Representation of the lac Repressor Headpiece

Kalman filter refinement of a protein structure in full atomic representation is made computationally feasible by the hierarchical approach of

[25] E. R. Zuiderweg, R. M. Scheek, and R. Kaptein, *Biopolymers* **24**, 2257 (1985).
[26] F. Frayman, "PROTO: An Approach for Determining Protein Structures from NMR Data," Ph.D. thesis. Department of Computer Science, Northwestern University, Evanston, Illinois, 1985.

our method. While in principle the Kalman filter can be used with full atomic representation from the outset, the calculation can become prohibitively expensive for larger molecules if random starting structures are used. If, on the other hand, the backbone topology is defined first, using the α-carbon approximation as described in the preceding section, the atomic level computation becomes tractable, and, in fact quite efficient with modern computational resources.

The starting structure for the full atomic calculation described here was an arbitrarily chosen member of the set of structures calculated by Dr. R. Scheek using restrained molecular dynamics. For simplicity of computation, hydrogen atoms, which have no independent degrees of motional freedom, have been included in pseudoatoms or "super" atoms, as is commonly done in other refinement methods.[27] The pseudoatom representation of the *lac* repressor headpiece protein has 399 objects, and the initial state vector consists of 399 mean positions or 1197 individual coordinates.

The set of constraints to be consistently updated by the Kalman filter mechanism include, for the *lac* repressor headpiece, 404 covalent bonds, 16 hydrogen bonds, 152 nonbonded distances in the backbone, and 150 short-range and 113 long-range NOEs. In addition, at the end of every cycle, all possible interatomic distances in the system are scanned for van der Waals radii violations. If such a violation is found, an additional distance constraint is added, and the mean and covariance matrices are dynamically updated.

Calculations were carried out with a FORTRAN 77 version of the program on both a CRAY X-MP/48 supercomputer and an Ardent TITAN minisupercomputer. This code was tested on all previously described experiments. The program uses a sparse matrix multiplication algorithm, which increases its performance since no multiplications involving a zero multiplicand are attempted. The total size of the covariance matrix at any stage of the update equals $9xn + 9x(nd^2 - nd)$, where n is the number of atoms and nd the number of atoms involved in previously introduced distance constraints.

The result is shown in Fig. 7 (see color insert). Figure 7A represents the uncertainty ellipsoid for the α carbons of helix-3 at the outset. Figure 7B shows the uncertainty in the position of a single atom before and after the full atomic Kalman filter iteration. Figure 7C and D shows the reduction of the uncertainty ellipsoid after one iteration for a segment containing helices-1 and -2 *with the side chains attached.*

It is noteworthy that the largest decrease in the residual error reflects a better placement of the mean atomic positions, with the decrease in the

[27] K. Wüthrich, M. Billeter, and W. Braun, *J. Mol. Biol.* **169**, 949–961 (1983).

size of the uncertainty ellipsoid being much smaller. The largest decrease in uncertainty is for the backbone atoms which are subject to the most constraints. This is exactly as expected, since the distance constraints have a relatively well defined mean value, but some are subject to appreciable uncertainty. It can be said as a general rule, that the uncertainty of the structure will depend on the uncertainties assigned to the individual constraints at the outset.

The program was also run on an Ardent TITAN, utilizing the state vector for 399 pseudoatoms, and the maximal length required for the covariance matrix for this problem. The Kalman filter was doubly iterated: for each of the four cycles the filter was iterated for all possible distance constraints (835). The average error for all constraints was decreased from 3.5 standard deviations to 0.4 standard deviations. The results of the error calculation at the end of each cycle are given in the table below, showing the doubly iterated Kalman filter error calculation for the *lac* repressor. [The average error for all constraints, as well as the maximum error of any single constraint (both in units of standard deviations), are shown.]

Average error (e)	Maximum error (e max)	Cycle
3.60	8.54	0
0.26	2.93	1
0.13	1.73	2
0.11	1.52	3
0.099	1.70	4

For each constraint $e = (D_{exp} - D_{obs})/\sigma$. D_{exp} is the expected distance, D_{obs} is the observed distance, and σ is the standard deviation of the observed distance.

In summary, the method can be used to obtain, in a single calculation, both the mean position and the uncertainty of the position of each atom (as defined by a given set of constraints) for a protein of moderate size.

Results Showing General Agreement of Independent Methods

For purposes of comparison, we used a sampled structure of the *lac* repressor as a starting point for the means and variances of each atom. The means and variances of each of the α carbons in the three helices were calculated by summarizing the locations of the carbons in the sampled clouds. These summary statistics were calculated only for the α carbons that are part of movable helices-2 and -3. The variance of the atoms ranged from 5 to 20 Å, since the sampled clouds were imprecise. The α carbons in the intervening random coils were initially placed at the origin of the coordinate system with a variance of 15 Å, the approximate volume of a polypeptide with 51 amino acids.

Each of the distance constraints used in the sampled system was represented as a distance distribution using the translation with empirical distances as described above. In addition, distance distributions for secondary structural constraints (derived from knowledge of the secondary structure and the empirical results of our secondary structure study) were added. Finally, the van der Waals constraints were also introduced between all α carbons.

The filter was run for 55 double-iterated cycles. Each cycle introduced all the constraints sequentially, followed by heating of the covariance matrix and repeated iteration. The final structure satisfied all the constraints an average of 1.1 standard deviations (the calculated value of the constraint is within 1.1 standard deviations of the expected value).

The uncertainties in the position of each atom clearly depend on the number and strength of constraints in which it is involved. In helical elements of the protein, ellipsoids drawn at 2 standard deviation confidence intervals have a maximum dimension of about 2.0 Å. In certain coil elements, however, the maximum dimension of uncertainty can have (at 2 standard deviations) an uncertainty of as much as 8.0 Å. This is expected because the coils are not well constrained by the NMR data in the first place, and when this is coupled with a simple α-carbon approximation to the backbone, there is still considerable freedom for these carbons.

In order to evaluate the structure further, we extracted the positions of the α carbons from the 10 structures provided by the Groningen group. We summarized these 10 structures with means and covariances and compared the mean positions as well as the size of the uncertainty ellipsoids. The mean positions of the Groningen structures match our mean positions to within 2.4 Å rms. If the relatively unstructured coils at the N- and C-termini are *not* considered, the rms deviations become 1.5 Å. The distributions of the two structures are shown in Fig. 8A and B (see color insert). The uncertainty ellipsoids of the Groningen structures are approximately half the size of our uncertainty clouds in each dimension, as would be expected. The areas of maximum and minimum deviation are the same in our structures and in the Groningen structures. Figure 8 also shows the relative uncertainty in the two solutions. The helical elements of the structure have the smallest variance in both solutions. The coil elements have greater variance. Most notably the eight amino acid coil between helix-2 and helix-3 has high variance in both solutions, as well as the N-termini.

When we compared each of the individual Groningen structures to our mean structures and variances, we found that 90% of the atoms fell within the 2 standard deviation ellipsoids, and essentially all the atoms (with the principal exception as noted above) fell within the 3 standard deviation ellipsoids.

A final comparison was conducted with the single structure of the *lac* repressor produced by Frayman.[26] Although a slightly different data set was used, we believe it was similar enough to allow a meaningful comparison. The single Frayman structure matches our structure to 2.6 Å rms and the average Groningen structure to 2.2 Å rms. In the case of Frayman's structures, there is no estimate of uncertainty in the atomic positions, so these could not be compared; 80% of the structure falls within the 2 standard deviation confidence ellipsoids, with 95% within 3 standard deviations. It is likely that the remaining atoms are positioned differently because different constraint sets were used. Figure 8C (see color insert) shows the mean structures as well as the ellipsoids of uncertainty.

lac Repressor Results Illustrating the Power of Hierarchical Representation

It is almost surprising that a simple α-carbon representation of the protein, with all atomic constraints abstracted to equivalent constraints between α carbons, produces an accurate structure with such precision. The structure produced by the probabilistic parametric system is accurate in the sense that it contains all the structures calculated by the group in Groningen and by Frayman. It is precise in the sense that the estimates of uncertainty are relatively tight. The uncertainties of the Kalman filter results are only twice those of the Groningen structures, even though the Groningen structures use full atomic representations. This result is important because it shows that hierarchical representations of protein structures, if used in a controlled way along with principled approximations for the constraints, can yield relatively high-resolution structures with limited computational resources.

Validation and Refinement Using Generalized Bloch Equations

The ultimate test of the "correctness" of an NMR structure is its ability to predict experimental data. Given a set of coordinates, it is possible to predict the original NOE spectrum by solving the full relaxation matrix, representing the generalized Bloch equations.[11] The method and its limitations are discussed in detail elsewhere in this series (Borgias and James [9], Vol. 176).

A second option is the iterative use of the relaxation matrix as a refinement procedure, in an attempt to define the "best" structure. Thus far this has only been used on relatively small structures such as the ST peptide, containing 17 amino acids,[28] and the *trp* operator.[29] It is, however, to be expected that this relatively expensive computational procedure will come into wider use. It remains to be established whether the limit of uncertainty that it will put on the solution structures derived from NMR data will be substantially narrower than those given by the probabilistic method described above.

Acknowledgments

The authors wish to thank Bruce Buchanan, Bruce Duncan, Jim Brinkley, Enrico Carrara, and Peter Cheeseman for their comments. We would like to acknowledge NIH Grants GM07365 and RR02300 and NSF Grant DMB8402348.

[28] J. Gariepy, A. Lane, F. Frayman, D. Wilbur, W. Robien, G. Schoolnik, and O. Jardetzky, *Biochemistry* 25, 7854 (1986).
[29] A. N. Lane, J.-F. Lefèvre, and O. Jardetzky, *J. Magn. Reson.* 66, 201 (1986).

[12] Proton Magnetic Resonance of Paramagnetic Metalloproteins

By Ivano Bertini, Lucia Banci, and Claudio Luchinat

Introduction

There is a class of biological compounds, mainly metalloproteins, that are paramagnetic and that give rise to isotropically shifted signals when investigated by ^{1}H NMR.[1-3] The isotropic shift is conceptually defined as the difference between the actual chemical shift and the shift of an "analogous" diamagnetic compound. It is due to the interaction between the resonating nuclei and the unpaired electrons residing on the metal ion. The same interaction is responsible for line-broadening effects, which

[1] I. Bertini and C. Luchinat, "NMR of Paramagnetic Molecules in Biological Systems." Benjamin/Cummings, Menlo Park, California, 1986.
[2] R. A. Dwek, "NMR in Biochemistry." Oxford Univ. Press, London and New York, 1973.
[3] G. N. La Mar, W. D. Horrocks, and R. H. Holm (eds.), "NMR of Paramagnetic Molecules." Academic Press, New York, 1973.

vary from modest to so large as to make the signals undetectable. The investigation of these compounds is often possible and fruitful, but requires a specific theoretical background and a specific methodological approach.[1]

Because the effect of a paramagnetic center in a molecule is that of shifting and broadening signals, to different extents, it is important that the broadening always be below the undetectability limit. Furthermore, it is important to be able to control the factors affecting the linewidth as much as possible. Although more details will be given in the next section, we mention here that nuclear relaxation, which gives rise to line broadening, depends on the type of metal ion, on its interaction with the nuclear probe, and on the magnetic field. Nuclear relaxation can be decomposed into several contributions, each one related to a correlation time τ_c. In a simple but fundamentally correct way, τ_c can be expressed as

$$\tau_c^{-1} = \tau_m^{-1} + \tau_r^{-1} + \tau_s^{-1} \tag{1}$$

where τ_s is the electron spin relaxation time, τ_r is the rotational correlation time, and τ_m is the lifetime of the nucleus under investigation in the paramagnetic environment, if there is chemical exchange. Information on these parameters can be obtained from NMR experiments.

In a protein, in the absence of chemical exchange, the contribution to nuclear relaxation due to τ_s is always present. The latter time is essentially a property of the metal ion, sometimes modulated by the number and kind of donor atoms. Some values are reported in Table I together with the contribution to the NMR linewidth at 100 MHz for a proton 5 Å away from the metal ion. It is apparent that a copper(II) or manganese(II) protein will broaden beyond detection the lines of many protons in the neighborhood of the paramagnetic center. On the contrary, iron(II) or low-spin iron(III) will provide spectra with well-shifted and relatively narrow signals. Depending on the purpose, it may even be useful to prepare particular paramagnetic metalloproteins through metal substitution. For example, zinc(II) is substituted by cobalt(II) in zinc proteins.[4]

Information on τ_s can be obtained independently through nuclear magnetic relaxation dispersion (NMRD) spectroscopy, also called relaxometry.[5-7]

When a metal ion with a short τ_s (metal ion 1) is magnetically coupled with a metal ion with a long τ_s (metal ion 2), the electron(s) of the latter

[4] I. Bertini and C. Luchinat, *Metal Ions Biol. Syst.* **15**, 101 (1983).
[5] I. Bertini, C. Luchinat, and L. Messori, *Metal Ions Biol. Syst.* **21**, 47 (1987).
[6] S. H. Koenig and R. D. Brown III, *Metal Ions Biol. Syst.* **21**, 229 (1987).
[7] R. E. London, this series, Vol. 176 [18].

TABLE I

ELECTRON SPIN RELAXATION RATES AND
NUCLEAR LINE-BROADENING EFFECTS
OF PARAMAGNETIC METAL IONS AT
ROOM TEMPERATURE

Metal ion[a]	τ_s^{-1} (sec^{-1})	Nuclear line broadening (Hz)[b]
Ti(III)	10^9–10^{10}	3,000–500
VO(IV)	10^8–10^9	20,000–3,000
V(III)	2×10^{11}	100
V(II)	2×10^9	9,000
Cr(III)	2×10^9	9,000
Cr(II)	10^{11}	300
Mn(III)	10^{10}–10^{11}	3,000–300
Mn(II)	10^8–10^9	200,000–40,000
Fe(III) (HS)	10^{10}–10^{11}	5,000–400
Fe(III) (LS)	10^{11}–10^{12}	40–10
Fe(II) (HS)	10^{12}	70
Co(II) (HS)	10^{11}–10^{12}	200–50
Co(II) (LS)	10^9–10^{10}	3,000–500
Ni(II)	10^{10}–10^{12}	1,000–25
Cu(II)	3×10^8–10^9	9,000–3,000
Ru(III)	10^{11}–10^{12}	10–40
Re(III)	10^{11}	100
Gd(III)	10^8–10^9	400,000–60,000
Ln(III)	10^{12}	30–100

[a] HS, High spin; LS, low spin.
[b] Calculated for ^1H, dipolar relaxation only, $r = 5$ Å, $B_0 = 2.35$ T.

will also relax through the relaxation mechanisms of the former metal ion, with the result that the electron relaxation rates of metal ion 2 will be shortened. Under such circumstances, the NMR spectra can be easily observed for protons of ligands of both ions.[8,9] Therefore, metal clusters in biochemistry with at least one fast-relaxing metal ion (typically iron–sulfur proteins with iron in mixed oxidation states[10]) are suitable systems to be investigated via NMR. Again, clusters of this kind can be prepared through chemical manipulation, either by replacing zinc with cobalt in any

[8] C. Owens, R. S. Drago, I. Bertini, C. Luchinat, and L. Banci, J. Am. Chem. Soc. 108, 3298 (1986).
[9] I. Bertini, C. Luchinat, C. Owens, and R. S. Drago, J. Am. Chem. Soc. 109, 5208 (1987).
[10] T. G. Spiro (ed.), "Iron–Sulfur Proteins." Wiley, New York, 1982.

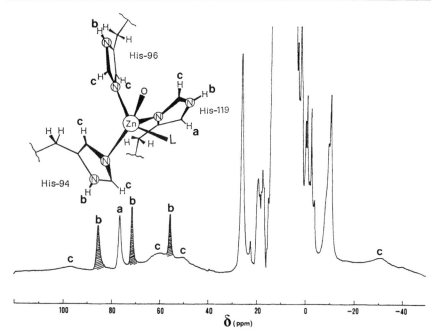

FIG. 1. ^1H NMR spectrum (room temperature, 200 MHz) of the NO_3^- adduct of cobalt-(II)-substituted bovine carbonate dehydratase II in unbuffered solution at pH 6.[12] The dashed signals disappear in D_2O. The structure shows the metal ligands: L is the NO_3^- ion and O is a water molecule. The correspondence between NMR signals and ligand protons is shown.

zinc plus other metal ion system,[11] or by artificially binding a metal ion, such as cobalt or a lanthanide, near a naturally occurring metal ion with a long τ_s.

^1H NMR Spectra of Some Typical Systems: Origin of Isotropic Shifts and Nuclear Relaxation

Some typical ^1H NMR spectra of paramagnetic metalloproteins are depicted in Figs. 1–4. Figure 1 shows the spectrum of an artificial high-spin cobalt(II) protein, i.e., cobalt(II) carbonate dehydratase II,[12] which contains zinc naturally. Figure 2 shows the spectrum of a natural heme protein containing low-spin iron(III).[13] Figure 3 shows the spectrum of a

[11] I. Bertini, G. Lanini, C. Luchinat, L. Messori, R. Monnanni, and A. Scozzafava, *J. Am. Chem. Soc.* **107**, 4391 (1985).

[12] I. Bertini, G. Canti, C. Luchinat, and F. Mani, *J. Am. Chem. Soc.* **103**, 7784 (1981).

[13] J. D. Cutnell, G. N. La Mar, and S. B. Kong, *J. Am. Chem. Soc.* **103**, 3567 (1981).

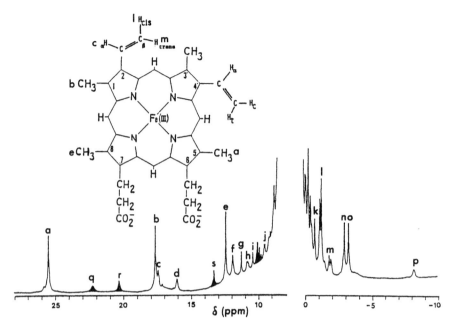

FIG. 2. ^{1}H NMR spectrum (40°, 360 MHz) of the cyanide adduct of sperm whale metmyoglobin at pH 8.6 in 0.2 M NaCl.[13] The shaded signals disappear in D_2O. The correspondence between some of the resonances and heme protons is shown.

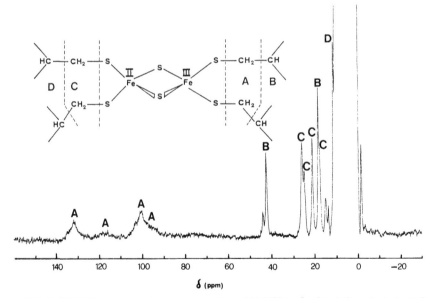

FIG. 3. ^{1}H NMR spectrum (room temperature, 200 MHz) of spinach ferredoxin in D_2O solution, 0.5 M Tris–HCl buffer, pH* 7.4 (*, the uncorrected pH meter reading).[14,31] The correspondence between NMR signals and ligand protons is shown. Note the much larger downfield shifts experienced by Fe(III) ligands.

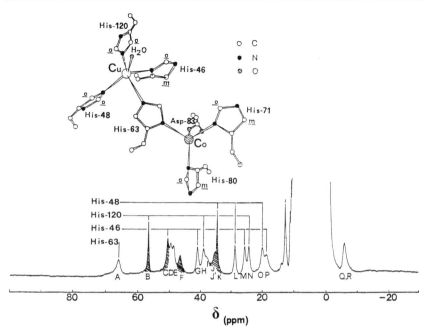

FIG. 4. ^{1}H NMR spectrum (300 MHz) of Cu_2Co_2SOD in H_2O at 303 K together with the signals' assignment for the protons of the copper domain. The shaded signals disappear when the spectrum is recorded in D_2O. The inset shows a schematic drawing of the active site.

2Fe2S-ferredoxin, containing a naturally occurring dimetallic cluster of iron(II) and iron(III) ions.[14] Figure 4 shows the spectrum of an artificial cobalt(II)–copper(II) dimetallic system.[15] In every case there are well-shifted signals with relatively short nuclear relaxation times, and slightly shifted signals with long nuclear relaxation times.

The first indications about the assignment come from the spectrum recorded in D_2O [in the case of metal-substituted proteins, it is better if the H_2O/D_2O exchange is performed on the demetallized protein and then the proper metal ion(s) are added]. All the signals due to exchangeable protons disappear. Typically they are due to histidine NH or peptide NH (these are seen in Figs. 1, 2, and 4 as shaded signals).

The nature of the isotropic shifts, when they are large, is essentially contact, i.e., scalar, in origin. This is due to the presence of unpaired

[14] I. Bertini, G. Lanini, and C. Luchinat, *Inorg. Chem.* **23**, 2729 (1984).
[15] L. Banci, I. Bertini, C. Luchinat, and A. Scozzafava, *J. Am. Chem. Soc.* **109**, 2328 (1987).

electron spin density at the resonating nuclei. Equation (2) gives the contact contribution[16,17]

$$\left(\frac{\Delta\nu}{\nu_0}\right)^{con} = -\frac{2\pi A_c}{h\gamma_N B_0}\langle S_z\rangle \tag{2}$$

where $\Delta\nu/\nu_0$ is the frequency shift relative to the spectrometer frequency (the shift in parts per million is, of course, 10^6 times $\Delta\nu/\nu_0$), A_c is the hyperfine coupling constant, h is Planck's constant, γ_N is the nuclear magnetogyric ratio, B_0 is the magnetic field, and $\langle S_z\rangle$ is the expectation value of S_z, in turn proportional to the z magnetization component of the electron spin magnetic moment, which in an isolated S manifold is given by

$$\langle S_z\rangle = -S(S+1)\frac{g_e\mu_B B_0}{3kT} \tag{3}$$

where S is the electron spin quantum number, g_e is the free-electron g factor, μ_B is the Bohr magneton, k is the Boltzmann constant, and T is the absolute temperature. Because $\langle S_z\rangle$ is proportional to B_0, the contact shift is independent of the magnetic field.

For some slightly shifted signals, such as those in the range 25 to -25 ppm in Fig. 1, the shift is often essentially dipolar in origin; it arises from a through-space dipolar interaction between the nucleus and the unpaired electron and from the presence of magnetic anisotropy, which, upon rapid rotation, averages the dipolar interaction to nonzero values depending on the coordinates of the resonating nucleus. The dipolar shifts are given in a simplified model by the following equation[17–19] (in the case that the magnetic susceptibility tensor is axially symmetric)

$$\left(\frac{\Delta\nu}{\nu_0}\right)^{dip} = \frac{1}{4\pi 3r^3}(3\cos^2\theta - 1)(\chi_\parallel - \chi_\perp) \tag{4}$$

where $\chi_\parallel - \chi_\perp$ is the magnetic susceptibility anisotropy, r is the length of the vector joining the metal ion and the resonating nucleus, and θ is the angle between the electron–nuclear vector and the z axis of the magnetic susceptibility tensor. The real unknowns in this type of analysis are the magnetic susceptibility tensor and the orientation of its principal axes. However, such shifts can be an important source of information if coupled to the nuclear relaxation properties. In the $S = \frac{1}{2}$ spin states, the χ

[16] H. M. McConnell and D. B. Chesnut, *J. Chem. Phys.* **28**, 107 (1958).
[17] R. J. Kurland and B. R. McGarvey, *J. Magn. Reson.* **2**, 286 (1970).
[18] W. D. Horrocks and D. D. Hall, *Inorg. Chem.* **10**, 2368 (1971).
[19] R. M. Golding, R. O. Pascual, and B. R. McGarvey, *J. Magn. Reson.* **46**, 30 (1982).

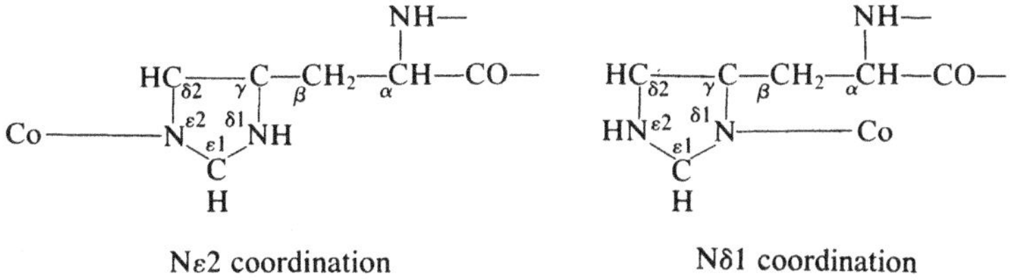

Nε2 coordination   Nδ1 coordination

SCHEME 1

values are proportional to the square of the g values obtained from EPR to first-order approximation; the dipolar shift can be expressed as[20]

$$\left(\frac{\Delta \nu}{\nu_0}\right)^{\mathrm{dip}} = \frac{\mu_0}{4\pi} \frac{\mu_B^2 S(S+1)}{9kTr^3} (3 \cos^2 \theta - 1)(g_\parallel^2 - g_\perp^2) \tag{5}$$

where μ_0 is the permeability of vacuum and the other symbols have been already defined.

The proton T_1 values of the spectrum in Fig. 1 are predominantly due to dipolar relaxation,[1] i.e., to through-space coupling between electrons and nuclei. The electron–nucleus dipolar contribution to the relaxation rate $(T_{1M}^{-1})^{\mathrm{dip}}$, is related to physical parameters through the following equation[21]:

$$(T_{1M}^{-1})^{\mathrm{dip}} = \frac{2}{15} \left(\frac{\mu_0}{4\pi}\right)^2 \frac{\gamma_N^2 g_e^2 \mu_B^2 S(S+1)}{r^6} \left(\frac{7\tau_c}{1+\omega_s^2\tau_c^2} + \frac{3\tau_c}{1+\omega_I^2\tau_c^2}\right) \tag{6}$$

where $g_e^2 \mu_B^2 S(S+1)$ is the square of the electronic magnetic moment, γ_N^2 is proportional to the square of the nuclear magnetic moment, r is the electron–nucleus distance, and the added terms in parentheses are a function of τ_c and of the electronic (ω_s) and nuclear (ω_I) Larmor frequencies. Equation (6) holds as long as the zero-field splitting is smaller than the Zeeman effect (3 cm^{-1} at 100 MHz); otherwise the predicted T_{1M}^{-1} can be shorter or longer than the actual value, depending on the external magnetic field and on the angle θ defined in Eq. (4).[22] The sharp signals about 50–90 ppm downfield have a T_1 typical of histidine metalike protons (either NH or CH)—it should be recalled that histidine NH protons are always in a metalike position with respect to the coordinating nitrogen, whereas metalike CH protons can only arise from Hδ2 protons when a histidine is coordinated through the Nδ1 nitrogen (Scheme 1). Other broad signals more downfield or in the same region can be assigned to His ortholike protons (Hε1 for Nδ1 coordination or both Hε1 and Hδ2 for Nε2

[20] H. M. McConnell and R. E. Robertson, *J. Chem. Phys.* **27**, 1361 (1958).
[21] I. Solomon, *Phys. Rev.* **99**, 559 (1955).
[22] I. Bertini, C. Luchinat, M. Mancini, and G. Spina, *J. Magn. Reson.* **59**, 213 (1984).

coordination). However, it soon becomes apparent, by looking at the T_1 values of the two sets of protons, that their ratio is much smaller than that expected on the basis of the $1/r^6$ ratios in Eq. (6) (which is about 25). This has been ascribed to the delocalization of the unpaired electrons on the imidazole ring: even a small fraction of unpaired electrons on a p orbital of a carbon to which a hydrogen atom is attached provides a large proton relaxation due to the short distance to the proton. This effect is particularly common in histidine ligands.

A further source of information derives from analysis of the T_2 values or linewidths. In the case of ^{1}H NMR, a dipolar contribution of the type described in Eq. (6) is always present.[1-3] Relaxation contributions due to contact interactions are usually negligible if the nucleus is not directly coordinated to the paramagnetic center. However, another contribution to the linewidth (but not to T_1) is operative, due to the different populations of the various electron spin M_s levels in a magnetic field. Such differences in populations give rise to a permanent moment associated with the molecule that is always aligned along the external magnetic field, which, upon molecular rotation, provides a further relaxation mechanism for the proton. Because the mechanism arises from the same physical phenomenon as that giving rise to magnetic susceptibility, this contribution is called Curie relaxation.[23,24] It is particularly relevant at high magnetic fields, because the electronic magnetic moment due to the difference in populations of the various M_s levels is proportional to the magnetic field, and nuclear relaxation depends on the square of the electronic magnetic moment. Its importance also increases with the size of the molecule, because the pertinent correlation time is the rotational correlation time τ_r defined in Eq. (1). The general equation for this contribution is

$$T_{2M}^{-1} = \frac{1}{5} \left(\frac{\mu_0}{4\pi}\right)^2 \frac{\omega_I^2 g_e^4 \mu_B^4 S^2 (S+1)^2}{(3kT)^2 r^6} \left(4\tau_r + \frac{3\tau_r}{1 + \omega_I^2 \tau_r^2}\right) \qquad (7)$$

where all the symbols have the usual meaning. Therefore the magnetic field dependence of the linewidth provides information on the distance: a plot of T_2^{-1} versus ω_I^2 gives a straight line with a slope, which is a function of S and proportional to $1/r^6$.

The spectrum of Fig. 2 is due to a cavity containing the low-spin heme iron(III) of cyanometmyoglobin. The $S = \frac{1}{2}$ spin state accounts for the relatively small isotropic shifts [Eqs. (2) and (3)]. The unpaired electron resides in an orbital of π symmetry, and therefore delocalizes onto the heme ring through π bonding. The heme methyl signals are easily identified from their larger intensity and have been assigned on myoglobin

[23] M. Gueron, *J. Magn. Reson.* **19**, 58 (1975).
[24] A. J. Vega and D. Fiat, *Mol. Phys.* **31**, 347 (1976).

samples artificially reconstituted with selectively deuterated heme rings.[25] Signals q, r, and s are due to exchangeable protons. Their selective T_1 values indicate that they belong to imidazole NH protons of the proximal and distal histidines and to the peptide NH proton of the proximal histidine, respectively.[13] In this system, homonuclear NOE experiments have been successfully performed on the upfield signals; through spectral difference they have been found to be related also to signals in the diamagnetic envelope. The buildup of NOE (dipolar interaction only) occurs according to Eq. (8).[26]

$$\text{NOE}_i(t) = \frac{\sigma_{ij}}{\rho_i} [1 - \exp(-\rho_i t)] \tag{8}$$

where ρ_i is the selective T_{1i}^{-1} value and σ_{ij} is the cross-relaxation term. Distances estimated from NOEs have allowed La Mar to assign signals n, o, and p to the Ile-99 protons and to recognize some vinyl signals.[27,28]

In the presence of magnetic coupling between two metal ions, when the absolute value of the magnetic coupling constant J (for the Hamiltonian $\mathcal{H} = J\mathbf{S}_1 \cdot \mathbf{S}_2$, with J positive for antiferromagnetic coupling) is not negligible with respect to kT, $\langle S_z \rangle$ is no longer given by Eq. (3), but should be evaluated for each metal ion i by taking into account the population distribution among the levels of the coupled system.[29]

$$\langle S_{iz} \rangle = \frac{\sum_{S'M_{s'}} \langle S'M_{s'}|S_{iz}|S'M_{s'}\rangle \exp\left(-\frac{E_{S'M_{s'}}}{kT}\right)}{\sum_{S'M_{s'}} \exp\left(-\frac{E_{S'M_{s'}}}{kT}\right)} \tag{9}$$

where the S' are the resulting electron spin levels in the pair and $E_{S'M_{s'}}$ is the energy of each $M_{s'}$ level. A nucleus interacting with both metal ions to a different extent would experience a contact shift given by an obvious extension of Eq. (2):

$$\left(\frac{\Delta\nu}{\nu_0}\right)^{con} = -\frac{2\pi}{h\gamma_N B_0} (A_{c1}\langle S_{1z}\rangle + A_{c2}\langle S_{2z}\rangle) \tag{10}$$

[25] A. Mayer, S. Ogawa, R. G. Shulman, T. Yamane, J. A. S. Cavaleiro, A. M. D. Rocha Gonsalves, G. W. Kenner, and K. M. Smith, *J. Mol. Biol.* **86**, 749 (1974).

[26] J. H. Noggle and R. E. Schirmer, "The Nuclear Overhauser Effect." Academic Press, New York, 1971.

[27] S. Ramaprasad, R. D. Johnson, and G. N. La Mar, *J. Am. Chem. Soc.* **106**, 3632 (1984).

[28] S. Ramaprasad, R. D. Johnson, and G. N. La Mar, *J. Am. Chem. Soc.* **106**, 5530 (1984).

[29] W. R. Dunham, G. Palmer, R. H. Sands, and A. J. Bearden, *Biochim. Biophys. Acta* **253**, 373 (1971).

where A_{c1} and A_{c2} are the hyperfine coupling constants with each metal ion (assumed to be invariant throughout the various $S'M_{s'}$ levels). By evaluating $\langle S_{1z} \rangle$ and $\langle S_{2z} \rangle$ in a simple case, Eq. (10) can be rewritten as

$$\left(\frac{\Delta\nu}{\nu_0}\right)^{con} = -\frac{2\pi g_e \mu_B}{h\gamma_N 3kT}\left[\frac{A_{c1}\sum_{S'} C_{1S'} S'(S'+1)(2S'+1)\exp\left(-\frac{E_{S'}}{kT}\right)}{\sum_{S'}(2S'+1)\exp\left(-\frac{E_{S'}}{kT}\right)}\right.$$

$$\left. +\frac{A_{c2}\sum_{S'} C_{2S'} S'(S'+1)(2S'+1)\exp\left(-\frac{E_{S'}}{kT}\right)}{\sum_{S'}(2S'+1)\exp\left(-\frac{E_{S'}}{kT}\right)}\right] \quad (11)$$

where $C_{1S'}$ and $C_{2S'}$ are appropriate coefficients related to the proportion of S_1 and S_2 in the S' wavefunctions[8,30]

$$C_{1S'} = \frac{S'(S'+1) + S_1(S_1+1) - S_2(S_2+1)}{2S'(S'+1)} \quad (12)$$

$$C_{2S'} = \frac{S'(S'+1) + S_2(S_2+1) - S_1(S_1+1)}{2S'(S'+1)} \quad (13)$$

When the nucleus feels only one metal ion (let us say 1), then the second term in parentheses in Eq. (10) or (11) vanishes. This case applies to reduced Fe_2S_2 proteins (Fig. 3), which contain iron(II) coupled with iron(III) with a coupling constant J of 100 cm^{-1}. As a result of the coupling, the electron spin relaxation rates of iron(III) approach those of iron(II), which are very short. The electron spin relaxation times of iron(III) in iron–sulfur proteins would be quite long, as demonstrated by very broad—or broad beyond detection—^{1}H NMR signals of the oxidized Fe_2S_2 proteins or oxidized rubredoxin, which contains an $Fe^{III}S_4$ moiety. As a result of magnetic coupling, sharp signals are obtained for all the protons feeling both iron(II) and iron(III) (Fig. 3). The shifts, however, are expressed by Eq. (11).[29,31] For J of the order of 100 cm^{-1}, by assuming that the hyperfine coupling is the same for both metal ions, a shift ratio of about $5:1$ is calculated for the β-CH_2 protons feeling iron(III) and iron(II), respectively. Furthermore, the shifts of protons of the iron(III) domain decrease with increasing temperature but those of the iron(II) domain increase with increasing temperature. This is what is actually

[30] J. Scaringe, D. J. Hodgson, and W. E. Hatfield, *Mol. Phys.* **35**, 701 (1978).
[31] L. Banci, I. Bertini, and C. Luchinat, "Structure and Bonding," in press (1989).

found and allows us to proceed with the assignment. The T_1 values depend on the square of the hyperfine coupling and therefore follow an equation of the type of Eq. (6), but contain the square of the coefficients[8,31]:

$$T_{1M}^{-1} = \frac{2}{15} \left(\frac{\mu_0}{4\pi}\right)^2 \gamma_N^2 g_e^2 \mu_B^2$$

$$\times \left[\frac{f(\tau_{c1}) \sum_{S'} C_{1S'}^2 S'(S' + 1) (2S' + 1) \exp\left(-\frac{E_{S'}}{kT}\right)}{r_1^6 \sum_{S'} (2S' + 1) \exp\left(-\frac{E_{S'}}{kT}\right)} \right.$$

$$\left. + \frac{f(\tau_{c2}) \sum_{S'} C_{2S'}^2 S'(S' + 1) (2S' + 1) \exp\left(-\frac{E_{S'}}{kT}\right)}{r_2^6 \sum_{S'}(2S' + 1) \exp\left(-\frac{E_{S'}}{kT}\right)} \right] \qquad (14)$$

where τ_{c1} and τ_{c2} are the correlation times for the coupling with each metal ion, possibly differing in τ_s, and $f(\tau_c)$ is the same function as that reported in Eq. (6). The analysis of the T_1 values confirms the assignment, and indicates that the electron spin relaxation times of iron(III) approach those of iron(II).[31]

If the absolute value of the magnetic coupling constant J is smaller than kT but larger than $\hbar\tau_s^{-1}$ of any uncoupled metal ion, then Eq. (11) shows that the shifts are purely additive with respect to the uncoupled system. However, nuclear relaxation rates are not, as shown by Eq. (14). Furthermore, the electronic relaxation times of the metal ion that by itself would have a longer τ_s are dramatically shortened by magnetic coupling. For example, in Cu_2Co_2 superoxide dismutase (Fig. 4), J has a value of 16 cm^{-1} (antiferromagnetic), and τ_s of copper decreases from 3×10^{-9} to about 10^{-11} sec.[15] As a result, the signals of protons of histidines bound to copper(II) are very sharp and have a long T_1. In qualitative agreement with Eq. (6), the T_1 values of ortholike protons of the cobalt domain would be the shortest due to the large value of $S(S + 1)$, followed by the corresponding protons of the copper domain; then the T_1 values of the metalike protons follow, respectively, for cobalt and copper.

The magnetic field dependence of the linewidth [Eq. (7)] between 60 and 400 MHz has also allowed us to distinguish between the protons which feel an $S = \frac{3}{2}$ magnetic moment (cobalt) and those which feel an $S = \frac{1}{2}$ magnetic moment (copper).[15]

Pitfalls for a quantitative agreement arise from large ligand-centered effects, which may be different from one proton to another. ^{1}H NOE experiments performed by saturating the His NH's have indicated which signals are due to protons vicinal to the NH's in the same His.[32]

NOEs have been measured for protons in the copper domain because such protons have the largest T_1 values. In the cobalt domain, NOEs have been observed between the signals assigned as geminal protons of the coordinated aspartate. A detailed assignment for each signal is reported in Table II.

Experimental Procedures

Recording the Spectra

^{1}H NMR spectra of metalloproteins in which one or more paramagnetic centers are present are characterized by (1) large chemical shift range, i.e., large range of isotropic shift values; (2) short T_1 and T_2 values; and (3) drastic field dependence of the linewidth (Curie contribution to T_2). In order to detect NMR signals under such conditions, some particular instrumental features are required.

To cover a large range of chemical shifts, the spectrometer should be able to irradiate a large band of frequencies with a good excitation profile. Therefore, short pulses must be used. Because for many purposes large tilt angles are also needed (*vide infra*), high radiofrequency power should ultimately reach the sample. This is achieved in the latest generation of commercial instruments either by a strong power output from the transmitter (up to 1–2 kW) or by the use of very high-Q coils. The latter choice, however, may make the tuning of the coil somewhat critical.

At high magnetic fields, fast digitizers are necessary in order to sample large spectral widths. Fast digitizers with still reasonable dynamic range are becoming available.

The short T_1 values of nuclei in these systems allow fast recycle times. On the other hand, very short T_2 values make the signals difficult to detect. If the instrumental dead time between the end of the pulse and the start of linear behavior of the receiver is of the same order of magnitude as the signal's T_2, a sizable part of the FID has decayed prior to detection, leading to a low signal-to-noise ratio. Furthermore, long delays prior to acquisition introduce signal phasing problems, which can be very serious with small broad signals in the presence of strong signals.

The best magnetic field value for recording ^{1}H NMR spectra is in general the highest one available to increase resolution and sensitivity, provided T_1 has not become too long. However, in paramagnetic macro-

[32] L. Banci, I. Bertini, C. Luchinat, M. Piccioli, A. Scozzafava, and P. Turano, *Inorg. Chem.*, submitted (1990).

TABLE II
ASSIGNMENT OF ISOTROPICALLY SHIFTED
SIGNALS IN Cu_2Co_2 SUPEROXIDE DISMUTASE[a]

Signal	Proton	Shift[b]
A	His-63 Hδ2	66.2
B	His-120 Hδ1	56.5
C	His-46 Hε2	50.3
D	His-80 Hδ2	49.4
	(His-71 Hδ2)	
E	His-71 Hδ2	48.8
	(His-80 Hδ2)	
F	His-80 Hε2	46.7
	(His-71 Hε2)	
G	His-46 Hδ2	40.6
H	His-120 Hε1	39.0
I	Asp-83 Hβ1 (Asp-83 Hβ2)	37.4
J'	Asp-83 Hβ2 (Asp-83 Hβ1)	35.6
J	His-71 Hε2	35.4
	(His-80 Hε2)	
K	His-48 Hε2	34.5
L	His-48 Hδ2	28.4
M	His-46 Hδ2	25.3
N	His-120 Hδ2	24.1
O	His-48 Hε1	19.6
P	His-46 Hβ1	18.7
Q, R	His-46 Hβ2, not assigned	−6.2

[a] At 300 MHz and 30°, in 50 mM acetate buffer,
pH 5.5.
[b] In parts per million from TMS.

molecules, Curie relaxation prevents the use of very high magnetic fields. For a six-coordinated cobalt(II) ion in a molecule of 100,000 Da, the best magnetic field to resolve signals of protons of residues bound to cobalt is 60 MHz!

With paramagnetic metalloproteins, there are broad, low-intensity signals together with high intensity signals due to the "diamagnetic" part of the protein, in addition to the solvent signal. It is often necessary to record the spectrum in H_2O rather than D_2O in order to detect signals of slowly (on the NMR time scale) exchangeable protons. The presence of signals with large differences in intensity may cause dynamic range problems, although modern instruments are often equipped with very high-resolution digitizers, and make baseline distortions more severe. The water signal can be suppressed through saturation either with the decoupler or with a presaturation pulse. This technique, however, induces a saturation transfer to the exchangeable protons, if they are present, reducing or

canceling their intensity. Solvent suppression techniques that do not ex-
cite the solvent are also applicable to paramagnetic samples. Good solvent suppression is achieved with the Redfield $2\bar{1}4\bar{1}2$ or the 1331 sequences[33]; however, they are not as good for simultaneously suppressing
the diamagnetic protein signals, unless severe distortions of the baseline
and intensity in the spectral region of interest are tolerable. The difference
in T_1 between isotropically shifted signals and the diamagnetic signals can
be used in some pulse sequences to eliminate (or reduce) both water and
diamagnetic signals.[33] The most commonly used sequences are WEFT,
super-WEFT, and a modification of the DEFT sequence. In our experience, the latter often turns out to be the most efficient. The modified
DEFT sequence uses a 90–τ–180–τ–90–AQ pulse train, with the τ value
and the recycle time chosen to be much smaller than the T_1 of the
diamagnetic signals but up to five times longer than the T_1 of the signals of
interest.[34] It is useful to cycle the pulse over the four different phases in
order to compensate for pulse imperfections and to set the transmitter on
the water resonance to eliminate strong ghosts in the spectrum.

The modified DEFT sequence can be easily used to measure nonselective T_1 values of fast-relaxing nuclei. The T_1 values are obtained by fitting
the signal intensity as a function of τ according to the following equation[34]:

$$I_\tau = I_\infty\left[1 - 2 \exp\left(\frac{-\tau}{T_1}\right) + (1 - a) \exp\left(\frac{-2\tau}{T_1}\right)\right] \qquad (15)$$

where I_τ is the signal intensity for the chosen τ value, a is cos θ, and θ is
the flip angle of the first pulse.

When T_1 values are not very short, and sizable diamagnetic contributions are operative, selective T_1 measurements are required. Selective
relaxation techniques are treated elsewhere in this series.[35]

NOE Experiments in Paramagnetic Metalloproteins

The nuclear Overhauser effect for a proton is the fractional variation
of the signal intensity when another resonance signal is saturated. Its
magnitude is proportional to the reciprocal of the sixth power of the
distance between the irradiated and the observed nucleus. The magnitude
of NOE depends on the applied magnetic field and on the correlation time
for the through-space interaction between the two protons.[36]

[33] P. J. Hore, this series, Vol. 176 [3].
[34] J. Hochmann and H. Kellerhals, J. Magn. Reson. 38, 23 (1980).
[35] N. Niccolai and C. Rossi, this series, Vol. 176 [10].
[36] J. A. Ferretti and G. H. Weiss, this series, Vol. 176 [1].

Theory predicts larger NOE values for macromolecules at high magnetic field (negative values) compared to those for small molecules in the condition of extreme motional narrowing (positive values) (cf. Borgias and James [9], Vol. 176, this series). In large macromolecules, however, considerable spin diffusion occurs, thus limiting the selectivity.

It has generally been assumed that unpaired electron(s) in a molecule would obviate any NOE effect; consequently, for a long time no investigation was attempted on paramagnetic systems. Recent reports on iron(III) (high and low spin) heme-containing proteins demonstrate, as can be predicted also from theoretical calculations, that NOE can be detected on fast-relaxing systems.[37] The amount of NOE depends on the intrinsic relaxation rate of the nucleus and on the correlation time for the interaction: for nuclei belonging to macromolecules and coupled with fast-relaxing unpaired electron(s), i.e., undergoing moderate nuclear relaxation enhancements, NOE effects up to 50% were detected. One great advantage of NOE in paramagnetic systems is that the large nuclear relaxation rates essentially eliminate spin diffusion effects, giving only primary NOE effects and providing selective structural information.

However, NOE detection in paramagnetic molecules presents experimental problems. First, the lines are difficult to saturate, so a decoupler with high rf power is necessary. A strong rf pulse produces large off-resonance effects over much of the spectrum. This situation could cause severe problems in NOE assignments due to difficulty in discriminating between true NOE and off-resonance effects. Therefore, attention is required in data analysis and, sometimes, more than one measurement with a different decoupler power (hence a different degree of saturation) is necessary for a safe assignment.

A second problem arises from the magnitude of the effect. Sometimes very small-intensity variations ($\leq 1\%$) are expected, which may fall below the dynamic range limit.[37] To alleviate this problem, suitable pulse sequences for suppressing residual water or intense diamagnetic signals are required. Sequences such as modified DEFT and super-WEFT on one side and $2\bar{1}4\bar{1}2$ and 1331 on the other side are or have been found useful in this respect.

The small expected NOEs require accumulation of a large number of scans. Although the total time may not be too long, owing to the short recycle times allowed by fast nuclear relaxation, the long-term instrumental stability is particularly critical in our case; the decoupler must often work at its maximum power, and a high stability in power and a symmetrical saturation profile are required to provide reliable data. The tempera-

---

[37] S. W. Unger, J. T. J. LeComte, and G. N. La Mar, *J. Magn. Reson.* **64,** 521 (1985).

ture must also be very stable ($\leq 0.1°$), because isotropic shifts are highly temperature dependent.

For obtaining good NOE difference spectra, it is important to appropriately choose the off-resonance positions in the spectrum that should be irradiated, in order to compensate for the strong off-resonance effects of the decoupler, especially in the diamagnetic region. The averaged spectrum, in which two symmetric positions with respect to the resonance line are irradiated, is found to be a good reference spectrum. Under these conditions, each point of the spectrum feels almost the same off-resonance effect both in the irradiated and in the reference spectrum, providing a good subtraction even in the diamagnetic region.

For t short enough compared to ρ_i^{-1}, Eq. (8) takes the form

$$\text{NOE}_i(t) = \sigma_{ij}t \tag{16}$$

This means the NOE is independent of the intrinsic nuclear relaxation, i.e., independent of the paramagnetism of the molecule. On the other hand, for saturation times long compared to ρ_i, we have a steady-state situation

$$\text{NOE}_i(t) = \sigma_{ij}/\rho_i \tag{17}$$

In the slow-motion limit, for a two-spin system, σ_{ij} has the form

$$\sigma_{ij} = (-h^2\gamma^4/10r_{ij}^6)\tau_c \tag{18}$$

If T_1 of the nucleus is not too short, we can monitor the NOE buildup described by Eq. (8) by measuring the NOE as a function of the irradiation time (truncated NOE). From this curve, the σ value can be extrapolated and structural information can be obtained. In the case of short T_1, steady state is quickly achieved and the NOE buildup curve is difficult to obtain. In any case, if the T_1 value is independently known, the cross-relaxation term can be estimated from a single NOE experiment. It is true, however, that NOE effects are small and sometimes difficult to detect. Because maximal differential NOEs are obtained for long τ_c values, it may be useful to increase the correlation time for molecular motion by dissolving the protein in a high-viscosity solvent or by increasing viscosity with appropriate solutes.

Perspectives and Concluding Remarks

Much structural information in a range of several angstroms from the metal can be garnered by analyzing the ^{1}H NMR spectra of paramagnetic metalloproteins. T_1, T_2, and NOE measurements provide information on the dynamics of the resonating nuclei. Furthermore, if T_1 and T_2 contain information on the distance of a nucleus from the metal, and NOE contains information on the distance between two nuclei, this approach may allow us to redesign the environment of metal ions in proteins, i.e., to design the active cavity of metalloenzymes. This information, coupled with molecular dynamics, is an independent, indispensable approach to the investigation of protein structures. Such studies can also be performed on ^{13}C, ^{19}F, and ^{31}P nuclei of the protein or of a substrate interacting with a protein.

The window of paramagnetic compounds that can be investigated is enlarged by magnetically coupled systems, either natural or artificial. A wealth of information may be obtained on the coupling and on the effects of the coupling on electron relaxation.

[13] Phosphorus-31 Nuclear Magnetic Resonance of Phosphoproteins

By Hans J. Vogel

Introduction

The ribosomal protein synthesis machinery utilizes the standard 20 amino acids as the building blocks for proteins. The five elements that are present in these amino acids are hydrogen, carbon, nitrogen, oxygen, and sulfur. Consequently, all proteins can in principle be studied by ^{1}H, ^{13}C, $^{14}N/^{15}N$, ^{17}O, or ^{33}S NMR techniques. Few studies of proteins, however, have utilized ^{14}N, ^{17}O, or ^{33}S NMR; the electric quadrupole moment of these nuclei renders resonances for the individual atoms very broad, leading to featureless unresolved spectra.[1] Because of their more favorable nuclear properties (spin $= \frac{1}{2}$), the study of the other three nuclei has provided considerable insight into structural, functional, and dynamic properties of proteins, as will be apparent from the various chapters in this volume.

As the element phosphorus is not incorporated during protein biosynthesis, it can only become incorporated into proteins through various posttranslational processes. Consequently only a few phosphorus atoms are found in most phosphoproteins. This has the advantage that the problem of resonance assignment, which can be quite formidable for ^{1}H, ^{13}C, and ^{15}N NMR studies of larger proteins, normally does not pose any

[1] S. Forsen and B. Lindman, *Methods Biochem. Anal.* **27**, 289 (1981).

METHODS IN ENZYMOLOGY, VOL. 177

TABLE I
PROPERTIES OF SOME COMMON NUCLEI

Nucleus	Resonance frequency (MHz)[a]	Natural abundance (%)	Sensitivity[b]	Chemical shift range for biomolecules (ppm)
^{1}H	400.1	99.98	1.00	15
^{13}C	100.6	1.11	0.016	200
^{15}N	40.5	0.37	0.001	100
^{31}P	162.0	100.0	0.066	20

[a] At 9.4 T.

[b] This is for an equal number of nuclei. The receptivity that actually governs the sensitivity of the NMR experiment (and which is usually defined as the isotopic abundance multiplied by the sensitivity) is less for nuclei with low isotopic abundance.

serious problems with ^{31}P NMR.[2] Furthermore, because of the paucity of resonances in ^{31}P NMR spectra of phosphoproteins, it has been possible to pursue studies of proteins with molecular weights up to 500,000.[2] In contrast, the problem of resonance overlap, which is inherent in ^{1}H, ^{13}C, and ^{15}N NMR studies of larger proteins, severely limits the size of protein that can be studied by these three techniques. On the other hand, the information that can be extracted by ^{31}P NMR is of a different nature than the overall structural detail that can potentially be obtained by the other techniques.

In addition to the different way in which phosphorus is incorporated into proteins, it is important to realize that the ^{31}P nucleus has a favorably high resonance frequency and a good sensitivity for NMR experiments (see Table I). It is also 100% naturally abundant, and thus can be studied without the need for introducing isotopic labels. Isotopic labeling can be expensive and experimentally cumbersome, and these factors have limited the use of nuclei with low natural abundance, as in ^{13}C or ^{15}N NMR.[3] The combination of the biological and physical factors mentioned above explains why ^{31}P NMR has become such a useful tool for the study of phosphoproteins. In this chapter we will focus on the practical aspects of the technique, which will be illustrated by some selected examples. It is

[2] H. J. Vogel, in "Phosphorus-31 NMR: Principles and Applications" (D. G. Gorenstein, ed.), pp. 104–154. Academic Press, Orlando, Florida, 1984.

[3] Modern NMR pulse methods can be used to study indirectly ^{15}N or ^{13}C nuclei, by utilizing heteronuclear quantum coherences and detecting the attached protons. These methods provide a great increase in sensitivity, which should overcome to some extent the need for incorporating isotopic labels.

not our intent to give an exhaustive review, and the reader wishing to obtain more details should consult other sources[2,4] or the original literature. Before discussing the experimental aspects of ^{31}P NMR of phosphoproteins, we will briefly review the biological mechanisms through which phosphate can be incorporated into proteins.

Phosphorus can become part of a protein through essentially three different mechanisms. The first mechanism is the incorporation of phosphorylated coenzyme into a protein. Such moieties are either extremely rigidly bound to the protein or in some cases they become covalently attached to the side chain of specific amino acids. An obvious example is the coenzyme pyridoxal phosphate, which occurs in the active site of a large number of enzymes.[5] Although the phosphate group of this coenzyme does not appear to be directly involved in enzymatic catalysis, the dephospho (pyridoxal) form generally does not support catalysis.[6,7] As such, the phosphate group presents a nonperturbing spectral reporter group for studying events that take place in the active site of the enzyme. Also, the cofactors FMN and FAD, which are part of many redox proteins, carry phosphate groups that can be studied by ^{31}P NMR.

The other two groups of phosphoproteins contain phosphorylated amino acids rather than coenzymes. Of particular interest is a large group of proteins and enzymes whose activity is regulated by reversible enzymatic phosphorylation/dephosphorylation reactions. Specialized protein kinases that can be activated by hormonal stimulation can transfer a phosphoryl group (PO_3^{2-}) from ATP onto unique serine, threonine, or tyrosine residues on selected target proteins.[8,9] In some nuclear proteins unique histidine, lysine, and arginine side chains can also become phosphorylated. Although these regulatory phosphorylation sites are often far removed from the active site of these enzymes, the phosphorylation has marked effects on the activity. The phosphoryl groups can subsequently be removed by specialized protein phosphatases that are also hormonally controlled. In addition to these regulatory phosphorylations there is a series of food storage proteins in milk and eggs (ovalbumin, casein, and phosvitin) and a series of polyelectrolyte proteins (bone, tooth, and saliva phosphoproteins) that are also phosphorylated by protein kinases. How-

[4] T. L. James, *CRC Crit. Rev. Biochem.* **18,** 1 (1985).
[5] K. D. Schnackerz and E. E. Snell, *J. Biol. Chem.* **258,** 4839 (1983).
[6] M. E. Mattingly, J. R. Mattingly, and M. Martinez-Carrion, *J. Biol. Chem.* **257,** 8872 (1982).
[7] S. G. Withers, N. B. Madsen, B. D. Sykes, M. Takagi, S. Shimomura, and T. Fukui, *J. Biol. Chem.* **256,** 10759 (1981).
[8] P. Cohen, *Eur. J. Biochem.* **151,** 439 (1985).
[9] A. M. Edelman, D. K. Blumenthal, and E. G. Krebs, *Annu. Rev. Biochem.* **56,** 567 (1987).

ever, in this case, the phosphorylation does not play a regulatory role and the proteinous phosphate group is either involved in the binding and precipitation of metal ions or simply functions as a phosphate store.

The second group of phosphoproteins, which contain phosphoamino acids, consists of enzymes in which the phosphorylated amino acid is an enzymatic catalytic intermediate.[10] The phosphoryl group is transferred from substrates (such as ATP, for example) onto an amino acid side chain in the active site. The residues that become phosphorylated are either a serine (as in the enzymes alkaline phosphatase and phosphoglucomutase), a histidine (as in HPr, succinate–CoA ligase, phosphoglycerate mutase, ATP citrate-lyase), or an aspartate (which has been observed for the Ca^{2+}-, Na^+,K^+-, and the H^+,K^+-ATPase). Because they are intermediates in catalysis, their formation is of a transient nature during active turnover. Nevertheless, many of these enzymes can be purified in a stable phosphorylated form that can be studied by ^{31}P NMR. In cases where the covalent phosphoamino acid is unstable, other techniques involving the use of the oxygen isotopes ^{18}O and ^{17}O may be useful to determine whether a phosphorylated intermediate exists[10,11] and whether it is an obligatory intermediate in catalysis.[12]

In the following discussion we examine the NMR parameters that can be measured by ^{31}P NMR. We will mainly focus on proteins that contain phosphoamino acids, but some examples involving bound coenzymes will also be mentioned.

Sample Preparations

For most proteins it will be necessary to make up a 1.5-ml sample with a protein concentration of ≥ 0.2 mM. This sample size is necessary to fill a 10-mm NMR tube to a level that adequately covers the receiver coil and to obtain a signal in a reasonable time (several hours). For proteins that give rise to relatively narrow resonances, it is possible to use protein concentrations that are somewhat lower. The relatively high protein concentration may sometimes result in protein aggregation. In such cases it is advisable to use a larger NMR tube and to reduce the protein concentration while keeping the same amount of protein within the area covered by the receiver coil. For all samples, it is usually advisable to use a Teflon plug, to avoid a vortex in the sample while spinning. Because most phosphorylated compounds carry a negative charge, precautions have to be

[10] J. R. Knowles, *Annu. Rev. Biochem.* **49,** 877 (1980).
[11] J. J. Villafranca, this volume [20].
[12] M. J. Wimmer and I. A. Rose, *Annu. Rev. Biochem.* **47,** 1031 (1978).

taken to exclude positively charged paramagnetic metal ions such as Mn^{2+} and Ni^{2+}. In order to avoid any such problems, it is therefore advisable to treat all the buffers—and even the 10–20% D_2O which is normally used for lock purposes—with a chelating agent such as Chelex 100.

It is also of extreme importance to ascertain that the sample does not contain any phosphorylated impurities. For example, nucleic acids or lipids can sometimes be difficult to extract and they can copurify with a protein. The phosphodiester linkages in these molecules give rise to peaks in the ^{31}P NMR spectrum. Extreme conditions, such as organic solvents, are sometimes necessary to remove such impurities. In one case we even were forced to precipitate the protein to prove that a phosphodiester linkage was not part of a protein, but rather a lipid impurity.[13] Not only nucleic acids and lipids can cause problems. For example, the extremely tight binding of the nucleotides AMP or IMP to glycogen phosphorylase necessitates a dialysis procedure against activated charcoal to remove these contaminants.[7] If such precautions are not taken, erroneous results can occur. For example, Fossel et al.[14] reported a highly unusual chemical shift for the catalytic phosphoaspartate in the active site of a Na^+, K^+-ATPase. The enthusiasm for this intriguing observation was severely tempered when subsequent studies showed that the resonance was in fact an ATP contamination and that the real intermediate had a very normal chemical shift.[15] Finally, it is important to ascertain that the sample is free from protease and phosphatase activity. As it may be in the spectrometer for several days at temperatures above 4°, it should preferably also be sterile to prevent growth of microorganisms. In some cases addition of a small amount of an antibiotic may be useful. If these conditions are not met, degradation reactions may take place in the course of the experiment, which will generally affect the experimental outcome. Some covalent enzymatic phosphoamino acid intermediates may be intrinsically unstable; this will lead to the gradual disappearance of their resonance and to the appearance of a P_i peak in the spectrum.

With respect to the pulsing conditions under which spectra should be collected, it is difficult to generalize. In our experience T_1 values for phosphoproteins can vary between 0.25 and 4 sec. Thus the optimal NMR parameters will need to be determined for each individual case.

[13] G. D. Armstrong, L. S. Frost, H. J. Vogel, and W. Paranchych, J. Bacteriol. **145**, 1167 (1981).
[14] E. T. Fossel, R. L. Post, D. S. O'Hara, and T. W. Smith, Biochemistry **20**, 7215 (1981).
[15] G. M. Sontheimer, H. R. Kalbitzer, and W. Hasselbach, Biochemistry **26**, 2701 (1987).

TABLE II
ACID–BASE STABILITY, pK_a VALUES, AND ^{31}P NMR PARAMETERS FOR PHOSPHOAMINO ACIDS[a]

Phosphoamino acid	Stable to strong acid	Stable to strong base	Chemical shift (ppm)		pK_a
			Deprotonated	Protonated	
Phosphomonoesters[b]					
Phosphoserine	+	−	4.6	0.6	5.8
Phosphothreonine	+	−	4.0	0.0	5.9
Phosphotyrosine	+	−	1.0	−3.3	5.8
Phosphodiesters[c]					
RNA, DNA, and phospholipids	+	−	0 to −1.5	No	No
Acyl phosphates					
Phosphoaspartate	−	−	−1.5	−6.5	4.8
Phosphoramidates					
N^3-Phosphohistidine	−	+	−4.5	No	No
N^1-Phosphohistidine	−	+	−5.5	No	No
Phospholysine	−	+	ND	ND	ND
Phosphoarginine	−	+	−3.0	−5.4	4.3

[a] All chemical shifts are reported with reference to 85% H_3PO_4. Upfield shifts are given a negative sign. ND, Not determined.

[b] The chemical shifts and pH titration behavior of the coenzymes pyridoxal and pyridoxamine phosphate are very similar to those of phosphothreonine, whereas for FMN, these parameters resemble those of phosphoserine.

[c] The diphosphodiester FAD (−10.8/−11.3 ppm) also does not show a titration behavior between pH 3 and 10.

Chemical Shift and Nature of Phosphorylated Amino Acids

The nature of a phosphoamino acid in a newly isolated phosphoprotein is generally determined from a study of its acid and base stability (see Table II).[16-18] This simple test normally provides a reasonable first indication but some caution has to be exercised, because the result is known to be dependent on adjacent amino acids; moreover, elimination reactions may occur.[16-18] Therefore the next step is usually the purification of a phosphopeptide, from which subsequently a phosphoamino acid is obtained and identified. ^{31}P NMR provides an alternative noninvasive method for determining the nature of phosphoamino acids.[2] The chemical shifts[19] measured for the various amino acids are quite different (see Table

[16] G. Taborsky, Adv. Protein Chem. **28,** 1 (1974).
[17] T. M. Mortensen, this series, Vol. 107, p. 1.
[18] J. M. Fujitaki and R. A. Smith, this series, Vol. 107, p. 23.
[19] Chemical shifts in NMR experiments are always measured with respect to a standard. The commonly used standard for ^{31}P NMR is 85% H_3PO_4. This standard is not convenient for

II) and thus they can be used directly for identification purposes. Naturally, chemical shifts are only a direct monitor of the chemical nature of a phosphoamino acid as long as the incorporation into the protein does not perturb the phosphate moiety. By and large no such deviations have been observed for all the regulatory sites where the phosphoryl group has been introduced by the action of a protein kinase.[2] With the exception of alkaline phosphatase, the majority of the covalent enzymatic intermediates have chemical shifts that are relatively close (<2.0 ppm) to those observed for the free amino acid in solution. Thus the chemical shifts for the histidyl-P of succinate–CoA ligase[20,21] and ATP citrate-lyase,[22] the aspartyl-P of the ATPases,[15] and the serine-P of phosphoglucomutase[23,24] were all close to their respective standards. Nevertheless, bond strain has been invoked to explain the 3.8-ppm downfield shift for the serine phosphate residue in the active site of alkaline phosphatase[25,26] and for the 1.5-ppm downfield shift of the active site N^1-phosphate of histidine in the HPr protein.[27,28] Thus it is often advisable to determine the chemical shift for the protein in both the native and a denatured state to be absolutely certain. Denaturation can be brought about by urea, pH extremes, etc. Particularly if one wants to differentiate between the N^1- and N^3-phosphohistidine, which have quite similar chemical shifts, this is a necessary step.[27,29] Further confirmation for the nature of the phosphoamino acid may be obtained by looking for a characteristic pH dependence of the shift (see below).

Although this step is useful for the identification of phosphomonoesters, unfortunately it does not provide help in identifying phosphohisti-

---

two reasons. First, it falls right in the middle of the spectrum. Second, and most important, the nature of the standard precludes its use as an internal standard. Although capillaries can be used, the high concentration of the standard also makes this impractical for samples that have weak signals. Therefore, we generally include compounds that have a well-characterized shift, such as phosphocreatine or methylene diphosphonate, as internal standards. The latter is conveniently downfield (~20 ppm) from the biologically occurring compounds, but the pH dependence of its chemical shift may cause some problems.

[20] H. J. Vogel, W. A. Bridger, and B. D. Sykes, *Biochemistry* **21**, 1126 (1982).
[21] H. J. Vogel and W. A. Bridger, *J. Biol. Chem.* **257**, 4834 (1982).
[22] S. P. Williams, B. D. Sykes, and W. A. Bridger, *Biochemistry* **24**, 5527 (1985).
[23] W. J. Ray, A. S. Mildvan, and J. B. Grutzner, *Arch. Biochem. Biophys.* **184**, 453 (1977).
[24] G. I. Rhyu, W. J. Ray, and J. L. Markley, *Biochemistry* **24**, 4746 (1985).
[25] J. L. Bock and B. Sheard, *Biochem. Biophys. Res. Commun.* **66**, 24 (1975).
[26] W. E. Hull and B. D. Sykes, *J. Mol. Biol.* **98**, 121 (1975).
[27] M. Gassner, D. Stehlik, O. Schrecker, W. Hengstenberg, W. Mauer, and H. Ruterjans, *Eur. J. Biochem.* **75**, 287 (1977).
[28] G. Dooijewaard, F. Roossien, and G. T. Robillard, *Biochemistry* **18**, 2996 (1979).
[29] J. M. Fujitaki, G. Fung, E. Y. Oh, and R. E. Smith, *Biochemistry* **20**, 3658 (1981).

dines, as a change in pH is not accompanied by a change in chemical shift.[27] In some highly phosphorylated phosphoproteins, such as phosvitin[30] or the neurofilament proteins,[31] phosphoserine and phosphothreonine residues can be present simultaneously. These can potentially be differentiated by their chemical shifts but further confirmation can be obtained if the coupling constants can be measured (see below).

For NMR studies of most nuclei, other than ^{31}P, there is often a good understanding for the manner in which the binding of certain agents should affect the chemical shift. For example, it would be useful if an upfield shift on the binding of a metal ion in a ^{31}P NMR spectrum could be interpreted as a direct effect of coordination. Unfortunately, however, this cannot be done. The phosphorus chemical shift is very dependent on the bond geometry, on the electronegativity of the substituents, and on the relative amount of π bonding. These three parameters are interdependent and their interrelationships are unknown.[32,33] Gorenstein and co-workers[32] have experimentally demonstrated the large effects that bond strain may exert on the chemical shift. Moreover, we have observed that the shifts induced by substituents with varying electronegativity were exactly opposite to what was theoretically predicted based on a simple deshielding model.[33] Thus, although the formation of salt linkages[34] and hydrogen bonds[35] undoubtedly will exert some effects on the chemical shifts, it is presently impossible to predict theoretically in what direction the shift should change. As a result, the interpretation of ^{31}P chemical shifts remains empirical.

Coupling Constants

Of all the phosphorylated amino acids, only phosphoserine and phosphothreonine display an easily detectable coupling to adjacent protons. In the case of phosphoserine, a three-bond coupling to the two β protons results in a triplet with a coupling constant $J_{POCH} \cong 6.5$ Hz. Phosphothreonine has a three-bond coupling to its single P proton and thus displays a doublet.[36] Broadband or composite pulse proton decoupling

[30] H. J. Vogel, *Biochemistry* **22,** 668 (1983).

[31] U. P. Zimmerman and W. W. Schlaepfer, *Biochemistry* **25,** 3533 (1986).

[32] D. G. Gorenstein (ed.), "Phosphorus-31 NMR: Principles and Applications," pp. 7–36. Academic Press, Orlando, Florida, 1984.

[33] H. J. Vogel and W. A. Bridger, *Biochemistry* **21,** 394 (1982).

[34] K. D. Schnackerz, K. Feldmann, and W. E. Hull, *Biochemistry* **18,** 1536 (1979).

[35] F. E. Evans and N. O. Kaplan, *FEBS Lett.* **105,** 11 (1979).

[36] C. Ho, J. A. Magnuson, J. B. Wilson, N. S. Magnuson, and R. J. Kurland, *Biochemistry* **8,** 2074 (1969).

can be used to remove these coupling effects. The resulting collapse of the signals is often useful, as it improves the signal-to-noise ratio. However, the J_{POCH} coupling-constant information contains valuable information because a Karplus relation exists describing the bond angles for these three-bond couplings.[37,38] Thus, a determination of the coupling constant may provide information about preferred side-chain conformations. In the case of ovalbumin a normal bond angle and rotational freedom were observed,[39] but for alkaline phosphatase an unusual rotamer population was involved.[40] For larger proteins with short T_2 relaxation times, the coupling is generally not observed. At a higher magnetic field they are also difficult to observe because of the effects of chemical shift anisotropy (see below) on the linewidths.

Relaxation Behavior

The ^{31}P nucleus in most biologically relevant molecules is primarily surrounded by ^{16}O and ^{12}C, which would not aid in the relaxation. In phosphoramidates the ^{14}N nucleus to which the ^{31}P is directly bonded could give rise to some scalar relaxation of the second kind; however, this effect is unlikely to be of great importance for macromolecules with long correlation times. As a result, the two major contributing mechanisms for relaxation in protein-bound phosphorous atoms is through dipole–dipole interactions with neighboring ^1H, or through chemical shift anisotropy (CSA) mechanisms.[41] In the limit of slow motion ($\omega_p\tau_c > 1$), which will apply for most proteins as long as the phosphate group does not rotate freely with respect to the protein surface, the formulas for the linewidth are as follows[2,20,42]:

Chemical shift anisotropy:
$$\Delta\nu = \frac{4}{45\pi} \omega_P^2 (\Delta\sigma)^2 \left(1 + \frac{\eta^2}{3}\right)\tau_c \quad (1)$$

Proton–phosphorus dipole–dipole:
$$\Delta\nu = \frac{1}{5\pi} \left(\frac{\gamma_P^2\gamma_H^2\hbar^2}{r_{PH}^6}\right)\tau_c \quad (2)$$

[37] L. D. Hall and R. B. Malcolm, *Can. J. Chem.* **50**, 2102 (1972).
[38] B. J. Blackburn, R. D. Lapper, and T. C. P. Smith, *J. Am. Chem. Soc.* **95**, 2873 (1973).
[39] H. J. Vogel and W. A. Bridger, *Biochemistry* **21**, 5825 (1982).
[40] J. F. Chlebowski, I. M. Armitage, P. P. Tusa, and J. E. Coleman, *J. Biol. Chem.* **251**, 1207 (1976).
[41] In small molecules, contributions of the spin–rotation mechanism to the relaxation cannot be ignored. For a discussion, see, for example, P. Bendel and T. L. James, *J. Magn. Reson.* **48**, 76 (1982) and references therein.
[42] B. D. Nageswara Rao, *in* "Phosphorus-31 NMR: Principles and Applications" (D. G. Gorenstein, ed.), pp. 57–103. Academic Press, Orlando, Florida, 1984.

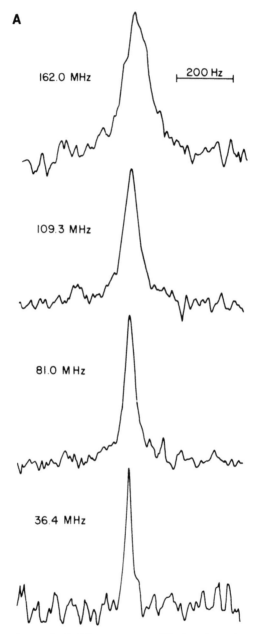

FIG. 1. (A) The N^3-phosphohistidine resonance of *Escherichia coli* succinate–CoA ligase as recorded at four spectrometers with different field strength. (B) The measured linewidth at half-height plotted against the square of the resonance frequency. The frequency-independent portion of the linewidth is indicated by the dotted line. The contribution of the chemical shift anisotropy is indicated by the solid line and clearly becomes dominant at higher fields. [Reprinted with permission from H. J. Vogel, W. A. Bridger, and B. D. Sykes,[20] *Biochemistry* **21,** 1126 (1982). Copyright 1982 American Chemical Society.]

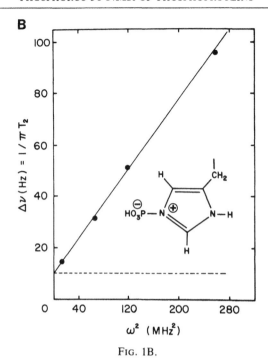

FIG. 1B.

where $\Delta\nu$ is the linewidth, τ_c is the correlation time, ω_P is the resonance frequency, $\Delta\sigma(1 + \eta^2/3)^{1/2}$ is the "anisotropy term," γ is the gyromagnetic ratio for different nuclei, and r_{PH} is the distance between phosphorus and proton nuclei.

Obviously for the chemical shift anisotropy mechanism the linewidth is linearly dependent on ω_P^2. If this relaxation mechanism contributes, the linewidth should increase with increasing resonance frequency. Thus, unlike ^{13}C and ^{1}H, for ^{31}P the highest magnetic field strength does not necessarily give the best resolution and sensitivity. As both dipolar and CSA relaxation mechanisms play a role in relaxation of protein-bound phosphorus nuclei, it will be necessary to determine their relative contributions to the relaxation before τ_c can be obtained from relaxation measurements. This can be accomplished by performing frequency-dependent measurements, where the linewidth is measured at various field strengths (see Fig. 1). Such an analysis has only been performed in a few instances; it has been demonstrated that at lower field (2.3 T) the dipolar mechanism is dominant, whereas at higher field (9.4 T) the linewidth is mainly determined by the chemical shift anisotropy mechanism.[20,39] Although this situation will not be as severe for relatively mobile phosphate groups, similar problems will be encountered for most phosphoproteins.

It should be noted that in contrast to the linewidth ($\sim 1/T_2$), the T_1 originating from CSA is not necessarily field dependent, but that the T_1 for dipolar mechanisms is inversely related to ω_P^2.[42] Thus an additional measurement of T_1 may be helpful in diagnosing the contributions from the various relaxation mechanisms. A combination of both T_1 and linewidth measurements should give greater confidence to the analysis; if only the latter is used, one has to consider the possibility that an increasing linewidth with ω_P^2 could also be caused by chemical exchange.[20]

Heteronuclear nuclear Overhauser effects (NOEs) from ^1H to ^{31}P for phosphoproteins have received relatively little attention.[43] Close to the maximum NOE can be observed in highly mobile phosphate groups under conditions where the heteronuclear dipolar relaxation mechanisms are still dominant. However, it should collapse relatively fast when measurements are done at higher magnetic field and motions are slow.[43]

pH Titrations

The data in Table II show that the majority of all phosphoamino acids display a pH dependence for the chemical shift, the only exception being the two phosphohistidine moieties.[44] Thus pH titration experiments provide a useful means for studying the solvent exposure of these residues. In addition, information about the pK_a values for individual groups can be obtained for proteins which give rise to ^{31}P NMR spectra in which the individual resonances can be resolved.[39] As indicated above, in addition to chemical shift the observation of a typical pH titration behavior provides further support for the identification of the chemical nature of a phosphoamino acid. Not only the pK_a values (see Table II) but also the changes in chemical shift (Δ ppm) with titration are diagnostic for the different classes of phosphoamino acids[45] (see Fig. 2).

The absence of a discernible pH titration for phosphoamino acids is usually caused by inaccessibility or by the presence of a strong linkage with a positively charged side chain of lysine or arginine, for example. However, the latter situation does not need to abolish the pH titration. Proximal positive charges may in fact lower the pK_a, or they may give rise to an increase in the Hill coefficient,[46] which describes the cooperativity

[43] P. A. Hart, in "Phosphorus-31 NMR: Principles and Applications" (D. G. Gorenstein, ed.), pp. 317–348. Academic Press, Orlando, Florida, 1984.
[44] Denaturants such as urea and high temperatures may have some effects on the chemical shifts and pK_a values determined in ^{31}P NMR pH titration experiments. These effects are generally quite small (<0.2 ppm or <0.2 pH units).
[45] H. J. Vogel and W. A. Bridger, Can. J. Biochem. Cell Biol. 61, 363 (1983).
[46] J. L. Markley, Acc. Chem. Res. 8, 70 (1975).

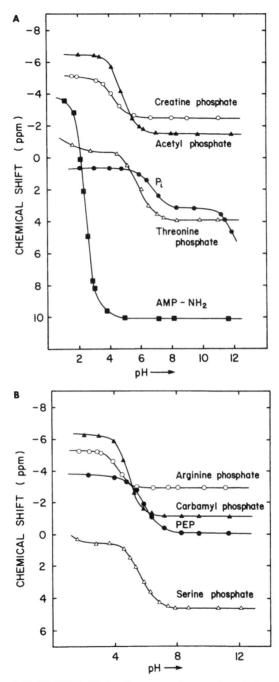

Fig. 2. (A and B) ^{31}P NMR pH titration curves for a series of phosphorus-containing model compounds that resemble phosphoamino acids. (Reproduced with permission from Vogel and Bridger.[45])

of the titration.[45] Alternatively, proximal negative charges could result in increases in the determined pK_a, or in a decrease in the Hill coefficient.[45] For the highly acidic protein phosvitin, for example, we have measured a fairly normal pK_a but a very low Hill coefficient, indicative of the fact that the many negative charges on this protein influence each others' titration behavior.[30] If the neighboring group has a $\Delta pK_a > 2$, a separate inflection in the titration curve may be observed.[47] From a comparison of titration data, it is sometimes even possible to estimate the maximum free energy that is involved in the formation of salt linkages.[45] Results obtained by ^{31}P NMR pH titration experiments may be further supported by FTIR experiments[48,49] in which the protonation state of the phosphate group can be directly followed. Thus, combining the two techniques allows one to distinguish readily between the direct effects of protonation or other conformational effects.

A final note of caution should be added: Observation of a change in chemical shift does not directly imply a protonation/deprotonation event on the phosphate group. For example, a phosphate moiety introduced as a modification into the active site of serine proteases displayed a clear pH titration, despite the fact that this compound does not have a pK_a within this range. These changes were attributed to the titration of neighboring groups.[50,51] Thus the possible effects of such interactions or the possibility of pH-induced conformational changes should always be taken into consideration.

A final question to be addressed in this section is whether, in the absence of a pH titration for a phosphomonoester moiety, a measurement of the chemical shift alone is sufficient to decide whether a residue is in its protonated or deprotonated form. This has been a contentious issue for some time. Naturally in the case of alkaline phosphatase, where a strained phosphoryl group gives rise to an unusual chemical shift, such information cannot be obtained.[25,26,40] However, in most other proteins and enzymes that have been studied, including all the pyridoxal phosphate-containing enzymes, the chemical shifts have been fairly close to those observed for model compounds, thus giving support to the notion that information about the chemical shift alone can be taken as support for the

[47] R. I. Shrager, J. S. Cohen, R. S. Heller, D. H. Sachs, and A. N. Schechter, *Biochemistry* **11**, 541 (1972).

[48] J. M. Sanchez-Ruiz and M. Martinez-Carrion, *Biochemistry* **25**, 2915 (1986).

[49] J. M. Sanchez-Ruiz and M. Martinez-Carrion, *Biochemistry* **27**, 3338 (1988).

[50] M. A. Porubcan, W. A. Westler, I. B. Ibanez, and J. L. Markley, *Biochemistry* **18**, 4108 (1979).

[51] A. C. M. Van der Drift, H. C. Beck, W. H. Dekker, A. G. Hulst, and E. R. J. Wils, *Biochemistry* **24**, 6894 (1985).

protonation state. Phosphodiester linkages can also readily be distinguished by ^{31}P NMR because they have no pK_a above pH 3.0 and as a result they should not titrate. Although the occurrence of such cross-links in proteins is not widespread, they have been found in some prokaryotic proteins.[52] ^{31}P NMR has been used to identify and study such linkages in the proteins flavodoxin[53,54] and glucose oxidase.[55]

Effects of Substrates and Other Ligands

The binding of substrates to the active site of enzymes that have a stable covalent phosphoamino acid catalytic intermediate may result in changes in the chemical shift and the linewidth for the phosphoamino acid. For example, for the Krebs cycle enzyme succinate–CoA ligase, it was observed that the addition of coenzyme A caused a large broadening of the N^3-phosphohistidyl resonance.[21] The subsequent addition of a nonmetabolizable analog of the second substrate succinate to form the ternary complex reduced the linewidth again to a value indicative of a tightly held residue, with no mobility with respect to the enzyme (see Fig. 3). These data have been interpreted in terms of a mechanism where the phosphohistidine can adopt two different conformations; exchange between these two is induced by the binding of coenzyme A. One of the conformations, which is dominant only when both coenzyme A and succinate are present, is thought to favor the in-line nucleophilic attack which generates the transient succinyl-P intermediate.[21] The other conformation was thought to facilitate the phosphorylation of the histidine by ATP. Also in the case of phosphoglucomutase, two exchanging conformations have been observed for the active-site phosphoserine.[24]

For the amino acids that are phosphorylated by kinases, little effect has been observed when ligands were added. Thus, in the case of the riboflavin-binding protein, the binding of riboflavin did not cause any changes for a series of phosphoserines.[56] However, in the case of xanthine oxidase, a phosphoserine residue was observed which broadened dramatically upon the binding of other ligands.[57] It was therefore suggested that this residue was in the active site of the enzyme and that it was possibly involved as a nucleophile in catalysis. Interestingly, in the case of glycogen phosphorylase, it was observed that addition of the inhibitor

[52] S. P. Adler, D. Purich, and E. R. Stadtman, *J. Biol. Chem.* **250**, 6264 (1975).

[53] D. E. Edmondson and T. L. James, *Proc. Natl. Acad. Sci. U.S.A.* **76**, 3786 (1979).

[54] C. T. Moonen and F. Muller, *Biochemistry* **21**, 408 (1982).

[55] T. L. James, D. E. Edmondson, and M. Husain, *Biochemistry* **20**, 617 (1981).

[56] M. S. Miller, M. T. Mas, and H. B. White, *Biochemistry* **23**, 569 (1984).

[57] M. D. Davis, D. E. Edmondson, and F. Muller, *Eur. J. Biochem.* **145**, 237 (1984).

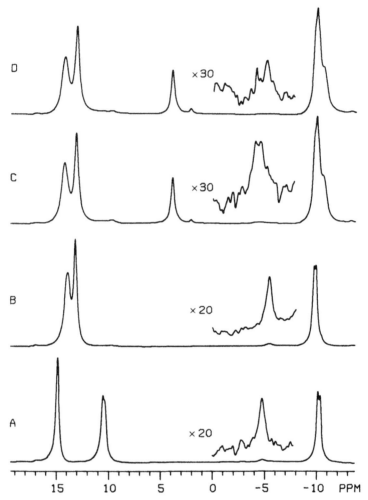

FIG. 3. ^{31}P NMR spectra of *Escherichia coli* succinate–CoA ligase (A) in the presence of the ATP analog AMPPCP, (B) plus Mg^{2+}, (C) plus coenzyme A, and (D) plus 2,2-difluoro-succinate, a competitive analog of the substrate succinate. The phosphohistidine resonance is around −5 ppm, shifts upfield on addition of Mg^{2+}, broadens and shifts downfield with coenzyme A, and narrows and moves upfield again when the difluorosuccinate is added. The ATP methylene analog was used here because it is nonhydrolyzable and its γ resonance does not overlap with the His-P resonance, unlike the one for ATPγ. The assignments of the resonances in these spectra are AMPPCP α (−10 ppm), AMPPCP β and γ (10 and 15 ppm, dependent on Mg^{2+}), coenzyme A phosphomonoester (4 ppm), and phosphodiester (−10 ppm). (From Vogel and Bridger.[21])

glucose increased the mobility for the regulatory phosphoserine-14.[58,59] Although glucose binds to the active site, which is far removed from this regulatory site, it was known that the hydrolysis of the phosphoserine by the protein phosphatase was drastically increased by the glucose addition.[60] Thus ^{31}P NMR provided the rationale for this observation: the binding of glucose to the remote active site increases the mobility and surface exposure of the serine phosphate, allowing the phosphatase to act.

One of the problems that can arise when substrates and other ligands are added is that these added components can be phosphorylated. Thus one may run into problems because of overlap of resonances for substrates and phosphoamino acids. Fortunately, these problems can be easily solved because a large number of commercially available analogs for compounds such as ATP, ADP, and AMP have substituted phosphorus groups. As these substitutions bring about characteristic changes in the chemical shift (see Table III),[33,59,61,62] overlap of resonances can be avoided through a judicious choice of analog (see Fig. 3). The use of these analogs has some further advantages. For example, the bridging methylene and imido substitutions provide for nonhydrolyzable ATP and ADP analogs. Not only does this limit the enzymatic turnover, which would normally complicate the lengthy NMR experiments, but if it can be demonstrated that the analog is a competitive inhibitor, one can also differentiate between events caused by the binding of ATP or phosphorylation by this nucleotide. Also, the slow rate at which thio analogs are used as substrates by many enzymes can be used to advantage, because one can potentially follow the slow turnover.

A somewhat different ligand than those discussed above is the electron, which can be introduced into the isoalloxazine ring of the FAD and FMN cofactors of redox proteins. The unpaired electron spin may cause broadening effects on the ^{31}P NMR resonances, which can subsequently be used to estimate the average distance between the electron and the phosphate group.[53-55]

Binding of Metal Ions

The majority of enzymatic phosphoryl-transfer reactions are strictly dependent on the presence of divalent cations such as Mg^{2+}.[63] Not only is

[58] M. Hoerl, K. Feldmann, K. D. Schnackerz, and E. J. M. Helmreich, *Biochemistry* **18**, 2457 (1979).

[59] S. G. Withers, N. B. Madsen, and B. D. Sykes, *Biochemistry* **20**, 1748 (1981).

[60] N. B. Madsen, *Enzymes* **17**, 365 (1986).

[61] E. K. Jaffe and M. Cohn, *Biochemistry* **17**, 652 (1978).

[62] S. Tran Dinh and M. Roux, *Eur. J. Biochem.* **76**, 245 (1977).

[63] A. S. Mildvan, *Adv. Enzymol.* **49**, 103 (1979).

TABLE III
^{31}P NMR PARAMETERS AND pK_a VALUES OF SUBSTITUTED PHOSPHORUS COMPOUNDS

Compound	Change in chemical shifts (ppm)[a]	Change in pK_a[b]	Analogs available	Reference
Thiophosphates	−40	−1.5	Nonbridging substitution, slowly hydrolyzable	61
Fluorophosphates	+11	No[c]	Terminal group of ATP, ADP, and AMP	33
Imidophosphates	−10	+0.9	Bridging substitution, nonhydrolyzable	62
Methylene phosphates	−25	+1.5	Bridging substitution, nonhydrolyzable	33
Cyclic phosphates	?	No[c]	Nonhydrolyzable, shift depends on bond angle	59

[a] Negative values indicate a downfield shift.
[b] Negative values indicate a decrease in pK_a.
[c] No pH titration occurs between pH 3 and 10; both the fluoro and cyclic phosphate analogs are monoanionic and therefore they do not closely resemble the natural compounds, which are generally mainly dianionic at pH 7.0.

the ATP–Mg^{2+} complex (rather than ATP) the preferred substrate, but some of the enzymes are metalloenzymes and contain tightly bound Zn^{2+} and Mg^{2+} ions. Thus there has been a considerable interest in using NMR techniques to determine the role of the metal ion. As Zn^{2+} does not have useful NMR properties, it has been replaced with ^{113}Cd^{2+}, which is a spin-$\frac{1}{2}$ nucleus.[64] For phosphoglucomutase[64] and alkaline phosphatase[65–67] (and with ^{112}Cd^{2+} or ^{114}Cd^{2+} as a no-spin control), ^{113}Cd–^{31}P couplings have been observed which provide evidence for the direct coordination of the metal ion to the phosphate group. Unfortunately, however, in our hands this approach has not worked well with enzymes wherein Mg^{2+} is not tightly bound to the protein. A second substitution that is often used is the paramagnetic Mn^{2+} in place of Mg^{2+}. Introduction of this cation generally

[64] G. I. Rhyu, W. J. Ray, and J. L. Markley, *Biochemistry* **23**, 252 (1984).
[65] J. D. Otvos, J. R. Alger, J. E. Coleman, and I. M. Armitage, *J. Biol. Chem.* **254**, 1778 (1979).
[66] P. Gettins and J. E. Coleman, *J. Biol. Chem.* **258**, 408 (1983).
[67] P. Gettins and J. E. Coleman, *J. Biol. Chem.* **259**, 4991 (1984).

produces a broadening that can be used to obtain distance information. Mn^{2+}, as well as the stable paramagnetic nitroxides, have also been used as general broadening reagents, to determine whether a residue is on the surface or is deeply buried within the protein. The latter residues in general do not experience any line broadening when such agents are added to the solution. If they are on the surface, broadening is generally observed.[13,55]

Conclusions and Other Experiments

In the foregoing discussions we have considered how the nature, mobility, and titratability of phosphoproteins can be determined and how these are affected by ligands and metal ions. However, some other types of experiments have also been performed. In the case of the enzyme alkaline phosphatase, both covalent E–P and noncovalent E · P intermediates can be detected in the ^{31}P NMR spectra. Because these were well separated, it was possible to perform saturation transfer experiments, which allowed the authors to determine the interconversion rates between these catalytic intermediates.[65] Solid-state NMR experiments have also been performed using phosphoproteins.[68] These have given support to the idea that there is little difference between the phosphate groups in the protein in solution or in a dried form. Finally, it should be mentioned that the technique of proton detection of nuclei such as ^{13}C and ^{15}N could also prove useful for studying ^{31}P.[69] Unfortunately, however, only in the case of phosphoserine and phosphothreonine can a direct coupling be observed between the proton and phosphorus nuclei; thus it is likely that this technique will remain limited to the study of these two phosphoamino acids.

In cases where information is available about the mobility and titratability of the phosphoamino acids, some interesting biochemical generalizations have come forward. Severe restrictions of motions usually occur in the active sites of enzymes.[2] This is consistent with the notion that the immobility facilitates the in-line nucleophilic attack[10] which is known to occur in all phosphoryl transfer enzymes. The fact that most of these sites do not titrate when ligands are bound suggests that these groups are generally shielded from the solvent. Conversely, flexibility and pH titratability were observed for all the sites that were covalently phosphorylated by protein kinases.[2,70,71] This flexibility may play an important role allow-

[68] L. J. Banaszak and J. Seelig, *Biochemistry* **21**, 2436 (1982).
[69] A. Bax, this series, Vol. 176 [8].
[70] S. P. Williams, W. A. Bridger, and M. N. G. James, *Biochemistry* **25**, 6655 (1986).
[71] M. W. Killimann, K. D. Schnackerz, and M. G. Heilmeyer, *Biochemistry* **23**, 112 (1984).

ing access of protein kinases and protein phosphatases to the regulatory site. All regulatory phosphorylation sites are dianionic at physiological pH. If salt linkages with basic amino acid side chains occur, the free energy involved[2,45] is $\Delta G° \leq -5.0$ kcal mol^{-1}. However, it is also possible that the highly charged and mobile phosphoryl group would prevent the interaction between two hydrophobic domains on a protein and exert its regulatory function in this fashion.

Acknowledgments

Research on regulatory proteins in the author's laboratory is presently sponsored by a grant from the Medical Research Council of Canada. The author is the recipient of a Scholarship from the Alberta Heritage Foundation for Medical Research. The secretarial assistance of Susan Clegg is greatly appreciated.

[14] Isotopic Labeling with Hydrogen-2 and Carbon-13 to Compare Conformations of Proteins and Mutants Generated by Site-Directed Mutagenesis, II

By JOYCE A. WILDE, PHILIP H. BOLTON, DAVID W. HIBLER, LYNN HARPOLD, TAYEBEH POURMOTABBED, MARK DELL'ACQUA, and JOHN A. GERLT

Introduction

Our preceding chapter in this volume [4] describes the preparation of isotopically labeled staphylococcal nuclease (Snase) samples as well as the characterization of these samples by one-dimensional NMR methods. The one-dimensional NMR spectra clearly show that Snase samples of high isotopic enrichment can be prepared in amounts suitable for NMR studies. In this chapter we describe the strategy for using isotopic labeling to determine the effects of site-specific amino acid replacements on the conformation of wild-type and mutant proteins.

A focus of our research has been the investigation of the effects of site-specific mutations on the activity and conformation of Snase.[1] At the beginning of this project we were confident that active-site mutants of Snase could be generated and that their enzymatic activity could be as-

[1] D. W. Hibler, N. J. Stolowich, M. A. Reynolds, J. A. Gerlt, J. A. Wilde, and P. H. Bolton, *Biochemistry* **26**, 6278 (1987).

sayed, as could the affinity of both wild-type and mutant proteins for calcium, which is essential for the activity of the wild-type protein as well as the binding of the active-site inhibitor pdTp. However, we were concerned that changes in specific amino acid residues could also lead to conformational changes in the protein and that these conformational changes might impact on the catalytic activity. The general approach was that if a particular mutant exhibited the same pdTp and calcium binding and activity as a wild type that it would be safe to assume that the wild-type amino acid is not essential for activity. However, if a change in either binding properties or enzymatic activity occurred following amino acid replacement, then independent evidence would be needed on the conformation of the mutant protein to assess the magnitude of conformational change.

As a general guideline we assumed that changes in internuclear distances on the order of 0.5 Å could be catalytically important. Distance changes of this order of magnitude can be detected by the nuclear Overhauser effect (NOE), which is described in detail in Chapters [6] and [9] in this volume. Because the NOE is sensitive to molecular motion and the internuclear distance between the pair of protons of interest, as well as the distances between the protons of interest and other protons, a *de novo* determination of internuclear distances in a particular protein to within 0.5 Å is extremely difficult due to the large number of unknowns. The comparison between two closely related systems for which the correlation times can be expected to be very similar and for which the overall arrangement of protons is similar can exploit the differences in the NOE magnitudes between wild-type and mutant proteins such that, while absolute proton–proton distances are not obtained, small differences in proton–proton distances can be obtained. For example, a change in a particular proton–proton NOE by a factor of two can be interpreted as an approximately 12% change in internuclear distance. Because interresidue NOEs are observed over the range of about 2.5–4.5 Å, this corresponds to a change in interproton distance on the order of 0.3–0.5 Å.

We recently presented a comparison of wild-type Snase with E43D and E43S, with the proteins in the presence of calcium and pdTp and, presumably, in a conformation near that of the productive complex. The conclusion was that there are conformational changes induced by amino acid replacement which result in changes in interproton distances on the order of 0.5 Å based on NOE data.

To characterize these changes more completely, we have begun to use isotopic labeling to obtain assignments of the NOE cross-peaks as to residue type, to position within a residue, and, whenever possible, to specific sites. In the discussion which follows, we will present the use of isotopic labeling to assign the aromatic partners of cross-peaks of both wild-type and mutant proteins through the use of deuteration of both aliphatic and aromatic amino acid side chains. Additional data will be shown which allow assignment of the position of the protons giving rise to a specific cross-peak through the use of labeled samples such as d_2-(3,5)-Y. Comparison of the data for isotopically labeled wild-type and mutant proteins allows independent assignment of the cross-peaks in these proteins. This contribution will conclude with a description of how the side-chain assignments can be extended to site-specific assignments.

It is noted that one very important feature of the isotopic labeling approach is that unambiguous results are obtained. For a protein the size of Snase, 149 amino acids plus a hexapeptide tail in the form we study, methods that depend on resolved scalar couplings or on sufficient resolution of the two-dimensional data (two-dimensional spectra of Snase are not completely resolved even at 620 MHz) will typically not be entirely adequate. Although isotopic labeling is both expensive and labor intensive, it does yield unambiguous information about both wild-type and mutant proteins. In favorable cases assignments can be made by comparing the NMR data of wild-type and mutant proteins. However, comparative methods are limited to instances of conformationally silent amino acids or when there is a localized probe of the protein structure.

Use of Deuteration to Obtain Residue-Type Assignments of NOE Cross-Peaks

We chose one region of the two-dimensional NOE map of Snase to begin our study; this region contains the cross-peaks between aromatic and upfield-shifted methyls, and it was chosen because it is relatively sparse—there are about 10 upfield-shifted methyls, and all NOEs in this region must be interresidue. The spectrum in Fig. 1 shows the NOE map for this region for wild-type Snase and d_8-V-labeled Snase obtained with a mixing time of 150 msec at a field strength of 400 MHz. For Snase at this field strength, there is little spin diffusion, as has been shown by comparison of data obtained with mixing times over the range of 50 to 500 msec.

To assign the residue types of the protons giving rise to the aromatic upfield-shifted methyl proton cross-peaks, we have examined the NOE obtained on a number of deuterated species of wild-type protein including d_5-F, d_4-Y, d_5-F + d_5-W, d_2-(3,5)-Y, d_2-(2,6)-Y, d_4-Y + d_5-W, d_5-F + d_2-(3,5)-Y, d_5-F + d_2-(2,6)-Y, d_2-(2,6)-F, d_3-(3,4,5)-F, and several of the above also containing d_5-W. The utility of such labeled samples in making residue-type assignments is indicated by the data in Fig. 2. The spectra shown are the slices through the two-dimensional NOE map of wild-type

Snase corresponding to the aromatic proton chemical shift of 7.03 ppm. The data show that there are NOE cross-peaks that are not present in the d_5-F data. The four signals, which can thus be assigned to F, can be further grouped into two pairs with the (-7.03 ppm, -0.1 ppm) and (-7.03 ppm, 0.4 ppm) being from V and the (-7.03 ppm, 0.1 ppm) and (-7.03 ppm, 0.8 ppm) being from L. The assignments to V and L are

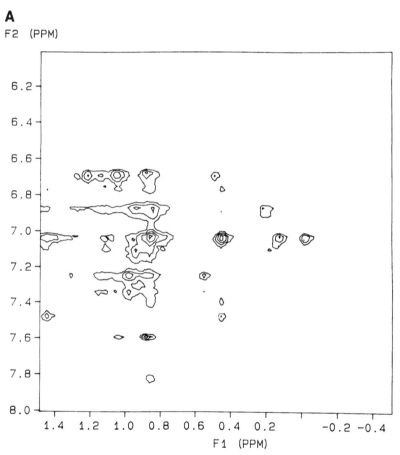

A

F2 (PPM)

F1 (PPM)

FIG. 1. (A) The two-dimensional map shown is for protio-Snase and contains the proton homonuclear NOE cross-peaks between the aromatic and upfield-shifted methyl regions. The two-dimensional data were obtained at 400 MHz using a mixing time of 150 msec and phase-sensitive detection. The sample was at 30° and the Snase was fully liganded. (B) The spectrum is for the same region as in A and contains the data for a sample of Snase labeled with d_8-V.

B

F2 (PPM)

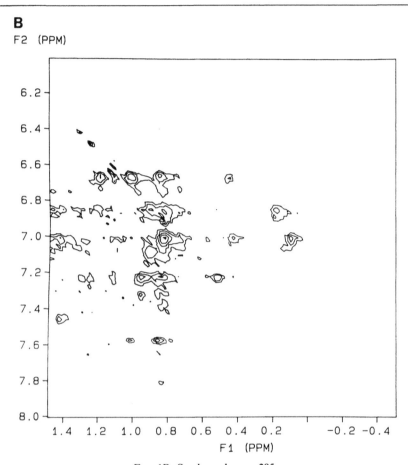

FIG. 1B. See legend on p. 285.

based on data obtained on samples labeled with either d_8-V or d_{10}-L. The examination of the chemical shift correlations or NOE correlations clearly shows that the two V methyls are on the same residue, as is the case for the two L methyls.

There is one NOE cross-peak on the 7.03-ppm line that is not assignable to F, because it is present for the d_5-F sample. Comparison of the d_5-F data with that obtained for d_5-F + d_2-(3,5)-Y and for d_5-F + d_2-(2,6)-Y shows that this additional NOE is from the 3,5 protons of a Y residue.

The above data show that deuteration can lead to results that are very simple to interpret with two-dimensional spectra, i.e., observation of the

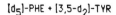

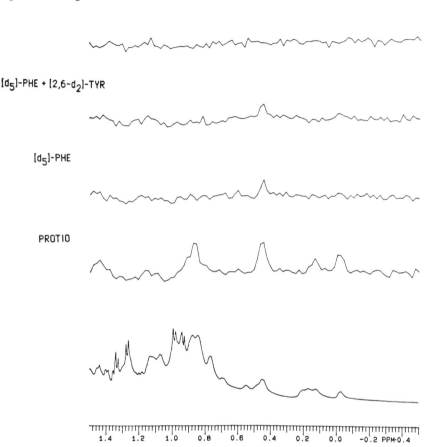

FIG. 2. The bottom spectrum is the normal one-dimensional proton spectrum of Snase. The other four spectra are the slices through the two-dimensional NOE maps at 7.03 ppm and contain the cross-peaks between aromatic protons having resonances at 7.03 ppm and the upfield-shifted methyls. The comparison of protio-Snase and labeled samples is given in the text.

disappearance of signals. The examples given above also show that deuteration is entirely satisfactory for dealing with cases of accidental overlap of cross-peaks, with the d_5-F giving partial loss of the (7.03 ppm, 0.4 ppm) cross-peak and further deuteration of Y giving complete elimination of the cross-peak. For analysis to be made in such a straightforward manner, high levels of deuteration are necessary and can be achieved using the methods given in Chapter [4] in this volume.

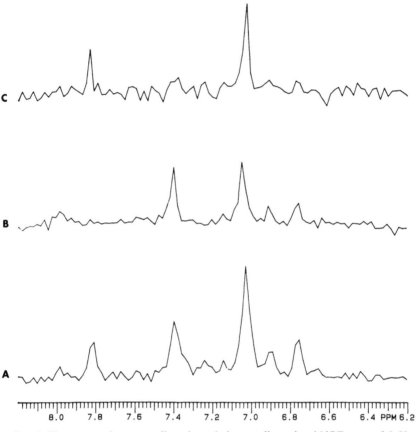

FIG. 3. The spectra shown are slices through the two-dimensional NOE maps of d_4-Y + d_5-W-labeled Snase (A), d_2-(2,6)-F + d_4-Y + d_5-W (B), and d_3-(3,4,5)-F + d_4-Y + d_5-W (C). The slices are vertical through the two-dimensional map at 0.4 ppm and contain the NOEs between valine and phenylalanine residues. The results are discussed in the text.

Use of Labeling to Obtain Positional Assignments within a Residue

Because the data cited above indicated that the (7.03 ppm, 0.4 ppm) cross-peak is due to valine and phenylalanine residues, it was of interest to determine which types of phenylalanine ring protons contribute to the NOE cross-peak. We therefore obtained two-dimensional NOE data on samples labeled with d_4-Y + d_5-W, d_2-(2,6)-F + d_5-W, and d_3-(3,4,5)-F + d_5-W. The slices at 0.4 ppm through the aromatic region are shown for these three samples in Fig. 3. The data on the protiophenylalanine sample indicate that there are NOE cross-peaks from at least two phenylalanines to this valine, because there are cross-peaks with five phenylalanine chemical shifts. When the ortho positions are deuterated, there are four distinct phenylalanine NOE cross-peaks to this valine; when the meta and para positions are deuterated, there are two chemical shifts. This indicates that there are two phenylalanines that have NOEs to this valine, whose ortho proton chemical shifts are 7.03 and 7.82 ppm, and that these two phenylalanines have meta and para chemical shifts of 7.4, 7.03, 6.92, and 6.78 ppm. Thus, the use of isotopic labeling allows the analysis of this region, which contains accidental degeneracy of phenylalanine chemical shifts. As will be presented elsewhere, this information, in conjunction with other results, has allowed assignment of the two phenylalanines to F34 and F76 and the valine to V74.

Use of Labeling to Aid in Comparing Results Obtained on Wild-Type and Mutant Proteins

We have previously shown that substitution of amino acids in Snase can give rise to mutant proteins that have proton chemical shifts quite different than those seen in the wild-type protein. Thus, the positions of NOE cross-peaks can change from wild-type to mutant protein even if the cross-peaks are from the same pair of protons or groups of protons. It is also possible that a NOE cross-peak could occur at the same chemical shift coordinates in both wild-type and mutant proteins but could arise from a different set of protons in each case. To be certain, independent assignment information for both wild-type and mutant proteins is needed. The use of labeling allows the determination of the partners of a specific cross-peak in both wild-type and mutant proteins and hence gives assurance that the same proton–proton cross-peak is being compared. The analysis of data on labeled mutant proteins proceeds as described above.

Use of Labeling to Aid in Assigning Inhibitor-Protein Contacts

We have used fast-exchange NOE experiments to probe the binding of the active-site inhibitor pdTp to wild-type and mutant Snase. As shown previously and discussed in Chapter [4] in this volume, under the proper conditions, using a mixture of an excess of low-molecular-weight inhibitor with protein, essentially all observed inhibitor NOEs arise from the bound inhibitor. These fast-exchange NOE data allow a detailed probing of the active-site region of the wild-type and mutant proteins; changes in internuclear distances in this region are likely to have catalytic consequences.

The structure of Snase determined on the basis of crystallographic information indicates that there are several amino acid residues of the wild-type protein that may be important in binding pdTp. Three of these are tyrosine-85, -113, and -115. Examination of the NOEs from the H1' proton of pdTp bound to wild type (E43D and E43S) showed that there are several aromatic residues spatially close to the H1' proton. The residue type of the aromatic protons could be assigned on the basis of comparing the NOE data obtained for protiated samples and for samples prepared with d_4-Y, d_2-(3,5)-Y and d_2-(2,6)-Y labeling. The results obtained for wild-type protein, with and without d_4-Y labeling, are shown in Fig. 4 and indicate that all of the H1'-to-aromatic proton NOEs are to tyrosines. Additional data obtained on Y-to-F mutant proteins have given site-specific assignments.

Site-Specific Assignments of Side-Chain Protons

The preceding discussion presents a variety of labeling approaches for assignments to residue type and position within a residue. The final assignment step is to obtain site-specific assignments. When independent structural information is available, such as that provided by a crystal structure, some working-model site-specific assignments can be made on the basis of interresidue NOE data; we have used this approach to obtain site-specific assignments of many residues. Another approach is the sequential assignment protocol developed by Wüthrich over the past several years.[2] Both of these approaches can be useful but both are limited either by resolution, incompletely resolved couplings, or insufficient or unreliable structural data. The most reliable assignment procedure is to utilize the $^1J_{^{13}C^{15}N}$ coupling across the amide bond as first demonstrated by Kainosho and later by Redfield and co-workers.[3,4] What is needed is an experiment to link the side-chain protons to either the carboxyl carbon or the amide nitrogen.

In most cases, establishment of linkage of the side-chain protons to the carboxyl carbon is of primary interest. There are couplings from the carboxyl carbon to the H_α and H_β protons. There are typically NOEs from the H_α and H_β protons to others in the side chain. For a protein the size of Snase there are readily observed NOEs from the ortho ring protons of F and Y to their H_α and H_β protons, for example. The linkage from the H_α and H_β to the carboxyl carbon can be made through the use of heteronuclear chemical shift correlation spectroscopy. Whereas the heteronuclear

[2] K. Wüthrich, "NMR of Proteins and Nucleic Acids." Wiley, New York, 1986.
[3] M. Kainosho and T. Tsuji, *Biochemistry* **21**, 6273 (1982).
[4] R. H. Griffey and A. G. Redfield, *Q. Rev. Biophys.* **19**, 51 (1987).

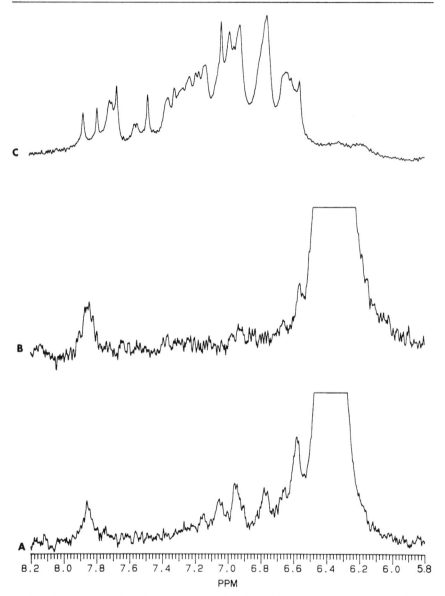

FIG. 4. (A) The spectrum is the one-dimensional NOE difference spectrum obtained from a mixture of liganded Snase upon irradiation of the H1' proton resonance of pdTp. (B) The spectrum is the same spectrum for d_4-Y-labeled Snase, and (C) the spectrum is the normal one-dimensional proton NMR spectrum of Snase.

couplings are small (3–6 Hz), the carboxyl carbon has a relatively narrow natural linewidth of typically 5 Hz. Thus, although the magnetization transfer is not highly efficient, the T_{1s} and T_{2s} of the nuclei involved are sufficiently long for transfer to be observed. Furthermore, the carboxyl carbon region is usually well resolved for a particular residue type—the 10 valines of Snase exhibit eight resolved resonances, only two of which overlap over a chemical shift range of almost 10 ppm. The three phenyl-alanine carboxyl resonances are well resolved with an almost equal spacing between resonances of 2.5 ppm. We are now performing heteronuclear experiments on Snase to link side-chain protons with the carboxyl carbon to be able to obtain unambiguous site-specific assignments using the previously demonstrated $^1J_{^{13}C^{15}N}$ coupling across the amide bond.

Acknowledgments

This research was supported by Grant GM-34573 from the National Institutes of Health to J. A. Gerlt and P. H. Bolton and Grant PCM-831422 from the National Science Foundation to P. H. Bolton.

Section III

Enzyme Mechanisms

[15] Phosphorus-31 Nuclear Magnetic Resonance of Enzyme Complexes: Bound Ligand Structure, Dynamics, and Environment

By DAVID G. GORENSTEIN

Introduction

The first ^{31}P NMR signals of phosphorus compounds were reported in 1951 by Dickenson[1] and Gutowsky and McCall.[2] By 1955, the availability of commercial multinuclear NMR spectrometers had led to the recognition that ^{31}P NMR could serve as an important analytical tool for structure elucidation. By 1956, chemical shifts on several hundred phosphorus-containing compounds had already been reported,[3] and some success had been achieved in correlating structure with ^{31}P chemical shifts.[4] Early spectrometers (pre-1963) generally required neat samples in large nonrotating tubes (8–12 mm o.d.). These spectrometers recorded only a single scan of the spectrum, and dilute samples gave very weak signals. In 1963, the introduction of signal averaging through a computer of average transients (CAT) and the availability in the middle 1960s of more sensitive, higher field electromagnets (2.3-T fields, equivalent to 100-MHz resonance frequency for protons) led to further rapid growth in the number of reported ^{31}P spectra. By 1967, Van Wazer and co-workers[5] published the first monograph entirely devoted to ^{31}P NMR spectroscopy.

With the introduction by 1970 of signal-averaging, Fourier transform (FT), and high-field superconducting-magnet NMR spectrometers, ^{31}P NMR spectroscopy finally expanded from the study of small organic and inorganic compounds to biological phosphorus compounds as well. The latest, routine, multinuclear FT NMR spectrometers (1.8–11.7 T) have reduced if not eliminated the one serious limitation to the widespread utilization of phosphorus NMR in biological systems, which is the low sensitivity of the phosphorus nucleus (6.6% at constant field compared to

[1] W. C. Dickenson, *Phys. Rev.* **81**, 717 (1951).

[2] H. S. Gutowsky and D. W. McCall, *Phys. Rev.* **82**, 748 (1951).

[3] J. R. Van Wazer, C. F. Callis, J. N. Shoolery, and R. C. Jones, *J. Am. Chem. Soc.* **78**, 5715 (1956).

[4] C. F. Callis, J. R. Van Wazer, J. N. Shoolery, and W. A. Anderson, *J. Am. Chem. Soc.* **79**, 2719 (1957).

[5] M. M. Crutchfield, C. H. Dungan, L. H. Letcher, V. Mark, J. R. Van Wazer, M. Grayson, and E. F. Griffin, *Top. Phosphorus Chem.* **5** (1967).

METHODS IN ENZYMOLOGY, VOL. 177

^{1}H NMR). Today, millimolar (or lower) concentrations of phosphorus nuclei in as little as 0.3 ml of solution are routinely and conveniently monitored. The ^{31}P nucleus has other convenient NMR properties suitable for FT NMR: spin-$\frac{1}{2}$ (which avoids problems associated with quadrupolar nuclei), 100% natural abundance, moderate relaxation times (providing relatively rapid signal averaging and sharp lines), and a wide range of chemical shifts (>600 ppm). As we will see, ^{31}P NMR can serve as a very convenient "reporter" probe of the structure, local active site environment, and dynamics of enzyme complexes.

Today's commercial FT NMR spectrometers cover the ^{31}P frequency range from 24 to 200 MHz. Generally for small, phosphorus-containing compounds, high signal-to-noise and resolution requirements dictate use of as high a magnetic field strength as possible, because both sensitivity and chemical shift dispersion increase at higher operating frequency (and field). However, consideration must be given to field-dependent relaxation mechanisms such as chemical shift anisotropy, which can lead to substantial line broadening of the ^{31}P NMR signal at high fields. Indeed, for moderate-sized biomolecules, sensitivity is often poorer at very high fields because the linewidths increase as the square of the field strength. An optimum compromise between increased sensitivity and resolution is often provided in medium-field spectrometers (200–300 MHz, ^{1}H).

Because signal intensity is proportional to the number of nuclei in the region of the observing receiver coil of the probe, high concentration and large volume provide minimal acquisition times. Probes for 20-mm o.d. sample tubes are available on wide-bore instruments, but the large sample volumes of 10–12 ml required often dictate against their use for any material of low availability, which is often the case in enzymatic studies. More typical are ^{31}P probes of 5- to 12-mm o.d. which require 0.5- to 2.5-ml sample volumes. Microcells can reduce the volume to less than 0.5 ml. The latest probe design provides a signal-to-noise ratio of 420 : 1 on a 1% trimethyl phosphite standard sample in a 10-mm tube in a single spectral acquisition (at 11.7 T on a GN-500 spectrometer).

For dilute samples, signal averaging increases the sensitivity of the NMR experiment because the coherent signal increases with the number of spectral accumulations whereas noise (incoherent) increases only with the square root of the number of scans. The signal-to-noise ratio will thus improve with the square root of the number of scans. The total time required to achieve acceptable sensitivity in the FT NMR experiment is determined by the nuclei concentration, the number of scans, the data acquisition time, and any delays introduced between successive radiofrequency pulses. With the tremendous increase in sensitivity in modern FT

NMR spectrometers, biological and medical applications of ^{31}P NMR spectroscopy grew dramatically during the 1970s, and numerous reviews on biological ^{31}P NMR dealt with this burgeoning field.[6-13] Gorenstein has edited a second monograph[14] entirely devoted to ^{31}P NMR, covering largely theory and the biochemical applications to this ever-expanding field. Verkade and Quin[15] and Burt[16] have recently edited additional books largely devoted to ^{31}P NMR of organophosphorus and metal complexes, and biological and medical aspects of the field, respectively.

In this chapter we will discuss the use of ^{31}P NMR spectroscopy and, in particular, ^{31}P chemical shifts to probe the structure and dynamics of phosphate ester enzyme complexes. More comprehensive reviews of the literature on ^{31}P NMR studies of enzyme complexes by Nageswara Rao[17] and on phosphoenzyme complexes by Vogel[18] are recommended. Additional excellent discussions of paramagnetic probes of enzyme complexes with phosphorus-containing compounds by Villafranca[19] and phosphorus relaxation methods by Hart[20] and James[21] are available.

[6] D. G. Gadian, G. K. Radda, R. E. Richards, P. J. Seeley, and R. G. Shulman, "Biological Applications of Magnetic Resonance." Academic Press, New York, 1979.

[7] M. Cohn and B. D. Nageswara Rao, *Bull. Magn. Reson.* **1**, 38 (1979).

[8] D. G. Gorenstein, *Jerusalem Symp. Quantum Chem. Biochem.* **11**, 1 (1978).

[9] D. G. Gorenstein, *Annu. Rev. Biophys. Bioeng.* **10**, 355 (1981).

[10] R. E. Gordon, P. E. Hanley, and D. Shaw, *Prog. Nucl. Magn. Reson. Spectrosc.* **15**, 1 (1982).

[11] M. Cohn, *Annu. Rev. Biophys. Bioeng.* **11**, 23 (1982).

[12] P. R. Cullis, S. B. Farren, and M. J. Hope, *Can. J. Spectrosc.* **26**, 89 (1981).

[13] B. D. Sykes, *Can. J. Biochem. Cell Biol.* **61**, 155 (1983).

[14] D. G. Gorenstein, in "Phosphorus-31 NMR: Principles and Applications" (D. G. Gorenstein, ed.), pp. 1-53. Academic Press, Orlando, Florida, 1984.

[15] J. G. Verkade and L. D. Quin (eds.), "Phosphorus-31 NMR Spectroscopy in Stereochemical Analysis—Organic Compounds and Metal Complexes," Vol. 8. VCH Publ., Deerfield Beach, Florida, 1987.

[16] C. T. Burt, "Phosphorus NMR in Biology." CRC Press, Boca Raton, Florida, 1987.

[17] B. D. Nageswara Rao, in "Phosphorus-31 NMR: Principles and Applications" (D. G. Gorenstein, ed.), pp. 57-103. Academic Press, Orlando, Florida, 1984.

[18] H. J. Vogel, in "Phosphorus-31 NMR: Principles and Applications" (D. G. Gorenstein, ed.), pp. 105-154. Academic Press, Orlando, Florida, 1984.

[19] J. J. Villafranca, in "Phosphorus-31 NMR: Principles and Applications" (D. G. Gorenstein, ed.), pp. 155-174. Academic Press, Orlando, Florida, 1984.

[20] P. A. Hart, in "Phosphorus-31 NMR: Principles and Applications" (D. G. Gorenstein, ed.), pp. 317-347. Academic Press, Orlando, Florida, 1984.

[21] T. L. James, in "Phosphorus-31 NMR: Principles and Applications" (D. G. Gorenstein, ed.), pp. 349-400. Academic Press, Orlando, Florida, 1984.

Phosphorus-31 Chemical Shifts

Introduction and Basic Principles

The interaction of the electron cloud surrounding the phosphorus nucleus with an external applied magnetic field B_0 gives rise to a local magnetic field. This induced field shields the nucleus, with a magnitude proportional to the field B_0, so that the effective field, B_{eff}, felt by the nucleus is given by

$$B_{eff} = B_0(1 - \sigma) \tag{1}$$

where σ is the shielding constant. The difference in chemical shift δ between two lines is simply the difference in shielding constants σ of the nuclei giving rise to the two lines. Note that throughout this chapter an attempt has been made to reference all ^{31}P shifts to 85% H_3PO_4 (or in some instances to trimethyl phosphate, which is 3.53 ppm downfield of 85% H_3PO_4) and always to follow the IUPAC convention[22] so that *positive values are to high frequency* (low field). One should cautiously interpret reported ^{31}P chemical shifts because the early literature (pre-1970s) and even many later papers use the opposite sign convention. In addition, with the introduction of superconducting magnets, sample geometry relative to the field differs from that of iron core magnets. Measured ^{31}P chemical shifts of a sample relative to a standard in a solvent of different diamagnetic susceptibility can differ by up to 1 ppm. To minimize this problem, it has been suggested that ^{31}P chemical shifts in aqueous samples be measured relative to 85% phosphoric acid in a spherical microcell inside the solution.[23]

Because the charge distribution in a phosphorus molecule will generally be far from spherically symmetrical, the ^{31}P chemical shift (or shielding constant) varies as a function of the orientation of the molecule relative to the external magnetic field.[24-28] This gives rise to a chemical shift anisotropy that can be defined by three principal components, σ_{11}, σ_{22}, and σ_{33}, of the shielding tensor.[24] For molecules that are axially symmetrical, with σ_{11} along the principal axis of symmetry, $\sigma_{11} = \sigma_\parallel$ (parallel

[22] Anonymous, *Pure Appl. Chem.* **45,** 217 (1976).

[23] M. Batley and J. W. Redmond, *J. Magn. Reson.* **49,** 172 (1982).

[24] R. K. Harris, "Nuclear Magnetic Resonance Spectroscopy." Pitman, London, 1983.

[25] H. Shindo, *in* "Phosphorus-31 NMR: Principles and Applications" (D. G. Gorenstein, ed.), Chap. 13, pp. 401–422. Academic Press, Orlando, Florida, 1984.

[26] J. Herzfeld, R. G. Griffin, and R. A. Haberkorn, *Biochemistry* **17,** 2711 (1978).

[27] S. J. Kohler and M. P. Klein, *Biochemistry* **15,** 967 (1976).

[28] P. Tutunjian, J. Tropp, and J. Waugh, *J. Am. Chem. Soc.* **105,** 4848 (1983).

component), and $\sigma_{22} = \sigma_{33} = \sigma_{\perp}$ (perpendicular component). These anisotropic chemical shifts are observed in solid samples[25-29] and liquid crystals,[30] whereas for small molecules in solution, rapid tumbling averages the shift. The average, isotropic chemical shielding σ_{iso} (which would be comparable to the solution chemical shift) is given by the trace of the shielding tensor or

$$\sigma_{iso} = \tfrac{1}{3}(\sigma_{11} + \sigma_{22} + \sigma_{33}) \tag{2}$$

and the anisotropy $\Delta\sigma$ is given by

$$\Delta\sigma = \sigma_{11} - \tfrac{1}{2}(\sigma_{22} + \sigma_{33})$$

or, for axial symmetry,

$$\Delta\sigma = \sigma_{\parallel} - \sigma_{\perp} \tag{3}$$

Theoretical ^{31}P Chemical Shift Calculations and Empirical Observations

Several attempts have been made to develop a unified theoretical foundation for ^{31}P chemical shifts of phosphorus compounds.[31-36] In one theoretical approach, Letcher and Van Wazer,[31,32] using approximate quantum-mechanical calculations, indicated that three factors appear to dominate ^{31}P chemical shift differences $\Delta\delta$, as shown by

$$\Delta\delta = -C\Delta\chi_X + k\Delta n_\pi + A\Delta\theta \tag{4}$$

where $\Delta\chi_X$ is the difference in electronegativity in the P–X bond, Δn_π is the change in the π-electron overlap, $\Delta\theta$ is the change in the σ-bond angle, and C, k, and A are constants.

As suggested by Eq. (4), electronegativity effects, bond angle changes, and π-electron overlap differences can all potentially contribute to ^{31}P shifts in a number of classes of phosphorus compounds. While these semiempirical isotropic chemical shift calculations are quite useful in providing a chemical and physical understanding for the factors affecting ^{31}P

[29] A. K. Cheetham, N. J. Clayden, C. M. Dobson, and R. J. B. Jakeman, *J. Chem. Soc. Chem. Commun.*, 195 (1986).
[30] H. Ye and B. M. Fung, *J. Magn. Reson.* **51**, 313 (1983).
[31] J. H. Letcher and J. R. Van Wazer, *J. Chem. Phys.* **44**, 815 (1966).
[32] J. H. Letcher and J. R. Van Wazer, *Top. Phosphorus Chem.* **5**, 75 (1967).
[33] M. Rajzmann and J. C. Simon, *Org. Magn. Reson.* **7**, 334 (1975).
[34] R. Wolff and V. R. Radeglia, *Z. Phys. Chem. (Leipzig)* **261**, 726 (1980).
[35] T. Weller, D. Deininger, and R. Lochman, *Z. Chem.* **21**, 105 (1981).
[36] P. Bernard-Moulin and G. Pouzard, *J. Chim. Phys.* **76**, 708 (1979).

chemical shifts, they represent severe theoretical approximations.[37] More exact *ab initio* chemical shift calculations of the shielding tensor are very difficult and very few have been reported on phosphorus compounds.[37-39] Whereas the semiempirical theoretical calculations have largely supported the importance of electronegativity, bond angle, and π-electron overlap on ^{31}P chemical shifts, the equations relating ^{31}P shift changes to structural and substituent changes unfortunately are not generally applicable. Also, because ^{31}P shifts are influenced by at least these three factors, empirical and semiempirical correlations can only be applied to classes of compounds that are similar in structure. It should also be emphasized again that structural perturbations will affect ^{31}P chemical shift tensors. Often variations in one of the tensor components will be compensated by an equally large variation in another tensor component with only a small net effect on the isotropic chemical shift. Interpretation of variations of isotropic ^{31}P chemical shifts should therefore be approached with great caution.

Within these limitations, a number of semiempirical and empirical observations and correlations, however, have been established and have proved useful in predicting ^{31}P chemical shift trends.[40] Unfortunately, however, no single factor can readily rationalize the observed range of ^{31}P chemical shifts.

Bond-Angle Effects. The Letcher and Van Wazer theory[32] suggested that changes in the σ-bond angles should make a negligible contribution $[|A| < 1$, Eq. (4)] to the ^{31}P chemical shifts of phosphoryl compounds, with electronegativity effects apparently predominating. Purdela,[41] in contrast, has suggested a correlation (although poor) between X–P–X bond angles and chemical shifts for a wide variety of phosphoryl compounds. Contrary to the theory of Letcher and Van Wazer, Kumamoto *et al.*[42] and Blackburn *et al.*[43] have argued on the basis of cyclic phosphate ester shifts that phosphorus bond angles must play some role in ^{31}P chemical shifts. A change in d_π–p_π bonding bond-angle changes was suggested

[37] B. D. Chesnut, *in* "Phosphorus-31 NMR Spectroscopy in Stereochemical Analysis— Organic Compounds and Metal Complexes" (J. G. Verkade and L. D. Quin, eds.), Vol. 8. VCH Publ., Deerfield Beach, Florida, 1987.

[38] F. R. Prado, C. Geissner-Prettre, B. Pullman, and J. P. Daudey, *J. Am. Chem. Soc.* **101**, 1737 (1979).

[39] C. Giessner-Pettre, B. Pullman, F. R. Prado, D. M. Cheng, V. Ivorno, and P. O. Ts'o, *Biopolymers* **23**, 377 (1984).

[40] D. G. Gorenstein, *in* "Phosphorus-31 NMR: Principles and Applications" (D. G. Gorenstein, ed.), pp. 7–36. Academic Press, Orlando, Florida, 1984.

[41] D. Purdela, *J. Magn. Reson.* **5**, 23 (1971).

[42] J. Kumamoto, J. R. Cox, Jr., and F. H. Westheimer, *J. Am. Chem. Soc.* **78**, 4858 (1956).

[43] G. M. Blackburn, J. S. Cohen, and I. Weatherall, *Tetrahedron* **27**, 2903 (1971).

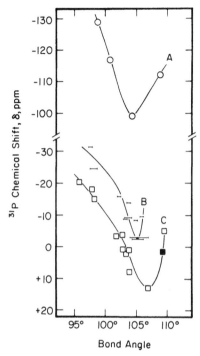

FIG. 1. (A) Phosphorus-31 chemical shifts of 2-thioxo-2-*tert*-butyl-1,3,2-dithiaphospha compounds versus S–P–S bond angle (◯). (Derived from Martin and Robert.[45]) (B) Phosphorus-31 chemical shifts of PO$_2$N$_2$ tetrahedra versus bond angle (|——|) (Derived from Contractor *et al.*[46]) (C) Phosphorus-31 chemical shifts of phosphate esters versus O–P–O bond angle [Derived from (▢) Gorenstein[44] and (■) Contractor *et al.*[46]]

as an explanation for these shifts. Blackburn *et al.*[43] compiled all of the known cyclic ester chemical shifts and concluded that these cyclic ester chemical shifts must arise from a complex stereoelectronic effect not explicable by present theory.

Gorenstein extended these observations and proposed an empirical correlation between ^{31}P chemical shifts and O–P–O bond angles in phosphates.[44] (Note that success here depends on the fact that we are dealing with only a limited structural variation: the number and chemical type of R groups attached to a tetrahedron of oxygen atoms surrounding the phosphorus nucleus.) As shown in Fig. 1C, for a wide variety of different alkyl phosphates (mono-, di-, and triesters; cyclic and acyclic neutral, monoanionic, and dianionic esters), a bond angle of ~108° gives maxi-

[44] D. G. Gorenstein, *J. Am. Chem. Soc.* **97**, 898 (1975).

mum shielding, while angles $<108°$ for the smallest O–P–O bond angle (obtained from X-ray data) or $>108°$ correlate with a deshielding (downfield shift) of the phosphorus nucleus.

Martin and Robert[45] have shown that a similar correlation of S–P–S bond angles and ^{31}P chemical shifts in 2-thioxo-2-tert-butyl-1,3,2-dithiaphospha compounds likely also exists (Fig. 1A). Contractor et al.[46] have observed a similar variation for PO_2N_2 tetrahedra, further confirming this bond-angle effect in tetracoordinated phosphorus compounds (Fig. 1B).

Bond-angle changes and hence distortion from tetrahedral symmetry in tetracoordinated phosphorus should affect the chemical shift anisotropy as well. Dutasta et al.[47] have verified experimentally this bond-angle effect in a solid-state ^{31}P NMR study on a series of cyclic thioxophosphonates. The shielding tensors are very sensitive to geometrical changes, and, in fact, a linear correlation appears to exist between the asymmetry parameter η and the intracyclic bond-angle α. The anistropy is also correlated to the bond angle whereas the average, isotropic chemical shielding shows a much poorer correlation.

Gorenstein and Kar[48] have attempted to calculate the ^{31}P chemical shifts for a model phosphate diester in various geometries to confirm theoretically the bond-angle correlation. Using CNDO/2 SCF molecular orbital calculations (for details, see Refs. 48 and 49), a correlation was drawn between calculated phosphorus electron densities and isotropic ^{31}P chemical shifts, and deshielding of the phosphorus atom with decreasing O–P–O bond angles was found. A slightly better correlation was achieved between observed and calculated ^{31}P chemical shifts using a Karplus–Das-type average excitation approximation in a semiempirical theoretical approach.[49]

Stereoelectronic Effects on ^{31}P Chemical Shifts. The semiempirical molecular orbital calculations of Gorenstein and Kar[48] suggested that ^{31}P chemical shifts are also dependent on P–O ester torsional angles ω, ω', which we define as a stereoelectronic effect. [The two torsional angles ω and ω' are defined by the R–O–P–O(R') dihedral angles (see Fig. 2).] These chemical shift calculations indicated that a phosphate diester in a gauche,gauche (g,g) conformation should have a ^{31}P chemical shift sev-

[45] J. Martin and J. B. Robert, *Org. Magn. Reson.* **15**, 87 (1981).

[46] S. R. Contractor, M. B. Hursthouse, L. S. Shaw, R. A. Shaw, and H. Yilmaz, personal communication.

[47] J. P. Dutasta, J. B. Robert, and L. Wiesenfeld, in "Phosphorus Chemistry" (J. G. Verkade and L. D. Quin, eds.). Amer. Chem. Soc., Washington, D.C., 1981.

[48] D. G. Gorenstein and D. Kar, *Biochem. Biophys. Res. Commun.* **65**, 1073 (1975).

[49] D. G. Gorenstein, *Prog. Nucl. Magn. Reson. Spectrosc.* **161**, 1 (1983).

A   B

FIG. 2. Definition of phosphate diester torsional angles ω,ω': $-g,-g$ (A) and $-g,t$ (B). R–O–P–O(R') torsional angles are $+60°$ ($+g$), $-60°$ ($-g$), and $180°$ (t).

eral parts per million upfield from a phosphate diester in more extended conformations, such as gauche,trans (g,t) or trans,trans (t,t) conformations. Chemical shifts of other conformations have been calculated and have been used to generate a ³¹P NMR chemical shift–torsional angle contour map (Fig. 3).

Pullman and co-workers[38,39] have also described a torsional-angle sensitivity to the ³¹P chemical shifts with generally more accurate *ab initio,* gauge-invariant-type, molecular orbital, chemical shift calculations. They found that the ³¹P chemical shift of a dimethyl phosphate in a g,g conformation is 3–6 ppm upfield from a phosphate in a g,t conformation. Interestingly, the C–O torsional angle also appears to influence ³¹P shifts, although not as much as the P–O ester torsional angle.[39]

Experimentally, this torsional angle, or stereoelectronic, effect on ³¹P shifts can be confirmed by using six-membered-ring model systems in which the torsional angles are rigidly defined by some molecular constraint, such as the diastereomeric phosphate triesters **1** and **2**.[50] Generally those diastereomeric phosphates with an axial ester group (structures **1a** and **2a**) have ³¹P chemical shifts as much as 6 ppm upfield from these

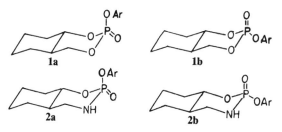

[50] D. G. Gorenstein and R. Rowell, *J. Am. Chem. Soc.* **101,** 4925 (1979).

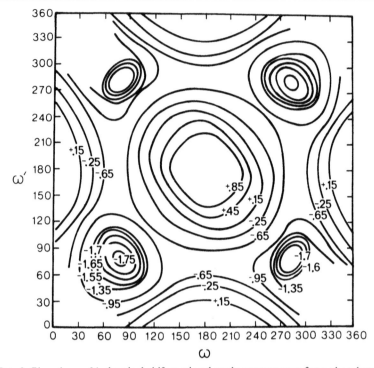

FIG. 3. Phosphorus-31 chemical shift torsional-angle contour map for a phosphate diester. Shift contours are in parts per million from 85% H_3PO_4. [Reprinted with permission
from D. G. Gorenstein and B. A. Luxon, *Biochemistry* **18**, 3796 (1979). Copyright 1979
American Chemical Society.]

isomeric phosphates with an equatorial ester group (**1b** and **2b**).[14,15,49,51] In
structures **1b** and **2b** the equatorial ester group ("equatorial" insofar as
the phosphorinane ring is viewed in a chair conformation) is locked into a
trans conformation relative to the endocyclic P–O ester bond. Thus trans
esters **1b** and **2b** are downfield of gauche esters such as **1a** and **2a**.

   The importance of the geometry about the phosphate tetrahedron in
influencing ^{31}P chemical shifts is illustrated in Fig. 4. Using magic-angle
spinning, solid-state NMR, Cheetham et al.[29] have measured the [31]P
shielding tensors, the chemical shift anisotropy, and the isotropic chemical shift [Eqs. (2) and (3)] for various crystalline orthophosphates and
higher phosphates. From the X-ray crystallographic structures of these
same phosphates, they calculated the summed bond strengths at oxygen
atoms, $\Sigma[S(O^{2-})]$, a term defined by Smith et al.,[52] and roughly related to

[51] B. E. Maryanoff, R. O. Hutchins, and C. A. Maryanoff, *Top. Stereochem.* **11**, 187 (1979).
[52] K. A. Smith, R. J. Kirkpatrick, E. Oldfield, and D. M. Henderson, *Am. Mineral.* **68**, 1206
   (1983).

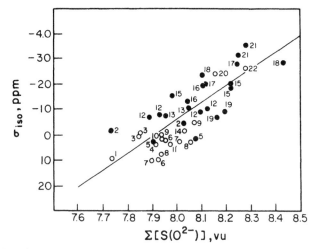

FIG. 4. Plot of the summed bond strengths at oxygen (in valence units, vu) following the method of I. D. Brown and R. D. Shannon [*Acta Crystallogr. Sect. A: Cryst. Phys. Diffr. Theor. Gen. Crystallogr.* **29**, 226 (1973)] versus the ³¹P magic-angle spinning NMR isotropic chemical shifts, σ_{iso} [Eqs. (1) and (2)] relative to 85% H_3PO_4. Open and closed circles are metal (M) orthophosphates (such as $M_nH_{3-n}PO_4$) and metal higher phosphates (such as $M_nP_2O_7$), respectively. Numbering and actual structures may be found in Ref. 29. (Copyright 1986 The Chemical Society.)

P–O bond lengths. As shown in Fig. 4 a reasonable correlation does appear to exist between the bond strengths at oxygen and isotropic ³¹P chemical shifts. The isotropic chemical shift moves upfield as the bond strength at oxygen increases. It should be noted that the variation in the individual shielding tensors is considerably larger than the isotropic shielding.[29,52]

These results are consistent with the stereoelectronic explanation for the variations in ³¹P chemical shifts of phosphates. In a series of CNDO and *ab initio* molecular orbital calculations, in our laboratory[53,54] and Perahia and Pullman's[55] laboratory, it has been shown that O–P–O bond angles, R–O–P–(OR) torsional angles, and P–O bond lengths (and strengths) are all highly coupled geometric parameters. Thus, these calculations, as well as X-ray structural comparisons, show that torsion about the P–O ester bond produces an 11° bond-angle distortion.[53] This coupling fits nicely with the proposed downfield shift in trans versus gauche conformations, because the ester RO–P–OR bond angle is reduced by ~5° in

[53] D. G. Gorenstein and D. Kar, *J. Am. Chem. Soc.* **99**, 672 (1977).

[54] D. G. Gorenstein, B. A. Luxon, and J. B. Findlay, *Biochim. Biophys. Acta* **475**, 184 (1977).

[55] D. Perahia and B. Pullman, *Biochim. Biophys. Acta* **435**, 282 (1976).

a g,t conformation relative to a g,g conformation[56] (the bond angle in a t,t conformation is reduced another $\sim5°$ from that in a g,t conformation). Molecular-orbital calculations have also shown that the P–O ester bond strengths are greatest in a g,g conformation and weakest in a t,t conformation.[57-59] These changes have been analyzed in terms of a stereoelectronic effect,[59] attributable to oxygen electron lone-pair orbital mixing with the adjacent polar P–O ester bonds. These orbital, stereoelectronic mixing effects are greatest when the lone-pair orbital is antiperiplanar to the adjacent P–O ester antibonding orbital in the gauche conformation.

Extrinsic and Other Effects on ^{31}P Chemical Shifts. Environmental effects on ^{31}P chemical shifts are generally smaller than the intrinsic effects discussed in previous sections. Lerner and Kearns[60] have shown the ^{31}P shifts of phosphate esters are sensitive to solvent effects [varying as much as 3 ppm from 100% H_2O to 70% DMSO (dimethyl sulfoxide in H_2O)], and Costello *et al.*[61] have noted similar sensitivity of ^{31}P shifts of orthophosphate, diethyl phosphate, and monoethyl orthophosphate to high salt (0–5 M added salt).

Gorenstein and co-workers[9,62,63] have noted that ^{31}P shifts are also sensitive to temperature, although they have analyzed this effect in terms of the stereoelectronic ^{31}P shift effect. In fact, it is possible to utilize this temperature sensitivity to design a ^{31}P chemical shift thermometer[62] (see below).

Under more modest changes in solvent and salt conditions, the intrinsic (and particularly stereoelectronic) effects appear to largely dominate ^{31}P chemical shifts of phosphate esters. Other possible factors that could affect ^{31}P shifts in phosphate esters have been found generally to be relatively unimportant. Thus, as discussed below, Gorenstein and co-workers have used this idea to probe the structure of double-helical nucleic acids,[8,14,40,59,62-65] and it is observed that ring-current effects associated

[56] R. O. Day, D. G. Gorenstein, and R. R. Holmes, *Inorg. Chem.* **22**, 2192 (1983).

[57] J. M. Lehn and G. Wipff, *J. Chem. Soc. Chem. Commun.*, 800 (1975).

[58] D. G. Gorenstein, J. B. Findlay, B. A. Luxon, and D. Kar, *J. Am. Chem. Soc.* **99**, 3473 (1977).

[59] D. G. Gorenstein, *Chem. Rev.* **87**, 1047 (1987).

[60] D. B. Lerner and D. R. Kearns, *J. Am. Chem. Soc.* **102**, 7612 (1980).

[61] A. J. R. Costello, T. Glonek, and J. R. Van Wazer, *J. Inorg. Chem. Soc.* **15**, 972 (1976).

[62] D. G. Gorenstein, J. B. Findlay, R. K. Momii, B. A. Luxon, and D. Kar, *Biochemistry* **15**, 3796 (1976).

[63] D. G. Gorenstein, E. M. Goldfield, R. Chen, K. Kovar, and B. A. Luxon, *Biochemistry* **20**, 2141 (1981).

[64] D. G. Gorenstein, S. A. Schroeder, M. Miyasaki, J. M. Fu, V. Roongta, P. Abuaf, J. T. Metz, and C. R. Jones, *Bull. Magn. Reson.* **8**, 137 (1986).

[65] S. A. Schroeder, J. M. Fu, C. R. Jones, and D. G. Gorenstein, *Biochemistry* **26**, 3812 (1987).

with the bases in double helical nucleic acids are expected and found[66] to have only small (<0.1 ppm) perturbations on the ^{31}P signals. This diamagnetic contribution to the ^{31}P chemical shift influences ^{1}H and heavy-atom chemical shifts to the same extent and is strongly distance dependent. The phosphorus nucleus is shielded by a tetrahedron of oxygens, and therefore aromatic groups such as nucleic acid bases can never approach close enough to cause any marked shielding or deshielding.

Most surprisingly, association of hydrogen-bonding donors often appears to have only a small effect on the ^{31}P chemical shift other than that explained by a shift in pK.[14,67,68] As described below in the section on ribonuclease A, secondary ionization of a phosphate monoester does produce a ~4-ppm downfield shift of the ^{31}P signal,[67–71] but this is possibly attributable to an O–P–O bond-angle effect.[44] Note that an increase in negative charge on the phosphate produces a downfield shift, opposite to the "normal" expectation in chemical shifts. This again reflects a subtle interplay of the various tensor components.

Experimental Considerations. Because of the low sensitivity of the ^{31}P experiment, high sample concentrations (0.1–5 mM in 0.5 ml) or larger sample volumes are generally required for obtaining a spectrum with an adequate signal-to-noise ratio in a reasonable time. As usual, several free induction decays (FIDs) are acquired and averaged to enhance spectral signal-to-noise ratios. Typical times for acquiring the ^{31}P FID following a 90° radiofrequency pulse are 1–8 sec depending on the required resolution (dictated by the linewidth of the signal, $1/\pi T_2^*$, where T_2^* is the time constant for the FID). Waiting longer than $2T_2^*$ will generally not improve the resolution or signal-to-noise ratio (S/N). Additional consideration for optimization of the S/N must be given to the time it takes for the ^{31}P spins to return to thermal equilibrium after a 90° radiofrequency pulse, which is roughly three times the spin–lattice relaxation time (T_1). If $T_1 \sim T_2^*$, as would be true for small, phosphorus-containing molecules wherein magnetic field inhomogeneity and paramagnetic impurities do not lead to any additional line broadening, then a waiting period between pulses of $3T_2^*$ provides a good compromise between adequate resolution and signal sensitivity. If $T_1 > T_2^*$, as is often the case in enzymatic complexes, then waiting only $3T_2^*$ does not allow the magnetization to return to equilibrium and an additional delay must generally be introduced so that the total time

[66] D. G. Gorenstein, S. A. Schroeder, J. M. Fu, J. T. Metz, V. Roongta, and C. R. Jones, *Biochemistry* **27**, 7223 (1988).

[67] D. G. Gorenstein and A. M. Wyrwicz, *Biochem. Biophys. Res. Commun.* **54**, 976 (1973).

[68] W. Haar, J. C. Thompson, W. Mauer, and H. Ruterjans, *Eur. J. Biochem.* **40**, 259 (1973).

[69] M. Cohn and T. R. Hughes, Jr., *J. Biol. Chem.* **237**, 3250 (1960).

[70] M. Blumenstein and M. A. Raftery, *Biochemistry* **11**, 1643 (1972).

[71] R. B. Moon and J. H. Richards, *J. Biol. Chem.* **248**, 7276 (1973).

between pulses is $\sim 3T_1$. This wait can be substantially shortened if the Ernst relationship[72] is used to set the pulse flip angles to $<90°$. At low field, 60–70° pulses, 4- to 8-K data points and 2.0- to 5.2-sec recycle times are generally used. The spectra are often broadband ^{1}H decoupled. The temperature of the sample can be controlled to within $\pm 1°$ by commercial spectrometer temperature control units. Broadband decoupling at higher superconducting fields can produce about 10–15° heating of the sample above the gas stream measured temperatures, even using a gated, two-level decoupling procedure.[62] As described below, this problem may be circumvented by using more modern decoupling methods (such as WALTZ decoupling).

^{31}P Thermometer. As indicated in the previous section, it is often necessary to correct for the solution heating by the decoupler, especially for ionic, aqueous samples. The temperature of the sample should be measured under the experimental decoupling conditions. It is incorrect to assume that the temperature of the sample, and the gas used to regulate the temperature of the sample, are the same, because the decoupler can heat the solution substantially above the gas temperature. This is especially true for biological solutions of high ionic strength. It is therefore necessary to be able to measure directly the solution temperature using either a nonmagnetic thermocouple sensor or thermistor inserted into the NMR solution, or better, a sample whose ^{31}P spectrum reflects the correct temperature. Several ^{31}P "thermometers" have been proposed.[62,73,74] Our laboratory has proposed the use of a sample containing a solution of trimethyl phosphate (10 mM), sodium hydrogen phosphate (10 mM), and EDTA (1 mM) in a Tris buffer (100 mM) and in the appropriate salt solution (e.g., 0.1 M NaCl) in 20% D_2O, pH 7.0 (pH meter reading calibrated against 100% H_2O buffers). The added salt is calculated to yield the approximate total ionic strength of the external NMR sample being studied, because decoupler heating is greater at higher salt concentrations. The frequency separation between the trimethyl phosphate and inorganic phosphate ^{31}P NMR signals is temperature sensitive (Fig. 5). Trimethyl phosphate shifts downfield with increasing temperature while the inorganic phosphate shifts upfield as the pH decreases with increasing temperature (because the Tris pK is very sensitive to temperature). The shift range (~ 1.5 ppm) and reproducibility of the measured shift difference from 0 to 90° is sufficient to calibrate the internal temperature of the ^{31}P thermometer to $\pm 1°$. Without decoupling and with a thermocouple in-

[72] M. L. Martin, J. J. Delduech, and G. J. Martin, "Practical NMR Spectroscopy." Heyden, London, 1980.

[73] F. L. Dickert and S. W. Hellman, *Anal. Chem.* **52**, 996 (1980).

[74] S. Lipson and A. Warshel, *J. Chem. Phys.* **49**, 5116 (1968).

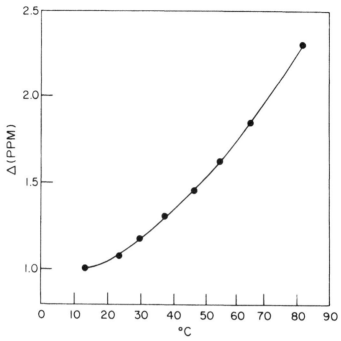

FIG. 5. A ³¹P thermometer. Calibration curve of chemical shift difference between the ³¹P signals of a solution of trimethyl phosphate (10 mM), sodium hydrogen phosphate (10 mM), EDTA (1 mM), Tris buffer (100 mM), 0.1 M NaCl, and 20% D_2O, pH 7.0, at indicated temperatures.

serted directly into the NMR sample, the ³¹P chemical shift difference between the trimethyl phosphate and inorganic phosphate signals is measured at various temperatures to produce a calibration curve (Fig. 5). In the absence of ¹H decoupling, the trimethyl phosphate signal is split into a multiplet and the center of the pattern can be used to determine the chemical shift difference. The decoupler is now turned on at the regulated temperature. Once the temperature of the sample has equilibrated, as evidenced by no further change with time in the ³¹P spectrum (it can take as long as 5–15 min for the sample to achieve its ultimate temperature as a result of the decoupler heating), the actual solution temperature is then determined from the shift difference and the calibration curve. On a GE/ Nicolet NT-200 NMR spectrometer the decoupler heated a biological sample of 0.2 M ionic strength from 10 to 26° and from 70 to 74° even with a two-level decoupling procedure (gated low, 1 W) at all times except during collection of the FID when it was gated high (2 W). Single level, 4-W decoupling produced greater than 25° heating and points out the importance of correction for this problem at higher fields. Properly run, WALTZ decoupling produces very little heating of the sample.

Stereoelectronic Effects on [31]P Chemical Shifts as Probe of DNA and DNA–Protein Structures

We have noted that ^{31}P chemical shifts can potentially provide a probe of the conformation of the phosphate ester backbone in nucleic acids and nucleic acid complexes.[8,9,75] As described in the previous section, these initial theoretical and simple model results suggested that we might be able to use this stereoelectronic effect on ^{31}P chemical shifts as a probe of nucleic acid conformations. If ^{31}P chemical shifts are sensitive to phosphate ester conformations, they potentially provide information on two of the most important torsional angles that define the nucleic acid deoxyribose phosphate backbone. One of these, the P–O3' torsional angle, is also the most variable one in the B-form of the double helix and the other P–O5' torsional angle is one of the most variable in the A-form of the duplex.[76] Indeed, following the original suggestion of Sundaralingam,[77] and based on recent X-ray crystallographic studies of oligonucleotides, Saenger[76] has noted that the P–O bonds may be considered the "major pivots affecting polynucleotide structure."

Our earlier ^{31}P NMR studies on poly- and oligonucleic acids[8] supported our suggestion that the base stacked and helical structure with a gauche,gauche phosphate ester torsional conformation should be upfield from the random-coil conformation, which contains a mixture of phosphate esters in other nongauche conformations as well (Fig. 2). More recently, we[64,78–80] and others[81–83] have used a [17]O/^{18}O labeling or two-dimensional (2D) ^{31}P/^{1}H heteronuclear correlated NMR[64,84,85] scheme

[75] D. G. Gorenstein, S. A. Schroeder, M. Miyasaki, J. M. Fu, V. Roongta, P. Abuaf, A. Chang, and J. C. Yang, in "Biophosphates and Their Analogues—Synthesis, Structure, Metabolism and Activity" (K. S. Bruzik and W. J. Stec, eds.), pp. 487–502. Elsevier, Amsterdam, 1987.

[76] W. Saenger, "Principles of Nucleic Acid Structure." Springer-Verlag, Berlin and New York, 1984.

[77] M. Sundaralingam, Biopolymers 7, 821 (1969).

[78] D. O. Shah, K. Lai, and D. G. Gorenstein, Biochemistry 23, 6717 (1984).

[79] K. Lai, D. O. Shah, E. Derose, and D. G. Gorenstein, Biochem. Biophys. Res. Commun. 121, 1021 (1984).

[80] D. G. Gorenstein, S. Schroeder, M. Miyasaki, J. Fu, C. Jones, V. Roongta, and P. Abuaf, Proc. Int. Conf. Phosphorus Chem., 10th, pp. 567–570 (1986).

[81] J. Ott and F. Eckstein, Nucleic Acids Res. 13, 6317 (1985).

[82] M. Petersheim, S. Mehdi, and J. A. Gerlt, J. Am. Chem. Soc. 106, 439 (1984).

[83] J. Ott and F. Eckstein, Biochemistry 24, 2530 (1985).

[84] V. Sklenář, H. Miyashiro, G. Zon, H. T. Miles, and A. Bax, FEBS Lett. 208, 94 (1986).

[85] J. M. Fu, S. A. Schroeder, C. R. Jones, R. Santini, and D. G. Gorenstein, J. Magn. Reson. 77, 577 (1988).

to identify the individual ³¹P resonances of oligonucleotides, which in B-DNA are observed to vary by ~0.7 ppm. These studies have allowed us to gain insight into the various factors responsible for ³¹P chemical shift variations in oligonucleotides.[78,79,81,86,87] As discussed above, one of the major contributing factors that determines ³¹P chemical shifts is the main-chain torsional angles of the individual phosphodiester groups along the oligonucleotide double helix. Phosphates located toward the middle of a B-DNA double helix assume the lower energy stereoelectronically favored g^-,g^- conformation,[88,89] while phosphodiester linkages located toward the two ends of the double helix tend to adopt a mixture of g^-,g^- and g^-,t conformations, where increased flexibility of the helix is more likely to occur. Because the g^-,g^- conformation is responsible for a more upfield ³¹P chemical shift, while a g^-,t conformation is associated with a lower field chemical shift, internal phosphates in oligonucleotides would be expected to be upfield of those nearer the ends. Although several exceptions have been observed, this positional relationship appears to be generally valid for oligonucleotides where ³¹P chemical shift assignments have been determined.[8,62,64,81,83] Thus position of the phosphorus (ends versus middle) within the oligonucleotide is one important factor responsible for variations in ³¹P chemical shifts.

Eckstein and co-workers[81,83,90] have recently noted that the occurrence of a 5'-pyrimidine–purine-3' base sequence (5'-PyPu-3') within the oligonucleotide has a ³¹P chemical shift further downfield than expected if based solely on the phosphate positional relationship. They have suggested an explanation for these anomolous chemical shifts based upon sequence-specific structural variations of the double helix as proposed by Calladine.[91] Local helical distortions arise along the DNA chain due to purine–purine steric clash on opposite strands of the double helix.[91] As a result, 5'-PyPu-3' sequences within the oligonucleotide represent positions where the largest helical distortions occur. Ott and Eckstein have proposed, based on the ³¹P assignments of two dodecamers and an octamer, that a correlation exists between the helical roll angle parameter[91,92] and ³¹P chemical shifts.[81,83] They have noted a considerably poorer corre-

[86] D. M. Cheng, L.-S. Kan, P. S. Miller, E. E. Leutzinger, and P. O. P. Ts'o, *Biopolymers* **21**, 697 (1982).

[87] D. M. Cheng, L. Kan, and P. O. P. Ts'o, *in* "Phosphorus NMR in Biology" (C. T. Burt, ed.), pp. 135–147. CRC Press, Boca Raton, Florida, 1987.

[88] The notation for the P–O ester torsion angles follows the convention of Seeman *et al.*[89] with the ζ P–O3' angle given first followed by the α P–O5' angle.

[89] N. C. Seeman, J. M. Rosenberg, F. L. Suddath, J. J. Park Kim, and A. Rich, *J. Mol. Biol.* **104**, 142 (1976).

[90] B. A. Connolly and F. Eckstein, *Biochemistry* **23**, 5523 (1984).

[91] C. R. Calladine, *J. Mol. Biol.* **161**, 343 (1982).

[92] R. E. Dickerson, *J. Mol. Biol.* **166**, 419 (1983).

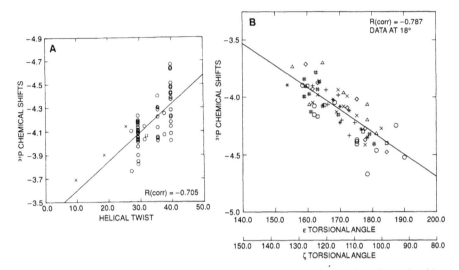

FIG. 6. (A) Plot of ^{31}P chemical shift versus calculated helix twist for nine oligonucleotide duplexes (an octamer, four dodecamers, and four tetradecamers).[85] Terminal phosphate shifts (○), drug-DNA (×), B-DNA (△), and A-DNA (□) shifts are shown. The correlation coefficient between ^{31}P shifts and calculated helical twist is 0.71. (See Ref. 66 for details.) (B) Comparison of ^{31}P chemical shifts of individually assigned phosphates with backbone torsional angles ζ and ε for various oligonucleotide duplexes: d(CCCGATCGGG) (△), G–T mismatch d(CGTGAATTCGCG) (○), G–U mismatch d(CGUGAATTCGCG) (□), and various 14-bp duplexes (×, +, #, ◇, *).

lation between ^{31}P chemical shifts of the oligonucleotides and other sequence-specific variations in duplex geometry (such as the helix twist).

Based on the ^{31}P assignments of oligonucleotides in our laboratory, we have found that there does appear to be a reasonable correlation between ^{31}P shifts and the helical twist at a particular base step.[64,65,85] In fact, if we plot the assigned ^{31}P shifts of the nine oligonucleotides from our laboratory and others, a reasonable correlation appears to exist between ^{31}P chemical shifts and calculated helical twist[66] (Fig. 6A).

These DNA studies have provided a means to test directly the hypothesis that ^{31}P chemical shifts are correlated with phosphate ester conformation. Thus, utilizing a heteronuclear proton flip or a selective version of the ^{1}H-detected heteronuclear pulse sequence,[66,92a,92b] a number of the backbone coupling constants (and through the Karplus relationship, the torsional angles) of over a dozen oligonucleotides have now been measured. It is thus most significant that there appears to be a strong

[92a] V. Sklenar and A. Bax, *J. Magn. Reson.* **71,** 379 (1987).

[92b] D. G. Gorenstein, C. R. Jones, S. A. Schroeder, J. T. Metz, J. M. Fu, V. A. Roongta, R. Powers, C. Karslake, E. Nikonowitz, and R. Santini, *in* "NMR and Macromolecular Structure" (N. Niccolai, ed.), in press. Birkhauser, Boston, Massachusetts, 1990.

correlation between the phosphate ester torsion angles and ^{31}P chemical shifts in the duplex oligonucleotides.

Using these measured $J_{H3'-P}$ coupling constants and the Karplus relationship, H3'–C3'–O3'–P (ε) torsional angles may be calculated. Because of a strong correlation, R = 0.92, shown by Dickerson,[92] between torsional angles C3'–O3'–P–O5' (ζ) and ε in the crystal structures of a 12-bp DNA fragment, ζ may also be calculated. Shown in Fig. 6B is a plot of the variation of ζ (and ε) versus ^{31}P chemical shifts for a number of the phosphates of these small 8- to 14-bp oligonucleotide duplexes. The correlation coefficient between ζ (or ε) and ^{31}P chemical shifts is −0.79 for all of the data in Fig. 6B, supporting the dependence of ^{31}P chemical shifts on phosphate ester conformation.[92b–92d]

Of potential significance, Klug and co-workers[93] have shown that the sites of DNase I-catalyzed hydrolysis of the palindromic 12-mer, d(CGCGAATTCGCG), correlates with the Calladine rules for sequence-specific local variation in duplex geometry. Drew and Travers[94] have suggested that these sequence-specific variations in the DNA structure could be responsible for the specificity of DNase I, DNase II, and copper-phenanthroline-catalyzed hydrolysis of DNA. As noted above, ^{31}P chemical shifts also appear to reflect these sequence-specific backbone structural changes and it will be interesting in the future to determine whether conformational changes in nucleic acids bound to enzymes can be conveniently monitored by ^{31}P NMR. Indeed, preliminary ^{31}P data on an *Eco*RI · dodecamer duplex complex suggests that several of the oligonucleotide ^{31}P signals are significantly perturbed in the duplex restriction enzyme complex (I. Russu, personal communication). In contrast, only small changes in the ^{31}P NMR spectra of a *lac* repressor headpiece protein with an operator DNA fragment have been observed.[92b]

^{31}P NMR Studies of Mononucleotide · Ribonuclease A Complexes

As noted above, association of hydrogen-bonding donors often appears to have little effect on the ^{31}P chemical shift other than that explained by a shift in the pK. Secondary ionization of a phosphate monoester does produce approximately a 4-ppm downfield shift of the ^{31}P signal.[67–71] Thus the pH dependence of the ^{31}P signal of pyrimidine nucleotides, both free in solution and when bound to bovine pancreatic ribonuclease A (RNase A), demonstrates this point (Fig. 7).[14,67,68]

[92c] D. G. Gorenstein, R. P. Meadows, J. T. Metz, E. Nikonowicz, and C. B. Post, *in* "Advances in Biophysical Chemistry" (C. A. Bush, ed.), in press. JAI Press, Greenwich, Connecticut, 1990.

[92d] S. A. Schroeder, V. A. Roongta, J. M. Fu, C. R. Jones, and D. G. Gorenstein, submitted (1990).

[93] G. P. Lomonossof, P. J. G. Butler, and A. Klug, *J. Mol. Biol.* **149**, 745 (1981).

[94] H. R. Drew and A. A. Travers, *Cell* **37**, 491 (1984).

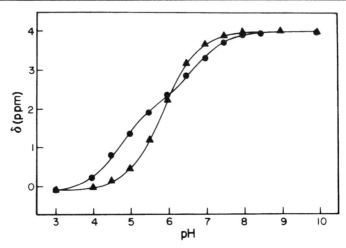

FIG. 7. Plot of ^{31}P chemical shift versus pH for 3′-CMP in 1 mM EDTA/H$_2$O. (▲) The shift of the inhibitor free in solution. (●) The chemical shift of the bovine pancreatic ribonuclease A complex (1:1), pH 7, 1 mM EDTA, 36.4 MHz. [Reprinted with permission from D. G. Gorenstein, A. M. Wyrwicz, and J. Bode,[95] *J. Am. Chem. Soc.* **98**, 2308 (1976). Copyright 1976 American Chemical Society.]

The ^{31}P chemical shift of cytidine 3′-monophosphate (3′-CMP) in free solution follows a simple titration curve, and the ionization constant derived from the ^{31}P shift variation agrees with potentiometric titration values. The ^{31}P chemical shift titration curve for the 3′-CMP · RNase A complex, however, cannot be analyzed in terms of a single ionization process. The two inflections in Fig. 7 suggest two ionizations with $pK_1 = 4.7$ and $pK_2 = 6.7$. The 5′-CMP, 2′-CMP, and 3′-UMP complexes of RNase A also exhibit similar biphasic ^{31}P titration curves.[68,95]

These results suggest that the nucleotides bind around neutral pH in the dianionic ionization state, as shown in Scheme 1. Thus the 3′-CMP · RNase A complex is shifted upfield less than 0.3 ppm from the free 3′-CMP between pH 6.5 and 7.5, whereas monoprotonation of the free dianion results in a 4-ppm upfield shift. Furthermore, the addition of the first proton to the nucleotide · protein complex ($pK_2 = 6.0$–6.7) must occur mainly on some site other than the dianionic phosphate because the ^{31}P signal is shifted upfield by only 1–2 ppm. The addition of a second proton ($pK_1 = 4.0$–5.7) to the complex shifts the ^{31}P signal further upfield so that at the lowest pH, the phosphate finally appears to be in the monoanionic ionization state.

On the basis of X-ray[96] and ^{1}H NMR[97,98] studies, it is known that the

[95] D. G. Gorenstein, A. M. Wyrwicz, and J. Bode, *J. Am. Chem. Soc.* **98**, 2308 (1976).

[96] R. F. M. Richards and H. W. Wyckoff, *Enzymes* **4** (1971).

[97] D. H. Meadows, G. C. K. Roberts, and O. Jardetzky, *J. Mol. Biol.* **45**, 491 (1969).

[98] J. L. Markley, *Acc. Chem. Res.* **8**, 70 (1975).

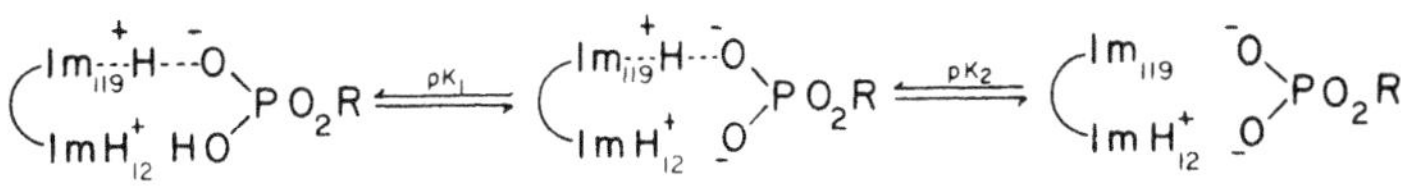

$$\left(\begin{matrix} Im_{119}^+ \text{-} H \cdots O \\ ImH_{12}^+ \quad HO \end{matrix} P O_2 R \right) \xrightleftharpoons{pK_1} \left(\begin{matrix} Im_{119}^+ \text{-} H \cdots O \\ ImH_{12}^+ \quad O^- \end{matrix} P O_2 R \right) \xrightleftharpoons{pK_2} \left(\begin{matrix} Im_{119} \quad O^- \\ ImH_{12}^+ \quad O^- \end{matrix} P O_2 R \right)$$

SCHEME 1. Ionization states for the 3'-CMP · RNase A complex. Only the imidazoles (Im) of His-12 and His-119 and the phosphate monoester inhibitor are shown.

TABLE I

CHEMICAL SHIFTS AND pH TITRATION DATA FOR
REPRESENTATIVE MODEL COMPOUNDS[a]

Compound	Chemical shift (ppm)[b]	Titratable[c]	pK_a	Reference(s)[d]
Phosphomonoesters				
Phosphoserine	4.6	+	5.8	1
Phosphothreonine	4.0	+	5.9	1
Pyridoxal phosphate	3.7	+	6.2	2, 3
Pyridoxamine phosphate	3.7	+	5.7	2, 3
Flavin mononucleotide	4.7	+	~6.0	4
Phosphodiesters				
RNA, DNA, phospholipids	0 to −1.5	−	—	5
Diphosphodiesters				
Flavin adenine dinucleotide	−10.8/−11.3	−	—	6
Phosphotriesters				
Dialkyl phosphoserine	0 to −3.0	−	—	6a
Phosphoramidates				
N^3-Phosphohistidine	−4.5	−	—	7
N^1-Phosphohistidine	−5.5	−	—	7
Phosphoarginine	−3.0	+	4.3	8
Phosphocreatine	−2.5	+	4.2	8
Acyl phosphates				
Acetyl phosphate	−1.5	+	4.8	8
Carbamoyl phosphate	−1.1	+	4.9	8

[a] Largely derived from Ref. 18.

[b] With respect to 85% H_3PO_4; upfield shifts are given a negative sign.

[c] Titratable: +, changes are observed in the chemical shift on pH titration; for phosphomonoesters this change is 4 ppm; for phosphoramidates, 2.5 ppm; and for acyl phosphates, 5.1 ppm; −, no significant changes are observed.

[d] References: (1) C. Ho, J. A. Magnuson, J. B. Wilson, N. S. Magnuson, and R. J. Kurland, *Biochemistry* **8**, 2074 (1969); (2) M. Martinez-Carrion, *Eur. J. Biochem.* **54**, 39 (1975); (3) K. Feldmann and E. J. M. Helmreich, *Biochemistry* **15**, 2394 (1976); (4) C. T. Monnen and F. Müller, *Biochemistry* **21**, 408 (1982); (5) G. D. Armstrong, L. S. Frost, H. J. Vogel, and W. Paranchych, *J. Bacteriol.* **145**, 1167 (1981); (6) T. L. James, D. E. Edmonson, and M. Husain, *Biochemistry* **20**, 617 (1981); (6a) D. G. Gorenstein, D. Shah, R. Chen, and D. Kallick, *Biochemistry* **89**, 2050 (1989); (7) M. Gassner, D. Stehlik, O. Schrecker, W. Hengstenberg, W. Maurer, and H. Rüterjans, *Eur. J. Biochem.* **75**, 287 (1977); (8) H. J. Vogel and W. A. Bridger, *Can. J. Biochem. Cell Biol.* **61**, 363 (1983).

nucleotides are located in a highly basic active site with protonated groups histidine-119, histidine-12, and, probably, lysine-41 quite close to the phosphate. This suggests that pK_1 is associated with ionization of a protonated histidine residue that hydrogen bonds to the phosphate (Scheme 1). This highly positive active site, which is capable of perturbing the pK of the phosphate from 6 to 4.7 (or to 4.0 for 2'-CMP), must have one or more hydrogen bonds to the phosphate over the entire pH region. Yet at the pH extrema, little if any perturbation of the ^{31}P chemical shift is found. Apparently, the ^{31}P chemical shift of the phosphate esters is mostly affected only by the protonation state and not by the highly positive local environment of the enzyme. Additional studies have supported this suggestion, although in some instances binding of an ionizable phosphate to a protein does produce ^{31}P shifts that cannot be rationalized in terms only of a shift in the pK of the phosphate.[9] Perhaps the most dramatic demonstration of enzyme-associated ^{31}P perturbation is the observation that the phosphate covalently bound to alkaline phosphatase[99–101] is shifted 6–8 ppm downfield from inorganic phosphate in solution. However, Chlebowski et al.[101] have suggested that this originates from bond-angle strain in the enzyme complex. Evans and Kaplan[102] have shown that ^{31}P chemical shifts in a series of 5'-nucleotides and related phosphate monoesters are only modestly sensitive to intramolecular hydrogen bonding between the base RNH and dianionic phosphate. The 0.4-ppm shielding resulting from this hydrogen-bonding interaction correlates nicely with the population of the g,g conformation about the C-4'–C-5' bond.

Nageswara Rao[17] and Vogel[18] have also recently reviewed ^{31}P NMR studies of various enzyme complexes. Table I, from Vogel,[18] provides an indication of the range of ^{31}P chemical shifts and the titration behavior of various phosphoprotein model compounds (see also Vogel [13], this volume).

Acknowledgments

Supported by the NIH (GM36281), the Purdue University Biochemical Magnetic Resonance Laboratory, which is supported by the NIH (Grant RR01077 from the Biotechnology Resources Program of the Division of Research Resources), and the NSF National Biological Facilities Center on Biomolecular NMR, Structure, and Design at Purdue (Grants BBS 8614177 and 8714258 from the Division of Biological Instrumentation). Support by the John Simon Guggenheim Memorial Foundation for a fellowship to D. G. Gorenstein is also gratefully acknowledged. The contributions of C. Jones, W. J. Ray, Jr., and R. Santini are much appreciated.

[99] J. L. Bock and B. Sheard, Biochem. Biophys. Res. Commun. 66, 24 (1975).
[100] W. E. Hull, S. E. Halford, H. Gutfreund, and B. D. Sykes, Biochemistry 15, 1547 (1976).
[101] J. F. Chlebowski, I. M. Armitage, P. P. Tusa, and J. E. Coleman, J. Biol. Chem. 254, 1207 (1976).
[102] F. E. Evans and N. O. Kaplan, FEBS Lett. 105, 11 (1979).

[16] Ligand–Protein Interactions via Nuclear Magnetic Resonance of Quadrupolar Nuclei

By CHARLES R. SANDERS II and MING-DAW TSAI

Introduction

A significant number of biologically relevant elements have isotopes that possess nuclear magnetic spin quantum numbers greater than $\frac{1}{2}$ and thereby possess nuclear quadrupole moments. This results in the powerful quadrupolar relaxation mechanism that normally dominates the relaxation and, therefore, the line shapes and widths of the NMR peaks observed from such nuclei. To a certain extent, quadrupolar NMR (QNMR) is advantageous because only a single relaxation mechanism is effectively operative. However, because of theoretical and experimental difficulties, QNMR is an underdeveloped and underutilized technique in the study of biological systems, particularly in enzymology. Keniry ([19], Vol. 176, this series) deals with ^{2}H NMR of ^{2}H-labeled proteins, where the useful information extracted regards the local dynamics of amino acid side chains. This chapter presents the use of QNMR to study ligand–protein or substrate–enzyme interactions. In addition to the possible determination of stoichiometry, binding constants, and exchange rates, QNMR can be a particularly effective method in the study of the motional dynamics (through determination of τ_c, the effective correlation time) and the electrical environment (through χ, the quadrupolar coupling constant) of bound ligands. The term QNMR should not be confused with a related physical technique: nuclear quadrupole resonance. Because the studies we are concerned with usually involve motion that is isotropic on the NMR time scale, we are technically concerned with high-resolution (HR) NMR. However, because QNMR lines tend to be rather broad, the practical study of quadrupolar species tends to involve a mixture of HR and "broad-line" NMR techniques.

Some aspects of QNMR have been discussed in a number of recent reviews.[1-6] This chapter emphasizes practical aspects and is written spe-

[1] P. Laszlo (ed.), "NMR of Newly Accessible Nuclei: Chemical and Biological Applications," Vols. 1 and 2. Academic Press, New York, 1983.
[2] S. Forsén, T. Drakenberg, and H. Wennerström, *Q. Rev. Biophys.* **19**, 83 (1987).
[3] S. Forsén and B. Lindman, *Methods Biochem. Anal.* **27**, 289 (1981).
[4] J. B. Lambert and F. G. Riddell (eds.), *NATO Adv. Study Ser., Ser. C* **103** (1983).
[5] M. Minelli, J. H. Enemark, R. T. C. Browlee, M. J. O'Connor, and A. G. Webb, *Coord. Chem. Rev.* **68**, 169 (1985).
[6] H. Mantsch, H. Saito, and I. P. Smith, *Prog. NMR Spectrosc.* **11**, 211 (1977).

cifically for biochemists with a general knowledge of HR NMR but without previous experience in QNMR.

Fundamental Considerations in Experimental Design

Sensitivity

Three factors contribute to the sensitivity problem in QNMR. (1) As shown in Table I, the intrinsic sensitivity of most quadrupolar nuclei is much lower than that for ^1H. Furthermore, many nuclei have low natural abundance. (2) Linewidths of quadrupolar species are often broad and the amount of sample is often limited in biochemically relevant studies. (3) Because T_1 and T_2 are usually very short for quadrupolar nuclei, the loss of signal during the "preacquisition delay" (DE; see Experimental Techniques) becomes a detrimental factor to sensitivity. On the other hand, short T_1 and T_2 allow rapid acquisition, and it is sometimes possible to collect as many as 10^6 scans within 24 hr.

To attain a rough estimate of the limit of detection of a sample and spectrometer system, it is often advisable to prepare a standard sample of the free ligand (or a structurally similar compound) at the same concentration as that to be used in the experiment of interest. If a signal can be

TABLE I
PARAMETERS FOR SELECTED QUADRUPOLAR NUCLEI OF BIOLOGICAL INTEREST

Nucleus	I	Natural abundance (%)	Sensitivity (relative to ^1H)	Q (10^{-28} m²)	Typical χ range (MHz)	Typical δ range (ppm)	Typical linewidth for $\tau_c = 0.1$ nsec (Hz)
$^6Li^a$	1	7.4	0.01	0.0007	<0.01	5	0.01
^7Li	3/2	92.6	0.29	−0.03	0.015–0.15	5	0.7
^{14}N	1	99.6	0.001	0.016	1–5	200	4,200
^{17}O	5/2	0.037	0.03	−0.026	1–15	700	3,000
$^2H^b$	1	0.015	0.01	0.0027	0.16–0.22	10	19
^{23}Na	3/2	100	0.1	0.14	0.2–2	20	125
^{25}Mg	5/2	10	0.003	0.22	1–4	—	120
^{33}S	3/2	0.76	0.002	−0.064	0.5–15	600	7,000
^{35}Cl	3/2	75	0.005	−0.079	1–4	700	500
^{39}K	3/2	93	0.0005	0.11	—	30	—
^{43}Ca	7/2	0.15	0.006	−0.065	0.5–4	70	29
^{59}Co	7/2	100	0.28	0.40	2–80	18,000	20,000

[a] Quadrupolar relaxation generally does not dominate in ^6Li NMR.

[b] Despite low Q, quadrupolar relaxation dominates due to nonspherical symmetry.

detected, the viscosity can be varied using glycerol and decreasing temperature to broaden the signal, until it reaches the limit of detection.

Chemical Shifts and Spin–Spin Coupling

As exemplified in Table I, the chemical shift range of each nucleus is variable. Observed chemical shifts are, of course, potential sources of information. However, the resolution of multiple signals is a problem when chemical shift differences do not exceed linewidths. This factor tends to limit the number of species observable in a single sample.

Because lines are often broad in QNMR, spin–spin coupling is frequently unobservable. Nevertheless, coupling constants are sometimes large enough to affect the spectra through broadening[7,8] or splitting[8,9] and must then be taken into account in determining line shape or linewidth ($\Delta\nu_{1/2}$).

Relaxation in the Absence of Chemical Exchange

Longitudinal (T_1) and transverse (T_2) relaxation for $I > \frac{1}{2}$ nuclei in the absence of chemical exchange is usually dominated by quadrupole-induced relaxation. Transverse relaxation largely dictates the shape and width of NMR peaks.

Quadrupolar relaxation properties are generally classified in two defined regions: extreme narrowing (where $\omega_0^2 \tau_c^2 \ll 1$) or nonextreme narrowing (where $\omega_0^2 \tau_c^2 \gtrsim 1$), where ω_0 is the angular Larmor frequency of the nuclei.

The rotational correlation time τ_c is related to the molecular tumbling rate: the faster the motion, the smaller the τ_c. This parameter can be roughly estimated for use in predicting the narrowing region using the Stokes–Einstein–Debye equation[10]:

$$\tau_c = 4\pi\eta r^3/3kT \tag{1}$$

where k is Boltzmann's constant, T is in degrees Kelvin, η is solution viscosity (~ 1 cP for H_2O at 20°), and r is the radius of the molecule. Yguerabide et al.[11] listed the τ_c values calculated from Eq. (1) and determined experimentally for a number of proteins of varying molecular weight. The τ_c value of a protein represents only the upper limit of the τ_c of the bound ligand, because the bound ligand is likely to possess some internal rotational freedom.

[7] S.-L. Huang and M.-D. Tsai, Biochemistry 21, 951 (1982).
[8] D. Sammons, P. A. Frey, K. Bruzik, and M.-D. Tsai, J. Am. Chem. Soc. 105, 5455 (1983).
[9] J. A. Gerlt, P. C. Demou, and S. Mehdi, J. Am. Chem. Soc. 104, 2848 (1982).
[10] W. Egan, J. Am. Chem. Soc. 98, 4091 (1976).
[11] J. Yguerabide, H. F. Epstein, and L. Stryer, J. Mol. Biol. 51, 573 (1970).

Relaxation in the Extreme Narrowing Limit. In this region transverse and longitudinal relaxations are both monoexponential decays characterized by single relaxation times related to the observed linewidth as follows:

$$\Delta\nu_{1/2} = \frac{1}{\pi T_2} = \frac{1}{\pi T_1} = \frac{3\pi}{10} \frac{2I + 3}{I^2(2I - 1)} \left(1 + \frac{\eta^2}{3}\right)\left(\frac{e^2 q_{zz} Q}{h}\right)^2 \tau_c \qquad (2)$$

The proportionality between τ_c and linewidth $\Delta\nu_{1/2}$ is evident: slow-moving species produce large linewidths. Equation (2) has been widely used to deduce τ_c from the observed linewidth in QNMR.

The term $(e^2 q_{zz} Q/h)$ is the quadrupolar coupling constant (χ) that describes the strength of the interaction of the quadrupole moment with its electric environment. This interaction is the physical basis for the quadrupolar relaxation mechanism. Q is the intrinsic quadrupole moment of the nucleus, e is the charge of the electron, and q_{zz} is the largest component of the electric field gradient interacting with Q and vanishes when electric symmetry around the nucleus becomes very high (e.g., T_d or O_h groups).

Table I shows that Q varies considerably from nucleus to nucleus. In the case of ⁶Li and ⁷Li, χ^2 can be so small that other relaxation mechanisms can actually compete, while for high-Q nuclei χ^2 is often so large that peaks become very broad, even for free ligands (unless q_{zz} is very low due to high symmetry).

The asymmetry parameter η varies from 0 to 1 and describes the deviation of the electric field gradient from axial symmetry ($\eta = 0$ for axial symmetry). Because even in the rare instance that $\eta = 1$ the term $1 + \eta^2/3$ is only 1.33, the asymmetry parameter can usually be neglected.

Relaxation Outside of Extreme Narrowing. As molecular motion becomes slower and $\omega_0^2 \tau_c^2$ approaches or exceeds 1, Eq. (2) no longer holds. T_2 and T_1 become nonequivalent: T_2 continues to get smaller (with corresponding broader linewidths), while T_1 starts to get larger. In the case of $I = 1$, nuclei lineshapes remain Lorentzian and T_1 and T_2 are described by Eqs. (3) and (4), respectively[12,13] [which reduce to Eq. (2) when $\omega_0^2 \tau_c^2 \ll 1$]:

$$\frac{1}{T_1} = \frac{3\pi^2}{100} \frac{2I + 3}{I^2(2I - 1)} \chi^2 \left(1 + \frac{\eta^2}{3}\right)\left(\frac{2\tau_c}{1 + \omega_0^2 \tau_c^2} + \frac{8\tau_c}{1 + 4\omega_0^2 \tau_c^2}\right) \qquad (3)$$

[12] A. Abragam, "The Principles of Nuclear Magnetism." Oxford Univ. Press (Clarendon), London and New York, 1961.
[13] S. Schramm and E. Oldfield, *Biochemistry* **22**, 2908 (1983).

$$\Delta \nu_{1/2} = \frac{1}{\pi T_2} = \frac{3\pi^2}{100} \frac{2I + 3}{I^2(2I - 1)} \chi^2 \left(1 + \frac{\eta^2}{3}\right)$$

$$\times \left(3\tau_c + \frac{5\tau_c}{1 + \omega_0^2\tau_c^2} + \frac{2\tau_c}{1 + 4\omega_0^2\tau_c^2}\right) \tag{4}$$

For nuclei with $I > 1$, Eqs. (3) and (4) are approximately accurate only when $\omega_0\tau_c \leq 1.5$.[14,15]

Above $\omega_0\tau_c \sim 1.5$, relaxations of $I > I$ nuclei become multiexponential, with line shapes being the superposition (thus non-Lorentzian) of contributions from each exponential. Data from such species (e.g., large proteins) can be very difficult to analyze. In solids and membranes τ_c becomes even larger, and the "quadrupolar splitting" becomes a measurable and useful parameter in QNMR (see Anisotropic Motion).

Because relaxation is so intimately associated with interpretability and observability, it is often helpful to predict likely behavior using estimated τ_c, χ, and ω_0 before running difficult and potentially fruitless experiments. Table I lists typical spectral behavior for different nuclei in extreme-narrowing conditions having the same τ_c.

Exchange

Direct observation of a quadrupolar species fully bound to an enzyme is often difficult. In such cases, the only way of getting NMR information about the bound species is to observe the spectra of the free component involved in exchange with the bound species. Some strategies for extracting the needed data are discussed later in this chapter (see Selective Examples of Applications).

Anisotropic Motion

As τ_c approaches or exceeds $1/\chi$, NMR begins to detect *orientational* anisotropy and peaks broaden and eventually split. Extreme examples are found in the NMR of solids and membranes when QNMR yields multiple resonances separated by "quadrupolar splittings."[16] For liquid samples of protein–ligand complexes τ_c is usually not large enough to induce quadrupolar splitting (if $\chi = 10^5$–10^6 Hz, τ_c would have to be at least 10^{-6}–10^{-5} sec). However, true *motional* isotropy rarely exists at a local

[14] B. Halle and H. Wennerström, *J. Magn. Reson.* **44**, 89 (1981).

[15] T. Andersson, T. Drakenberg, S. Forsén, E. Thulin, and M. Swärd, *Eur. J. Biochem.* **126**, 501 (1982).

[16] C. A. Fyfe, "Solid State NMR for Chemists." CFC Press, Guelph, Ontario, Canada, 1983.

level in macromolecular systems, and the τ_c deduced from Eqs. (2)–(4) should be regarded as "effective" correlation times resulting from a combination of slower, generally isotropic overall tumbling and faster, anisotropic internal motions.

Determination of τ_c and/or χ

Correlation times and χ provide information of interest to biochemists and chemists. Because ω_0 and I are always known and η is usually negligible, relaxation times are straightforward functions of τ_c and χ according to Eqs. (2)–(4). If one of these values is known, the other can be determined from the relaxation time or linewidth using algebra (in extreme narrowing) or the Newton–Raphson (or some other) numerical method.[17]

The correlation time is strictly a motional parameter and is affected by chemistry, solvation, temperature, etc., only to the extent that these affect molecular motion. On the other hand, χ is dependent on chemical factors. When the electrical environment of the quadrupolar nuclei is defined by stable covalent bonds, χ is likely to be constant in a variety of environments provided that no change in bonding occurs. Thus, χ for a C–D species is likely to be nearly constant free and bound to an enzyme. On the other hand, χ for $^{39}K^+$ is likely to change on binding to an anion or a protein.

In ^2H NMR studies χ can be estimated by chemical analogy with similar compounds for which χ has been determined by independent techniques,[6,13,18,19] and can be used to determine changes in τ_c upon binding. In observing ligands bound to proteins, the simultaneous solution of τ_c and χ based on Eqs. (3) and (4) can be accomplished outside of extreme narrowing by measuring both T_1 and T_2 or by measuring either value at two different magnetic field strengths.

Experimental Techniques

Special Problems in QNMR

Experimental difficulties in QNMR are sometimes different from those commonly encountered in other HR NMR or solid-state NMR. In this section the nature of some of these problems is described. The discussion includes hardware and software techniques used to minimize these and other problems.

[17] R. L. Burden, J. D. Faires, and A. C. Reynolds, "Numerical Analysis," 2nd ed. Prindle, Weber, & Schmidt, Boston, 1981.
[18] P. Tsang, R. R. Vold, and R. L. Vold, *J. Magn. Reson.* **71**, 276 (1987).
[19] H. Saito, H. Mantsch, and I. C. P. Smith, *J. Am. Chem. Soc.* **95**, 8453 (1973).

Acoustic Ringing. Acoustic ringing refers to the acoustic waves generated by electromagnetic interaction of the radiofrequency pulse with probe materials. These waves result in spurious rolling in the spectral baseline and can also be manifested as a sharp spike at the beginning of a free induction decay (FID). This phenomenon is present in all NMR experiments but is typically a serious problem in QNMR. In other HR NMR, acoustic ringing can be minimized by employing a sufficiently long preacquisition delay (DE), but in QNMR the length of this DE may cause other problems (see later). Furthermore, in usual HR NMR, signals are often sharp and are easily distinguished from a rolling baseline, and the spectral width is often so small that the entire "ring" is not observed. In QNMR large spectral widths make ringing very apparent and broad peaks may be difficult to differentiate from the baseline rolling, as illustrated previously.[20,21] Acoustic ringing is also more severe at relatively low observation frequencies, a fact of some importance because many quadrupolar nuclei have low Larmor frequencies due to low magnetogyric ratios (γ). Gerothanassis recently published an excellent review[21] on the phenomenon of acoustic ringing and how to deal with it.

Need and Consequences of "Hard" Pulsing. Another consequence of the small γ possessed by many nuclei is that, under a given condition, a relatively long pulsewidth is needed. A long 90° pulse may present difficulties in complex pulse sequences and limits the breadth of spectral excitation. This latter effect is in conflict with the frequent need in QNMR to cover a large spectral range including broad signals. In order to reduce the length of the 90° pulse, higher transmitter power (to produce a "hard" pulse) must be utilized, which worsens acoustic ringing.

Preacquisition Delay (Deadtime). Many spectrometers automatically set DE equal to the dwell time to facilitate phasing. In pulsed FT NMR, the dwell time (DW) is the inverse of the sampling rate that is determined by the spectral width (SW), i.e., DW = 1/(2SW). In a typical ^{1}H NMR experiment at 300 MHz, SW = 3000 Hz and DW = 167 μsec. Because T_1 and T_2 are on the order of 0.1–10 sec (for protons), the data lost during DE are usually negligible. However, in QNMR there are a number of conflicting problems. (1) A large SW is generally needed, resulting in smaller DW (and thus a smaller default value of DE). (2) Acoustic ringing is often more serious (as described previously), which would require a longer DE to minimize. (3) T_1 and T_2 are often so short (e.g., <1 msec) that a significant portion of their signal can be lost during DE. This results in a loss of S/N,

[20] M.-D. Tsai, this series, Vol. 87, p. 235.
[21] I. P. Gerothanassis, *Prog. NMR Spectrosc.* **19**, 267 (1987).

distortion of peaks,[22] and problems in intensity/integration measurement. As a result, the T_1 data should be treated with caution, and the integrated signals with different linewidths cannot be compared.

Hardware Considerations

Magnet and Console. Large superconducting magnets have the advantages of superior sensitivity and stability and result in higher Larmor frequencies. However, because of the broad signals usually encountered in QNMR, resolution and field stability requirements are often less stringent than in typical HR work.

Rapid pulsing and data sampling of a great number of scans require a high dynamic range, which puts high demands on the electronic and computational components of the spectrometer system.

The Probe. The performance of the probe is a key variable in QNMR. In QNMR, as for solid-state NMR, high-power probes are frequently used.[21,22] These probes generally utilize a single horizontal solenoid transmitter/receiver coil which is inherently more sensitive than Helmholtz saddle coils. These coils produce a short 90° pulse compared to saddle coils and also handle higher pulse power more efficiently. High-power probes are also built with materials chosen to reduce acoustic ringing (a negative factor in terms of sensitivity).

The disadvantages of high-power probes lie in the difficulty of shimming and achieving high resolution. Furthermore, they often possess no locking or spinning capability. Fortunately, these factors are usually less important in QNMR. Sample change is also laborious because the entire probe has to be removed from the magnet.

High-resolution probes are often quite suitable when linewidths are not too broad, and these probes are a necessity when accurate linewidths are needed in the high-resolution region.

Software Considerations

Pulse Sequences. When initiating a study, a standard single-pulse experiment can be attempted to evaluate the system at hand (sensitivity, baseline distortion, linewidth, etc.). At this point, DE could be manually adjusted to either reduce acoustic ringing (larger DE) or to increase S/N (smaller DE).[20] Alternatively, one can employ a pulse sequence which will reduce acoustic ringing. Many such pulse sequences have recently become available and have been reviewed by Gerothanassis.[21] In our

[22] E. Fukushima and S. B. W. Roeder, "Experimental Pulse NMR—A Nuts and Bolts Approach." Addison-Wesley, Reading, Massachusetts, 1981.

experience the RIDE (ring-down elimination) sequence[4,21,23] worked effectively.

Some pulse sequences involving use of spin echoes have also been developed for QNMR.[21] When T_2 is very short but T_1 is much longer, spin-echo techniques should allow the true spectra to be observed and also can increase S/N.[21,24,25] However, when T_1 and T_2 are both very short (0–100 μsec), acquisition becomes very difficult due (among other reasons) to the significant relaxation that occurs during the actual pulse sequence. To our knowledge, there is no way around this problem in pulsed FT NMR and in this case CW NMR can actually be advantageous if sensitivity is not a big problem.[21,26] (CW NMR also eliminates the problems associated with generating adequate pulse power to cover a very broad spectral width.)

Although the sampling rate could be as high as 100 scans per sec, it should be kept in mind that the recycling time needs to be $\geq 3\,T_2$ to avoid truncation of the FID (which will cause a regular series of spikes at the bottom of the peak),[27] and needs to be $\geq 5\,T_2$ when the quantitative intensity is of interest.[28] Becker et al.[29] presented a lucid description of the optimization of pulse repetition times as well as pulse angles.

T_1/T_2 Determination. The relative merits of the various T_1/T_2 determination methods have been discussed extensively.[26] These methods can be integrated into the sequences used to minimize acoustic ringing. Data taken for very short T_1/T_2 should be interpreted with caution, because these data may deviate from true values due to significant relaxation during pulsing and DE.

Selective Examples of Applications

Metal Ion Binding to Proteins

As shown in Table I, many biologically relevant metal ions are quadrupolar nuclei. QNMR has been extensively utilized to study binding of these ions to various proteins, as described by several reviews.[2,3,23,30] The

[23] H. J. Vogel and S. Forsén, *Biol. Magn. Reson.* **7**, 247 (1986).
[24] A. Allerhand and D. W. Cochran, *J. Am. Chem. Soc.* **92**, 4482 (1970).
[25] E. D. Becker, J. A. Ferretti, and T. C. Ferrar, *J. Am. Chem. Soc.* **91**, 7784 (1969).
[26] M. Llinas and A. DeMarco, *J. Am. Chem. Soc.* **102**, 2226 (1980).
[27] A. E. Derome, "Modern NMR Techniques for Chemical Research," Vol. 6 of the Organic Chemistry Series. Pergamon, Oxford, 1987.
[28] M. L. Martin, J.-J. Delpuech, and G. J. Martin, "Practical NMR Spectroscopy." Heyden, London, 1980.
[29] E. D. Becker, J. A. Ferretti, and P. N. Gambhir, *Anal. Chem.* **51**, 1413 (1979).
[30] W. H. Braunlin, T. Drakenberg, and S. Forsén, *Curr. Top. Bioenerg.* **14**, 97 (1985).

most successful study in this category is binding of Ca^{2+} to calmodulin (CaM), which is used to illustrate the procedures, problems, and interpretations of the actual experiments.

CaM (M_r 16,800) has four Ca^{2+}-binding sites, numbered I–IV starting from the N-terminus. Various studies have suggested that the four sites can be grouped into two high-affinity sites (III and IV, $K_a \approx 10^6$–$10^7\,M^{-1}$) and two lower affinity sites (I and II, $K_a \approx 10^5$–$10^6\,M^{-1}$).[31-34] The ^{45}Ca NMR signals of Ca^{2+} fully bound to the high-affinity sites have been observed at 2.8 mM CaM and 4.8 mM Ca^{2+}, with $\Delta\nu_{1/2} = 770 \pm 50$ Hz ($T_2 = 0.42$ msec). An inversion–recovery experiment determined $T_1 = 0.86$ msec. The nonequivalence of T_1 and T_2 suggests that the system is not in the extreme narrowing limit and Eq. (2) cannot be applied. Because CaM is a small protein and satisfies $\omega_0\tau_c \leq 1.5$, even if binding is totally rigid, Eqs. (3) and (4) can be used to solve for χ (1.1 MHz) and τ_c (8.2 nsec) simultaneously from T_1 and T_2.[15]

Additional information concerning binding of Ca^{2+} and Mg^{2+} to CaM have been obtained by titration and temperature-dependence experiments, which are illustrated using ^{25}Mg NMR.[35] Because the affinity of Mg^{2+} to CaM is much weaker than that of Ca^{2+}, it is not possible to observe fully bound Mg^{2+}. Even if a large excess of CaM can be used to fully bind Mg^{2+}, the larger quadrupole moment of Mg^{2+} could result in a $\Delta\nu_{1/2}$ too large to be measured at the concentration used. Figure 1 shows ^{25}Mg NMR spectra at different ratios of CaM $\cdot$ Ca_2^{2+} (which represents CaM when sites III and IV are saturated with Ca^{2+} and sites I and II are open for binding Mg^{2+}). The plots of $\Delta\nu_{1/2}$ versus $[Mg^{2+}]/[CaM \cdot Ca_2^{2+}]$ are shown in Fig. 2A. All spectra are Lorentzian, as exemplified in spectra I and J (fittings of A and H, respectively). The Lorentzian shape and the successive changes in linewidths rule out the possibility that the observed signal is due to overlap of the signals of free and bound Mg^{2+} in *very slow exchange* relative to the NMR time scale (Case A in Table II). There are, however, three other possible interpretations. (1) The observed signal is due to free Mg^{2+}, which is in *slow exchange* with bound Mg^{2+} [Case B(a) in Table II]. The signal of bound Mg^{2+} is either too broad to be detected or is well separated from the observed signal of free Mg^{2+} so that the signal

[31] J. D. Potter, P. Strang-Brown, P. L. Walker, and S. Iida, this series, Vol. 102, p. 135.

[32] J. A. Cox, M. Comte, A. Malone, D. Burger, and E. A. Stein, *Metal Ions Biol. Syst.* **17**, 215 (1984).

[33] S. Forsén, H. J. Vogel, and T. Drakenberg, in "Calcium and Cell Function" (W. Y. Cheung, ed.), Vol. 6, pp. 113–157. Academic Press, New York, 1986.

[34] C.-L. A. Wang, *Biochem. Biophys. Res. Commun.* **130**, 426 (1985).

[35] M.-D. Tsai, T. Drakenberg, E. Thulin, and S. Forsén, *Biochemistry* **26**, 3635 (1987).

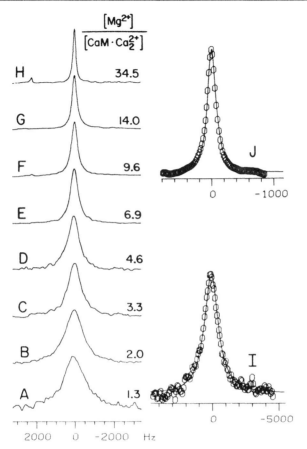

FIG. 1. ^{25}Mg NMR spectra (at 22.15 MHz, 25°) of Mg^{2+} at various ratios of $[Mg^{2+}]/$ $[CaM \cdot Ca_2^{2+}]$ (A–H). Spectra I and J are the Lorentzian fits of spectra A and H, respectively. The circles in spectra I and J represent experimental data, and the solid lines represent the fitted spectra. The RIDE sequence was used to reduce acoustic ringing. The line broadening used is 100 Hz. (Reproduced from Tsai *et al.*[35] with permission.)

remains Lorentzian. The broadening is caused by chemical exchange as shown in Table II. (2) The observed signal is due to the *fast exchange* average of free and bound Mg^{2+} [Case C(a) in Table II] and the successive broadening is mainly due to the rapid relaxation (large $\Delta\nu_{1/2}$) of bound Mg^{2+}. (3) The system is in the *intermediate exchange* region (Case D in Table II), and the broadening is caused by chemical exchange and by averaging with bound Mg^{2+}.

A temperature-dependence study can differentiate the above situa-

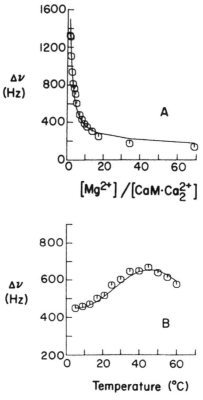

Fig. 2. Titration curve (A) and temperature curve (B) of Mg^{2+} binding to $CaM \cdot Ca_2^{2+}$. The titration curve was obtained at 25°. The temperature curve was obtained at $[Mg^{2+}]/[CaM \cdot Ca_2^{2+}] = 5.24$. Circles are experimental points and solid curves are obtained by connecting the calculated points from the iterative fitting. The $\Delta\nu$ values shown include a line broadening of 100 Hz. (Reproduced from Tsai et al.[35] with permission.)

tions. In the case of slow exchange, increasing temperature will increase the exchange rate (i.e., increase $1/\tau_{ex}$) and cause an increase in the observed $\Delta\nu$ if the increase due to exchange rate is large enough to outweigh the decrease in the intrinsic linewidth (due to increased molecular motion). In the case of fast exchange, the intrinsic linewidths of both free and bound Mg^{2+} decrease due to decreasing τ_c, which will result in a decrease in the observed $\Delta\nu_{1/2}$ upon increasing temperature. Figure 2B shows that the $Mg^{2+}/CaM \cdot Ca_2^{2+}$ system is in slow exchange at 25° and gets into the intermediate region at 30–50°.

When such a complete set of data can be obtained, the data can be analyzed by "total line shape analysis." The details of such analysis have

TABLE II
EFFECTS OF EXCHANGE ON TITRATION EXPERIMENTS AND
TEMPERATURE DEPENDENCE OF $\Delta\nu_{1/2}{}^a$

Cases	Titration experiments	Temperature dependence
A. Very slow exchange $\dfrac{1}{\tau_{ex}} \ll \dfrac{1}{(T_{1,2})_b}$	(a) Bound species not detectable: one Lorentzian signal with successive decrease in intensity, but with constant $\Delta\nu_{1/2}$ (b) Bound species detectable: two signals, or one apparently non-Lorentzian signal, if the two signals overlap	$\Delta\nu_{1/2}$ decreases with increasing temperature due to decreasing τ_c
B. Slow exchange $\dfrac{1}{\tau_{ex}} < \dfrac{1}{(T_{1,2})_b}$	(a) Bound species not detectable: one Lorentzian signal, successively broadened $$\frac{1}{T_{1,2}} = \frac{1}{(T_{1,2})_f} + \frac{1}{\tau_{ex}} \frac{P_b}{P_f} \qquad (6)$$ or $$\Delta\nu_{1/2} = \Delta\nu_f + \frac{1}{\pi\tau_{ex}} \frac{P_b}{P_f}$$ (b) Bound species detectable: two signals, each broadened by exchange, or one apparently non-Lorentzian signal, if the two signals overlap	$\Delta\nu_{1/2}$ increases with increasing temperature due to decreasing τ_{ex}
C. Fast exchange $\dfrac{1}{\tau_{ex}} \gg \dfrac{1}{(T_{1,2})_b}$	(a) Exchange narrowing ($1/\tau_{ex} \gg \Delta\omega$): one signal, $$\frac{1}{T_{1,2}} = \frac{P_f}{(T_{1,2})_f} + \frac{P_b}{(T_{1,2})_b} \qquad (7)$$ (b) Not exchange narrowing ($1/\tau_{ex} \ll \Delta\omega$): two signals separated by $\Delta\omega$	Same as A
D. Intermediate exchange $\dfrac{1}{\tau_{ex}} \sim \dfrac{1}{(T_{1,2})_b}$	In between B and C	Irregular or insensitive

[a] $1/\tau_{ex}$ is the ligand–protein on/off exchange rate (k_{ex}), b and f designate free and bound states, respectively, $\Delta\omega$ represents the separation in resonance frequencies of free and bound ligands, and P_f and P_b represent fractions of ligand in free and bound states, respectively. Only one ligand site per molecule is assumed as is the relationship $1/(T_{1,2})_b \approx 1/(T_{1,2})_b - 1/(T_{1,2})_f$.

been described elsewhere[36] and are beyond the scope of this article. The curves in Fig. 2 resulted from iterative fittings, which yielded the following information: $K_a = 2000 \ M^{-1}$ (assuming two equivalent sites), k_{off} ($\approx k_{ex}$) = 2300 sec^{-1}, and $\Delta\nu_{1/2}$ = 3.5 kHz for bound Mg^{2+}. In this case, τ_c and χ cannot be obtained simultaneously, but if τ_c of Mg^{2+} is assumed to be the same as τ_c of Ca^{2+} when bound to CaM, χ = 1.6 MHz can be obtained. The result unequivocally establishes that Mg^{2+} binds to sites III and IV of CaM, as described in detail by Tsai et al.[35]

In many other studies, complete line shape analysis by the above method is not possible because the complete temperature range may not be obtainable, and the protein available for experiments may be limited in quantity and stability. Even if the protein is stable enough for a variable temperature study, the system may remain in slow (or fast) exchange throughout the entire region. The titration data of these systems are frequently analyzed by the "Swift–Connick" equation[37] [Eq. (5)], which is a combination of Eqs. (6) and (7) described in Table II when $P_f \approx 1$ and $P_b \ll 1$:

$$\frac{1}{T_{1,2}} = \frac{1}{(T_{1,2})_f} + \frac{P_b}{\tau_{ex} + (T_{1,2})_b} \tag{5}$$

Equation (5) is applicable for slow, intermediate, and fast exchange regions. It allows a straightforward measurement of $1/(\tau_{ex} + T_{2b})$ from the slope of the plot of $(\pi\Delta\nu_{1/2})$ versus P_b. Because $P_b \ll 1$ (e.g., 0–5%), it can often be assumed that $P_b = [protein]/[ligand]_{total}$. The main problem in this approach is to determine whether the system is in slow exchange (slope = $1/\tau_{ex}$ because $\tau_{ex} \gg T_{2b}$), fast exchange (slope $\approx \pi\Delta\nu_b$ because $T_{2b} \gg \tau_{ex}$), or intermediate exchange (slope is not directly interpretable).

Halide Ion Binding to Proteins

The technical problems for halide ion binding are very similar to those of metal ion binding to proteins, but the systems amenable to such studies are more limited. Well-studied proteins include, among others, hemoglobin[38] and Cl$^-$ transport proteins.[39,40] One advantage of ^{35}Cl NMR is that fast exchange can be established by the following criterion[3]:

[36] T. Drakenberg, S. Forsén, and H. Lilja, *J. Magn. Reson.* **53**, 412 (1983).
[37] T. J. Swift and R. E. Connick, *J. Chem. Phys.* **37**, 307 (1962).
[38] E. Chiancone, J.-E. Norne, and S. Forsén, this series, Vol. 76, p. 552.
[39] J. J. Falke and S. I. Chan, *J. Biol. Chem.* **260**, 9537 (1985).
[40] J. J. Falke, K. J. Kanes, and S. I. Chan, *J. Biol. Chem.* **260**, 9544 (1985).

$$\frac{(\Delta\nu - \Delta\nu_f) \quad \text{for} \quad {}^{35}\text{Cl}}{(\Delta\nu - \Delta\nu_f) \quad \text{for} \quad {}^{37}\text{Cl}} = \left(\frac{\chi \quad \text{for} \quad {}^{35}\text{Cl}}{\chi \quad \text{for} \quad {}^{37}\text{Cl}}\right)^2 = 1.60 \tag{8}$$

where $\Delta\nu$ and $\Delta\nu_f$ represent the observed linewidth and the linewidth of free Cl^-, respectively. In Cl^- binding studies, fast exchange is observed in most cases, and the detailed ${}^{35}Cl$ NMR analysis under various conditions (e.g., pH dependence and effect of other ligands) has been used to deduce the number and the environment of Cl^- binding sites.

Substrate Binding to Enzymes

The most desirable information in this category is the "internal motions of bound substrates." It is widely accepted that enzyme-bound substrates possess a certain extent of internal rotational freedom relative to the enzyme. However, it is difficult to describe the dynamics of bound substrates quantitatively, because a single measurement can yield only the *effective* τ_c, which may include contributions from several motional components. Indeed, more studies have been done on the dynamics of local segments of proteins (e.g., London [18] and Keniry [19], Vol. 176, this series) than that of bound substrates. The latter studies are often complicated by exchange processes as described previously herein.

In enzyme–substrate binding studies, ${}^2H$ NMR has been used to show that the adenine portion of NAD^+ bound to lactate dehydrogenase has a longer τ_c than the τ_c of the pyridine ring,[41] and that when linoleic acid is bound to lipoxygenase, the internal motions of the substrate increase at positions away from the carboxylic acid group.[42] We describe in detail a binding study of adenylate kinase (AK) using ${}^2H$ NMR.[43]

Chicken muscle AK was titrated with adenylyl(β,γ-methylene)diphosphonate (AMPPCP) deuterated on the phosphonate chain and on the adenine ring and followed by measuring the ${}^2H$ NMR linewidth of the single peak that results from the average of the bound and free AMPPCP. Plots of $\Delta\nu_{1/2}$ versus $[AMPPCP]_{bound}/[AMPPCP]_{total}$ were linear, as shown in Fig. 3. The line shapes of the observed peaks were usually Lorentzian, and limited T_1 inversion recovery data taken were always monoexponential. Thus, the data meet "fast-exchange" criteria and the linewidths of the fully bound species can be determined from linear extrapolations to the fraction bound equal to 1.[44] The effective τ_c values were then calcu-

[41] A. P. Zens, P. T. Fogle, T. A. Bryson, R. B. Dunlap, R. R. Fisher, and P. D. Ellis, *J. Am. Chem. Soc.* **98**, 3760 (1976).
[42] T. S. Viswanathan and R. J. Cushley, *J. Biol. Chem.* **256**, 7155 (1981).
[43] C. R. Sanders II and M.-D. Tsai, *J. Am. Chem. Soc.* **110**, 3323 (1988).
[44] A. C. McLaughlin and J. S. Leigh, *J. Magn. Reson.* **9**, 296 (1973).

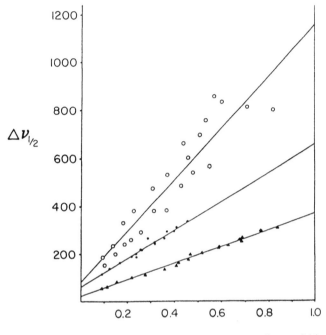

Fraction of AMPPCP bound to AK

FIG. 3. High-resolution ^{2}H NMR (46.1 MHz) linewidths ($\Delta\nu_{1/2}$ in hertz) of ^{2}H-labeled AMPPCP and MgAMPPCP as a function of fractions bound to AK. The experiments were carried out by titrating 1–2 mM AK with small aliquots of AMPPCP or MgAMPPCP ([Mg^{2+}]/[AMPPCP] = 4). Sample conditions: pH 7.0 in ^{2}H-depleted H$_2$O with 45 mM HEPES–K$^+$ or imidazole–HCl, 117 mM KCl, 1–8 mM dithiothreitol, and 0.1 mM EDTA in a 10-mm NMR tube (starting volume of 1.75 ml). Spectral conditions: the probe temperature was 10°; the digital resolution varied from 1 Hz/point for narrow signals to 15 Hz/point for very broad signals; the 90° pulsewidth was 12 μsec. The CYCLOPS pulse sequence was used. The reported $\Delta\nu_{1/2}$ have been corrected for line broadening (1–10 Hz). The fraction of AMPPCP bound to AK was calculated using K_d values of 210 μM for AMPPCP and 190 μM for MgAMPPCP. (○) [8-^{2}H]AMPPCP, (●) MgAMPPCD$_2$P, and (▲) AMPPCD$_2$P. (From Sanders and Tsai.[43])

lated from Eq. (4) using values of χ from structurally related compounds. The contributions of two-bond ^{2}H–^{14}N and ^{2}H–^{31}P scalar and dipolar couplings to the observed $\Delta\nu_{1/2}$ were insignificant. The validity of deducing τ_c from linewidths was further supported by limited T_1 experiments.

The τ_c values obtained indicate that the adenine ring of bound AMPPCP is motionally rigid (τ_c = 27 nsec) and approaches the overall τ_c of AK. The β–γ region of the phosphonate chain (τ_c = 7 nsec) possesses considerable local mobility, consistent with more qualitative ^{17}O NMR studies.[45] This motional freedom is reduced upon binding of Mg^{2+} ($\tau_c = 16$ nsec). The implication of these dynamic properties on the catalytic mechanisms of AK has been discussed.[43]

[45] D. A. Wisner, C. A. Steginsky, Y.-J. Shyy, and M.-D. Tsai, *J. Am. Chem. Soc.* **107**, 2814 (1985).

[17] Ligand Conformations and Ligand–Enzyme Interactions as Studied by the Nuclear Overhauser Effect

By PAUL R. ROSEVEAR and ALBERT S. MILDVAN

Introduction

The nuclear Overhauser effect (NOE) has proved to be a powerful tool for the elucidation of molecular structure.[1] Application of this methodology for studying the solution structure and environment of flexible ligands bound to macromolecules has provided knowledge of the conformation of nucleotides bound to peptides and proteins,[2-6] hormones bound to proteins,[7] nucleic acids bound to proteins,[8-10] and peptides bound to phospholipids.[11,12] These studies usually involve measurement of interproton NOEs using one- or two-dimensional NMR techniques, qualitative estimation or quantitative calculation of interproton distances, and hand or computer model building to evaluate the conformation of the bound li-

[1] J. H. Noggle and R. E. Schirmer, "The Nuclear Overhauser Effect." Academic Press, New York, 1971.
[2] D. C. Fry, S. A. Kuby, and A. S. Mildvan, *Biochemistry* **24**, 4680 (1985).
[3] P. R. Rosevear, H. N. Bramson, C. O'Brien, E. T. Kaiser, and A. S. Mildvan, *Biochemistry* **22**, 3439 (1983).
[4] P. R. Rosevear, T. L. Fox, and A. S. Mildvan, *Biochemistry* **26**, 3487 (1987).
[5] P. R. Rosevear, V. M. Powers, D. Dowhan, A. S. Mildvan, and G. L. Kenyon, *Biochemistry* **26**, 5338 (1987).
[6] R. S. Ehrlich and R. F. Colman, *Biochemistry* **24**, 5378 (1985).
[7] D. H. Live, D. Cowburn, and E. Breslow, *Biochemistry* **26**, 6415 (1987).
[8] C. W. Hilbers, A. Heerschap, J. A. L. I. Walters, and C. A. G. Haasnoot, "Nucleic Acids: The Vectors of Life," p. 427. Reidel, Dordrecht, The Netherlands, 1983.
[9] P. R. Rosevear and A. G. Redfield, unpublished results.
[10] L. J. Ferrin and A. S. Mildvan, *Biochemistry* **25**, 5131 (1986).
[11] K. Wakamatsu, A. Okada, T. Miyazawa, Y. Masui, S. Sakakibara, and T. Higashijima, *Eur. J. Biochem.* **163**, 331 (1987).
[12] K. Wakamatsu, A. Okada, T. Higashijima, and T. Miyazawa, *Biopolymers* **25**, S193 (1986).

METHODS IN ENZYMOLOGY, VOL. 177

gand. Hand model building is usually sufficient for determining the bound conformation of small ligands, but methods based on computer algorithms are more efficient for determining the conformational space, defined by the NOE constraints, available to larger, more flexible ligands. NOEs from protons of a protein to those of a ligand may be used to obtain information on amino acids of the protein that are near the bound ligand.

Limitations of the NOE method for determining the conformation of a bound ligand result from the assumptions of fixed interproton distances and a single correlation time for all internuclear vectors. In many cases, the validity of these assumptions can be tested by experiment. However, if several bound conformations coexist with different interproton distances, the NOE method will yield a nonlinear average conformation skewed toward those conformations having the shortest interproton distances. Multiple conformations for the bound ligand can usually be detected by observing two or more interproton NOEs that are not mutually consistent with a single conformation. An alternative approach to evaluate the uniqueness of the bound conformation is the combined use of the NOE and paramagnetic probe–T_1 methods.[3,4]

The recent development of one- and two-dimensional isotope-filtered NOEs should greatly extend the usefulness of this methodology for studying the conformations of ligands bound to macromolecules.[13–15] In these techniques, either the proton to be saturated is selectively irradiated, or the proton receiving the transfer of saturation is selectively detected through heteroatom decoupling. The advantage of selective irradiation or detection of the NOE is that it can thereby be unequivocally demonstrated that the NOE exists between a given pair of protons. This is particularly important when studying the conformation of a ligand bound to a large macromolecule containing numerous protons with chemical shifts similar to those of the ligand under study. Detailed reviews of the theory of the nuclear Overhauser effect have been published elsewhere.[1,16–18] This chapter will concentrate on the practical aspects and

[13] R. H. Griffey, M. A. Jarema, S. Kunz, P. R. Rosevear, and A. G. Redfield, *J. Am. Chem. Soc.* **107**, 711 (1985).

[14] M. A. Weiss, A. G. Redfield, and R. H. Griffey, *Proc. Natl. Acad. Sci. U.S.A.* **83**, 1325 (1986).

[15] O. Jardetzky and G. C. K. Roberts, "NMR in Molecular Biology." Academic Press, New York, 1981.

[16] A. S. Mildvan, P. R. Rosevear, J. Granot, C. O'Brien, H. N. Bramson, and E. T. Kaiser, this series, Vol. 99, p. 93.

[17] A. A. Bothner-By and R. Gassend, *Ann. N.Y. Acad. Sci.* **222**, 668 (1973).

[18] B. A. Borgias and T. L. James, this series, Vol. 176 [9].

pitfalls in using the NOE method to determine the conformations and environment of ligands bound to macromolecules.

Theory

Steady-State NOE

The NOE is defined as the change in intensity or area of a nuclear resonance when another resonance is preirradiated with a radiofrequency field. The steady-state equation for the fractional change in the intensity of resonance A on preirradiation of resonance B [$f_A(B)$], where A and B are protons, is given by Eq. (1). The parameter σ_{AB} (sec^{-1}) is the cross-relaxation rate and ρ_A (sec^{-1}) is the longitudinal relaxation rate ($1/T_{1A}$) of proton A.

$$f_A(B) = \sigma_{AB}/\rho_A \tag{1}$$

The cross-relaxation rate between two nuclei is given by Eq. (2)

$$\sigma_{AB} = (D/r_{AB})^6 f(\tau_r) \tag{2}$$

where D is a constant ($\gamma^4 \hbar^2/10)^{1/6}$, numerically equal to 62.02 Å sec$^{-1/3}$, if A and B are protons, r_{AB} is the interproton distance in angstroms, and $f(\tau_r)$ is a function of the correlation time τ_r, or the time constant for motion of the interproton vector r_{AB}. Evaluation of the cross-relaxation rate σ_{AB} from the experimentally determined $f_A(B)$ and ρ_A permits the internuclear distance r_{AB} to be calculated from Eq. (2). The correlation function $f(\tau_r)$ is given by Eq. (3), where ω_I is the nuclear precession frequency. Equations (2) and (3) assume isotropic rotation of the molecule as a whole with a time constant τ_r shorter than any of the time constants of internal motion.

$$f(\tau_r) = \frac{6\tau_r}{1 + 4\omega_I^2\tau_r^2} - \tau_r \tag{3}$$

Application of these basic equations to exchanging systems has proved to be extremely powerful in determining the conformation of ligands bound to macromolecules. The theory for measuring NOEs in exchanging systems [transferred NOE (TRNOEs)] has previously been extensively reviewed.[19] The basis of the TRNOE is that a change in magnetization, as a result of cross-relaxation between two protons of a bound ligand, is transferred to the free or averaged ligand resonance by chemical exchange. Therefore, the cross-relaxation rate between the two protons in the bound

[19] G. M. Clore and A. M. Groneborn, *J. Magn. Reson.* **53**, 423 (1983).

ligand can be measured by observation of changes in the intensity of the free or averaged ligand resonance, following preirradiation of either the free, averaged, or bound resonance of the ligand, provided that exchange is much faster than the longitudinal and cross-relaxation rates of the bound ligand protons. Under these conditions, negative TRNOEs are observed because $f(\tau_r)$ is approximated by $-\tau_r$, in Eq. (3), when the molecular weight becomes large enough (i.e., when $\omega_1\tau_r \gg 1$).

Time Dependence of NOE

In practice, measurement of the steady-state TRNOE between protons of the ligand in the presence of a macromolecule is not selective due to the phenomenon of spin diffusion. Spin diffusion arises from the indirect transfer of magnetization from resonance A to B through proton resonances in close proximity to A or B. In order to overcome this difficulty and to evaluate the spatial proximity of the pair of protons more accurately, the cross-relaxation rate is best determined from the early time dependence of $f_A(B)$ or the approach to the steady-state NOE, using Eq. (4).

$$f_A(B)_t = \frac{\sigma_{AB}}{\rho_A}\left(1 - e^{-\rho_A t}\right) + \frac{\sigma_{AB}}{\rho_A - c}\left(e^{-\rho_A t} - e^{-ct}\right) \qquad (4)$$

The approach to the steady-state NOE thus depends on σ_{AB}, the cross-relaxation rate from irradiated spin B to A; on ρ_A, the spin–lattice relaxation rate of A; on t, the duration of irradiation of spin B; and on c, a phenomenological rate constant for saturation of the irradiated resonance which can be approximated by $1/2(1/T_{1B} + 1/T_{2B})$.[20] Alternatively, the rate constant for saturation of the irradiated resonance can be obtained directly from measurement of the rate of saturation of spin B as a function of the preirradiation time. For instantaneous saturation of irradiated spin B, c is very large, and the time dependence of the NOE reduces to Eq. (5).

$$f_A(B)_t = \sigma_{AB}/\rho_A(1 - e^{-\rho_A t}) \qquad (5)$$

Note that as $t \to \infty$ the NOE approaches the steady-state value σ_{AB}/ρ_A. This equation assumes an isolated pair of protons with cross-relaxation occurring only between protons A and B. This is almost never the case in the multiproton environment of a macromolecule. To approximate this multiproton environment, a three-spin equation of the form of Eq. (4) has been developed where the positioning of the third spin relative to the irradiated spin B and observed spin A can be varied.[20] Multispin methods have also been developed which take into account the complete spin

[20] A. Dubs, G. Wagner, and K. Wüthrich, *Biochim. Biophys. Acta* **577**, 177 (1979).

TABLE I
FIXED INTERNUCLEAR DISTANCES USED TO ESTIMATE
CORRELATION TIME OF LIGAND–PROTEIN COMPLEXES

Ligand	Proton pair	Assumed distance (Å)
ATP	H1'–H2'	2.9 ± 0.2
NADP	H1'–H2'	2.9 ± 0.2
	HN4–HN5	2.5 ± 0.1
	HN5–HN6	2.5 ± 0.1
Deoxynucleotides	H2"–H1'	2.4 ± 0.1
Uridine nucleotides	H5–H6	2.4 ± 0.1
Thymidine nucleotides	H^5-methyl–H6	2.8 ± 0.1

system.[18,21] Due to the large number of spins involved, these methods are usually impractical for analysis of complex ligand–protein complexes. However, at relatively short times under favorable conditions, the two-spin approximation can be used to calculate the cross-relaxation rate between spins A and B.

Estimation of Correlation Time

Calculation of σ_{AB} from Eq. (4) and estimation of the correlation time τ_r permit the determination of the internuclear distance between protons A and B. The correlation time τ_r may be estimated from a known fixed distance on the ligand which is independent of conformation.[3,19] This approach makes the assumption that τ_r is the same for all interproton vectors in the bound ligand molecule. Fixed internuclear distances which have been used to estimate the correlation time are listed in Table I.

The upper limit of the correlation time can be estimated from the ratio (T_1/T_2) of the longitudinal and transverse relaxation rates of the ligand protons with Eq. (6).[3]

$$\frac{T_1}{T_2} = \frac{12\omega_I^4\tau_r^4 + 37\omega_I^2\tau_r^2 + 10}{16\omega_I^2\tau_r^2 + 10} \qquad (6)$$

An alternative method for determining the τ_r values that makes no assumption of the equality of τ_r for each interproton vector is to study the frequency dependence of the cross-relaxation rate according to Eq. (7).[2]

$$\frac{\sigma_{AB} \text{ (Freq. 1)}}{\sigma_{AB} \text{ (Freq. 2)}} = \frac{f(\tau_r)\text{(Freq. 1)}}{f(\tau_r)\text{(Freq. 2)}} \qquad (7)$$

[21] A. A. Bothner-By and J. H. Noggle, *J. Am. Chem. Soc.* **101**, 5152 (1979).

This approach has been successfully used to evaluate the correlation time of MgATP bound to a peptide.[2] However, it will be useful only for relatively small systems where $f(\tau_r)$ is not approximated by $-\tau_r$, but is frequency dependent as shown in Eq. (3). For a comparison between 250 and 500 MHz, a suitable range for τ_r would be $(1-9) \times 10^{-10}$ sec. Provided the correlation time of the system is of the correct magnitude, this approach will be particularly valuable for ligands which lack an internal reference distance. This approach permits the individual correlation times for all internuclear vectors to be experimentally determined and directly tests the validity of the assumption usually made that all internuclear vectors have the same correlation time.

Exchange Rate of Ligand

Recently, the time dependence of the TRNOE was solved for a ligand in slow exchange between the free and macromolecule bound state according to the following scheme:

$$E + L \underset{k_{off}}{\overset{k_{on}}{\rightleftharpoons}} EL \tag{8}$$

assuming instantaneous saturation of spin B in the free state.[22] With instantaneous saturation of free B, an initial lag phase is observed due to the time required to transfer the saturation from spin A in the free state to spin A in the bound state. This type of lag phase has been named an exchange lag phase.

Lag phases in the initial slope of the time dependence of the NOE can thus result from slow exchange between the free and bound forms of spin A (exchange lag phase), from noninstantaneous saturation of the irradiated resonance as shown in Eq. (4) (saturation lag phase), and from indirect cross-relaxation with other spins (spin diffusion lag phase). Therefore, before the cross-relaxation rate is calculated from the time dependence of the NOE buildup curve, it is important to evaluate the role, if any, of the three types of lag phases. It has previously been shown that in the fast-exchange condition, when the rate constant for saturation is large enough to be neglected, the cross-relaxation rate, σ_{AB}, can be estimated from the initial slope of the time-dependent NOE buildup curves, as may be seen by differentiation of Eq. (5) with respect to time, and setting $t = 0$. However, for the slow-exchange condition,

$$\left\{ \frac{d[f_A(B)_t]}{dt} \right\}_{t=0} = \sigma_{AB} \tag{9}$$

[22] D. Kohda, G. Kawai, S. Yokoyama, M. Kawakami, S. Mizushima, and T. Miyazawa, *Biochemistry* **26,** 6531 (1987).

the maximal slope in the NOE buildup curve occurs just after the "exchange lag phase." This "apparent initial slope" has been shown to yield an apparent cross-relaxation rate σ'_{AB}, which is approximately proportional to the actual cross-relaxation rate σ_{AB} and can therefore be used to estimate interproton distances and for determining the conformation of the bound ligand.[22] This approach extends the TRNOE method to systems in which the ligand is more tightly bound and not in fast exchange with respect to σ_{AB} and ρ_A. However, because only a limited number of data points are available, in all such cases the use of initial or early slopes is only an approximate method, and a study of the approach to the steady state, fitting the data to Eq. (4), is preferable.

Sample Preparation for NOE Experiments

Usually 400 μl of 0.2–2.0 mM protein is used in the NOE experiment. The ligand to be observed is typically 5 to 20 times more concentrated than the protein, but it is most conveniently added after a reference spectrum of the protein has been obtained. If nonexchangeable protons are being observed, protein samples are deuterated using phosphate or deuterated Tris–Cl buffer in ^{2}H$_2$O. Deuterated Tris base is available from Merck (St. Louis, Missouri 63116). All components of the mixture should be passed through a microcolumn containing neutralized Chelex 100 (Bio-Rad) to remove trace paramagnetic metals before preparing the sample. Such treatment is especially necessary if phosphate buffer is used.

Concentration and deuteration of the protein solution can be carried out by repeated lyophilization and redissolving in ^{2}H$_2$O, or by pressure filtration in an Amicon-type cell. For easily denatured proteins, the concentration should be done by vacuum filtration using Millipore cartridge filters, or by vacuum dialysis in a Schleicher Scheull collodion bag, followed by repeated redilution with ^{2}H$_2$O. Alternatively, deuteration of a concentrated protein solution may be accomplished by gel filtration on a column of Sephadex G-25 (medium grade, Pharmacia) that has previously been equilibrated with deuterated buffer. This deuteration procedure results in a 2- to 3-fold dilution of the protein and requires larger volumes of deuterated buffer. The concentrated and deuterated sample, approximately 400 μl, is placed in a 5-mm o.d. NMR tube for the NMR experiment. Depending on the probe being used, it may be possible to use a smaller volume without loss of resolution.

Measurements of the time dependence of the NOE, as well as selective T_1 values of protons receiving the NOE, usually require at least 2 days of data collection. The specific activity of enzyme samples should therefore be determined before and after the data collection, as well as periodically during the experiment, depending on the stability of the enzyme. Data collected after a significant loss of enzymatic activity ($\geq 20\%$) should not be used for conformational analysis, because the conformation of the bound ligand may change.[3] Under these circumstances, the experiment should be continued with fresh protein samples. The instability of some proteins, under conditions of prolonged NMR data collection, preclude measurement of the time dependence of the NOE buildup. Relative cross-relaxation rates have been estimated in such cases from a single time point. However, because of the existence of lags as well as the possibility of spin diffusion, such single-point determinations of relative distances are, at best, approximations. Measurements of the NOE buildup rates by two-dimensional NOE spectroscopy require much longer data acquisition periods, usually 12 to 48 hr per mixing time, and higher concentrations of protein, 1.0–2.0 mM. Thus, two-dimensional measurements of the time dependence of the NOE buildup are suitable only for proteins that are readily available and reasonably stable for long periods of time.

Instrumentation

^{1}H NMR spectra are normally obtained at frequencies between 250 and 600 MHz. Because most experiments are performed in aqueous solution, where the residual HDO signal, even after suppression (cf. Hore [3], Vol. 176, this series), can contribute significantly to the total signal, 16-bit analog-to-digital conversion provides better observable dynamic range, which is important in observing the small intensity changes of the ligand resonances in the biological sample. The spectrometer frequency is normally set at the frequency of the residual HDO signal and a spectral width is used sufficient to cover -1 to 10 ppm. Experiments are typically performed at temperatures between 5 and 40°, depending on the ligand exchange rates and the stability of the protein under study. Chemical shifts are usually referenced to external sodium 4,4-dimethyl-4-silapentane 1-sulfonate (DSS).

Location of Ligand Resonances and Estimation of
  Ligand Exchange Rate

Initially, a ^{1}H NMR spectrum of the protein is acquired under the conditions of the NOE experiment. The protein is then titrated with the ligand under study and NMR spectra obtained under identical conditions, i.e., the same number of FIDs, recycle times, and pulsewidths. Typically, 0.2–2.0 mM protein is titrated with ligand to a final ligand concentration of 5–15 mM. At low ligand/protein ratios, difference spectra (FID of protein alone minus FID of ligand–protein) can be used to extract the chemical shifts for the ligand. Changes in the chemical shifts of the ligand resonances as a function of the ligand/protein ratio can be used to estimate the exchange rate of the ligand from the protein,[23] which is usually limited by k_{off}, the rate constant for dissociation of the ligand–protein complex. The rate constant k_{off} can also be estimated from the dependence of linewidth, obtained from peak widths at half-maximum height ($\Delta\nu_{1/2}$), on the ligand/protein ratio.[24] The largest transverse relaxation rate ($1/T_2$) estimated from the linewidth ($\pi\Delta\nu_{1/2}$) sets a lower limit on the exchange rate of the ligand from the protein. Because of the low values of interproton cross-relaxation rates and longitudinal relaxation rates in diamagnetic systems (≤ 5 sec^{-1}), the fast-exchange case generally holds. However, if necessary, the exchange rate can usually be increased by raising the temperature, provided the protein is stable.

Titration of the protein with ligand also permits the unequivocal identification of the chemical shifts of the ligand resonances. This is necessary before these resonances can be selectively irradiated in the NOE experiment. Identification of the ligand resonances is not always straightforward due to resonance overlap, because there are usually many more resonances from the protein with similar or identical chemical shifts to those of the ligand. In such cases, a technique which may be extremely useful in identifying the chemical shifts of the ligand resonances in the presence of protein is heteronuclear multiple-quantum NMR[25] (cf. Markley [2] and Rance et al. [6], Vol. 176, this series). Such techniques utilize a 180° heteronuclear carbon pulse in combination with a 180° proton pulse to produce ^{13}C-labeled proton magnetization opposite to the unlabeled magnetization at the time of the echo. The delay in the echo is set equal to $1/2J$ which is approximately 3.2 msec for $^1J_{^{13}C-^1H}$ coupling constants. Because the ligand proton resonances are at a concentration approximately 10 to 20 times that of the protein resonances, only the proton resonances from the ligand are thereby detected under the appropriate conditions. The intensities of the resonances from the protein are also decreased further due to their faster relaxation rates. A conventional spin-echo sequence is not as effective as the heteronuclear experiment in identifying the ligand resonances. In extreme cases the carbon-bound protons of the protein can be deuterated by cloning and by appropriate biosynthetic incorporation of deuterated amino acids.[26]

[23] J. Feeney, J. G. Batchelor, J. P. Albrand, and G. C. K. Roberts, *J. Magn. Reson.* **33**, 519 (1979).

[24] J. A. Pople, W. G. Schneider, and H. J. Berstein, "High Resolution Nuclear Magnetic Resonance." McGraw-Hill, New York, 1959.

[25] A. Bax, R. H. Griffey, and B. L. Hawkins, *J. Magn. Reson.* **55**, 301 (1983).

[26] S. H. Seholzer, M. Cohn, J. A. Putkey, A. R. Means, and H. L. Crespi, *Proc. Natl. Acad. Sci. U.S.A.* **83**, 3634 (1986).

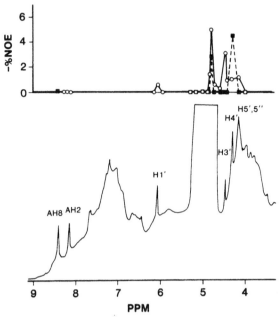

FIG. 1. Action spectra of interproton nuclear Overhauser effects to adenine H8 and ribose H1' of MgATP in the presence of creatine kinase. Effects on adenine H8 (O) and ribose H1' (■) were monitored while the ribose proton region was sequentially preirradiated. A control spectrum is shown on the bottom. The sample contained 1.3 mM rabbit muscle creatine kinase sites and 14.5 mM MgATP in 10 mM Tris–d_{11} buffer containing 0.1 mM DTT (dithiothreitol) at pH 8.0. NOE difference spectra were obtained at 500 MHz from sequentially preirradiating the ribose proton region for 300 msec with 256 scans, 16-bit A/D conversion, 8192 data points, an acquisition time of 0.7 sec, a spectral width of 3000 Hz, a relaxation delay of 5 sec, and a 90° pulse. [Reprinted with permission from P. R. Rosevear, V. M. Powers, D. Dowhan, A. S. Mildvan, and G. L. Kenyon,[5] *Biochemistry* **26,** 5338 (1987). Copyright 1987 American Chemical Society.]

An action spectrum (Fig. 1) can also be used to identify the chemical shifts of the ligand resonances and to evaluate the selectivity of preirradiation.[5,27] Figure 1 shows the effect of preirradiating various resonances in the spectrum of creatine kinase in the presence of MgATP.[5] The NOEs observed in either adenine H8 or ribose H1' of ATP are selective because each NOE exhibits a clear maximum in the action spectrum. Thus, by varying the duration and power of the preirradiation pulse, selective NOEs can be measured under conditions in which spin diffusion does not significantly contribute to the observed NOEs. Selectivity of the preirra-

[27] G. M. Smith and A. S. Mildvan, *Biochemistry* **21,** 6119 (1982).

diation pulse is an important criterion to satisfy before beginning to measure the detailed time dependence of the NOE buildup.

Collection of the NOE Data

NOE data are usually obtained by selectively preirradiating the resonance of interest, B, for a varying period of time, t, before applying the observation pulse using the sequence $\{[RD–preirradiate(T_1, \omega_1)–observation pulse]_{64}–[RD–preirradiate(t_1, \omega_{off-res})–observation pulse]_{64}\}_n$, where ω_1 is the frequency of resonance B, $\omega_{off-res}$ is a control frequency away from any of the ligand resonances, RD is the relaxation delay, which is typically in the range of 5 to 7 T_1s, and n is the number of times through which the entire sequence is cycled. The selective preirradiation pulse is applied using the proton decoupler for 50 to 750 msec. The power of the preirradiation pulse is adjusted to provide selective saturation of the resonance at the shortest possible preirradiation time. These conditions are sufficient to produce selective NOEs at short preirradiation times. At longer preirradiation times, 500–1000 msec, indirect or secondary NOEs are frequently observed.

Due to instabilities in instrumentation, it was previously necessary to insert a control frequency between each pair of the test frequencies.[16] However, using the spectrometers currently available, instabilities that led to poor subtraction of the two FIDs are not a major problem and a single control frequency needs to be collected for every 8 to 15 test frequencies. Routinely, 64 scans at each frequency are alternatively collected until the total number of scans necessary to obtain the desired signal-to-noise ratio is obtained. The FIDs obtained in this manner are stored on a magnetic disk for processing at a later time. Commercial software currently available enables both the frequencies and preirradiation times to be varied under computer control.

The magnitude of the NOE is measured by subtraction of the FID of the control frequency, $\omega_{off-res}$ from the FID of the test frequency (ω_1). The difference FID is then multiplied by an exponential filtering function to improve the signal-to-noise ratio and is Fourier transformed. Negative NOEs are usually observed in the presence of large macromolecules and at high field strengths, $\omega_1\tau_r \gg 1$. However, difference spectra are more conveniently phased with such NOEs positive. Importantly, the control spectrum must be processed with the same parameters used to obtain the difference spectrum. Calculation of the NOE to a given resonance is achieved by integrating the difference peak for that resonance in the difference spectrum and dividing it by the integrated intensity of the same resonance in the control spectrum. If irradiation does not alter the line shape, then the amplitude in the difference spectrum divided by the control amplitude also yields the NOE. The magnitude of the NOE versus preirradiation time, together with the longitudinal relaxation rate of the resonance receiving the NOE (ρ_A), is fitted to Eq. (4) to obtain the cross-relaxation rate (σ_{AB}).

Determination of ρ_A by Selective Saturation Recovery

The longitudinal relaxation rate is measured by using selective saturation recovery to avoid spin-diffusion effects from other resonances.[28] The pulse sequence used for this is $[(t_1, \omega_A)-\tau-observation\ pulse-RD]_n$, where t_1 is the time required for saturation of resonance A. The recovery time (τ) is a variable delay between the saturating pulse and the observation pulse, the observation pulse is a 90° pulse, and RD is the relaxation delay, usually $>5\ T_1$s. Recovery of the magnetization over the time period of approximately $1/\rho_A$ is a first-order process and a plot of $\ln[I_0 - I(\tau)]$ versus τ yields ρ_A, where I_0 is the maximal intensity of the resonance and $I(\tau)$ is the observed intensity at time τ after the saturating pulse.

Collection of Two-Dimensional NOE Data

One-dimensional NOE experiments measuring the initial buildup rate of the NOE are often limited by poor selectivity in preirradiating the resonance of interest in crowded spectral regions.[29] Two-dimensional phase-sensitive NOESY spectra provide more selective NOEs in crowded spectral regions and can yield a complete set of cross-relaxation rates.[20] Phase-sensitive NOESY spectra at several mixing times are collected using the pulse sequence $(RD-90°-t_1-90°-\tau_m-90°-t_2)_n$, where RD is the relaxation delay between pulses, 90° represents a 90° pulse, t_1 is the incremented period during which the spins become frequency labeled, τ_m is the fixed mixing time during which cross-relaxation occurs, and t_2 is the acquisition time. Data are collected as described by States et al.,[30] which is a method of phase cycling permitting quadrature phase detection in both dimensions, and is generally part of the program package provided with commercial instruments. A 180° pulse is used during the mixing period in combination with symmetrization during data processing to eliminate the contribution of multiple-quantum J cross-peaks to the NOE cross-peaks.[31] The two-dimensional (2D) spectra are usually acquired with 4096 points in t_2, 512 points in t_1, and a spectral width of 10,000 Hz in

[28] J. Tropp and A. G. Redfield, *Biochemistry* **20**, 2133 (1981).
[29] A. Kumar, G. Wagner, R. R. Ernst, and K. Wüthrich, *J. Am. Chem. Soc.* **103**, 3654 (1981).
[30] D. J. States, R. A. Haberkorn, and D. J. Ruben, *J. Magn. Reson.* **48**, 286 (1982).
[31] S. Macura, Y. Huang, D. Suter, and R. R. Ernst, *J. Magn. Reson.* **43**, 259 (1982).

both dimensions. The time domain in t_1 is expanded to 1024 points by zero filling, giving 10 Hz resolution in both dimensions. The smallest possible spectral width is preferable because it will give higher resolution in both dimensions.

Cross-relaxation rates are calculated from time-dependent NOE buildup curves obtained by collecting phase-sensitive NOESY spectra at several mixing times (typically 75, 150, 250, 500, and 700 msec). Auto- and cross-peak intensities are calculated by the volume integral using the software package provided with the spectrometer or by calculation of the sum of the areas of the individual slices composing the auto or cross-peak.[32,33] Alternately, and much more rapidly, one may monitor the time-dependent buildup of the area under the peak in an individual slice of the 2D spectrum, approximated by the product of peak height and half-width. The magnitudes of the NOEs determined by this simpler and faster method agree to with 10–15% with those measured by the more elaborate volume integrals, and summations of areas.[34] Cross-relaxation rates are then estimated from the initial slopes of the time-dependent NOE buildup curves according to Eq. (9), with the limitations described.

Data Reduction

Calculation of Internuclear Distances

Cross-relaxation rates are calculated by fitting the time dependence of the NOE buildup curve to Eq. (4) with a nonlinear least-squares fitting algorithm that can vary σ_{AB}, c, and ρ_A. Usually ρ_A is kept constant at the value independently measured and the time dependence of the NOE buildup is fit by varying σ_{AB} and c.

Saturation of the resonance of interest is usually not instantaneous, and the rate constant for saturation of the irradiated resonance c is given by

$$I(t) = I_0 \exp[-ct] \cos(\gamma B_2 t) \qquad (10)$$

where c is a function of the spin–lattice relaxation rate, spin–spin relaxation rate, and B_0 inhomogeneities, and $\cos(\gamma B_2 t)$ is the Torrey oscillation term.[35] However, the rate constant c is usually approximated by $1/2(1/T_{1B} + 1/T_{2B})$. Transverse relaxation rates ($1/T_{2B}$) can be estimated from the linewidth at half-height ($\Delta\nu$) according to the relation $1/T_2 = \pi\Delta\nu$. In the presence of complex spin–spin coupling, which prevents the direct and

[32] G. H. Weiss and J. A. Ferretti, *J. Magn. Reson.* **55**, 397 (1983).
[33] M. S. Broido, T. L. James, G. Zon, and J. W. Keepers, *Eur. J. Biochem.* **150**, 117 (1985).
[34] J. W. Keepers and T. L. James, *J. Magn. Reson.* **53**, 104 (1984).
[35] H. C. Torrey, *Phys. Rev.* **76**, 1059 (1949).

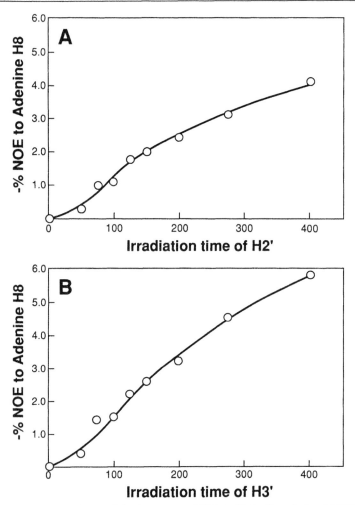

FIG. 2. Time dependence of the interproton NOEs in Mg(α,β-methylene)ATP interacting with truncated methionine–tRNA ligase. Time dependence of the NOEs to adenine H8 on preirradiation of ribose H2' (A) or ribose H3' (B). The sample contained 3.1 mM Mg(α,β-methylene)ATP, 0.36 mM *Escherichia coli*-truncated methionine–tRNA ligase in 20 mM Tris–d_{11} buffer containing 50 mM NaCl and 0.1 mM DTT at pH 7.6. NMR spectra were obtained at 500 MHz with 2,048 scans, 16-bit A/D conversion, 16,384 data points, a spectral width of 8,000 Hz, and a relaxation delay of 2 sec.

accurate measurement of the linewidth, the Carr–Purcell–Meiboom–Gill pulse sequence is preferable for measurement of T_2.

To avoid higher order effects, NOEs observed at early preirradiation times should be used in the curve-fitting procedure. As an example, Fig. 2 shows the time dependence of the NOE buildup to adenine H8 of (α,β-

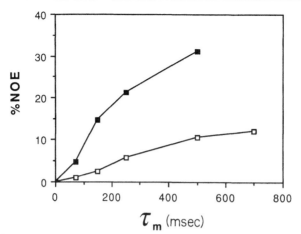

FIG. 3. The NOE buildup of cross-peaks as a function of the mixing time in phase-sensitive NOESY spectra. The data were measured in the cross-sections through the diagonal peak of the nicotinamide ribose N1' proton. The time course of NOE buildup to the nicotinamide N2 and N6 protons are shown. Two-dimensional spectra were acquired with 4,096 points in t_2, 512 points in t_1, and a spectral width of 10,000 Hz in both dimensions. The time domain in t_1 was expanded to 1,024 points by zero filling. Samples used to collect the NOESY spectra contained 2 mM type II dihydrofolate reductase subunits, 9 mM NADP$^+$, and 10 mM potassium phosphate buffer, pH 5.9, in ^2H_2O. (□) N2/N1' and (■) N6/N1'.

methylene) ATP upon preirradiation of ribose H2' or H3' in the presence of the methionine–tRNA ligase.[36] The solid curves represent theoretical fits to the data using Eq. (4) with ρ_A fixed at 4.0 sec^{-1} and varying σ_{AB} and c. The best fit of the time dependence of the NOE from ribose H2' to adenine H8 was found using a $\sigma_{AB} = -0.21$ sec^{-1} and $c = 21$ sec^{-1}. The NOE buildup curve from ribose H3' to adenine H8 was best fit using a $\sigma_{AB} = -0.31$ sec^{-1} and $c = 21$ sec^{-1}. The cross-relaxation rates together with the correlation function are used to calculate internuclear distances. The correlation function $f(\tau_r)$ can be estimated using Eq. (2) and the ribose H1' to H2' distance that lies within the limits of 2.9 ± 0.2 Å, regardless of either the ribose conformation or the glycosyl torsional angle.[3,37,38] As pointed out above, calculation of internuclear distances using this approach assumes rigidly fixed internuclear distances and that all internuclear vectors have the same correlation time.

The time dependence of the NOE buildup from two-dimensional NOESY data can also be used to measure the cross-relaxation rate. Figure 3 shows the time dependence of the NOE buildup, obtained from 2D

[36] J. S. Williams and P. R. Rosevear, unpublished results.
[37] H. P. M. DeLeeuw, C. A. G. Haasnoot, and C. Altona, *Isr. J. Chem.* **20,** 108 (1980).
[38] M. Levitt and A. Warshel, *J. Am. Chem. Soc.* **100,** 2067 (1978).

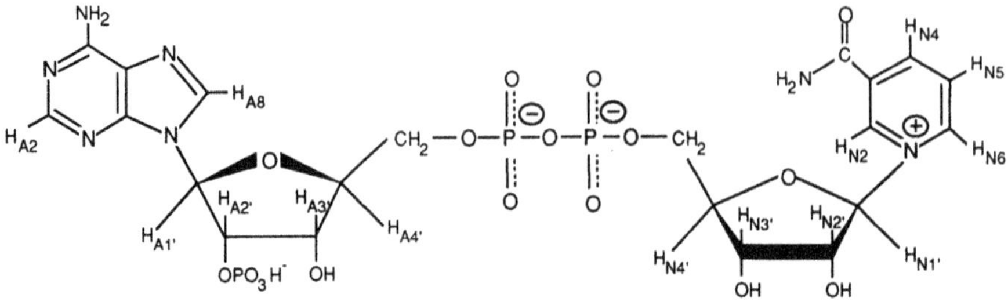

FIG. 4. Numbering system for the protons of NADP$^+$. Protons from the adenosine and nicotinamide moieties are labeled A and N, respectively. NADP is drawn with the adenine in the anti conformation and nicotinamide in the syn conformation.

NOESY data, for the nicotinamide ribose H1' to the nicotinamide N2 and N6 protons of NADP$^+$ in the presence of a type II dihydrofolate reductase.[39] The numbering system for the protons of NADP$^+$ are shown in Fig. 4. Cross-relaxation rates are calculated from the initial slope of the data shown in Fig. 3 using Eq. (9).[19] Provided that the effective correlation time of all internuclear vectors is approximately equal to that of a vector of known distance, that from N5 to N6 for NADP$^+$ (Table I), internuclear distances can then be calculated from Eq. (11),

$$r_{CD}/r_{AB} = (\sigma_{AB}/\sigma_{CD})^{1/6} \tag{11}$$

where r_{CD} and r_{AB} are the known and unknown distances, respectively, and σ_{CD} and σ_{AB} are the respective cross-relaxation rates.[19,34,40,41] See Borgias and James [9] in Vol. 176 of this series for caveats regarding use of Eq. (11).

Model Building

In principle, interproton distances can be obtained with an accuracy of ±0.2 Å from the time dependence of the NOE buildup, providing structural information comparable to that of protein X-ray crystallography.[18] However, there are several difficulties involved in the quantitative interpretation of NOE data.[34,42] Therefore, qualitative interpretation of the

[39] R. Britto, P. S. Vermersch, G. N. Bennett, F. Rudolph, and P. R. Rosevear, unpublished results.
[40] G. Wagner and K. Wüthrich, *J. Magn. Reson.* **33**, 675 (1979).
[41] C. M. Dobson, E. T. Olejniezak, F. M. Paulsen, and R. G. Ratcliffe, *J. Magn. Reson.* **48**, 87 (1982).
[42] G. M. Clore and A. M. Gronenborn, *J. Magn. Reson.* **61**, 158 (1985).

NOE data is preferable and, in many cases, sufficient for determining the conformation of the protein-bound ligand. This is particularly true for small ligands, such as nucleotides.

Initially a hand-built stick model of the ligand is made and its conformation is adjusted to be consistent with the observed NOEs and calculated internuclear distances. Model kits provided by Prentice Hall (Englewood Cliffs, New Jersey) are particularly useful for this stage of conformational analysis. The conformation of the ligand is adjusted with internuclear distances calculated from the time dependence of the NOE. This process usually also permits a rough evaluation of the uniqueness of the bound ligand conformation. For example, hand-built models showed that interproton distances on MgATP bound to the active site of pyruvate kinase were not consistent with a single nucleotide conformation, but required multiple adenine–ribose conformations to fit the internuclear distances.[4] In some cases, in which the number of rotatable bonds is small, this stage of model building is sufficient to evaluate the conformation of the bound ligand.

With free, flexible ligands and occasionally with protein-bound ligands, the measured interproton distances cannot be accommodated by a single ligand conformation. In such cases the choice of a basis set of individual conformations to fit the average distances is somewhat arbitrary. We have suggested that, under these circumstances, it is most appropriate to seek the minimum number of low-energy conformations, ideally two, which satisfy the interproton distances measured by the NOE method. Low-energy conformations are experimentally detected from X-ray data on a large number of related ligand structures that show a clustering of stable conformations. By this criterion, the low-energy nucleotide conformations are low anti, C-3' endo; high anti, C-2' endo; high anti, O-1' endo; and syn, C-2' endo. The interproton distances in these conformations are given in Table II,[3] and Fig. 5 defines the conformational angles in a nucleotide.

Interproton distances in each member of the basis set of conformations to be used (I, II, III, ...) are obtained from the crystallographic coordinates or by model building (Table II). The fractional contribution $(f_I, f_{II}, \text{and } f_{III})$ of each conformation to the observed average interproton distance is calculated from simultaneous equations of the following form:

$$\langle r_{AB}^{-6} \rangle = f_I(r_{I,AB})^{-6} + f_{II}(r_{II,AB})^{-6} + f_{III}(r_{III,AB})^{-6} \tag{12}$$

$$\langle r_{CD}^{-6} \rangle = f_I(r_{I,CD})^{-6} + f_{II}(r_{II,CD})^{-6} + f_{III}(r_{III,CD})^{-6} \tag{13}$$

$$\langle r_{EF}^{-6} \rangle = f_I(r_{I,EF})^{-6} + f_{II}(r_{II,EF})^{-6} + f_{III}(r_{III,EF})^{-6} \tag{14}$$

$$\sum f_i = 1.0 \tag{15}$$

TABLE II

INTERPROTON DISTANCES IN LOW-ENERGY PURINE NUCLEOTIDE
CONFORMATIONS FOR USE AS BASIS SET IN FITTING NOE DATA[a]

		Distances (Å)			
		From purine H8 to ribose			From ribose
Conformation		H1'	H2'	H3'	H1' to H4'
I	Low anti-C-3'-endo (χ = 15°, δ = 83°)[b]	3.7	4.1	3.3	3.2
II	High anti-C-2'-endo (χ = 48°, δ = 143°)	3.8	2.5	3.8	3.5
III	High anti-O-1'-endo (χ = 55°, δ = 97°)	3.8	2.4	3.2	2.8
IV	Syn-C-2'-endo (χ = 217°, δ = 143°)	2.3	3.7	5.6	3.5

[a] Data from Rosevear et al.[3]

[b] The glycosidic angle χ ($\pm 10°$) is defined by O-4'–C-1'–N-9–C-8, and
the sugar pucker angle δ ($\pm 10°$) is defined by C-5'–C-4'–C-3'–O-3'.
See also Fig. 5.

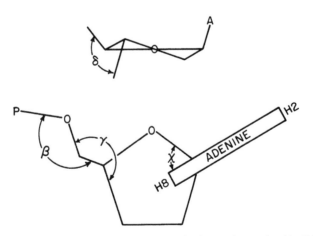

FIG. 5. Definition of selected conformational angles in a purine nucleotide. [M. Sundaralingam, *Biopolymers* **7**, 821 (1969); W. Saenger, "Principles of Nucleic Acid Structure," pp.
14–24. Springer-Verlag, Berlin and New York, 1984.]

With free $Co(III)(NH_3)_4ATP$ and in the pyruvate kinase–MgATP complex (Table III),[2-5,10,43-45] contributions from three of these low-energy conformations of ATP were required to fit the interproton distances. These multiple conformations were also consistent with Mn^{2+}-to-proton distances measured in similar complexes by paramagnetic effects on T_1.[4] In the DNA polymerase I–dGTP complex, two low-energy conformations of dGTP were sufficient to fit the interproton distances (Table III).

It is emphasized that the choice of a basis set of conformations is arbitrary. While it is reasonable to assume that a protein-bound ligand that retains flexibility will assume low-energy conformations, this may not always be the case. Indeed, in most cases, a single conformation of a bound ligand suffices to fit the interproton distances, indicating that the protein has restricted the conformational space of the ligand, thus raising its free energy by lowering its entropy. In enzyme–substrate interactions, binding energy may also be used, not only to restrict but also to distort the substrate conformation to one of higher energy, approaching that of the transition state. Hence, the use of energy minimization methods to adjust and refine conformations of enzyme-bound substrates when a single conformation satisfies the data is inappropriate, unless all of the enzyme–substrate interactions can be properly considered (see below).

A further stage of model building uses the NOE constraints in a computer algorithm to search systematically the conformational space available to the bound ligand. Such an analysis permits an objective evaluation of the completeness of the data, as well as the uniqueness of the ligand conformation. In computer searches, the detection of multiple ligand conformations usually results from incomplete data rather than from a true multiplicity of conformations. This distinction, which must be made in interpreting the results, is usually obvious from the distribution of the measured distances. Two computer algorithms have been widely used for this analysis, i.e., distance geometry[5,46-49] and constrained energy minimization together with molecular dynamics.[50-52] In principle, each method

[43] D. C. Fry, S. A. Kuby, and A. S. Mildvan, *Biochemistry* **26**, 1645 (1987).
[44] L. J. Ferrin and A. S. Mildvan, *Biochemistry* **24**, 6904 (1985).
[45] G. P. Mullen and A. S. Mildvan, *Fed. Proc., Fed. Am. Soc. Exp. Biol.* **2**, A588 (1988).
[46] I. D. Kuntz, G. M. Crippen, and P. A. Kollman, *Biopolymers* **18**, 939 (1979).
[47] T. F. Havel, I. D. Kuntz, and G. M. Crippen, *Bull. Math. Biol.* **45**, 665 (1983).
[48] P. R. Rosevear, S. Sellin, B. Mannervik, I. D. Kuntz, and A. S. Mildvan, *J. Biol. Chem.* **259**, 11436 (1984).
[49] W. Braun and N. Gō, *J. Mol. Biol.* **186**, 611 (1985).
[50] B. R. Brooks, R. E. Bruccoleri, B. D. Olason, D. J. States, S. Swaminathan, and M. Karplus, *J. Comput. Chem.* **4**, 187 (1983).
[51] G. M. Clore, A. M. Groneborn, A. T. Brünger, and M. Karplus, *J. Mol. Biol.* **186**, 435 (1985).

TABLE III

CONFORMATIONS OF PROTEIN-BOUND NUCLEOTIDE LIGANDS DETERMINED BY NOE METHOD

Protein	Ligand	Glycosidic bond	Conformation[a]: sugar pucker	Other	Reference(s)
cAMP-dependent protein kinase	MgATP	High anti (χ = 78 ± 10°)	O-1' endo	—	3
Adenylate kinase	AMP	High anti (χ = 110 ± 10°)	C-3' endo C-2' exo (δ = 105 ± 10°)	γ = 180 ± 10°, β = 170 ± 20°	43
	MgATP	High anti (χ = 65 ± 10°)	C-3' endo–O-1' endo (δ = 97 ± 10°)	γ = 170 ± 10°	2
Adenylate kinase peptide 1–45	MgATP	High anti (χ = 60 ± 10°)	C-3' endo–O-1' endo (δ = 94 ± 10°)	γ = 50 ± 10°	2
Pyruvate kinase	MgATP[b]	Multiple: (62% I, 20% III, 18% IV)[c]			4
Creatine kinase	MgATP[d]	High anti (χ = 68 ± 10°)	C-3' endo (δ = 10 ± 10°)	—	4
	MgATP	High anti (χ = 78 ± 10°)	O-1' endo or C-4' exo	γ = 140 ± 10° or 70 ± 10°	5
DNA Polymerase I large fragment	MgdATP	High anti (χ = 50 ± 10°)	O-1' endo (δ = 95 ± 10°)		44
	MgdATP[e]	High anti (χ = 62 ± 10°)	O-1' endo–C-3' endo (δ = 90 ± 10°)		10

MgdTTP[f]	High anti (χ = 40 ± 10°)	O-1' endo (δ = 100 ± 10°)	10, 44
MgdGTP	Multiple	O-1' endo–C-3' endo	
	60% low anti (χ = 32°)	(δ = 90°)	
	40% syn (χ = 222°)	C-2' endo (δ = 144°)	45
MgdGTP[g]	High anti (χ = 60 ± 15°)	O-1' endo (δ = 100 ± 20°)	45
MgdddGTP	Multiple		
	30% high anti (χ = 45°)	C-2' endo (δ = 135°)	
	30% high anti (χ = 95°)	C-3' endo (δ = 85°)	
	40% syn (χ = 212°)	C-2' endo (δ = 135°)	
MgdddGTP[g]	High anti (χ = 45 ± 10°)	C-2' endo (δ = 135 ± 10°)	45
(rU)$_{54}$	High anti (χ = 60 ± 10°)	O-1' endo (δ = 105 ± 10°)	10

[a] The dihedral angles χ and δ are defined in Table II and Fig. 5; γ is defined by O-5'–C-5'–C-4'–C-3' and β is defined by P–O-5'–C-5'–C-4'.

[b] MgATP at active site.

[c] Active-site conformations I, II, and IV are defined in Table II.

[d] MgATP at secondary site.

[e] In presence of (rU)$_{54}$ template.

[f] In absence or presence of (rA)$_{50}$ template.

[g] In presence of (rC)$_{37}$ template.

allows the conformational space available to the bound ligand to be systematically explored within the distance constraints determined from the NOE experiment. Constraints can be entered either qualitatively as upper or lower limit values or quantitatively as absolute internuclear distances with associated errors. Primary NOEs when observed can be used as qualitative constraints by setting lower limit distances equal to van der Waals contact, 2.4 Å, and an upper bound distance of the order of 3.5 to 4.5 Å.[53] Upper bound distances depend on the system under study and can be estimated from the noise level in the data.

Constrained energy minimization and molecular dynamics have been shown to have poor convergent properties due to their inability to overcome local energy minima[51,54,55] (but see Scheek et al. [10], this volume). Therefore, these methods are most satisfactory for generating structures from NMR data when the starting structure is close to a final structure that obeys all of the NMR constraints. Another more serious problem is that these methods assume that the final structure is at an energy minimum. This is generally not the case for ligands bound to proteins, because binding energy is often used to limit the structure of a flexible ligand to a single conformation or even to distort the structure of the bound ligand to a higher energy form.[56] Therefore, distance geometry (cf. Kuntz et al. [9], this volume), which uses only distance constraints from covalent bonding and NMR data, provides the least biased search of the conformational space available to the bound ligand.[5,48,54] The final structure generated by distance geometry is not dependent on the starting structure. It has been suggested that acceptable structures generated by the distance geometry algorithm might be then energy minimized to eliminate unfavorable bonded and nonbonded interactions.[57] If this is done, the energy-minimized structures should be carefully checked to assure that they still obey the NMR distance constraints (e.g., using CORMA; see Borgias and James [9], Vol. 176, this series), and that they do not produce unfavorable ligand–protein interactions, if the protein structure is known.

In the conformational search procedure, using the distance geometry program, the lower bound distance between any two atoms not bonded to each other, and not known, is set to the sum of the van der Waals radii.

[52] M. Nilges, G. M. Clore, A. M. Groneborn, A. T. Brünger, M. Karplus, and L. Nilsson, Biochemistry 26, 3718 (1987).
[53] K. Wüthrich, "NMR of Proteins and Nucleic Acids." Wiley, New York, 1986.
[54] S. W. Fesik, G. Bolis, H. L. Sham, and E. T. Olejniczak, Biochemistry 26, 1851 (1987).
[55] E. R. P. Zuiderweg, R. M. Scheek, R. Boelens, W. F. van Gunsteren, and R. Kaptein, Biochimie 67, 707 (1985).
[56] W. Jencks, Adv. Enzymol. Relat. Areas Mol. Biol. 43, 219 (1975).
[57] M. P. Williamson, T. F. Havel, and K. Wüthrich, J. Mol. Biol. 182, 295 (1985).

The upper bound is set to a value sufficiently large to permit the conformation of the molecule to vary freely. Bond lengths and bond angles are obtained from a crystallographic structure of the ligand or are calculated using standard bond lengths and angles. Bond lengths and bond angles are expressed as 1–2 and 1–3 distances, respectively. Errors in the 1–2 and 1–3 distances are usually set at 0.01 and 0.1 Å, respectively.[5,48] All dihedral angles are expressed as 1–4 distances and are allowed to vary by at least 1.5 Å to permit a complete conformational search. Asymmetric carbons are constrained to have the correct chirality by an additional penalty term based on quantitative deviations from the chiral volume. The chiral volume V_{ch} for each asymmetric carbon atom is calculated according to Eq. (16).

$$V_{ch} = (\mathbf{V}_1 - \mathbf{V}_4) \cdot [(\mathbf{V}_2 - \mathbf{V}_4) \times (\mathbf{V}_3 - \mathbf{V}_4)] \qquad (16)$$

where $\mathbf{V}_1$ through $\mathbf{V}_4$ are the position vectors of the atoms directly bonded to the asymmetric carbon atom, labeled in ascending rank according to the Cahn–Ingold–Prelog system.[5,47,48] Planarity of peptide bonds, sp^2 carbons, and aromatic rings can be maintained by addition of a subroutine that minimizes the weighted sum of the squares of the distances from the best mean plane through the atoms defined to be in that plane.[48]

The conformational search is carried out 25 to 50 times and the resulting structures are examined for their consistency with the original data. Conformations which satisfy all of the measured distances are next examined for their mutual consistency. Variations in these conformations usually reflect limitations in the data and indicate the need for additional measured distances. The consistent failure of the resulting conformations to satisfy all of the experimentally determined distances within their experimental errors usually indicates the presence of more than one conformation of the bound ligand, a situation which may be treated by assuming the averaging of several conformations, as discussed above. Table III gives the conformations of several enzyme-bound nucleotide substrates determined by the NOE method.

Intermolecular Protein–Ligand NOE Studies

Intermolecular NOEs from protons of a protein to those of a ligand provide valuable qualitative information on the types of amino acids of the protein which are within 5 Å of the bound ligand. This approach was first utilized to study the interaction of peptide ligands with the protein neurophysin.[58] NOE studies of enzyme–substrate interactions have pro-

[58] P. Balaram, A. A. Bothner-By, and E. Breslow, *J. Am. Chem. Soc.* **94**, 417 (1972).

vided evidence for Lys, Arg, and an aromatic residue at the MgATP binding site of creatine kinase[59,60]; His-36, Leu and Ile residues at the MgATP binding site of adenylate kinase[2,27]; Ile, Leu and Tyr residues at the dATP binding site of DNA polymerase I (Fig. 6)[10,44]; and cationic Arg or Lys residues at the template binding site of DNA polymerase I.[10]

As in the determination of ligand conformations, typical samples contain 0.2 to 2.0 mM protein (or enzyme) sites, 2.0 to 10 mM ligand (or substrate), buffer, and salts in ^2H_2O. The proton decoupler is used to preirradiate the protein resonances, typically at twice the power (20 μW) used to determine ligand conformations, and NOEs to the ligand resonances are sought. A single preirradiation time of 0.5 to 1.0 sec, which is short enough to minimize secondary NOEs and spin diffusion yet long enough to permit the detection of primary intermolecular NOEs, is chosen by preliminary studies. A coarse NOE action spectrum is initially obtained by preirradiation of the major peaks in the aliphatic and aromatic regions of the protein, searching for NOEs to the ligand. The same pulse program as that described above for intermolecular NOEs is used. At regular intervals, typically every third to every twelfth run, depending on the stability of the protein and the ligand, an empty region of the spectrum is preirradiated in order to generate a control spectrum. Regions of the protein spectrum that yield intermolecular NOEs to protons of the ligand are then examined in greater detail, preirradiating at every 0.03 to 0.1 ppm in order to obtain a higher resolution NOE action spectrum in a long 1- to 2-day run. If the protein and ligand are stable enough, the coarse search may be omitted, and after preliminary studies to determine the appropriate preirradiation power and time, the high-resolution NOE action spectrum is directly obtained by a fine search throughout the entire protein spectrum. The resulting NOE action spectrum is a subspectrum of the protein that gives the chemical shifts of those amino acid residues of the protein that are within 5 Å of the ligand. A study of the substrate-binding site of DNA polymerase I (Fig. 6) suggested that at least two hydrophobic residues, most likely Ile and Tyr, were located at the substrate-binding site.[10,44] While such data provide valuable clues to amino acid types, the absolute assignment to specific residues require separate experiments. In the case of DNA polymerase, the photoaffinity probe 8-azido-dATP was found to label Tyr-766 in the sequence Leu-Ile-Tyr,[61] in agreement with the NOE studies.

[59] T. L. James, *Biochemistry* **15**, 4724 (1976).
[60] M. Vasak, K. Nagoyama, K. Wüthrich, M. L. Mertens, and J. H. R. Kagi, *Biochemistry* **18**, 5050 (1979).
[61] C. M. Joyce, D. L. Ollis, J. Rush, T. A. Steitz, W. H. Konigsberg, and N. D. F. Grindley, "UCLA Symposium on Protein Structure Folding and Design," pp. 197–205. Liss, New York, 1986.

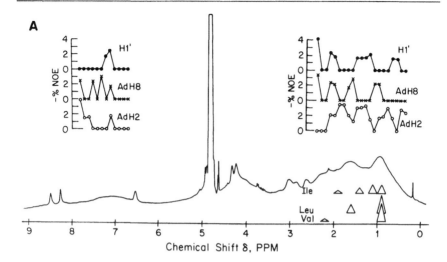

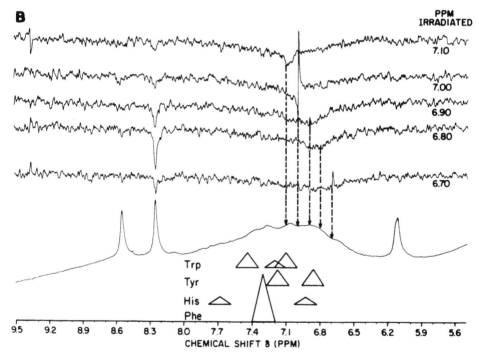

FIG. 6. NOE action spectra.[10,44] (A) From protons of the large fragment of Pol I to protons of Mg²⁺dATP together with the chemical shift and linewidths of Ile, Leu, and Val residues.[44] Preirradiation pulse was 20 μW for 1.0 sec. (B) From aromatic portions of the large fragment of Pol I to protons of MgAMPCPP. A control spectrum is shown below, together with chemical shifts and linewidths of aromatic protons. Preirradiation pulse was 20 μW for 0.5 sec.

Future Developments

There most certainly will be continued effort in quantitatively analyzing one- and two-dimensional NOE data taking multispin effects into account. Computer programs will become more readily available that automatically pick the peaks of interest, integrate their volumes, and store the data in a matrix for complete analysis. Advances will continue to be made in computational methods that determine the conformation of the bound ligand based on NMR constraints. Molecular graphics will also be more readily available for viewing and docking the computed structures into the crystallographic structure of the protein.

However, a major advance in biological NMR will be the application of one- and two-dimensional isotope-filtered NOEs along with the selective labeling of the protein and ligand. This methodology will greatly increase the selectivity, identification, and resolution of the NOE experiment and thus our ability to determine the conformations and arrangements of bound ligands and the amino acid environment provided by the protein. The major limitation, preventing wider use of this method at present, is the unavailability of appropriately labeled ligands. Absolute identification of the protein amino acids that are near bound ligands will be greatly facilitated by directed mutagenesis of specific residues showing intermolecular NOEs.

[18] Determination of Equilibrium Constants of Enzyme-Bound Reactants and Products by Nuclear Magnetic Resonance

By B. D. Nageswara Rao

Introduction

In an NMR spectrum the area enclosed by a nonoverlapping resonance, or a group of resonances in a spin-coupled multiplet, belonging to a specific nucleus, is directly proportional to the concentration of the molecular species containing the nucleus in the sample, provided that (1) all the nuclei are in thermal equilibrium, i.e., fully relaxed and (2) in cases where spin–spin coupling is present, the coupling is weak, i.e., the coupling constant between any pair of spins is small compared to their chemical shift difference. If appropriate precautions are taken to ensure these conditions, then the NMR spectrum provides a straightforward and non-

METHODS IN ENZYMOLOGY, VOL. 177

invasive method for the accurate determination of ratios of concentrations of the different species containing a chosen type of nucleus. Such measurements are of considerable value in a variety of investigations in which either the variations in the concentrations as a function of time or their values at equilibrium are of interest. Thus, if an adequate number of reactants and products in an enzymatic reaction contain the same nuclear species, the NMR signals readily allow the determination of the equilibrium constant of the reaction and the dependence of this constant on externally controllable sample conditions such as pH and temperature. Such measurements are easily performed at catalytic concentrations of enzymes.[1] On the other hand, if it is possible to prepare the equilibrium mixture with enzyme concentrations in sufficient excess over those of the reactants and products, such that all the substrates are in their enzyme-bound complexes, then the NMR signals may be used to determine the equilibrium constant (and kinetic parameters) of the interconversion step of the reaction.[1-4] This point becomes evident by considering a minimal kinetic scheme of an enzymatic reaction of two reactants S_1 and S_2, yielding two products P_1 and P_2 as shown in Fig. 1. If the concentration of E is in sufficient excess over S_1, S_2, P_1, and P_2, it will be possible to isolate and monitor the reaction $E \cdot S_1 \cdot S_2 \rightleftharpoons E \cdot P_1 \cdot P_2$. A measurement of the areas enclosed by the resonances of these complexes yields the equilibrium constant, and an analysis of the line shapes of the resonances allows the determination of the exchange rates of this step.[5]

On the basis of the minimal kinetic scheme in Fig. 1, the catalytic and enzyme-bound equilibrium constants may be written as

$$K_{eq} = \frac{[P_1][P_2]}{[S_1][S_2]} \quad \text{and} \quad K'_{eq} = \frac{[E \cdot P_1 \cdot P_2]}{[E \cdot S_1 \cdot S_2]} \tag{1}$$

and the dissociation constants involving S_1 as

$$K_{S_1} = \frac{[E][S_1]}{[E \cdot S_1]} \quad \text{and} \quad K'_{S_1} = \frac{[E \cdot S_2][S_1]}{[E \cdot S_1 \cdot S_2]} \tag{2}$$

Expressions similar to Eq. (2) may be given for K_{S_2} and K'_{S_2}, K_{P_1} and K'_{P_1} and K_{P_2} and K'_{P_2}. With the help of these definitions it may be readily shown that

[1] B. D. Nageswara Rao, D. Buttlaire, and M. Cohn, *J. Biol. Chem.* **251,** 6981 (1976).
[2] B. D. Nageswara Rao, "NMR and Biochemistry: A Symposium Honoring Mildred Cohn" (S. J. Opella and P. Lu, eds.), p. 371. Dekker, New York, 1979.
[3] M. Cohn and B. D. Nageswara Rao, *Bull. Magn. Reson.* **1,** 38 (1979).
[4] B. D. Nageswara Rao, in "Phosphorus-31 NMR: Principles and Applications" (D. G. Gorenstein, ed.), p. 57. Academic Press, Orlando, Florida, 1984.
[5] B. D. Nageswara Rao, this series, Vol. 176 [14].

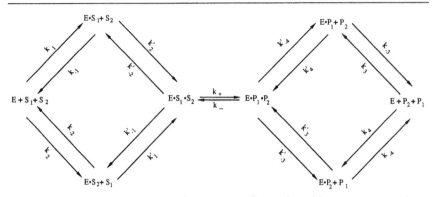

FIG. 1. A minimal kinetic scheme for an enzymatic reaction with two reactants and two products.

$$K'_{eq}/K_{eq} = K_{S_1} K'_{S_2}/K_{P_1} K'_{P_2} \qquad (3)$$

Equation (3) may be written in other equivalent forms by using such identities as

$$K_{S_1} K'_{S_2} = K'_{S_1} K_{S_2} \qquad (4)$$

Clearly, K_{eq} is determined by thermodynamics, and remains unaffected by the enzyme. On the other hand, K'_{eq} depends on the enzyme, and the dependence is signified by the ratios of dissociation constants occurring on the right-hand side of Eq. (3). These ratios may be altered, thus affecting a change in K'_{eq}, whereas K_{eq} remains invariant. This implicit variability of K'_{eq} provides, to some degree, the motivation for the measurement and understanding of this equilibrium constant.

The various experimental aspects relevant for making the measurements of equilibrium constants of enzyme-bound reactants and products by the NMR method are elaborated below. Following this, experimental results obtained from such measurements on a number of ATP-utilizing enzymes are presented and discussed.

Experimental Considerations

In order to measure the equilibrium constants with all the reactants and products in enzyme-bound forms, the enzyme concentration is required to be in sufficient excess of all the substrate concentrations such that, on the basis of known dissociation constants, the substrates are predominantly found in the enzyme-bound complexes. A second important factor is that the concentrations of these complexes must be large enough ($\geq$1–2 mM) for detection by NMR with reasonable sensitivity. For dissociation constants of 100–200 μM this will require enzyme concentrations of about 4–5 mM. The sample volumes are typically 1.5–2.5 ml for nuclei other than protons. Thus, the measurements require large quantities of enzyme (~200 to 300 mg per sample, for a molecular of weight of about 40,000) and favorable dissociation constants. If all the substrates contain at least one nucleus of the species being detected giving rise to identifiable and nonoverlapping resonances, it is convenient for the measurement. For a two-substrate reaction, it is necessary to have at least one substrate on either side of the reaction giving rise to a distinct resonance in the spectrum. The concentrations of the other substrates, which do not possess resonances, may be inferred on the basis of the stoichiometry of the reaction. Because the goal of the measurements is to exclusively monitor the step $E \cdot S_1 \cdot S_2 \rightleftharpoons E \cdot P_1 \cdot P_2$ in the overall reaction, the presence of any of S_1, S_2, P_1, and P_2 in other nonparticipating complexes such as $E \cdot S_1$ and $E \cdot S_1 \cdot P_2$, although enzyme bound, will impinge on the accuracy of the equilibrium constant determined. The presence of such unproductive complexes is minimized if all the four substrates possess similar dissociation constants. However, in a number of cases the near equality of dissociation constants is not likely to obtain, and therefore the accuracy of the measured equilibrium constant will be affected.

There are a number of factors to be recognized from the NMR point of view regarding reliable measurements of enzyme-bound equilibrium constants. In order that the integrated areas enclosed by the resonances in the spectrum are proportional to the concentrations, it is necessary that if spin–spin splittings are present in the spectrum, it must be a weakly coupled spin system (pairwise spin–spin coupling constants negligible in comparison with the corresponding chemical shift differences) and the integration must include all the resonances in the multiplet. Clearly, nonoverlapping resonances are preferable, and if overlapping resonances are unavoidable only the sums of concentrations of different species contributing to the overlapping region will be measureable. Because spectra are normally obtained using the Fourier transform (FT) mode of operation, aside from acquiring sufficient numbers of transients for the best signal-to-noise ratio, it is important to allow sufficient delay [compared to the spin–lattice relaxation times (T_1) of all the nuclei] between pulses so that the spin system returns to thermal equilibrium. Furthermore, it is useful to avoid signals of interest occurring too close to the edges of audio filters used in the detection, and to use as large a data size as possible (for a given sweep width). Line-broadening or apodization methods should be used with care lest they alter the integrated areas to be measured. Finally, additional radiofrequency fields such as those used for decoupling or double-resonance experiments should not be used, as they produce intensity changes due to nuclear Overhauser effects.

Experimental Results for Specific Enzymes

Arginine Kinase

The determination of K'_{eq} by the use of ^{31}P NMR was first accomplished for the arginine kinase reaction.[1] The source of the enzyme was the tail muscle of the American lobster. Arginine kinase reversibly catalyzes the reaction[6]

$$ATP + arginine \overset{Mg(II)}{\rightleftharpoons} ADP + P\text{-arginine} \tag{5}$$

Mg(II) is an obligatory component of the reaction. The ^{31}P NMR spectrum of an equilibrium mixture of the arginine kinase reaction set up with catalytic concentrations of the enzyme, shown in Fig. 2A, exhibits resonances that were readily assigned to six phosphate groups, three in ATP, two in ADP, and one in P-arginine. The apparent equilibrium constant, K_{eq}, may be readily determined by integration of the signals. It is found that at 12°, $K_{eq} = 0.1 \pm 0.02$ at pH 7.25, and 0.22 ± 0.04 at pH 8.0.

The ^{31}P NMR spectrum of an equilibrium mixture of enzyme-bound reactants and products of arginine kinase is shown in Fig. 2B. The sample used for this spectrum contained the enzyme at concentrations in sufficient excess over that of the substrates such that over 90% of the substrates was in enzyme-bound complexes. It was shown that the line shapes of the three groups of resonances in Fig. 2B arose primarily from the interconversion of bound substrates and products on the surface of the enzyme, by adding EDTA to the sample of Fig. 2B to sequester Mg(II) from the reaction complexes; thus the reaction was stopped but the substrates still remain bound to the enzyme. The ^{31}P NMR spectrum of the sample after the addition of EDTA is shown in Fig. 2C, in which the resonances are well defined without the complex line-shape pattern in Fig. 2B. The assignment of the three groups of resonances in Fig. 2B to the six (interconverting) phosphate groups was thus confirmed. The analysis of line shapes such as those in Fig. 2B for the determination of interconversion rates is presented elsewhere in this series.[5] For the purpose of determination of the equilibrium constant of the interconversion step, it is clear that the area enclosed by the signal at ~ −19 ppm in Fig. 2B is proportional to the concentration of enzyme-bound ATP and that at ~ −10 ppm to the sum of the concentration of enzyme-bound ATP and

[6] J. F. Morrison, *Enzymes* **8**, 457 (1973).

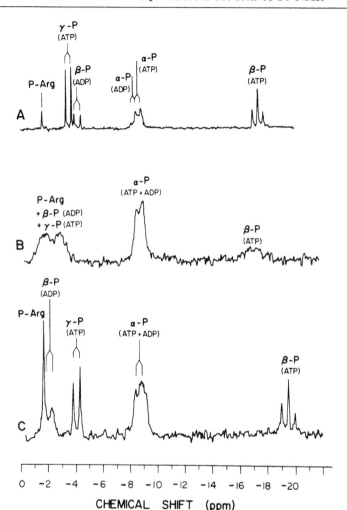

FIG. 2. ³¹P NMR spectra (at 40.3 MHz) of equilibrium mixtures of the arginine kinase reaction. (A) Catalytic concentrations of enzyme; substrate concentrations are in the range of 3–12 mM. (B) Enzyme concentrations in excess of substrate concentrations: enzyme, 4.4 mM; substrate concentrations 2–3 mM. (C) Spectrum obtained after the addition of excess EDTA to the sample in B to stop the reaction (see Ref. 1).

ADP. From the areas of these two signals the ratio of the concentrations of enzyme-bound ADP and ATP is readily obtained. However, in order to calculate K'_{eq} the ratio of E · MgADP · P-arginine and E · MgATP · arginine is required. In these experiments it is assumed that this ratio is nearly equal to that of the enzyme-bound ADP and ATP concentrations in Fig.

2B. K'_{eq} may then be evaluated from the areas of the signals at ~ -10 and ~ -19 ppm. What are the implications of such an assumption? The experiment was initiated with nearly equal concentrations of ATP and arginine. Equal concentrations of ADP and P-arginine are produced by the reaction. Although the dissociation constants for the four substrates, estimated from various types of measurements, do clearly show that the substrates are almost entirely in enzyme-bound complexes, it is difficult to estimate the concentrations of unproductive complexes (such as $E \cdot MgADP$ and $E \cdot MgADP \cdot arginine$). Nevertheless, the following relation holds:

$$\frac{[\text{Bound ADP}]_{\text{Total}}}{[\text{Bound ATP}]_{\text{Total}}} = \frac{[\text{Bound ADP}]_{\text{Reaction complex}} + [\text{Bound ADP}]_{\text{Other complexes}}}{[\text{Bound ATP}]_{\text{Reaction complex}} + [\text{Bound ATP}]_{\text{Other complexes}}} \quad (6)$$

The ratio of the first terms in the numerator and denominator on the right-hand side of Eq. (6) is K'_{eq}, whereas the ratio on the left-hand side of Eq. (6) is its calculated value. Remembering that $(a + b)/(c + d) = a/c$, if $a/c = b/d$, the calculated value will be exactly equal to K'_{eq} if the ratio of the second terms on the right-hand side of Eq. (6) has the same value as that of the first terms, i.e., when equilibrium is established in the sample the concentrations of the unproductive complexes are also governed by K'_{eq}. This condition may actually obtain in some cases, but the accuracy of its validity is difficult to assess unless precise information is available about the dissociation constants involved in all the multiple equilibria involved. It is important to recognize that the equality of the ratios of first terms and second terms on the right-hand side is implicit in the measurements of K'_{eq} by using the NMR method.

On the basis of the integrated areas under the signals at ~ -10 and ~ -19 ppm and with the assumption discussed above, K'_{eq} for arginine kinase was found to be 0.8 ± 0.2, at $12°$ and pH 7.25, significantly larger than K_{eq} (= 0.1 ± 0.02) under the same conditions.[1,7]

Creatine Kinase

Creatine kinase, which reversibly catalyzes the reaction[8]

$$\text{ATP} + \text{creatine} \underset{}{\overset{\text{Mg(II)}}{\rightleftharpoons}} \text{ADP} + \text{P-creatine} \quad (7)$$

is similar to arginine kinase in that both these enzymes catalyze phosphoryl transfer to a guanidino nitrogen. Arginine kinase is sometimes referred to as the invertebrate analog of creatine kinase.[6] ^{31}P NMR experiments

---

[7] K'_{eq} was incorrectly defined in Ref. 1. This mistake was corrected in B. D. Nageswara Rao and M. Cohn, *J. Biol. Chem.* **252**, 3344 (1977).

[8] D. C. Watts, *Enzymes* **8**, 383 (1973).

similar to those described above for arginine kinase were performed for rabbit muscle creatine kinase.[9] At catalytic enzyme concentrations the spectrum for the equilibrium mixture was similar to that in Fig. 2A, with the signal for P-creatine occurring approximately at the chemical shift value of P-arginine. The value of the catalytic equilibrium constant K_{eq}, at 4° and pH 8.0, was determined to be 0.08. The ^{31}P NMR spectrum with excess enzyme concentrations shows signals in the same three regions as in Fig. 2B with similar line shapes leading to similar assignments for the signals, as well as a similar value for K'_{eq}.[5] However, the evaluation of K'_{eq} for creatine kinase was impaired by two factors: (1) during the experiments there is an irreversible accumulation of P_i due to a contaminant ATPase activity, and (2) the substrates do not bind creatine kinase as well as they bind arginine kinase, resulting in the possibility of the presence of appreciable concentrations of free substrates as well as in errors arising from the implications of Eq. (6) discussed earlier. The value of K'_{eq} was found to be approximately unity on the basis of integrated areas proportional to enzyme-bound ATP and ADP. However, in view of the above, the range of uncertainty is expected to be rather large with a value somewhere between 0.5 and 2.5.

Adenylate Kinase

Adenylate kinase catalyzes the reversible reaction[10]

$$MgATP + AMP \rightleftharpoons MgADP + ADP \tag{8}$$

The enzyme is most abundant in tissues, such as muscle, in which the energy turnover is considerable, and is essential for production of adenine nucleotides beyond the monophosphate level. ^{31}P NMR experiments were performed on the enzyme-bound substrate complexes of this enzyme from two different sources, namely, porcine and carp muscle, and among the goals of these experiments was the determination of K'_{eq}.[11] While all the four substrates in Eq. (8) give rise to clearly identifiable ^{31}P NMR signals at catalytic enzyme concentrations,[5] the spectra of the equilibrium mixtures of enzyme-bound reactants and products for both carp and porcine enzymes gave just two broad resonances in the regions -4 to -6 ppm and -9 to -11 ppm.[5] Before any attempt was made to evaluate K'_{eq} it was necessary to assign the resonances observed to specific phosphate groups in the substrates and account for the missing signals in the AMP (4 ppm) and β-P(ATP)(-19 ppm) regions. One experiment, which was

[9] B. D. Nageswara Rao and M. Cohn, *J. Biol. Chem.* **256**, 1716 (1981).
[10] L. Noda, *Enzymes* **8**, 279 (1973).
[11] B. D. Nageswara Rao, M. Cohn, and L. Noda, *J. Biol. Chem.* **253**, 1149 (1978).

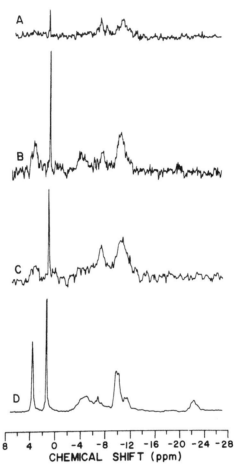

8   4   0   -4   -8   -12 -16 -20 -24 -28
CHEMICAL SHIFT (ppm)

FIG. 3. ^{31}P NMR spectra (at 40.3 MHz) of an equilibrium mixture of reactants and products bound to porcine adenylate kinase at 15° pH 7.0. (A) After the ~4-hr accumulation of NMR signal. (B) As in A, after a ~12-hr accumulation; note the appearance of two distinct resonances of β-P(ADP). (C) As in B, but with the addition of a small amount of ATP. (D) As in C, but with excess EDTA added to stop the reaction. The ^{31}P exchanges in the reaction are shown at the top (see Ref. 11).

serendipitously performed, turned out to be the key for the interpretation of these spectra.[11] The experiment is briefly described below.

In Fig. 3A the spectrum of a stoichiometric equilibrium mixture of porcine adenylate kinase[5] is reproduced. This spectrum was obtained with ~4 hr of signal averaging by the spectrometer; note the presence of a small P_i signal, ~2 ppm, which accumulated due to some ATPase activity present in the sample as a contaminant. When the signal averaging was continued for ~12 hr on the sample (with the intention of obtaining a better signal-to-noise ratio), the spectrum in Fig. 3B was obtained. Clearly, the irreversible buildup of the effect of ATP hydrolysis has altered the nature of the spectrum with time. In Fig. 3B not only is there an increase in P_i, there is a significant AMP signal at 4 ppm, and two signals are now obtained in the region of -3 to -7 ppm where there was only one in Fig. 3A. The spectrum in Fig. 3B may be explained as follows: as the irreversible hydrolysis of ATP slowly progresses along with the reversible adenylate kinase activity, AMP and P_i increase, and ATP is progressively depleted. Thus the β-P(MgADP) that exchanges with β-P(MgATP) gives rise to an observable signal; the signal is no longer broadened by the exchange because there is not enough ATP to exchange with.[11] On this basis, the signal at -4 ppm in Fig. 3B is assigned to the β-P of acceptor ADP. This assignment is verified in Fig. 3C, in which a small amount of ATP is added to the sample of Fig. 3B, and the resonance at -4 ppm is now broadened due to the onset of appreciable exchange arising from the addition of ATP. No signal is visible for β-P(MgATP) at -19 ppm in Fig. 3C because it is too broad to be seen. If now the reaction is stopped by the addition of EDTA to the sample of Fig. 3C, the ATP signal appears (see Fig. 3D). Another noteworthy feature of Fig. 3D is the coalescence of the two ADP signals into one in the absence of Mg(II). This increase in the exchange rate and evidence from other experiments was used to show that only metal-free nucleotides bind at the AMP or donor ADP site, whereas ADP and ATP can bind at the acceptor ADP site with or without metal.[11,12] The metal ion thus plays the additional role of providing the distinction between the acceptor and donor ADP molecules on adenylate kinase besides facilitating catalysis as in other kinases.

The experiment described above thus made it clear that the two signals observed at -4 to -6 ppm and -9 to -11 ppm in Fig. 3A belong to the phosphate groups that are in fast exchange (exchange rate large compared to the chemical shift difference) among the interconverting enzyme-bound reactants and products, namely, γ-P(MgATP) and β-P (donor ADP) giving rise to the signal at -4 to -6 ppm, and α-P(MgATP) and α-P (acceptor MgADP) giving rise to the signal at -9 to -11 ppm. For the same value of the exchange rate the other two pairs of exchanging signals, namely, AMP exchanging with α-P (donor ADP), and β-P(MgATP) exchanging with β-P (acceptor MgADP), are in slow exchange because the corresponding chemical shift differences are large compared to the exchange rate. These signals in slow exchange are thus too broad to be

[12] K. V. Vasavada, J. I. Kaplan, and B. D. Nageswara Rao, *Biochemistry* **23**, 961 (1984).

observed. Density-matrix analyses of line shapes were used to extract information on the interconversion rate on this enzyme, as well as the rate at which the acceptor and donor ADP molecules interchange their roles (binding sites) on the enzyme, together with the influence of the cation Mg(II) on such an interchange.[5,12] However, from the point of evaluating K'_{eq} and the assignment of the two observed signals, the spectrum is bereft of adequate information, because the integrated areas under the resonance at -4 to -6 ppm are proportional to the sum of the concentrations of enzyme-bound ATP and donor ADP while those for the signal at -9 to -11 ppm are proportional to the sum of enzyme-bound ATP and acceptor ADP. The ratio of the concentrations of $E \cdot ADP \cdot MgADP$ and $E \cdot AMP \cdot MgATP$ (i.e., K'_{eq}) cannot be determined from just these integrated areas.

In order to obtain an estimate of K'_{eq}, a ^{31}P NMR spectrum of the equilibrium mixture of enzyme-bound reactants was recorded at a suboptimal value of Mg(II) concentration.[11] If the reaction was initiated with the addition ADP and Mg(II) to the enzyme, the optimum Mg(II) concentration is $\frac{1}{2}[ADP]$ because the enzyme contains just one Mg(II) ion per reaction complex. A spectrum recorded with Mg(II) at one-third the optimum value does show signals in the AMP and β-P(ATP) regions because the signals form substrates present in enzyme complexes devoid of Mg(II), are not undergoing the reaction, and therefore are not subject to exchange. The superposition of the broad resonances from reaction complexes [with Mg(II)] and sharp resonances from nonreacting complexes [without Mg(II)] in exchange because of dissociation and association results in a signal of intermediate width, which becomes observable. Assuming that all the bound substrates are present in the four complexes $E \cdot ADP \cdot ADP$, $E \cdot ADP \cdot MgADP$, $E \cdot AMP \cdot ATP$, and $E \cdot AMP \cdot MgATP$, the integrated areas yield

$$\frac{[E \cdot ADP \cdot MgADP] + [E \cdot ADP \cdot ADP]}{[E \cdot AMP \cdot MgATP] + [E \cdot AMP \cdot ATP]} = 1.6 \pm 0.5 \qquad (9)$$

If it is further assumed that in the reaction mixture

$$\frac{[E \cdot ADP \cdot MgADP]}{[E \cdot AMP \cdot MgATP]} = \frac{[E \cdot ADP \cdot ADP]}{[E \cdot AMP \cdot ATP]} \qquad (10)$$

the ratio of 1.6 ± 0.5 in Eq. (9) becomes equal to K'_{eq} at $4°$, pH 7.0. [The reasoning is similar to that used in discussing Eq. (6) earlier.] On the other hand, integrating areas in the ^{31}P NMR spectrum of the equilibrium mixture obtained with catalytic enzyme concentrations[5] gives $K_{eq} = 0.38 \pm 0.05$ at $4°$ and pH 7.0.[11] It should be noted that the evaluation of K'_{eq} is subject to the validity of Eqs. (9) and (10), and, therefore, the uncertainty in K'_{eq} is expected to be larger than ± 0.5 quoted for the ratio in Eq. (9).

Pyruvate Kinase

Pyruvate kinase catalyzes the reaction[13]

$$\text{ATP} + \text{pyruvate} \rightleftharpoons \text{ADP} + \text{P-enolpyruvate.} \qquad (11)$$

The enzyme requires a divalent cation, usually Mg(II), and a monovalent cation, normally K^+, as obligatory components. At catalytic enzyme concentrations, the equilibrium of the reaction is predominantly in favor of ATP production. The enzyme thus strongly forces the unidirectional flow of the coupled reactions in the glycolytic pathway toward pyruvate formation.

^{31}P NMR spectra of equilibrium mixtures of pyruvate kinase at catalytic and stoichiometric enzyme levels are shown in Fig. 4A.[14] The catalytic equilibrium mixture was driven in the reverse direction by adding a 15-fold excess of pyruvate over ATP to see if a signal would be observed for P-enolpyruvate. A weak signal, barely discernible at about 0.9 ppm, is probably due to P-enolpyruvate. As is evident from the spectrum, this signal would not have been considered at all but for the fact that one is looking for a signal at that position in the spectrum. The spectrum is thus clearly indicative of the one-sided nature of the equilibrium. An approximate estimate of the equilibrium constant yields $K_{eq} \approx 3 \times 10^{-4}$.

The ^{31}P NMR spectrum obtained at excess enzyme concentrations shows both ATP and ADP signals as well as a signal at ~4 ppm, assigned to P_i and P-enolpyruvate. A closer examination revealed that this signal is also contributed to by AMP. (Note that these assignments imply substantial chemical shifts for both P_i and P-enolpyruvate.) P_i is produced because of a contaminant ATPase activity of the enzyme. AMP is produced because of contaminant adenylate kinase coupling with the ATPase activity. The quantity of P-enolpyruvate contributing to the area of this composite signal is of relevance for the evaluation of K'_{eq}. In order to make an approximate estimate of K'_{eq}, the accumulation of AMP and P_i were estimated on the basis of experiments with E · MgATP, E · MgADP, and E · P-enolpyruvate. Furthermore, it was assumed that the instantaneous concentrations of the reactants and products were primarily governed by the equilibrium associated with the pyruvate kinase reaction. This procedure yielded [E · MgADP · P-enolpyruvate] ≤ 0.9 mM and

[13] F. J. Kayne, *Enzymes* **8**, 352 (1973).
[14] B. D. Nageswara Rao, F. J. Kayne, and M. Cohn, *J. Biol. Chem.* **254**, 2689 (1979).

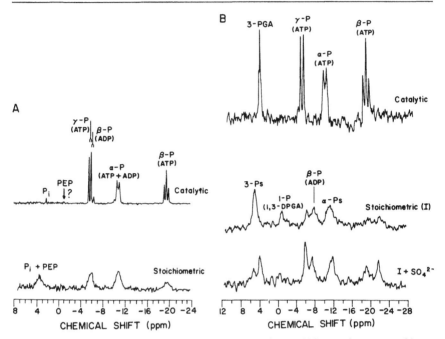

FIG. 4. ^{31}P NMR spectra of equilibrium mixtures of (A) rabbit muscle pyruvate kinase (see Ref. 14) and (B) yeast 3-P-glycerate kinase (see Ref. 18) at catalytic and stoichiometric enzyme concentrations. Spectra for pyruvate kinase were obtained at 40.3 MHz, 15°, and pH 8.0; for 3-P-glycerate kinase catalytic equilibrium mixture was recorded at 24 MHz and stoichiometric equilibrium mixtures at 109.3 MHz, 4°, and pH 7.0. The appearance of a small P-enolpyruvate signal in the catalytic equilibrium mixture of pyruvate kinase occurs in the presence of a 15-fold excess of pyruvate over ATP.

[E · MgATP · pyruvate] ≤ 1.2 mM. The ratio of these two upper limits of concentrations gives $K'_{eq} \approx 0.8$. It must be recognized that this value is, at best, a crude estimate of K'_{eq}.

The value of K'_{eq} for pyruvate kinase was redetermined recently by Stackhouse et al., who used an enzyme with a much-reduced ATPase activity compared to that used by Nageswara Rao et al.[14] Stackhouse et al.[15] estimate a value of ~0.1 for K'_{eq}. The discrepancy between this value and the rather crude estimate of 0.8 made by Nageswara Rao et al.[14] is not surprising. It may be noted that the estimate of Stackhouse et al.[15] also implicitly assumes that the concentrations of bound substrates are determined exclusively by K'_{eq} [see earlier discussion following Eq. (6)]. Pyru-

[15] J. Stackhouse, K. P. Nambiar, J. J. Burbaum, D. M. Stauffer, and S. A. Brenner, J. Am. Chem. Soc. 107, 2757 (1985).

vate kinase binds three of the substrates (excluding P-enolpyruvate) rather poorly and the assumption of all the substrates being in the fully bound form is most likely to be violated. All these factors contribute to uncertainties in K'_{eq}. The correct value of K'_{eq} may thus be elusive in spite of a well-controlled NMR experiment with a highly pure enzyme. Thus, while it is reasonable to argue that K'_{eq} differs by a factor of ~ 1000 from K_{eq}, its value is difficult to ascertain with precision.

3-Phosphoglycerate Kinase

3-Phosphoglycerate kinase catalyzes the first ATP-generating reaction in anaerobic glycolysis[16]:

$$\text{ATP} + \text{3-P-glycerate} \overset{\text{Mg(II)}}{\rightleftharpoons} \text{ADP} + \text{1,3-bis-P-glycerate} \qquad (12)$$

The equilibrium of the overall reaction is predominantly in favor of ATP production and a value of 3×10^{-4} was estimated for K_{eq}.[17] ^{31}P NMR spectra of equilibrium mixtures of this reaction obtained at catalytic and stoichiometric enzyme concentrations are shown in Fig. 4B.[18] The source of the enzyme used in the experiments was yeast. The catalytic equilibrium mixture, prepared with ATP, Mg(II), and 3-P-glycerate, reveals no other signals on the addition of the enzyme, because the equilibrium is highly one-sided in favor of ATP. The equilibrium mixture at excess enzyme concentrations, however, shows signals attributed to all the four substrates. Integration of areas under the signals allowed an estimate of $K'_{eq} = 0.8 \pm 0.3$. This value is, once again, subject to the assumption regarding equilibrium concentrations [see Eq. (6)]. In addition, 3-P-glycerate kinase has two ATP-binding sites, and at one of these, presumably the noncatalytic site, ATP has poor affinity for Mg(II).[18–20] This disparity in affinity for Mg(II) readily appears in the shape of the β-P(ATP) resonance in the region of -19 to -21 ppm. The presence of the nucleotide in such unproductive complexes is an additional factor, unique to this enzyme, which might contribute to departures from the assumption regarding the equilibrium concentrations of enzyme-bound substrates.

Sulfate ion is known to affect the kinetics of the 3-P-glycerate kinase reaction.[18,20,21] Since the spectrum of the equilibrium mixture of bound substrate exclusively monitors the interconversion step of the reaction

[16] R. K. Scopes, *Enzymes* **8**, 335 (1973).
[17] W. K. G. Krietsch and T. Bücher, *Eur. J. Biochem.* **17**, 568 (1970).
[18] B. D. Nageswara Rao, M. Cohn, and R. K. Scopes, *J. Biol. Chem.* **253**, 8056 (1978).
[19] R. K. Scopes, *Eur. J. Biochem.* **91**, 119 (1978).
[20] B. D. Ray and B. D. Nageswara Rao, *Biochemistry* **27**, 5574 (1988).
[21] R. K. Scopes, *Eur. J. Biochem.* **85**, 503 (1978).

(see Fig. 1), it can be used to assess the effect of sulfate solely on this step. The ^{31}P NMR spectrum of the stoichiometric equilibrium mixture recorded in the presence of ~50 mM sulfate concentration is also shown in Fig. 4B. The spectrum indicates[18] (1) a change in the chemical shift 3-P resonance of 3-P-glycerate by ~1.5 ppm, (2) a reduction in the linewidths of β-P(ATP) resonances indicating a reduction by ~10–15% the rate of phosphoryl transfer from ATP on the enzyme (compared to the rate in absence of sulfate), and (3) an apparent shift in K'_{eq} in favor of E · MgATP · 3-P-glycerate. The shift in K'_{eq} implies changes in the ratios of the dissociation constants of the reactants and products from their enzyme-bound complexes [see Eq. (3)].

Methionyl-tRNA Synthetase

Methionyl-tRNA synthetase catalyzes the aminoacylation of tRNA^{Met} in a two-step reaction, the first step (activation reaction) of which involves nucleotidyl transfer[22,23]

$$E + \text{methionine} + ATP \overset{\text{Mg(II)}}{\rightleftharpoons} E \cdot \text{Met-AMP} + PP_i \tag{13}$$
$$E \cdot \text{Met-AMP} + \text{tRNA}^{Met} \rightleftharpoons E + AMP + \text{Met-tRNA}^{Met} \tag{14}$$

Mg(II) is an obligatory component of the first step of this reaction. The ^{31}P NMR experiments were performed[24] with a monomeric trypsin-modified fragment of the enzyme (from *Escherichia coli*), which fully retains the activity of the first step of the reaction.[25] A ^{31}P NMR spectrum of an equilibrium mixture of the activation reaction is shown in Fig. 5A. An ambiguity existed initially regarding the position of enzyme-bound Met-AMP signal in the spectrum. The ^{31}P resonance of chemically synthesized Met-AMP at pH 4.1 was found to be at −7.5 ppm. In order to arrive at the assignment of signals shown in Fig. 5A, a saturation transfer experiment was performed.[24] A spectrum of the sample of Fig. 5A was recorded while the β-P(ATP) resonance was irradiated with a saturating radiofrequency field. When the reaction proceeds at a rate faster than the relaxation rates of ^{31}P nuclei in the different complexes in equilibrium, the saturation of the β-P(ATP) resonance is transferred to ^{31}P resonances of the chemical species into which this nucleus converts during reaction, i.e., MgPP$_i$ in the forward direction and subsequently γ-P(ATP) on reversal. The result

[22] G. Fayat and J. P. Waller, *Eur. J. Biochem.* **44**, 335 (1974).
[23] S. Blanquet, G. Fayat, and J. P. Waller, *Eur. J. Biochem.* **44**, 343 (1974).
[24] G. Fayat, S. Blanquet, B. D. Nageswara Rao, and M. Cohn, *J. Biol. Chem.* **255**, 8164 (1980).
[25] F. Hyafil, Y. Jacques, G. Fayat, M. Fromant, P. Dessen, and S. Blanquet, *Biochemistry* **15**, 3678 (1976).

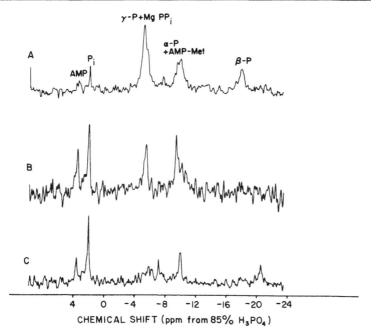

FIG. 5. ³¹P NMR spectra (at 145.7 MHz) of (A) an equilibrium mixture of the activation reaction of methionyl-tRNA synthetase with reactants and products primarily bound to the enzyme at 20° and pH 7.6. (B) As in A, after saturating the β-P(ATP) resonance by radio-frequency irradiation. (C) As in B, but with excess EDTA added to stop the reaction (see Ref. 24).

of the experiment is shown in Fig. 5B. The group of resonances in the region -4 to -6 ppm is now significantly reduced in intensity and the total intensity is the same as that of the group in the region -9 to -11 ppm. These results clearly indicate that the PP$_i$ and γ-P(ATP) resonances superpose, and the Met-AMP resonances superpose that of α-P(ATP). The spectrum in Fig. 5A also contains both P$_i$ and AMP resonances owing to contamination of inorganic pyrophosphatase in the enzyme preparation and to decomposition of Met-AMP. After the spectrum in Fig. 5B was recorded, EDTA was added to the sample to remove Mg^{2+} from PP$_i$- and ATP-containing complexes, and a ³¹P NMR spectrum of the resultant sample was recorded (Fig. 5C). The spectrum shows depletion of ATP because of the inorganic pyrophosphatase. However, a distinct resonance is observed at about -10 ppm for the residual E · Met-AMP. The results depicted by Fig. 5B and C confirm the assignment shown in Fig. 5A.

The equilibrium constant for the interconversion of the enzyme-bound substrates to enzyme-bound products for the methionyl-tRNA synthetase reaction, given by [E · AMP-Met · MgPP$_i$]/[E · Met · MgATP], was found to have a value between 1 and 2 at pH 7.65 and 20°. This value is in good agreement with that determined from stopped-flow experiments in which equilibrium constants of 1.5 and 1.1 were found for the native and trypsin-modified enzyme, respectively.[25]

Discussion

The NMR method of determination of equilibrium constants and exchange rates of interconverting reactants and products on the surface of the enzyme offers the advantage of being direct and straightforward, if a sample can be prepared in which the predominant species are E · S$_1$ · S$_2$ and E · P$_1$ · P$_2$, and the different substrates give rise to an adequate number of nonoverlapping resonances that can be detected with good sensitivity in a reasonable amount of time. The measurements can be made while the interconversion is in progress, and no intervention, such as stopping the reaction to measure concentrations of one species or the other, is required. In procedures in which stopping the reaction is necessary, there is always the question of the effect of the method of stoppage on the equilibrium that was intended to be monitored. However, the NMR method is likely to require larger quantities of enzymes and substrates than any other technique, which limits its applicability to relatively few enzymatic systems. Even in cases where all the substrates are in the desired complexes, and give rise to identifiable resonances in the absence of interconversion, the onset of the reaction might make some of the resonances undetectable if they are excessively broadened due to the exchange process (see, for example, the experiments with adenylate kinase[11]). Although the measurements may be made in a satisfactory manner, there is one factor that might cause a departure between the NMR-estimated value and the correct value of K'_{eq}, namely, the extent of the presence of unproductive enzyme complexes in the sample [see Eq. (6)]. The contribution of this factor is difficult to assess quantitatively. It is, nevertheless, most likely to vitiate the measurements in cases where there are large disparities in the affinities of the different substrates to enzyme, especially if appreciable concentrations of free substrates may be present in the sample, such as in the case of pyruvate kinase. The first-published measurements on pyruvate kinase[14] were also plagued by the interference from a sizeable ATPase activity, which led to what appears to be an incorrect estimate of K'_{eq}, in view of the recent work of Stackhouse *et al.*,[15] who used an enzyme with much less ATPase interference. However, because Stackhouse *et al.*[15] used the same sample conditions, even the value they estimated is subject to the same questions regarding the contribution of unproductive complexes to the departure between the estimated value and the true value of K'_{eq}. This example is mentioned here to suggest that although the NMR measurements in cases of this kind may be used to arrive at some general and qualitative features related to the interconversion step, it is perhaps not wise to reach incisive conclusions about the mechanism of the enzymatic reaction on the exclusive basis of the estimates of K'_{eq}.

With the exception of pyruvate kinase, for all other enzymes studied by ^{31}P NMR the value of K'_{eq} is in the range 0.5 to 2.5. Values of K'_{eq} close to unity were obtained for yeast hexokinase[26] and rabbit muscle triose-phosphate isomerase,[27] by biochemical methods. Albery and Knowles,[28–30] in connection with their results on triose-phosphate isomerase, argued that K'_{eq} should approach unity for a well-evolved enzyme. A corollary to this argument is that the interconversion should not be rate limiting in the overall reaction, which is also the case for all the enzymes discussed here, with the possible exception of pyruvate kinase.[15] The thermodynamic arguments of Albery and Knowles are the subject of some debate. Furthermore, their suggestion implicitly assumes that evolutionary pressure on the enzymes in the cell manifests itself primarily through the kinetics of the various steps of the respective reactions. Nevertheless, the fact is that for a number of enzymes for which the estimates are made the values of K'_{eq} do hover around unity. Whether there is a deeper significance to this is yet to be assessed in an incontrovertible manner.

Acknowledgments

The writing of this chapter was supported in part by the NSF Grant DMB 8608185. Thanks are due to Ms. Margo Page for typing the manuscript.

[26] K. D. Wilkinson and I. A. Rose, *J. Biol. Chem.* **254**, 12567 (1979).
[27] R. Iyengar and I. A. Rose, *Biochemistry* **20**, 1223 (1980).
[28] W. J. Albery and J. R. Knowles, *Biochemistry* **15**, 5627 (1976).
[29] W. J. Albery and J. R. Knowles, *Biochemistry* **15**, 5631 (1976).
[30] J. R. Knowles and W. J. Albery, *Acc. Chem. Res.* **10**, 105 (1977).

[19] Mechanistic Studies Utilizing Oxygen-18 Analyzed by Carbon-13 and Nitrogen-15 Nuclear Magnetic Resonance Spectroscopy

By JOHN M. RISLEY and ROBERT L. VAN ETTEN

Introduction

Nuclei to which different isotopes are covalently attached exhibit different NMR chemical shifts. Within a series, for example, the isotopes of hydrogen, of carbon, of nitrogen, and of oxygen, the chemical shifts of nuclei bearing the heavier isotopes are generally upfield with respect to those with the lighter isotopes. The magnitudes of the differences in chemical shifts, called the isotope-induced shift, are generally small. They vary considerably and are a function, in part, of the chemical shift range of the nucleus being observed. Although oxygen-18 was reported to exert an isotope shift (relative to oxygen-16) in ¹H NMR[1] and in ⁵⁵Mn and ⁹⁵Mo NMR,[2,3] it was not until reports of the ¹⁸O isotope shifts in ³¹P NMR,[4,5] ¹³C NMR,[6] and ¹⁵N NMR[7] that use of the oxygen isotope-induced shift for mechanistic studies was established to be of considerable practical importance. The use of the isotopes of oxygen in mechanistic studies using ³¹P NMR is discussed elsewhere in this volume.[8] The utilization of the ¹⁸O shift on ¹³C NMR to elucidate the source of oxygen in the biosynthesis of natural products will not be discussed here, because this particular topic has been reviewed[9] and there are a number of recent examples of its continued application.[10,11]

In the present chapter, we restrict our discussion to mechanistic studies utilizing ¹⁸O, as analyzed by ¹³C and ¹⁵N NMR spectroscopy. Specifically, the mechanistic studies covered include rates of hydrolysis reac-

[1] S. Pinchas and E. Meshulan, *J. Chem. Soc. D,* 1147 (1970).
[2] K. U. Buckler, A. R. Haase, O. Lutz, M. Müller, and A. Nolle, *Z. Naturforsch. A* **32,** 126 (1977).
[3] A. R. Haase, O. Lutz, M. Müller, and A. Nolle, *Z. Naturforsch. A* **31,** 1427 (1976).
[4] M. Cohn and A. Hu, *Proc. Natl. Acad. Sci. U.S.A.* **75,** 200 (1978).
[5] O. Lutz, A. Nolle, and D. Staschewski, *Z. Naturforsch. A* **33,** 380 (1978).
[6] J. M. Risley and R. L. Van Etten, *J. Am. Chem. Soc.* **101,** 252 (1979).
[7] R. L. Van Etten and J. M. Risley, *J. Am. Chem. Soc.* **103,** 5633 (1981).
[8] J. J. Villafranca, this volume [20].
[9] J. C. Vederas, *Can. J. Chem.* **60,** 1637 (1982).
[10] A. A. Ajaz, J. A. Robinson, and D. L. Turner, *J. Chem. Soc., Perkin Trans. 1,* 27 (1987).
[11] V. C. Emery and M. Akhtar, *Biochemistry* **26,** 1200 (1987).

tions, the point of bond cleavage, and oxygen exchange reactions. These studies are conducted in aqueous solutions or in mixed solvents, although analysis of some reactions in organic solvents is possible. Utilizing NMR to study these reactions has the advantage that the reactions are assayed in a nearly continuous mode, sample handling is minimized, conversion to volatile derivatives is not necessary, and positional information is obtained directly. As a consequence, it is often possible to answer more than one question with one experiment. Although the majority of our examples will deal with ^{13}C NMR, it should be kept in mind that most of the descriptions and characteristics apply to ^{15}N NMR studies as well.

Experimental Design and Acquisition of Data

The natural abundance of ^{18}O is 0.20%, but enrichments of ^{18}O up to 99 atom % are commercially available in the forms of normalized water ($H_2{}^{18}O$), unnormalized water ($D_2{}^{18}O$), and oxygen ($^{18}O_2$). Although the design of experiments involving reactions at carbon–oxygen bonds is largely dictated by the reaction being studied, considerable flexibility in the experimental design does exist. Reactions in which the carbon–oxygen bond is broken and reformed permit complementary experimental designs wherein the oxygen-18 isotope may be incorporated into the substrate or it may be present in the aqueous solvent. When ^{18}O is incorporated into the substrate, the ^{18}O label is replaced by ^{16}O from the normal aqueous solvent as the reaction proceeds, and the label is essentially lost "irreversibly" to the bulk solvent, here termed an out-exchange. An advantage to this experimental design is the conservation of the ^{18}O label. The label is used only during the synthesis of the substrate. A disadvantage is the (possible) difficulty of a synthesis that must result in the incorporation of the oxygen-18 label into the substrate at sufficiently high atom percent excess enrichment.

When the ^{18}O is present in the bulk solvent, the ^{16}O in the substrate is replaced by the ^{18}O label from the solvent as the reaction proceeds, and the label is eventually equilibrated at the atom percent of the ^{18}O enrichment in the aqueous solvent, an in-exchange. An advantage to this experimental design is the relative availability of unlabeled substrate. A disadvantage is the generally larger quantity of ^{18}O-enriched solvent necessary for each experiment (this disadvantage is somewhat offset by the fact that the ^{18}O-labeled water may be recovered and used in subsequent experiments). The atomic percent of ^{18}O in the solvent is easily determined by reaction of the ^{18}O-labeled water with phosphorus pentachloride to form [$^{18}O_4$]phosphoric acid, followed by analysis using ^{31}P NMR or derivatization with diazomethane to form trimethyl[$^{18}O_4$]phosphate and analysis by mass spectrometry (MS) or GC-MS.[12]

There are some kinds of experiments, e.g., point of bond cleavage reactions, where the carbon-oxygen bond is not broken. Although the carbon–oxygen bond may not be broken, the reactants and products may have distinctive spectral characteristics that are measurable. For example, the hydrolytic cleavage of benzyl phosphate by an acid phosphatase from human prostate occurs with P–O bond cleavage and not with C–O bond cleavage. The ^{18}O isotope shifts on the benzyl carbon atom are slightly different in the ester and alcohol and are measurable.[13] The experimental design for these reactions requires that the ^{18}O label be incorporated into the substrate.

As a general rule, we have found that it is preferable to work with substrates or solutions that have an ^{18}O enrichment between 15 and 85%. The level of enrichment depends on the exact experiment. The kinetics of an oxygen exchange reaction are more precisely followed using a higher level of ^{18}O enrichment. On the other hand, a point of bond cleavage reaction may be observed just as easily at a lower level as at a higher level of ^{18}O labeling.

One additional consideration is the use of ^{13}C or ^{15}N enrichment in the substrate. The low natural abundance of ^{13}C (1.1%) and ^{15}N (0.37%) results in the simplification of the NMR spectra by eliminating complicated coupling patterns but at the cost of sensitivity. This is a significant disadvantage in these studies due to the quantity of substrate necessary to obtain an NMR spectrum within a reasonable time. Limitations due to the solubility of the substrate may also arise. Isotopic enrichments of up to 90–99 at. % for ^{13}C and of up to 95–99 at. % for ^{15}N for a number of compounds are commercially available. In general, this makes it possible to purchase ^{13}C-enriched and ^{15}N-enriched compounds, which may be used to synthesize the desired substrates with specific enrichments. The increased enrichment in ^{13}C or ^{15}N that can be incorporated into the substrate decreases by a power of two the quantity of substrate necessary to achieve the same signal-to-noise ratio in the NMR spectrum, and also significantly reduces the complexity of the NMR spectrum, usually to a single signal—that of the enriched site. This can simplify data acquisition and data analysis and can also decrease instrumental time. We recommend the use of ^{18}O-labeled, ^{13}C- or ^{15}N-enriched substrates when at all feasible.

The magnitudes of the ^{18}O isotope effect in ^{13}C NMR and ^{15}N NMR are small and are significantly dependent on the functional group. In ^{13}C

[12] J. M. Risley and R. L. Van Etten, *J. Labelled Compd. Radiopharm.* **15**, 533 (1978).
[13] J. E. Parente, J. M. Risley, and R. L. Van Etten, *J. Am. Chem. Soc.* **106**, 8156 (1984).

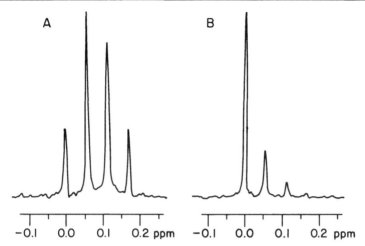

FIG. 1. ^{15}N NMR spectra of the [^{15}N,^{18}O]nitrate ion. Sodium [^{15}N,^{18}O]nitrate (95 at. % ^{15}N) was synthesized. (A) The at. % ^{18}O in this synthesis was 50%. (B) The at. % ^{18}O in this synthesis was 15%. The ^{18}O isotope-induced shift is 56 ppb upfield for each ^{18}O label. The spectra were recorded on a Varian XL-200 NMR spectrometer operating at 20.28 MHz at ambient temperature, using D_2O as a solvent.

NMR, the value of the isotope effect ranges from approximately 15 parts per billion (ppb) for primary alcohols to approximately 54 ppb for ketones. For multiple equivalent isotopic substitutions the isotope shifts are additive. There is also a positional dependence of the isotope effect that is potentially advantageous. The magnitude of the ^{18}O isotope shift on ^{13}C NMR has been reviewed.[14-15] An example of the ^{18}O isotope effect in ^{15}N NMR is shown in Fig. 1. In ^{15}N NMR, the isotope effect is somewhat larger and ranges from 27 ppb for a nitrile oxide to 159 ppb for an isoxazole, per ^{18}O label[16]; the isotope effect is 138 ppb for nitrite ion and 56 ppb for nitrate ion, per ^{18}O[7] (Fig. 1; note also the additivity of the effect).

The relatively small magnitudes of the isotope shifts affect the choice of the magnetic field strength for these studies. While we originally observed the small isotope effects in ^{13}C NMR at 1.9 T (80-MHz proton),[6] we have found that, practically speaking, a 4.7-T (200-MHz proton) magnetic field strength has been preferable for almost all of these studies. Although higher magnetic field strengths may be used, a 200-MHz NMR instrument equipped with a tunable probe is now more routinely available and there is

[14] P. E. Hansen, *Prog. NMR Spectrosc.* **20**, 207 (1988); see also Ref. 14a.

[14a] P. E. Hansen, *Annu. Rep. NMR Spectrosc.* **15**, 105 (1983).

[15] D. A. Forsyth, *in* "Isotopes in Organic Chemistry" (E. Buncel and C. C. Lee, eds.), Vol. 6, p. 23. Elsevier, Amsterdam, 1984.

[16] G. Rajendran, R. E. Santini, and R. L. Van Etten, *J. Am. Chem. Soc.* **109**, 4357 (1987).

seldom a necessity to use higher magnetic field strengths in studies with small molecules and ^{13}C or ^{15}N NMR. (However, natural abundance studies of larger molecules may require the added signal dispersion afforded by higher field instruments.)

To observe the small isotope-induced shifts, and for quantitation, the magnet must be shimmed to a much higher precision than is necessary for routine spectral analysis. It has been possible to shim a 4.7-T magnet (200-MHz instrument) for ^{13}C NMR to achieve baseline resolution between signals for isotope effects of 15–20 ppb (and in some instances 10 ppb). Although considerable time might initially be spent to attain symmetrical peaks with this resolution, it is well worth the effort and we recommend this high degree of resolution. Of course, the necessary resolution also depends on the isotope effect being observed and would generally not be necessary for observation of larger isotope effects. For shimming, we recommend the use of a standard solution such as 40% (v/v) dioxane/D_2O having the same *volume* as that of the reaction solutions; an isotopically substituted standard may also be used. The advantages of this are numerous: (1) knowledge of what the spectrum should be in a good NMR tube, (2) location of the NMR signal with respect to the field, (3) faster shimming using one-pulse spectra, and (4) the ready availability of a standard solution to determine whether an experiment was unsuccessful due to a sample problem or an instrumental problem.

The instrumental setup for these types of experiments is straightforward and generally quite simple. For reactions wherein enriched substrates are used, only the NMR signals for the reactant and product(s) are of concern. The choices of spectral window, data block size, and acquisition time are interrelated. The size of the spectral window (whether quadrature detection or not) will depend on the chemical shifts for the reactant and product(s). These types of experiments generally involve reactions in which the chemical shifts of the reactant and product(s) do not differ significantly. Indeed, for exchange reactions the chemical shifts differ by the magnitude of the total isotope-induced shift, so a small spectral window is used. The data block should be of sufficient size to define each NMR signal with a minimum of approximately 10 data points; the subsequent use of zero-filling in the analysis of the data may affect the choice of the size of the data block. Because it is often possible to use a 512- or 1K-size data block for these experiments, the storage disks should be configured to accommodate as many data blocks of this size as possible. The acquisition time is a function of the spectral window and the size of the data block, and should be used in conjunction with the spin–lattice relaxation (T_1) times of the reactant and product(s) to obtain the necessary total delay time. The T_1 times for the reactant and product(s) should be known under the conditions of the reactions. The presence of an ^{18}O label does not change the T_1 time of the ^{13}C or ^{15}N nucleus; this has been demonstrated experimentally for a number of ^{18}O-labeled, ^{13}C-enriched compounds,[17] and confirms what is expected theoretically because ^{18}O has no spin. Consequently, it is possible to use natural abundance, unlabeled compounds to measure the T_1 times for the reactant and product(s). In order to attain full relaxation of the nucleus, a minimum total delay time of five times the longest T_1 should be used between pulses. Thus, knowing the longest T_1 time will permit an optimal selection of the spectral window, data block size, acquisition time, and delay time. A 90° pulse should be used and protons should be gated decoupled to maximize the ability to quantitate accurately the reaction.[18] These are the most general experimental conditions and they may be modified depending on the specific experiment being done.

The probe should be allowed to equilibrate at the desired reaction temperature for at least 30 min. The temperature should not be expected to be maintained at better than ±1° and should be checked using a chemical shift thermometer[19–21] or a thermometer that can be inserted directly into the solution in the NMR tube. The minimal number of pulses needed for a spectrum is determined by the concentration of the substrate, the sensitivity of the instrument (as measured by the signal-to-noise ratio), and the phase cycling of the pulse sequence. A high signal-to-noise ratio is necessary for quantitation of the spectrum because the baseline detection limit is generally approximately 5% of the largest signal in a spectrum with a high signal-to-noise ratio, but, of course, is much poorer in a spectrum with a low signal-to-noise ratio.

Once the parameters for the acquisition of each spectrum are established, the number of spectra required to follow three to four half-lives of the reaction and the total experimental time may be estimated. The acquisition of the spectra should be under computer control in order to maintain a consistent set of parameters, and each half-life of the reaction is probably best defined by approximately 10 data points (NMR spectra). For reactions wherein substrates at natural abundance are used, it is particularly important to remember that fold-back of other NMR signals in the spectrum may interfere or distort the NMR signal of interest if the offset and spectral window are not carefully chosen; otherwise, the instrumental setup is the same.

[17] T. L. Mega, Ph.D. thesis, "Quantitative and Mechanistic Studies Using the Oxygen-18 Isotope Shift in Carbon-13 Nuclear Magnetic Resonance Spectroscopy." Purdue University, West Lafayette, Indiana, 1989.

[18] J. N. Schoolery, *Prog. Nucl. Magn. Reson. Spectrosc.* **11**, 79 (1977).

[19] A. L. Van Geet, *Anal. Chem.* **40**, 2227 (1968).

[20] A. L. Van Geet, *Anal. Chem.* **42**, 679 (1970).

[21] D. S. Raiford, C. L. Fisk, and E. D. Becker, *Anal. Chem.* **51**, 2050 (1979).

The instrument specifications should be consulted to determine the sample volume recommended for optimal resolution. For small volumes a vortex plug may be required to prevent vortexing into the detection coil while the sample is spinning. An alternative is to use a bulb insert centered in the detection coil. The concentration of the substrate that is needed is determined by the ^{13}C or ^{15}N isotopic enrichment of the compound, by the sensitivity of the instrument, by the acquisition time for each spectrum, and by the half-life of the reaction. This concentration must be determined for each reaction system. A minimum of 10% (v/v) deuterium solvent must be present for instrumental lock. A coaxial or a bulb insert in the NMR tube may be used to keep the lock solvent and reaction solution separate, thus avoiding kinetic deuterium isotope effects. If an enzyme is a part of the reaction solution, the stability of the enzyme must also be established. Solutions used to initiate the reaction, e.g., addition of enzyme, substrate, acid, or base, should be equilibrated at the temperature for the reaction prior to addition (at least 30 min is recommended).

Analysis and Interpretation of Data

The isotope shift is contained in the "tail" of the FID. Thus, the choice of a weighting function to be applied to the FID must be made carefully. The application of an exponential function to the FID permits observation of the isotope effect and quantitation of the NMR signals. The use of other weighting functions, such as a Gaussian function, permits observation of the isotope effect, but can present difficulties in the quantitation of the spectra; thus, if a weighting function other than an exponential one is used, we suggest that stringent controls be used to assure that the NMR signals are quantitated accurately. Often it is advantageous to zero-fill the data block once before Fourier transform of the FID (note, however, that it is possible zero-fill the data block more than once, but doing so generally does little to improve the spectral quantitation). An absolute intensity (amplitude) calculation of the same magnitude should be applied to each spectrum in order to observe correctly the relative intensities of the signals between spectra and to quantitate accurately (proportionally) the spectra.

The area under each peak of interest is needed and may be determined in three ways: by integration, a curve simulation (deconvolution) routine, and peak height. The areas under the peaks are not absolute quantities, but are all relative values. In order to determine areas by integration or by a curve simulation routine, it is necessary to define an area as some number. Typical peak printout routines assign arbitrary height values to the peaks in the spectrum relative to the digitization of the spectrum;

a relative area for the peaks may be calculated knowing the resolution (width at half-height) of the signals. It is frequently most convenient to define the total relative area as 100 ("100%"). Thus, for example, the sum of the areas of the NMR signals for the ^{18}O and ^{16}O isotopic species is defined as 100 and the relative areas ("quantities") of each species are easily calculated. As an oxygen exchange reaction proceeds, the relative areas of each species may be expected to change as a function of time, but the total relative area is considered to remain at 100. As may be shown mathematically, when relative areas computed from peak height measurements are converted to a percentage of the total area, the resultant number is dependent only on the peak height of the NMR signal, providing that the linewidth of the NMR signal for each isotopic species is the same, as is expected to be the case, especially for oxygen exchange reactions. Therefore, the peak heights may be used directly. Using the three methods of peak area measurement, we have found that the relative area under each peak may be calculated to an accuracy of ±3%. The accuracy of these calculations is for the measurement of the relative peak area above the baseline of the spectrum. Near the baseline of the spectrum, the measurement of the relative area of the individual species is highly dependent on the signal-to-noise ratio of the spectrum, but the accuracy tends to decrease to approximately ±7 to ±10%. The accuracy with which the relative area of an NMR signal may be determined is dependent on the detection limits for the different isotopic species. These are the reasons why we have found it preferable to work with substrates or solutions that have an ^{18}O enrichment between 15 and 85%. Although an isotopic species present at 5% may be detected, we have found that in general a practical lower limit of 15% may be reliably detected with minimal error. Furthermore, a 5% change may be quantitated more accurately for changes of 85–80% or 15–20% than for 100–95% or 0–5%. Therefore, excellent experimental data may be obtained for oxygen exchange reactions using an ^{18}O-enriched substrate in normal water by following the loss of the ^{18}O label where the greatest changes in the spectra are observed for the first three half-lives of the reaction. The quantitation of the spectra support the design of these type of experiments using an ^{18}O-enriched substrate and following the reaction in normal water, although we have followed the incorporation of the ^{18}O from the solvent into the substrate with similar accuracy.

The interpretation of the resulting data involves testing proposed models for the reaction and subsequently refining the model according to the experimental data. Based on the model, a rate constant for the reaction is computed from the data using an appropriate plot of the data as a function of time. Time points for each spectrum are computed from the midpoint of the acquisition of each FID. A hydrolysis reaction is easy to analyze using standard models to obtain a rate constant. For oxygen exchange reactions, a valid assumption that often may be made to simplify a model is that the exchange reaction is effectively irreversible due to the large difference in concentration between the aqueous solvent and the substrate. Exchange reactions that involve two isotopic species (e.g., alcohols, aldehydes, and ketones) are analogous to hydrolysis reactions and therefore may be analyzed using standard models to obtain a rate constant.

Two basic models may be proposed for exchange reactions which involve more than two isotopically substituted species [e.g., three isotopic species of carboxylic acids, three isotopic species of nitrite, and four isotopic species of nitrate (cf. Fig. 1)]. These basic models are random sequential exchange and multiple exchange.

Random sequential exchange reactions are simply the exchange of one oxygen per event such that the exchange reaction appears as a step exchange. For example, an exchange reaction for carboxylic acids would proceed from an isotopic species containing two ^{18}O to an intermediate form with one ^{18}O and one ^{16}O and finally to a species with two ^{16}O. The differential equations describing this model have been solved.[22,23] In order to calculate a pseudo-first-order rate constant for this model, the set of equations is reduced to a first-order process by computing the total percentage ^{18}O in each spectrum and plotting it as a function of time, from which the pseudo-first-order rate constant is easily obtained.

Multiple exchange reactions are those where more than one exchange reaction occurs per event. An entire range of exchange reactions is possible. For example, any number of exchange reactions for carboxylic acids between one and two is possible; although two is the maximum number that can be detected, additional exchange reactions are possible but would not be detected.

The set of equations describing random sequential exchange and multiple exchange for carboxylic acids may be combined to yield the following set of solved differential equations

$$C_2(t) = C_2(0)e^{-kt} \tag{1}$$
$$C_1(t) = [C_1(0) + 2C_2(0)]e^{-kt/(2-R)} - 2C_2(0)e^{-kt} \tag{2}$$
$$C_0(t) = [C_0(0) + C_1(0) + C_2(0)]$$
$$\qquad - [C_1(0) + 2C_2(0)]e^{-kt/(2-R)} + C_2(0)e^{-kt} \tag{3}$$

where C_2 represents $[RC^{18}O_2H]$, C_1 represents $[RC^{18}O^{16}OH]$, C_0 represents $[RC^{16}O_2H]$, $C_n(0)$ for $n = 0-2$ is the initial concentration of each

[22] J. M. Risley, Ph.D. thesis, "Medium Phosphate(Oxygen)–Water Exchange Reaction Catalyzed by Acid Phosphatases: The Oxygen-18-Isotope Effect in Carbon-13 Nuclear Magnetic Resonance Spectroscopy." Purdue University, West Lafayette, Indiana, 1980.
[23] J. M. Risley and R. L. Van Etten, J. Am. Chem. Soc. 103, 4389 (1981).

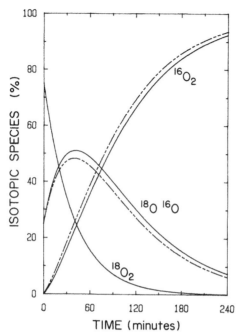

FIG. 2. Theoretical curves for the oxygen exchange reaction of carboxylic acids. The oxygen exchange reaction of carboxylic acids might proceed either by random sequential exchange of the oxygen atoms or by multiple exchange of the oxygen atoms. This plot illustrates the differences observed between a random sequential exchange (——) and a reaction that proceeds by 10% multiple (coupled) exchange (----). There is no difference between the curves for the exchange reactions of the carboxylic acid labeled with two ^{18}O atoms. The differences between the exchange reactions arise in the species containing one ^{18}O and no ^{18}O. Although 10% multiple exchange is not large, the difference between the curves for the two reactions should be readily detectable.

isotopic species, k is the pseudo-first-order rate constant for the exchange reaction, and R is the ratio of multiple exchange to random sequential exchange. The criteria that are satisfied are that as $R \rightarrow 0$ the set of equations reduces to the set of equations describing random sequential exchange and as $R \rightarrow 1$ the set of equations reduces to the set of equations describing total multiple (coupled) exchange. Figure 2 illustrates the differences expected for a reaction undergoing random sequential exchange and for a reaction where there is 10% multiple exchange, i.e., $R = 0.1$. The exchange reaction of the doubly labeled species is the same in both exchange reactions, and the differences arise in the reactions of the other two species. The plot was made with $C_2(0) = 75$, $C_1(0) = 25$, and $C_0(0) = 0$. It is apparent that there is a detectable difference between these two exchange reactions, which becomes even more dramatic as the percentage of multiple exchange increases. The equations for multiple exchange reactions (defined slightly differently) for the phosphate(oxygen)–water exchange reaction have been solved.[22,24] By analogy to these examples, the equations describing a model for multiple exchange in other reactions may be solved and the rate constants for the reactions may be calculated from the experimental data.

Theoretical curves for the different isotopic species in the exchange reaction are calculated using the equations describing the model and compared to the actual experimental data. Refinements to the model may be made by iterations of the computed curves to minimize the error between the experimental data and the model.

Problems and Cautionary Notes

An accurate measurement of the spin–lattice relaxation time (T_1) for the NMR signals being observed under the experimental conditions of the reaction is necessary to assure that sufficient time is allowed for the nucleus to relax fully between pulses (approximately $5T_1$). Because many things may affect the quality of the spectra, care should be exercised to minimize paramagnetic ion contamination (by passing the solution through a small column of a chelating agent, bubbling nitrogen through the solution to remove dissolved oxygen, etc.), to remove particulate matter in the solution (by filtering the solution), to eliminate thermal gradients, and to prevent the formation of bubbles in solution at elevated temperatures (by equilibration of the solution to the approximate temperature of the probe using an external bath before the sample is inserted into the probe). The probe temperature should be calibrated with a thermometer or with a chemical thermometer. Note that the probe temperature and the temperature of the reaction mixture in the NMR tube should be the same, but experimental conditions, such as high ionic strength solutions studied at high decoupling power, can lead to large differences between the probe temperature and the solution temperature. Thus, considerable effort must be expended to determine an accurate temperature and to maintain it throughout the reaction solution. A standard NMR sample should be prepared for which the spectrum is known. The value of a standard sample is that it is possible to deduce quickly whether problems with a spectrum are due to the sample or the instrument.

Care must be taken when analyzing spectra, since the presence of different isotopic species can result in changes of spectral properties. For example, the NMR spectrum of [1,4-^{13}C$_2$]succinic acid partially labeled

[24] D. D. Hackney, K. E. Stempel, and P. D. Boyer, this series, Vol. 64, p. 60.

with ^{18}O is the summation of the NMR signals for three symmetric isotopic species plus three AB spectra for three unsymmetric isotopic species, which results in a very complicated spectrum.[17] The NMR spectrum
of [^{15}N$_2$]dimethylglyoxime partially labeled with ^{18}O shows single NMR
signals for the symmetric isotopomers and an AB pattern for the unsymmetric isotopomer.[16] Also, the isotope shift of the carboxyl carbon in
carboxylic acids is a function of the ratio of the acid to the anion and is a
maximum at the pK_a of the acid.[17,25]

Finally, it is possible that under some reaction conditions, exchange
broadening of the NMR signal may occur, which may obscure the isotope
shift.

Examples

The hydrolysis of 2,2-dimethyloxirane by the enzyme epoxide hydrolase (EC 3.3.2.3) is a classic example of the use of the ^{18}O shift in ^{13}C
NMR. The 2,2-dimethyl[3-^{13}C,^{18}O]oxirane was synthesized containing 90
at. % ^{13}C and 48 at. % ^{18}O.[26] A preparation of mouse liver microsomes, which contains the enzyme epoxide hydrolase, and the ^{13}C-en-
riched, ^{18}O-labeled oxirane was incubated at pH 7.9 and 35°. ^{13}C NMR
spectra were recorded as the epoxide hydrolase catalyzed the hydrolysis
of the oxirane to 2-methylpropane-1,2-diol; selected spectra from the reaction are shown in Fig. 3 in a stacked plot with the time point for each
spectrum given on the right. Under these conditions there is no nonenzy-
matic hydrolysis of the oxirane, so only the enzyme-catalyzed reaction is
observed. The magnitude of the ^{18}O isotope effect on the primary carbon
in the oxirane is 31 ppb. Figure 3 shows that the enzyme catalyzes the
hydrolytic cleavage of the primary carbon–oxygen bond and not the tertiary carbon–oxygen bond. This is clearly evident by comparison of the
^{18}O isotope shift in the substrate (oxirane) with the absence of an ^{18}O
isotope shift in the ^{13}C NMR signal for the primary carbon atom in the
product diol, which appears upfield from the ^{13}C NMR signals for the
substrate. (In the complementary experiment involving incubation of
mouse liver microsomes with 2,2-dimethyl[3-^{13}C]oxirane in ^{18}O-labeled
water, an ^{18}O-isotope effect of 19 ppb is observed for the primary carbon
atom in the product diol.) Thus, the data from this experiment are used to
answer simultaneously two questions: what is the site of bond cleavage
and what is the rate of the enzyme-catalyzed hydrolysis reaction? The

[25] S. L. R. Ellison and M. J. T. Robinson, *J. Chem. Soc., Chem. Commun,* 745 (1983).
[26] J. M. Risley, F. Kuo, and R. L. Van Etten, *J. Am. Chem. Soc.* **105,** 1647 (1983).

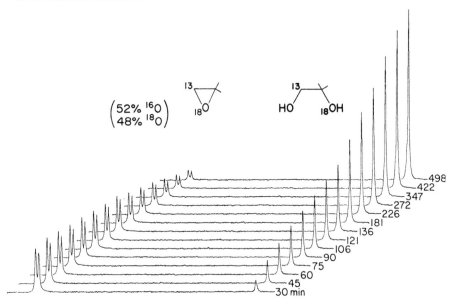

FIG. 3. Hydrolysis of 2,2-dimethyl[3-^{13}C,^{18}O]oxirane at pH 7.9 and 35° by mouse liver microsomes. Epoxide hydrolase in mouse liver microsomes catalyzes the hydrolysis of the oxirane at the primary carbon–oxygen bond to yield the diol product. The ^{18}O isotope shift on the primary carbon atom in the oxirane can be seen, but is absent in the product diol. Not only do these data show the point of bond cleavage, but from these data the pseudo-first-order rate constant for the hydrolysis may be calculated. The spectra were recorded at 50.31 MHz on a 200-MHz spectrometer.

answer to the first question is clearly indicated by the spectra. The answer to the second question may be calculated directly from the NMR data ($k = 7.21 \times 10^{-5}$ sec^{-1}).

D-Erythrose will undergo an oxygen exchange reaction at the anomeric carbon atom, and this reaction may be followed very easily by using the ^{18}O isotope shift on ^{13}C NMR.[27] The reaction system was chosen to follow the in-exchange of ^{18}O from the solvent water. D-[1-^{13}C]Erythrose was dissolved in unbuffered 60% H$_2$^{18}O at 23°, and ^{13}C NMR spectra were collected as a function of time. The series of spectra recorded at 10-min intervals for the β-D-[1-^{13}C]erythrofuranose anomer are shown in Fig. 4. As the exchange reaction proceeds, the ^{13}C NMR signal for the anomeric carbon with the ^{18}O label appears 19 ppb upfield, and the reaction eventually reaches equilibrium at the atomic percent ^{18}O of the solvent (60%).

[27] J. M. Risley and R. L. Van Etten, *Biochemistry* **21**, 6360 (1982).

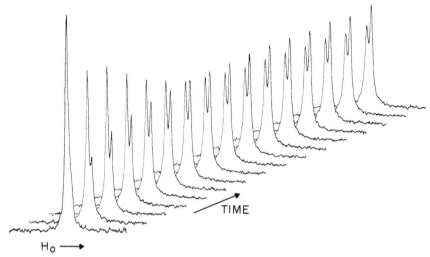

$H_0 \longrightarrow$

FIG. 4. Oxygen exchange at the anomeric carbon in D-erythrose. The series of spectra were taken of the β-D-[1-^{13}C]erythrofuranose anomer of D-erythrose during the oxygen exchange reaction in unbuffered 60% $H_2{}^{18}O$. The oxygen exchange at the anomeric carbon atom is easily followed by ^{13}C NMR and these data are easily quantitated to give a pseudo-first-order rate constant for the reaction. The spectra were recorded at 50.31 MHz on a 200-MHz spectrometer.

From these data, the rate of oxygen exchange at the anomeric carbon is calculated using standard first-order equations to obtain the pseudo-first-order rate constant (4.81×10^{-4} sec^{-1}).

Numerous other examples from organic chemistry and enzymology could be cited, but many of them can easily be found by referring to earlier publications from this laboratory,[28] or to a forthcoming review.[29]

Acknowledgments

Preparation of this paper and work in this laboratory was supported in part by DHHS Grant GM 27003 from the National Institute of General Medical Sciences and by DHHS Grant RR01077 and NSF Grant BBS-8714258 in support of the spectrometer facility.

[28] K. H. Röhm and R. L. Van Etten, *Eur. J. Biochem.* **160,** 327 (1986).
[29] J. M. Risley and R. L. Van Etten, *in* "NMR—Basic Principles and Progress" (H. Günther, ed.). Springer-Verlag, Berlin and New York, 1989.

[20] Positional Isotope Exchange Using Phosphorus-31 Nuclear Magnetic Resonance

By Joseph J. Villafranca

Introduction

The chemical shift of a phosphorus atom relative to other phosphorus atoms depends on the induced magnetic fields of the circulating electrons in the atomic environment. The immediate environment obviously includes the directly attached atoms and the substitution of one atomic isotope for another would be expected to alter the electronic distribution in the bond joining the two atoms. This phenomenon is the basis for the isotopic shift difference observed in ^{31}P NMR spectra of phosphorus-containing compounds when ^{18}O is substituted for ^{16}O.[1] A spectrum of inorganic phosphate is presented in Fig. 1 illustrating the $^{18}O/^{16}O$ isotopic shift.

From the initial observations of a ~0.02-ppm upfield shift in the ^{31}P NMR resonance per oxygen ^{18}O substituted for ^{16}O in inorganic phosphate,[2,3] the isotopic shift in several phosphate-containing species has been studied.[1,4,5] Table I lists ranges of chemical shifts for phosphate esters as well as specific values measured for bridge versus nonbridge positions in selected adenine nucleotides. For simple esters, the ^{18}O-induced chemical shift follows bond order, with a larger shift found for a $P{=}O$ double bond than for a $P{-}O$ single bond; the shift is generally additive. There is also a difference for the ^{18}O-induced shift for ^{18}O in a bridge versus nonbridge position in nucleotides (Table I).

Based on the chemical shift difference for ^{18}O- and ^{16}O-containing phosphate species, ^{31}P NMR investigations of several biochemical systems have been reported and summarized.[1,4] The main use of the NMR experiments discussed in this chapter exploits the method known as positional isotope exchange, or PIX, with enzymatic substrates such as ATP. For enzymatic reactions, if a mechanism can be written involving transfer of an atom from one molecule to another, then the movement of this atom

[1] C. W. DeBrosse and J. J. Villafranca, *in* "Magnetic Resonance in Biology" (J. S. Cohen, ed.), p. 1. Wiley, New York, 1983.

[2] M. Cohn and A. Hu, *Proc. Natl. Acad. Sci. U.S.A.* **75**, 200 (1978).

[3] O. Lutz, A. Nelles, and D. Staschewski, *Z. Naturforsch. A* **33**, 380 (1978).

[4] F. M. Raushel and J. J. Villafranca, *CRC Crit. Rev. Biochem.* **23**, 1 (1988).

[5] W. von der Saal and J. J. Villafranca, *Bioorg. Chem.* **14**, 28 (1986).

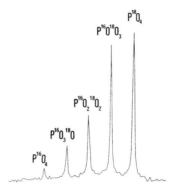

FIG. 1. ^{31}P NMR spectrum of inorganic phosphate at pH 9.0. The data were recorded at 145 MHz. Each labeled peak represents the ^{18}O and ^{16}O composition of the species present in solution. The sample contained 78% ^{18}O and 22% ^{16}O, statistically distributed among the five species for this isotopic content.

among enzyme-bound substrates, intermediates, or products may be followed, in favorable cases, by the substitution of one isotope for another, e.g., ^{18}O for ^{16}O. Several conditions have to be met for the successful observation of this phenomenon and these are discussed in other reports.[1,4] The main limitations for observing PIX in an enzymatic reaction occur when (1) the release of the substrate undergoing the PIX reaction (e.g., ATP) is slow relative to the overall reaction, (2) the phosphate group does not have rotational mobility on the enzyme, and (3) all substrates have to be present for the chemical reaction to occur and PIX is not observable because of condition 1 or 2. More detailed explanations of these limitations are given elsewhere.[4] The main thrust of this chapter deals with the details of sample preparation, experimental conditions for NMR observation of the isotopic shift, and, finally, data analysis.

Sample Preparation: Experimental Conditions

The limitations of the ^{31}P NMR method for observing PIX reactions arise from two primary instrumental considerations: (1) resolution and (2) sensitivity. For most configurations of NMR probes (10 mm) designed to observe the ^{31}P nucleus, about 1 ml of sample is required. The current sensitivity limitations require that the sample contain 0.5 to 10 μmol of phosphorus-containing material. Data gathered with 1.7 mM [γ-^{18}O$_4$]ATP illustrates typical signal-to-noise (S/N) conditions (Fig. 2).

To begin a PIX experiment, the ^{18}O-containing substrates must be prepared. $H_3P[^{18}O_4]$ is readily synthesized from PCl_5 and $H_2^{18}O$ following

TABLE I
CHEMICAL SHIFTS OF ^{31}P NMR RESONANCES DUE TO
[18]O IN VARIOUS PHOSPHATES[a]

Compound	$\Delta\delta/^{18}O$ (ppm)
$HP^{18}\bullet_4^{2-}$	0.021
$(CH_3O)_3P{=}^{18}\bullet$	0.036
$(CH_3O)_2P^{18}\bullet_2^-$	0.029
$(CH_3O)P^{18}\bullet_3^{2-}$	0.023

$$
\begin{array}{c}
\text{O} \\
\| \\
\text{AMP}{-}^{18}\bullet{-}\text{P}{-}\text{O} \\
| \\
\text{O}
\end{array}
\qquad 0.0205
$$

$$
\begin{array}{c}
^{18}\bullet \\
| \\
\text{AMP}{-}\text{O}{-}\text{P}{-}\text{O} \\
| \\
^{18}\bullet
\end{array}
\qquad 0.0227
$$

$$
\begin{array}{c}
\text{O} \\
\| \\
\text{ADP}{-}^{18}\bullet{-}\text{P}{-}\text{O} \\
| \\
\text{O}
\end{array}
\qquad 0.0206
$$

$$
\begin{array}{c}
^{18}\bullet \\
| \\
\text{ADP}{-}\text{O}{-}\text{P}{-}\text{O} \\
| \\
\text{O}
\end{array}
\qquad 0.0225
$$

$$
\begin{array}{c}
^{18}\bullet \\
| \\
\text{AMP}{-}^{18}\bullet{-}\text{P}{-}^{18}\bullet \\
| \\
^{18}\bullet
\end{array}
\qquad
\begin{array}{l}
0.0166\ (\alpha,\beta\ \text{bridge}); \\
0.0215\ (\text{nonbridge})
\end{array}
$$

$$
\begin{array}{c}
^{18}\bullet \qquad\quad \text{O} \\
| \qquad\qquad \| \\
\text{AMP}{-}^{18}\bullet{-}\text{P}{-}\text{O}{-}\text{P}{-}\text{O} \\
| \qquad\qquad | \\
^{18}\bullet \qquad\quad \text{O}
\end{array}
\qquad
\begin{array}{l}
0.0172\ (\alpha,\beta\ \text{bridge},\ \alpha\text{-P}); \\
0.0281\ (\text{nonbridge})
\end{array}
$$

[a] Data from M. Cohn [*Annu. Rev. Biophys. Bioeng.* **11**, 23 (1982)] and from W. von der Saal and J. J. Villafranca [*Bioorg. Chem.* **14**, 28 (1986)].

the method of Risley and Van Etten.[6] The $H_3P[^{18}O_4]$ is then used in several synthetic procedures to produce the ^{18}O-labeled ATP of choice. The reaction below illustrates the reaction monitored by ^{31}P-NMR in Fig. 2 for the enzyme CTP synthase.

[6] J. M. Risley and R. L. Van Etten, *J. Labelled Compd. Radiopharm.* **15**, 533 (1978).

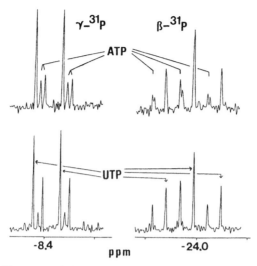

FIG. 2. ^{31}P NMR spectra of the γ-P and β-P regions of a solution containing UTP and [γ-^{18}O$_4$]ATP. The bottom spectrum represents a solution of 2 mM UTP and 1.8 mM [γ-^{18}O$_4$]ATP (95 at. % ^{18}O) prepared as described by von der Saal et al.[8] at $t = 0$ before reaction with CTP synthase. The distribution of γ-P[^{18}O$_4$] : P^{16}O[^{18}O$_3$] was 81 : 19 as observed at −8.4 ppm. The top spectrum is at $t = 120$ min after the PIX reaction has progressed. The γ-P[^{18}O$_4$] : P^{16}O[^{18}O$_3$] region of the spectrum has changed, as has the β-P region (−24 ppm). Upfield peaks are now found corresponding to ^{18}O in the β-P nonbridging positions (see Reynolds et al.[13] for an excellent spectrum demonstrating β-P bridge and nonbridge ^{18}O shifts in ATP and also the β-P region of ATP in Fig. 4 of this chapter).

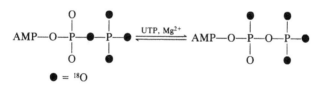

Because well-resolved peaks are required for quantification of the PIX rate, the instrument should first be shimmed on a concentrated sample of $H_3P^{18}O_4$. The concentration of $H_3P^{18}O_4$ has to be at least 1 M so that the shimming can be done using single pulses. This sample should be dissolved in a 50/50 mixture of D_2O/H_2O and the pH adjusted to 9.0. EDTA (10–100 mM) is added to chelate metal ions. Figure 1 shows a sample prepared to illustrate the ~0.02-ppm (2.9 Hz at 145 MHz) difference among the various ^{18}O- and ^{16}O-containing species in a sample of H_3PO_4 (78% ^{18}O, 22% ^{16}O). The spectrum also illustrates that a well-shimmed sample can have linewidths at half-height of ≤0.2 Hz. For an NMR spectrometer operating at 80 MHz, the peak-to-peak separation is 1.6 Hz, whereas at 160 MHz the separation is 3.2 Hz. Thus a wide range of spectrometers can be used to obtain useful data. After the instrument is shimmed on the $H_3P^{18}O_4$ sample the spectrum of $[\gamma\text{-}^{18}O_4]ATP$ can be recorded. The $[\gamma\text{-}^{18}O_4]ATP$ can be synthesized by the method of Midelfort and Rose.[7] The $[\gamma\text{-}^{18}O_4]ATP$ sample should also be dissolved in a 50/50 mixture of D_2O/H_2O and the pH should be adjusted to 8.5. EDTA can be added to chelate metal ions or the $[\gamma\text{-}^{18}O_4]ATP$ sample can be passed through Chelex.

The enzymatic reaction is carried out using $[\gamma\text{-}^{18}O_4]ATP$ as the substrate.[8] The reaction is quenched by the addition of a solution of 1 M Tris buffer, pH 8.5, and EDTA at a concentration that is sufficient to chelate the metal present in the reaction buffer. Protein is removed by the addition of several drops of CCl_4 followed by vortexing and centrifuging. D_2O is added to the supernatant so that the final concentration of D_2O is 50% and then the sample is filtered through glass wool into an NMR tube.

Alternatively, EDTA and Tris buffer, pH 8.5, can be added to the solution, followed by CCl_4 and vortexing. In place of H_2O/D_2O, $[^2H_6]$acetone (400 μl) can be added and the solution passed through glass wool into a 10-mm NMR tube. The linewidths are generally sharper when the sample is shimmed on $[^2H_6]$acetone rather than on D_2O.

For the data with CTP synthase shown in Fig. 2, the ^{31}P NMR spectra were recorded at 145 MHz on a Brüker WM 360 spectrometer using a sweep width of 5 kHz and an acquisition time of 1.6 sec. The digital resolution can be markedly improved if zero-filling is used, and typically zero-filling to 64 K can be done before Fourier transformation. With good digital resolution the areas under the peaks can be used to measure the relative amounts of each $^{18}O/^{16}O$-containing species for use in calculating the PIX rate. Alternatively, the peak heights can be used to determine the amount of each species when peaks overlap. However, extreme caution must be employed when using this method due to the artifacts that can be introduced when peak-smoothing routines are used to improve the S/N ratio of spectra.

Taking the spectra in Fig. 2 as an example, the PIX reaction catalyzed by CTP synthase was followed using $[\gamma\text{-}^{18}O_4]ATP$ in the presence of UTP. Each phosphorus resonance of UTP and ATP can be resolved at 145 MHz[9]; the γ (about -8.4 ppm) and β (about -24.0 ppm) resonances are presented in Fig. 2. In this experiment the ^{18}O content was 95 at. %

[7] C. F. Midelfort and I. A. Rose, *J. Biol. Chem.* **251**, 5881 (1976).

[8] W. von der Saal, P. M. Anderson, and J. J. Villafranca, *J. Biol. Chem.* **260**, 14993 (1985).

[9] W. von der Saal, J. J. Villafranca, and P. M. Anderson, *J. Am. Chem. Soc.* **107**, 703 (1985).

in each of the four positions of the γ-P of ATP, and this results in an $81:19$ distribution of $P[^{18}O_4]:P^{16}O[^{18}O_3]$. If the PIX reaction went to complete equilibrium, then the starting ^{18}O in the β,γ-P–O–P bridge position would be two-thirds in the β nonbridge position and one-third in the β,γ-P–O–P bridge position at equilibrium. The beginning distribution of $81:19$ ($P[^{18}O_4]:P^{16}O[^{18}O_3]$) would be $26:74$ at equilibrium. The bottom spectrum in Fig. 2 was taken at $t = 0$ at the top spectrum at $t = 120$ min. The top spectrum shows a distribution of $55:45$ after 120 min and during this period of time 35% of the ATP was hydrolyzed (measured relative to the UTP peaks, because UTP is not consumed during this reaction).

The PIX rate is calculated using the equation of Litwin and Wimmer[10]

$$V_{PIX} = \frac{X}{\ln(1 - X)} \frac{A_0}{t} \ln(1 - F)$$

where A_0 is the initial concentration of ATP, X is the fraction of substrate consumed at time t, and F is the fraction of the equilibrium value for PIX. The experiment in Fig. 2 was begun with 1.8 μmol of ATP and thus

$$V_{PIX} = \frac{0.35}{\ln(0.65)} \times \frac{1.8}{120} \ln(1 - 0.46) = 0.0075 \ \mu mol/min$$

Quite often the PIX rate is given in units of micromoles/minute/milligram of enzyme so that it can be compared to V_{max} (micromoles/minute) or the rate of a partial reaction catalyzed by the enzyme. When this is desired each rate is normalized by the amount of enzyme used so that V_{PIX}/V_{max} can be expressed as a dimensionless number. For many enzymatic systems individual rate constants are not known and the ratio V_{PIX}/V_{max} is a useful kinetic measure of the relative extent of formation of an intermediate in the enzymatic reaction. The ratio V_{PIX}/V_{max} is a combination of rate constants relating not only the off rates of substrate (e.g., ATP undergoing PIX) and products, but also the partitioning of the intermediate. Specific examples will follow to illustrate this.

Data Analysis

Using the PIX experimental approach, evidence for intermediates in enzymatic reactions can be obtained. The nature of the experimental design requires that a bond be broken and reformed (e.g., in ATP as given above) to observe the phenomenon of positional isotope exchange. Below are given mathematical expressions for several enzyme mechanisms to

---

[10] S. Litwin and M. J. Wimmer, *J. Biol. Chem.* **254**, 1859 (1979).

illustrate the diversity of questions that can be approached using PIX methodology. The derivation of each kinetic model follows the approach of Cleland[11] using net rate constants.

Simple System

A simple kinetic scheme involving enzyme, E, and substrate, S, that can undergo PIX and product, P, is given here:

$$E \underset{k_2}{\overset{k_1S}{\rightleftharpoons}} ES \underset{k_4}{\overset{k_3}{\rightleftharpoons}} EP \overset{k_5}{\rightarrow} E + P$$

In a general reaction involving ATP

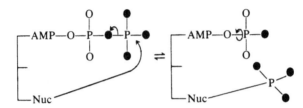

the attack by a nucleophile on the γ-P gives an enzyme-bound intermediate. Reversal of the reaction and release of ATP gives the observed PIX and the rate is measured as presented in the previous section. The rate of PIX is determined by the net rate constant k'_4 for reversal of the reaction from the EP complex and includes ATP molecules that may have, but not necessarily must have, undergone PIX (remember that limitations of the rotational freedom of the bound phosphate may reduce or prevent observation of PIX even though an intermediate is formed). Thus,

$$V_{PIX} = k'_4[EP] = k_2 k_4[EP]/(k_2 + k_3)$$

and using the steady-state assumption and conservation equation for enzyme,[1] one can arrive at

$$\frac{V_{PIX}}{E_T} = \frac{k_2 k_3 k_4}{(k_2 + k_3)(k_3 + k_4 + k_5)}$$

With $V_{max}/E_T = k_5[EP] = k_3 k_5/(k_3 + k_4 + k_5)$, one can derive the ratio

$$\frac{V_{PIX}}{V_{max}} = \frac{k'_4}{k_5} = \frac{k_2 k_4}{k_5(k_2 + k_3)}$$

Without additional data, these experiments provide the ratio of rate constants relating the partitioning of the EP complex (i.e., k'_4/k_5) between

---

[11] W. W. Cleland, *Biochemistry* **14**, 3220 (1975).

PIX and net turnover. The individual values of each rate constant can be obtained if rapid-quench kinetic data are combined with the PIX data, and this has been done for several enzymes.[4] Lacking these experiments assumptions can be made to determine a range of reasonable rate constants from the PIX and V_{max} data.

For the case $k_2 > k_3$, $V_{PIX}/V_{max} = k_4/k_5$. If the measured value of V_{PIX}/V_{max} is 0.03 (low extent of PIX) and V_{max} is 100 min^{-1}, then $k_4 = 3$ min^{-1} and $k_5 = 100$ min^{-1}. Limits can be placed on the other rate constant, i.e., $k_3 > k_5$ because k_5 must be rate limiting. Assuming that $k_3 = 300$ min^{-1} and $k_2 = 3000$ min^{-1} ($k_2 > k_3$) for the case under consideration, one can easily understand why PIX is observed but at only 3% of the rate of V_{max}. When ES $\rightarrow$ EP occurs, the isotope-containing bonds are broken; reformation of S, EP $\rightarrow$ ES, and consequently positional isotope exchange takes place only 3% of the time, even though escape of S from ES is highly probable ($k_2 > k_3$). The alternative $k_3 > k_2$ would preclude observation of PIX and need not be considered.

More Complex Systems

In the past several years many extensions of the isotope exchange methodology have been reported to further explore enzyme mechanisms.[8,12–14] Data about the partitioning of enzyme-bound intermediates, about the order of addition of substrates or release of products, and about establishing the timing of formation of reaction intermediates have been obtained with PIX studies using ^{31}P NMR.

The order of addition of substrates in the enzyme CTP synthase was studied using [γ-^{18}O$_4$]ATP. Figure 2 shows data from these experiments. The question addressed in this study was whether ATP or UTP bound in a preferred order or randomly to the enzyme. Scheme 1 shows these possibilities.

For the ordered addition of ATP before UTP

$$\frac{V_{PIX}}{V_{max}} = \frac{k_2 k_4 k_{10}}{k_{prod}(k' + k_3 k_9[\text{UTP}])(1 + k_3 k_9 k_{prod}[\text{UTP}]/k'')}$$

where k_{prod} is the net rate constant for release of product and k' and k'' are given in Ref. 8.

[12] D. D. Clark and J. J. Villafranca, *Biochemistry* **24**, 5147 (1985).
[13] M. A. Reynolds, N. J. Oppenheimer, and G. L. Kenyon, *J. Am. Chem. Soc.* **105**, 6663 (1983).
[14] H. C. Wang, L. Ciskanik, D. Dunaway-Mariano, W. von der Saal, and J. J. Villafranca, *Biochemistry* **27**, 625 (1988).

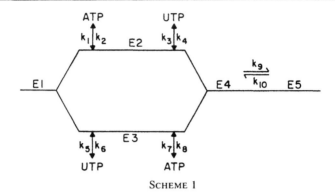

SCHEME 1

For the ordered addition of UTP before ATP

$$\frac{V_{PIX}}{V_{max}} = \frac{k_8 k_{10}}{(k_8 + k_9)k_{prod}}$$

In one kinetic expression the PIX rate depends on the concentration of the other substrate, UTP, whereas the other expression does not. The design of the experiment is simple and involves measuring the PIX rate for $[\gamma\text{-}^{18}O_4]$ATP at various UTP levels that bracket the binding constant for UTP. If ATP binds before UTP, then the ratio V_{PIX}/V_{max} will be zero at low and high levels of UTP, reaching a maximum at an intermediate concentration of UTP. If UTP binds before ATP, then the V_{PIX}/V_{max} ratio will be independent of the UTP level except at very low levels of UTP. For CTP synthase, the experimental results show that UTP binds before ATP.[8]

The inhibition of the PIX reaction rate is another experimental method that can be used to explore kinetic mechanisms of enzymes. This method has been applied to study the forward and reverse reactions catalyzed by UTP–glucose-1-phosphate uridylyltransferase (UDPglucose pyrophosphorylase).[15]

UTP + glucose 1-phosphate $\rightleftharpoons$ PP$_i$ + UDPglucose

As shown below, the reaction occurs by nucleophilic attack at the α-P of UTP by glucose 1-phosphate. Thus, two PIX reactions occur and each can be followed by ^{31}P NMR. This was done and PIX was observed.

An additional set of experiments led to following the *inhibition* of the PIX rate by varying the concentration of the nonlabeled substrate, e.g.,

[15] L. S. Hester and F. M. Raushel, *Biochemistry* **26**, 6465 (1987).

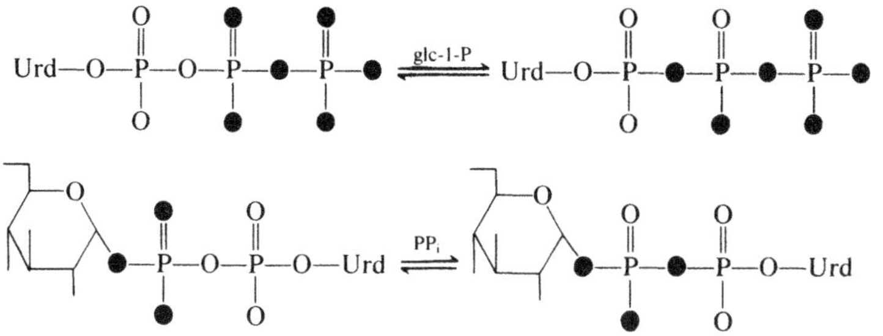

glucose 1-phosphate in an experiment with ^{18}O-labeled UTP. The full kinetic expressions are given in Ref. 15 and are not reproduced here. The results of this thorough study are given in Scheme 2 and demonstrate that with the proper experimental design PIX experiments using ^{31}P NMR in combination with other kinetic methods can provide a complete description of the kinetic steps in an enzyme-catalyzed reaction. Indeed the method is quite powerful.

Finally, rather complex reaction mechanisms can be followed by PIX methods provided the proper substrate molecules can be synthesized with the isotopes in the bonds to be cleaved. A complex system that has been studied in this way involves the reaction catalyzed by pyruvate,phosphate dikinase (PPDK). The overall reaction is

$$ATP + P_i + pyruvate \rightleftharpoons PEP + AMP + PP_i$$

The reaction has been proposed to occur in three distinct steps with the intermediacy of two covalent enzyme forms, a pyrophosphorylated enzyme and a phosphorylated enzyme. These steps are given here:

$$E^{his} + ATP \rightleftharpoons E^{his}\text{-}\overset{\beta\gamma}{PP} + AMP \tag{1}$$

$$E^{his}\text{-}\overset{\beta\gamma}{PP} + P_i \rightleftharpoons E^{his}\text{-}\overset{\beta}{P} + \overset{\gamma}{PP_i} \tag{2}$$

$$E^{his}\text{-}\overset{\beta}{P} + pyruvate \rightleftharpoons E^{his} + \overset{\beta}{PEP} \tag{3}$$

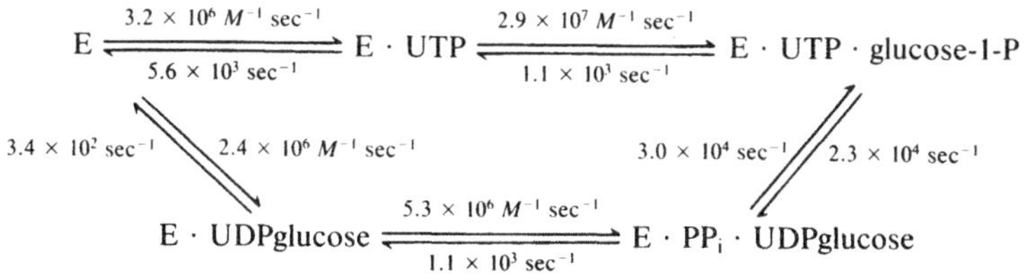

SCHEME 2

According to this mechanism, cleavage of ATP between the α- and β-P [partial reaction (1)] should take place in the absence of P_i and cleavage at the $\beta\gamma$ position [partial reaction (2)] should require the participation of P_i. Other kinetic results indicate that P_i is required for partial reaction (1) as well as partial reaction (2). However, these results do not distinguish between the P_i requirement for AMP formation on the enzyme and the P_i requirement for release of AMP from the enzyme. In order to make this distinction, we tested the P_i dependency of the α,β-P and β,γ-P cleavage reactions by testing the ability of PPDK to catalyze PIX in [α,β-$^{18}O,\beta,\beta$-$^{18}O_2$]ATP (1) and [β,γ-$^{18}O,\gamma,\gamma,\gamma$-$^{18}O_3$]ATP (5). As indicated, reversible α,β-P cleavage accompanied by torsional equilibration of the AMP α-P oxygen atoms will lead to the formation of [α-$^{18}O,\beta,\beta$-$^{18}O_2$]ATP (2). If β,γ-P cleavage follows, as would be expected if P_i were present, [β,γ-$^{18}O,\beta$-$^{18}O,\alpha,\beta$-^{18}O]ATP (3) and [β,γ-$^{18}O,\beta$-$^{18}O,\alpha$-^{18}O]ATP (4) will be formed. PIX in [β,γ-$^{18}O,\gamma,\gamma,\gamma$-$^{18}O_3$]ATP (5) can only take place if reversible cleavage of the β,γ-P position takes place. Accordingly, 5 should be converted to [β-$^{18}O,\gamma,\gamma,\gamma$-$^{18}O_3$]ATP (6).

α,β-P cleavage

β,γ-P cleavage

The conversions of 1 to 2, 3, and 4 and of 5 to 6 were monitored by ^{31}P NMR. Shown in Figs. 3 and 4 are the ^{31}P NMR spectra of the [α,β-$^{18}O,\beta,\beta$-$^{18}O_2$]ATP (1) and the [β,γ-$^{18}O,\gamma,\gamma,\gamma$-$^{18}O_3$]ATP (5) reaction mixtures prior to and after the addition of P_i and PPDK. The $\alpha\beta \rightarrow \alpha$ and $\beta \rightarrow$

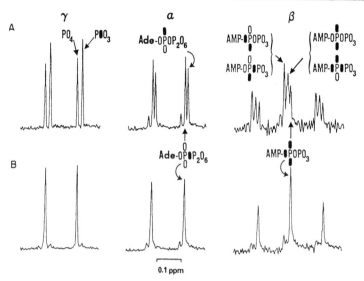

FIG. 3. ^{31}P NMR spectrum at 145.81 MHz of (B) $[\alpha,\beta\text{-}^{18}O,\beta,\beta\text{-}^{18}O_2]$ATP (1) and (A) PIX reaction mixture of 1 generated from incubation of PPDK with 5 mM 1, 12 mM P$_i$, at 30° for 90 min. (Data from Wang et al.[14])

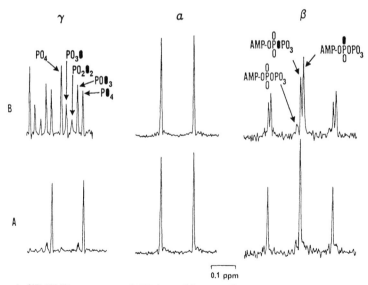

FIG. 4. ^{31}P NMR spectrum of (A) $[\beta,\gamma\text{-}^{18}O,\gamma,\gamma,\gamma\text{-}^{18}O_3]$ATP (5) and (B) PIX reaction mixture of 5 generated from incubation of PPDK with 5 mM 5, 12 mM P$_i$, at 30° for 90 min. (Data from Wang et al.[14])

$\beta\gamma$ PIX reactions of **1** are evident from the ^{31}P NMR spectrum shown in Fig. 3A. Equilibration of $[\alpha,\beta\text{-}^{18}O,\beta,\beta\text{-}^{18}O_2]$ATP should partition ^{18}O between the α,β-bridge and α-nonbridge positions in a $1:2$ ratio and between the β-nonbridge position and β,γ-bridge position in a $1:2$ ratio. The ^{31}P NMR spectrum shows these ratios are $1.0:1.8$, close to the theoretical values.

The ^{31}P NMR spectrum of **5** is shown in Fig. 4A. In the equilibrium mixture (Fig. 4B), the remainder of the ATP had ^{18}O present either in the β,γ-P bridge position (31% of the total ATP) or in the β-P nonbridge position (62% of the total ATP). The ATP that had ^{18}O in the β,γ-P bridge position consisted of **5** and $[\beta,\gamma\text{-}^{18}O]$ATP, whereas the ATP that had ^{18}O in the β-P nonbridge position consisted of **6** and $[\beta\text{-}^{18}O]$ATP. The $[\beta,\gamma\text{-}^{18}O]$ATP and $[\beta\text{-}^{18}O]$ATP are formed as a result of the rotation (presumably while dissociated from the enzyme) of the PP$_i$ formed from **5** or **6** followed by transfer of the $[^{16}O_3]$phosphoryl moiety of the PP$_i$ during the back reaction. Also, formation of ^{18}O-labeled P$_i$ accompanied the formation of the $[\beta\text{-}^{18}O]$ATP and $[\beta,\gamma\text{-}^{18}O]$ATP.

In other experiments, the dependency of the PIX rates (namely, $\alpha,\beta\text{-}^{18}O \rightarrow \alpha\text{-}^{18}O$; $\beta\text{-}^{18}O \rightarrow \beta,\gamma\text{-}^{18}O$) of **1** were studied in the presence of P$_i$. Addition of P$_i$ to the reaction mixture increased both exchange rates equally, increasing them to 4% of the maximum velocity of the forward (AMP-forming) reaction. In view of the ^{18}O washout that was observed with the $[\beta,\gamma\text{-}^{18}O,\gamma,\gamma,\gamma\text{-}^{18}O_3]$ATP (**5**), the slow PIX rate of **1** is probably due to PP$_i$ release from the enzyme and subsequent rebinding. Not surprisingly, the inclusion of PP$_i$ in the reaction increased the PIX rate 10-fold by providing a pathway to recycle the phosphorylated enzyme back to free enzyme. Likewise, inclusion of yeast inorganic pyrophosphatase in the reaction prevented the PIX by destroying all PP$_i$ released from the enzyme and thereby drawing all of the enzyme into the phosphorylated form. The PIX experiments described above support a Bi Bi Uni Uni mechanism in which ATP and P$_i$ must be bound to PPDK before AMP formation can take place and in which AMP and PP$_i$ must be bound to the phosphorylated form of the enzyme for P$_i$ formation. Thus the complicated reaction catalyzed by PPDK was studied by PIX methods and the stepwise nature of the phosphoryl transfer steps was revealed directly by ^{31}P NMR.

Summary

This chapter has presented the basic methods involved in the use of ^{31}P NMR to study positional isotope exchange in enzyme-catalyzed reactions involving phosphorus-containing substrates. The method is straightforward but requires synthesis of specific ^{18}O isotopically labeled substrates at the site of bond cleavage. Analysis of the kinetic consequences of the PIX reaction depends on the nature of the enzymatic reaction and the number of other substrates involved in the kinetic reaction mechanism. This may be simple or formidable. From the examples described in this article one can appreciate that PIX experiments, when combined with other kinetic methods, can in favorable cases unravel many (if not all) of the rate constants in an enzyme-catalyzed reaction.

[21] Paramagnetic Probes of Macromolecules

By JOSEPH J. VILLAFRANCA

Throughout this series, NMR has been used to explore the structure of macromolecules of modest size (1,000–40,000 Da) by direct observation of nuclei, e.g., ^{1}H and ^{13}C of proteins and ^{1}H and ^{31}P of nucleic acids. High-resolution information is available and these studies are aided by spectrometers that operate using high magnetic fields. For large proteins and polynucleotides (>50,000 Da) other approaches must be used. A general and useful method has been to observe directly or indirectly small-molecule probes that when bound to macromolecules function as "reporters" of structural features. Among such probes are paramagnetic metal ions, nitroxyl spin labels, diamagnetic species such as $^{23}Na^{+}$, $^{25}Mg^{2+}$, and $^{43}Ca^{2+}$, and ^{2}H- or ^{19}F-labeled probes.

Consideration of which probe should be used is fundamentally dictated by (1) the macromolecular system being studied (protein, nucleic acid, membrane, etc.) and (2) the type of information desired (binding constants, distances between distinct sites, etc.). For example, a paramagnetic metal ion may be an integral part of the macromolecular system, such as Fe^{3+} in a heme, Cu^{2+} in a metalloenzyme, or a paramagnetic ion that substitutes for naturally occurring metal ions (e.g., Mn^{2+} for Mg^{2+} or Gd^{3+} for Ca^{2+}). In other cases a paramagnetic probe (nitroxyl spin label) may be introduced into a system (perhaps covalently) often without serious perturbation of structure or function and may thus serve as a reporter of molecular dynamics and/or structure. This chapter focuses on the theory and application of paramagnetic probes in NMR experiments with macromolecules by highlighting several examples from the literature.

Numerous experiments have been conducted with paramagnetic probes. First, the influence of the electron–nuclear dipolar interactions on the NMR spectra of nuclei in macromolecules has been reported for sev-

METHODS IN ENZYMOLOGY, VOL. 177

eral systems[1]; second, the paramagnetic influence of metal ions on the protons of water in enzyme–metal ion and nucleic acid–metal ion complexes have been reported[2]; third, the influence of paramagnetic probes on the nuclei of small molecules (substrates and substrate analogs) interacting with the macromolecule–probe complex has been used to determine structures of active-site complexes.[3] The description of the electron–nuclear dipolar relaxation that follows applies to all the above situations. Each case cited above has individual restrictions and these will be discussed where appropriate.

The first case to be discussed is the interaction of a paramagnetic metal ion with solvent water molecules. Observations are made by monitoring the ^1H NMR relaxation rates of bulk water molecules. The aim of many of the studies is to determine the number of water molecules in a macromolecular complex and to evaluate whether changes occur in the number of bound water molecules as the system is perturbed. For example, if a substrate molecule binds to the metal ion in an enzyme–Mn^{2+} complex, there may be a change (reduction?) in the number of rapidly exchanging water molecules when substrate is added. How would this change be measured?

The basic parameter that is measured is the longitudinal $(1/T_1)$ relaxation rate of water protons. Occasionally, the transverse $(1/T_2)$ relaxation rate is also measured. Addition of a paramagnetic ion to water produces an increase in the solvent water proton relaxation rates by virtue of the large electronic magnetic moment associated with the unpaired electrons of the paramagnetic ion. Subsequent addition of a macromolecule to which the paramagnetic ion binds in a specific binding site can produce a further increase in water proton relaxation rates. This phenomenon is due to a correlation time change(s) in the presence of the macromolecule and this second relaxation enhancement is the essence of the proton relaxation enhancement (PRE) method. Because the observed enhancement is proportional to the amount of bound metal ion, PRE can be simply used as a titration indicator to yield metal binding information.

The magnitude of the PRE effect has been widely analyzed in terms of a number of equations collectively known as the Solomon–Bloembergen–Morgan (SBM) equations.[2] The equations contain several variables that determine the magnitude of the observed enhancement: a term related to symmetry (B) and a correlation time (τ_v) determining the electron spin-

[1] R. A. Dwek, "NMR and Biochemistry." Oxford Univ. Press (Clarendon), London and New York, 1973.

[2] D. R. Burton, S. Forsén, G. Karlström, and R. A. Dwek, *Prog. Nucl. Magn. Reson. Spectrosc.* **13**, 1 (1979).

[3] J. J. Villafranca, *in* "Phosphorus-31 NMR: Principles and Applications" (D. Gorenstein, ed.), p. 155. Academic Press, Orlando, Florida, 1984.

relaxation time (τ_s) of the paramagnetic ion; a rotational correlation time (τ_r) for the ion–macromolecule complex; the lifetime (τ_m) of a water molecule in the metal ion coordination sphere; the number (q) of exchangeable water molecules in the metal ion coordination sphere; the metal ion–water proton distance (r); and other variables that relate the activation energies corresponding to the correlation times and the temperature of the measurement. Along with temperature, an additional parameter at the discretion of the investigator is the NMR frequency (ω) of the relaxation rate measurement. The quantitative application of PRE has been to obtain relaxation data at varying temperature and frequency followed by a multiparameter fit to the data using the above parameters as variables in the SBM equations. In view of the large number of parameters (up to 9) involved in this fitting procedure, considerable problems in obtaining good definition of these parameters can arise. References 2–5 present a critical analysis of the limitations of these methods and a brief overview will be presented here to evaluate the number of rapidly exchanging water molecules in a macromolecular complex.

Measurement of the paramagnetic (P) contribution to $1/T_1$ (or $1/T_2$) is done by subtracting the relaxation rate of a diamagnetic control solution (everything but the paramagnetic metal ion) from the total relaxation rate $[1/T_{1p} = 1/T_1(\text{total}) - 1/T_2(\text{diamagnetic})]$.

Because the PRE method measures the relaxation effect in *bulk* water, consideration of the exchange of water molecules from the macromolecule–ion complex is essential. The overall correlation time that modulates the electron-nuclear interaction is given as $1/\tau_c = 1/\tau_s + 1/\tau_m + 1/\tau_r$. For paramagnetic ions in solution such as Mn^{2+} and Gd^{3+}, $\tau_s \simeq 10^{-9}$ sec, $\tau_m \simeq 10^{-9}$ sec, and $\tau_r \simeq 10^{-11}$ sec. When these ions are bound to a macromolecule, $\tau_r \simeq 10^{-7}$ sec for proteins of MW $\simeq 100,000$ and therefore $1/\tau_c$ will change in magnitude, having contributions from both τ_s and τ_m. These ions will have the largest PRE change when bound to a macromolecule in contrast to ions such as Co^{2+}. For Co^{2+}, $\tau_s \sim 10^{-12}$ sec, and only a change in this parameter will produce a change in PRE value in the presence of a macromolecule. Thus, ions with a long τ_s value are most suitable for investigations because PRE changes are large.

Once corrections have been made for diamagnetic contributions to $1/T_1$, the lifetime (τ_m) of the ligand (water) in the enzyme complex can be evaluated. Equation (1) describes the relationship between the experimentally obtained paramagnetic relaxation rate $(1/T_{1p})$, the paramagnetic

4 S. C. Ransom, Ph.D. thesis, "Magnetic Resonance Studies of *E. coli* Glutamine Synthetase." The Pennsylvania State University, University Park, Pennsylvania, 1984.
5 C. D. Eads, Ph.D. thesis, "Spectroscopic Studies of the Metal Ion Sites of *E. coli* Glutamine Synthetase." The Pennsylvania State University, University Park, Pennsylvania, 1985.

dipolar electron–nuclear relaxation (T_{1m}) in the enzyme complex, and the lifetime of this complex (τ_m). In this expression,

$$1/T_{1p} = fq/(T_{1m} + \tau_m) \tag{1}$$

f is the [paramagnetic complex]/[55.5] and q is the number of water molecules in the complex.

Three conditions are important: (1) fast exchange, where $T_{1m} \gg \tau_m$ and $1/T_{1p} = fq/T_{1m}$; (2) slow exchange, where $\tau_m \gg T_{1m}$ and $1/T_{1p} = fq/\tau_m$; and (3) intermediate exchange, where $T_{1m} \simeq \tau_m$.

The SBM equation [Eq. (2)] relates $1/T_{1m}$ to the distance (r) between the paramagnetic center and the nucleus (nuclei) being investigated (2.80 Å for Mn^{2+}–H of H_2O); $f(\tau_c)$ is the correlation function that describes the dipolar electron–nuclear interaction, and C relates the

$$1/T_{1m} = Cf(\tau_c)/r^6 \tag{2}$$

physical properties of the nucleus and the electron(s) according to Eq. (3). In this expression γ_I is the magnetogyric ratio of the nucleus (2.675×10^4 rad/sec G for ^{1}H) and μ_{eff} is the effective magnetic

$$C = \tfrac{2}{15}\gamma_I^2\mu_{eff}^2 \tag{3}$$

moment of the electron. For simple paramagnetic species, Eq. (4) relates S, the spin quantum number ($S = \tfrac{1}{2}, 1, \tfrac{3}{2}$, etc., for 1, 2, 3, etc., electrons), g the electronic g factor (usually 2), and β the Bohr magneton (9.274×10^{-21} erg/G). The expression in Eq. (3) does not hold for ions such as Cu^{2+} (anisotropic g) or for Co^{2+} (spin-orbit coupling and $g \geq 2$). Equation (4) is valid for "S-state" ions such as Cr^{3+}, Mn^{2+}, and Gd^{3+}.

$$\mu_{eff}^2 = S(S + 1)g^2\beta^2 \tag{4}$$

For most macromolecular complexes, Eq. (5) applies (ω_I equals the Larmor precession frequency of the nucleus).

$$f(\tau_c) = 3\tau_c/(1 + \omega_I^2\tau_c^2) \tag{5}$$

Rearrangement of Eq. (2) gives Eq. (6), with $C' = C^{1/6}$ for fast-exchange conditions ($T_{1m} = fT_{1p}$).

$$r = C'[T_{1m}f(\tau_c)]^{1/6} \tag{6}$$

Values for C' are given for various metal ions and nuclei other than ^{1}H in Ref. 6.

To evaluate the number of rapidly exchanging water molecules in a complex, $1/T_{1m}$ must be measured at several frequencies, and ideally at several temperatures. Lacking this rigorous approach a few simple experiments can be conducted, but the reliability of the parameters falls off. For

6 J. J. Villafranca, this series, Vol. 87 [12].

TABLE I
PRE PARAMETERS[a]

Parameter	Alone	GLUT	MSOX	APBA	ACPS
B (10^{18} rad/sec)	5.65	3.60	1.99	4.31	4.65
τ_v (10^{-11} sec)	1.86	1.38	9.52	1.46	1.13
τ_m (10^{-9} sec)	4.11	4.76	4.94	4.49	5.12
q	1.79	1.27	0.76	1.25	0.88

[a] From analysis of the data in Fig. 1. All solutions contained GS–Mn^{2+} either alone or with the molecules as indicated. (Data from Ref. 7.)

example, only two frequencies could be used, or one frequency and two nuclei (H$_2$O and D$_2$O), or measurement of $1/T_{1m}$ and $1/T_{2m}$ at one frequency. All of these "shortcuts" have their drawbacks and these have been amply discussed in reviews.[2,6] Best results are obtained with a multiparameter fit to the above equations using an iterative fitting routine, several of which have been described in the literature.[2,7,8]

Because τ_s often contributes to τ_c, another equation must be introduced to fit the frequency dependence of τ_s. This is the Bloembergen–Morgan equation

$$1/\tau_s = Bf(\tau_v) \tag{7}$$

where B is a constant related to the zero-field splitting parameter of an $S > \frac{1}{2}$ ion such as Mn^{2+} and τ_v is a correlation time related to the modulation of the zero-field splitting, often a result of solvent impact on the complex. In general, for the multiparameter fitting procedure, r is held constant while B, τ_v, q, and τ_m are allowed to vary; the computed $1/T_{1p}$ values are compared to the measured values and the goodness of the fit is assessed.

Examples of such multiparameter fits are given in Table I for glutamine synthetase (glutamate–ammonia ligase) complexes with Mn^{2+} as the paramagnetic probe. Structures of the molecules studied as analogs of the substrate glutamate and the tetrahedral adduct are given here.

MSOX   ACPS   APBA   TETRAHEDRAL ADDUCT

[7] C. D. Eads and J. J. Villafranca, *Arch. Biochem. Biophys.* **252**, 382 (1987).
[8] P. P. Chuknyisky, *J. Magn. Reson.* **84**, 153 (1989).

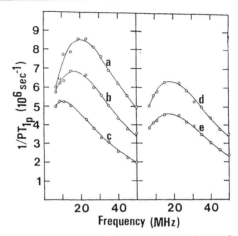

FIG. 1. Frequency dependence of $1/PT_{1p}$ values for complexes of glutamine synthetase: (a) GS–Mn^{2+} alone, (b) GS–Mn^{2+} plus glutamate, (c) GS–Mn^{2+} plus MSOX, (d) GS–Mn^{2+} plus APBA, and (e) GS–Mn^{2+} plus ACPS. (Data from Ref. 7.)

Figure 1 shows how the theoretical curves closely fit the data for the cases given in Table I. Figure 2 shows the temperature dependence of $1/PT_{1p}$ data at 27 MHz for the five complexes, GS–Mn^{2+} (a), and GS–Mn^{2+} complexes with glutamate (b), methionine sulfoximine (MSOX) (c), aminophosphonobutyric acid (APBA) (d), and 3-amino-3-carboxypropanesulfonamide (ACPS) (e).

Changes in the effect of enzyme-bound paramagnetic metals ions on the solvent proton relaxation rates reflect changes in the structure of the metal ion environment. A quantitative analysis of these data was at-

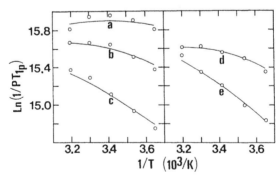

FIG. 2. Temperature dependence of $1/PT_{1p}$ data for glutamine synthetase Mn^{2+} complexes at 27 MHz. The letters refer to the same complexes described in Fig. 1. (Data from Ref. 7.)

tempted to identify such structural changes based on theoretical expressions relating the measured relaxation rates to parameters describing the motion and geometry of the interacting molecules. For each complex, two fits to the data were made, one in which only the frequency-dependent data were considered, and one in which both frequency- and temperature-dependent data were included.[7] Table I lists the parameters used to fit the data in Fig. 1.

Although the fits to the data are reasonably good, the physical significance of the derived parameters is unclear. For example, consider the value of the rotational correlation time, τ_r. It was possible to obtain a good fit to the temperature-independent data (see Table II of Ref. 7) by fixing the value of this parameter at a value of 200 nsec for the enzyme complex (MW $\sim$ 600,000). Under this constraint the major contribution to the overall correlation time, τ_c, comes from τ_m, which characterizes exchange of water molecules from the inner coordination sphere of the metal ion. In contrast, to get an acceptable fit to the temperature-dependent data, it was necessary to treat τ_r as an adjustable parameter. In this case it is τ_r that makes the major contribution to τ_c, and τ_m assumes a value corresponding to intermediate exchange. This value of τ_r is too short to represent rotation of a molecule as large as the glutamine synthetase dodecamer. Thus it is unclear what physical process τ_r represents. However, the differences in the relaxation data for the various inhibitors do reflect structural differences in the metal ion site that are most likely related to water molecules at this site.

Thus, there appear to be two water molecules exchanging in the enzyme–Mn^{2+} complex. When glutamate is added to this complex, a change occurs in the structure of the metal ion environment. Because the relaxation rates decrease upon substrate addition, one reasonable interpretation is that the binding of glutamate diminishes the accessibility of the metal ion to solvent water. The observed change in solvent accessibility may therefore represent the immobilization of a bridging water molecule between the metal ion and the substrate.

When MSOX (an analog of the tetrahedral adduct) is added to GS–Mn^{2+}, a greater reduction than that found with glutamate is observed in the relaxation rate data. Recent spin-echo EPR data from our laboratory demonstrated that the sulfoximine portion of MSOX lies in the inner coordination sphere of the metal ion and that a single water molecule is coordinated to the metal ion in this complex.[9] Therefore, the effect of MSOX on the relaxation data probably reflects the direct displacement of a water molecule from the metal ion by the inhibitor. The difference in the

[9] C. D. Eads, R. LoBrutto, A. Kumar, and J. J. Villafranca, *Biochemistry* **27**, 165 (1988).

relaxation behavior of the solutions containing MSOX and glutamate indicates that the structure of the metal ion environment differs for the two complexes. This suggests that the metal ion plays a direct role in catalysis and is consistent with our previous hypothesis that the metal ion serves to bind to and stabilize the formation of the tetrahedral adduct.

The data for the glutamate and MSOX complexes can be compared to the data for the APBA and ACPS complexes. The APBA behavior is similar to the effect of glutamate, while the effect of ACPS is similar to MSOX. To determine a structural basis for this effect, the distance between the metal ion and the phosphorus of APBA can be determined by ^{31}P NMR observations. The distance between a nucleus and a paramagnetic center can be estimated from Eq. (6) under fast-exchange conditions. For this situation analysis of the paramagnetic effects is the same as described in Eqs. (1)–(6) with the modification that $q = 1$ in Eq. (1), because the ligand is the analog APBA. To estimate the correlation time, the effect of enzyme–bound Mn^{2+} on the longitudinal and transverse relaxation rates of the phosphorus atom of APBA at 145.8 MHz was measured at three temperatures. Using the dissociation constant of 0.75 mM determined by following the water proton relaxation rate during titration with APBA, the normalized relaxation rates $1/PT_{1p}$ and $1/PT_{2p}$ (in sec^{-1}) were found to be 60.9 and 1560 at 293 K, 83.0 and 2440 at 303 K, and 99.6 and 3480 at 313 K, respectively.[7]

If neither the longitudinal nor the transverse relaxation rates are exchange limited, then the correlation time can be determined from the ratio of these rates using Eq. (8):

$$\tau_c^2 = (T_{1p}/T_{2p} - \tfrac{7}{6})/\tfrac{2}{3}\omega^2 \qquad (8)$$

This analysis gives only a lower limit on the correlation time of $\tau_c > 7.8$ nsec. The tumbling time of the protein places an upper limit on the correlation time, $\tau_c < 200$ nsec. Using Eq. (6) and the longitudinal relaxation time at 293 K, this range of correlation times corresponds to a range of distances of $5 < r < 8$ Å between the metal ion and the phosphorus atom of APBA. This distance is too large for APBA to be an inner-sphere ligand to the metal ion and APBA may be interacting with the metal ion as a glutamate analog based on the similarity in the water proton relaxation behavior and on the similarity in the distances to the metal determined by NMR. ACPS may be interacting with the metal ion as a transition-state analog based on the similarity in the water proton relaxation behavior compared to that obtained with MSOX. Thus, the PRE and ^{31}P NMR data have indicated that the metal ion is involved in binding substrates and has a direct role in the catalytic mechanism.

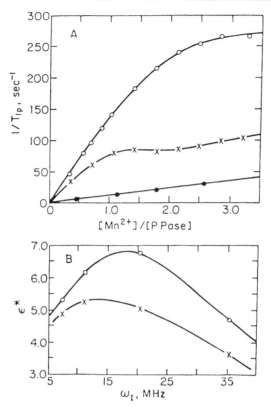

FIG. 3. PRE data for yeast inorganic pyrophosphatase. (A) Titration with Mn^{2+} of enzyme alone (○) or of enzyme in the presence of Co^{3+} $(NH_3)_4PNP$ (×). Mn^{2+} solution in buffer (●). (B) PRE data for inorganic pyrophosphatase as a function of NMR frequency. The symbols are the same as in A. (Data from Ref. 10.)

Another example of the use of the PRE method to explore metal ion sites of enzymes is the study of yeast inorganic pyrophosphatase.[10] This enzyme has multiple metal ion sites, and stable Co^{3+} and Cr^{3+} analogs were used to explore one site and Mn^{2+} was used as a paramagnetic probe of the other sites. Figure 3A presents PRE titration data for solutions containing Mn^{2+} and pyrophosphatase. The relaxation rate of the solvent water protons was monitored at 20 MHz. These data show that at least two Mn^{2+} bind per subunit and that each bound Mn^{2+} produces a pro-

[10] W. B. Knight, D. Dunaway-Mariano, S. C. Ransom, and J. J. Villafranca, *J. Biol. Chem.* **259**, 2886 (1984).

nounced paramagnetic effect on the relaxation properties of water protons. The $1/T_{1p}$ values are not enhanced beyond two bound Mn^{2+} per subunit.

The PRE values for a solution of $[Mn^{2+}]/[enzyme] = 0.26$ (at which 98% of the Mn^{2+} is present as $E \cdot Mn$ and 2% is present as $E \cdot Mn \cdot Mn$) were measured at four frequencies and the results are provided in Fig. 3B. The data are expressed as PRE enhancements (ε^*) that were calculated by dividing the $1/T_{1p}$ values for enzyme solutions by the $1/T_{1p}$ values for an equivalent amount of Mn^{2+} in the absence of enzyme. The relaxation rate data obtained for the solution in which $[Mn^{2+}]/[enzyme] = 0.26$ (open circles, Fig. 3B) were fit to Eqs. (1)–(7), using a pattern search routine as described before.

The titration of pyrophosphatase with Mn^{2+} was repeated in the presence of saturating amounts of the substrate analog, $Co^{3+}(NH_3)_4PNP$. The results are provided in Fig. 3A. The PRE data with this analog are given in Fig. 3B. Fits to the data were as described before. For $E–Mn^{2+}$, $q = 1.2$ and $\tau_v = 1.3 \times 10^{-11}$ sec, while for the complex with the Co^{3+} analog, $q = 0.9$ and $\tau_v = 7.7 \times 10^{-11}$ sec. Thus, no significant change in hydration number is found for the bound Mn^{2+} when another metal–substrate site is filled, but the PRE values change in magnitude. This is primarily due to a

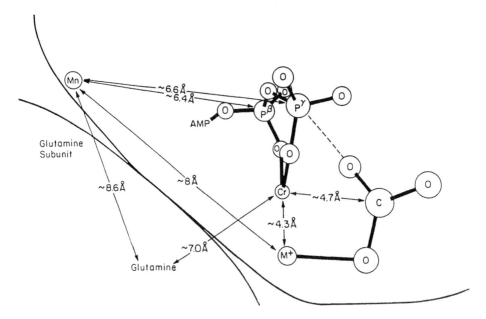

FIG. 4. Spatial relationships among the metal ion and substrate sites of carbamoylphosphate synthase obtained from NMR studies. (Data from Refs. 11 and 12.)

change in symmetry of the bound Mn^{2+} when the substrate analog complex is bound, and demonstrates that without an extensive PRE analysis, incorrect conclusions about the changes in PRE values (i.e., reduction in bound water molecules) may have been made.

Finally, a combination of PRE of solvent and direct NMR observation of metal nuclei ($^6Li^+$, $^7Li^+$, and $^{133}Cs^+$) was used to explore the active and allosteric activator sites of carbamoyl-phosphate synthase.[11] The paramagnetic probes were Mn^{2+}- and Cr^{3+}-ATP and the distances from these paramagnetic probes to various nuclei were calculated.[11] These studies, along with paramagnetic probe studies of ^1H, ^{31}P, and ^{13}C nuclei of substrates for this enzyme,[12] provide a "topographic map" of the various sites on this enzyme. The analysis of these data closely follows the outline given in this chapter and the reader is referred to these papers for an extensive discussion of the use of these methods in the computation of paramagnetic probe–nuclei distances. Figure 4 shows a schematic drawing of the spatial relationship among these sites obtained by using the paramagnetic Mn^{2+} and Cr^{3+} probes.

In conclusion, this chapter has presented a brief overview of the use of paramagnetic probes to study structural features of macromolecules. When applied with caution and intelligent experimental design, these methods can provide structural information that can be a substantial aid in relating structure to function in biological macromolecules.

---

[11] P. G. Kasprzyk, E. Whalen-Pederson, P. M. Anderson, and J. J. Villafranca, *Bioorg. Chem.* **13**, 98 (1985).

[12] F. M. Raushel, P. M. Anderson, and J. J. Villafranca, *Biochemistry* **22**, 1872 (1983).

Section IV

In Vivo Studies of Enzymatic Activity

[22] Enzyme Regulation of Metabolic Flux

By SHEILA M. COHEN

Introduction

Carbon-13 NMR spectroscopy provides a convenient way to study metabolism in cells, isolated perfused organs, and whole animals.[1-4] The low natural abundance of ^{13}C nuclei (~1.1%) is actually advantageous for metabolic studies. In the applications of the ^{13}C NMR method discussed here, the ^{13}C label is introduced by administration of specifically enriched substrates to the living cell; signals are measured, repetitively, from substrate, intermediate, and end products. The distribution of ^{13}C label at individual carbon atoms in these metabolites is then used to elucidate pathways and kinetics. Our understanding of the regulation of the enzymes of intermediary metabolism ultimately must be tested within the context of the physiology of the whole cell. The nondestructive ^{13}C NMR method has considerable potential for providing information that may yield new insights into cellular metabolism, while also serving as a complement to more familiar *in vitro* studies with purified enzymes or ^{14}C tracer investigations of metabolism. We examine here some of the practical considerations encountered when applying ^{13}C NMR to investigations of the regulation of metabolic flux in liver preparations. Criteria defining circumstances under which a ^{13}C NMR approach is advantageous should unfold during our examinations. Perfused liver preparations have proved to be particularly convenient for these studies because of their long-term stability and the ease and speed with which perfusion conditions can be controlled and modified. As a consequence of these properties, it is fairly straightforward to construct mathematical models describing metabolic flux in this preparation. Although the examples discussed here are specific, the repertoire developed to handle these particular problems of metabolic regulation has more general application. The ^{13}C NMR method is fairly versatile, and its extension to new problems is usually both direct and rich in opportunities for exploiting unique features of the new model.

[1] S. M. Cohen, P. Glynn, and R. G. Shulman, *Proc. Natl. Acad. Sci. U.S.A.* **78,** 60 (1981).
[2] S. M. Cohen, *Biochemistry* **26,** 563 (1987).
[3] N. V. Reo, B. A. Siegfried, and J. J. H. Ackerman, *J. Biol. Chem.* **259,** 13664 (1984).
[4] S. M. Cohen (ed.), *Ann. N.Y. Acad. Sci.* **508** (1987).

General Considerations of Experimental Design

¹³C, Not ¹⁴C

Because the radioactive ¹⁴C isotope has been widely used in metabolic studies for over 40 years, we mention that, contrary to what might be conjectured, a well-designed ¹⁴C tracer experiment is usually not a good model for a ¹³C NMR approach. Such a model would be severely limited because, for example, it would not include provision for use of the information conveyed by ¹³C–¹³C *J* splitting, nor would it usually incorporate the kinetics information available in time-resolved sequentially acquired ¹³C spectra of metabolizing cells. Because ¹³C is a stable isotope and because NMR is a relatively insensitive method, a tracer approach is neither necessary nor in most cases even appropriate for ¹³C NMR at all. Rather, substrates highly enriched with ¹³C (>90% at a specific site) are used. That is, cells are presented with the ¹³C label at about substrate levels, typically 2 to 10 mM.

Stability of Preparation

To obtain metabolically meaningful information from an NMR experiment, it is important that the preparation studied meet all necessary criteria of long-term stability and metabolic responsiveness. The ability of a perfused liver preparation to carry out gluconeogenesis at acceptable rates from substrates that enter the pathway at the level of pyruvate is a particularly stringent test, as is responsiveness to physiological levels of hormones. Liver preparations meeting these criteria have been routinely obtained with use of the perfusion apparatus modified for NMR and conditions described in detail previously.[2,5] The importance of constructing the perfusion apparatus (especially the cell holding the liver) of plastic materials and of using water of the highest purity for making physiological buffers is emphasized. Also of importance is use of a perfusion medium containing either fresh, washed erythrocytes plus bovine serum albumin[5] (as an oncotic pressure agent) or an appropriate fluorocarbon emulsion.[6] It is worthwhile initially to measure ³¹P spectra of liver preparations repetitively over a period exceeding the longest lasting ¹³C NMR experiment anticipated and under the exact conditions to be used for ¹³C NMR, including the same substrate, sample temperature, and age and physiological state of the donor rat. In ³¹P spectra of well-perfused liver, the signals arising from ATP are relatively sharp and intense and the ratio ATP : P$_i$ is

[5] S. M. Cohen, *J. Biol. Chem.* **258,** 14294 (1983).
[6] A. W. Wolkoff, K. L. Johansen, and T. Goeser, *Anal. Biochem.* **167,** 1 (1987).

unchanged over several hours under our perfusion conditions.[2,5] Subsequently, even when acquisition of ^{13}C spectra is the major interest, measurement of several ^{31}P spectra, both before and after collection of the ^{13}C spectra,[2] or measurement of ^{31}P and ^{13}C spectra simultaneously by an alternate scan technique[5] is advisable. In this way, the utility of the ^{13}C NMR data will not be compromised by inclusion of any preparation of dubious quality initially or which deteriorated during the experiments. (We note that rf heating of perfused liver samples due to broad band ^{1}H decoupling has not been a problem when gated two-level decoupling is used, either with or without employment of multiple pulse sequences such as MLEV.)

Freshly isolated rat hepatocytes, because of their relative fragility, are much more difficult to incubate in the NMR probe, especially at cell densities useful for measuring intracellular metabolites by ^{13}C NMR. Several investigators have described various cell perfusion systems for NMR[7]; however, to date none of these has been demonstrated to be appropriate for the fragile liver cell. Earlier, we measured ^{31}P and ^{13}C spectra of isolated rat hepatocytes at 25° while depending on endogenous catalase activity to provide oxygen as small quantities of H_2O_2 in saline were gently mixed into the cellular suspension at regular intervals via computer control.[8,9] Clearly, these important but difficult cells require the development of an optimized scheme for incubation/oxygenation in the NMR probe. Partially for this reason, most of the ^{13}C NMR methods described here apply to perfused liver preparations.

Before Addition of ^{13}C-Labeled Substrate

In the typical experiment considered here, steady-state conditions are established in the presence of a ^{13}C-labeled substrate. Reproducibility is improved and interpretation of results is more straightforward if endogenous substrates remaining in the perfused liver preparation are first washed out and depleted by an initial 30- to 40-min period during which perfusate is not recirculated, even when the donor rat has been starved. This step would be omitted if the competition of flux from endogenous and labeled exogenous substrates were being examined by the methods described below.

[7] W. M. Egan, in "NMR Spectroscopy of Cells and Organisms" (R. K. Gupta, ed.), Vol. 1, p. 135. CRC Press, Boca Raton, Florida, 1987.

[8] S. M. Cohen, S. Ogawa, and R. G. Shulman, *Proc. Natl. Acad. Sci. U.S.A.* **76**, 1603 (1979).

[9] S. M. Cohen and R. G. Shulman, in "Noninvasive Probes of Tissue Metabolism" (J. S. Cohen, ed.), p. 119. Wiley, New York, 1982.

After the wash-out period, the simple substrate-free Krebs bicarbonate medium is exchanged for one containing fresh, washed erythrocytes; this medium is then recirculated. One ^{13}C spectrum is measured at this point before addition of any substrate. For ease of interpretation, this ^{13}C natural abundance background spectrum of a given liver is subtracted from each spectrum accumulated after addition of ^{13}C-labeled substrate. ^{13}C natural abundance spectra of liver are typically dominated by signals from triacylglycerols. Because of the sensitivity of ^{13}C relaxation rates of carbons in lipids to temperature, the cleanness of the subtraction procedure depends on careful equilibration of the temperature of liver and medium in the NMR perfusion apparatus prior to measurement of the natural abundance spectrum. Even in challenging cases, such as subtraction of the intense ^{13}C natural abundance background of liver from the genetically obese (*ob/ob*) mouse,[10] the subtraction can be clean and essentially complete if adequate care is given to thermal equilibrium and if other possible changes are minimized. For example, use of a properly constituted perfusion medium, described above, is required if liver swelling is to be discouraged.

Calibration of ^{13}C Intensities

An efficient method for measuring ^{13}C spectra is required to achieve good time resolution. For this reason, a low flip angle and a pulse repetition time much less than 5 T_1 are used. ^{13}C spectra are usually measured with broadband ^1H decoupling to improve the signal-to-noise ratio and to facilitate interpretation. Consequently, all spectral intensities used in calculations must be corrected for the nuclear Overhauser effect (NOE) and T_1, or saturation, effects. Only corrected values are used in this chapter. The most convenient approach for making these corrections usually depends on whether absolute quantities or simply ratios of intensities within a given spectrum are required.[2,5] When ^{13}C spectra are measured under nonsaturating conditions or when appropriate intensity corrections are made, specific label distributions measured in the living cell by ^{13}C NMR agree closely with the ^{14}C isotopic distribution measured by standard isolation procedures in extracts of the same doubly labeled samples.[11] In processing ^{13}C spectra, the signal-to-noise ratio is improved and resolution is not hurt by use of optimum filtering, i.e., by use of a line broadening on the order of the observed linewidth. At 90.56 MHz, application of 8–12 Hz line broadening is typical for perfused liver.

[10] S. M. Cohen, *Ann. N.Y. Acad. Sci.* **508**, 109 (1987).
[11] S. M. Cohen, R. Rognstad, R. G. Shulman, and J. Katz, *J. Biol. Chem.* **256**, 3428 (1981).

[13C] Assays of Enzyme Regulation of Metabolic Flux

Metabolism of [13C]-Labeled Glycerol: Modeling of Pentose Cycle

Use of Steady-State Assumption. ^{13}C NMR measurements of fluxes through enzymes are demonstrated most succinctly by suspensions of rat liver cells carrying out gluconeogenesis from [2-^{13}C]glycerol. The pathway from [2-^{13}C]glycerol to L-glycerol-3-P and into the C-2 and C-5 positions of glucose in cells from fasted donor rats is readily followed by ^{13}C NMR.[8,11] In the oxidative part of the pathway involving glucose-6-P dehydrogenase (GPD), [1,4-^{13}C]pentoses are formed from [2,5-^{13}C]glucose-6-P with loss of CO_2. Further reactions in the pentose phosphate pathway involve transaldolase (TA) and transketolase (TK):

$$P + P \stackrel{TK}{\rightleftharpoons} GAP + S \tag{1}$$

$$GAP + S \stackrel{TA}{\rightleftharpoons} E + \text{fructose-6-P} \tag{2}$$

$$E + P \stackrel{TK}{\rightleftharpoons} GAP + \text{fructose-6-P} \tag{3}$$

in which P denotes either xylulose-5-P or ribose-5-P, GAP is glyceraldehyde-3-P, S is sedoheptulose-7-P, and E is erythrose-4-P. The sum of these three reactions, where it is considered that the pentoses are all in equilibrium, is

$$3P \rightleftharpoons GAP + 2\text{fructose-6-P}$$

All the flux through the oxidative pentose path is assumed to be returned to hexoses; that is, the purpose of the cycle here is the regeneration of NADPH.

Calculation of the activity of the pentose cycle is a canonical example of use of the steady-state assumption in physiological ^{13}C NMR. If the three [1,4-^{13}C]pentoses formed in the oxidative part of the pathway from [2,5-^{13}C]glucose-6-P go through Eqs. (1)–(3) irreversibly, the two hexoses formed are [1,3,5-^{13}C]fructose-6-P and [1,5-^{13}C]fructose-6-P. With the assumption of metabolic and isotopic steady state,[12] the rate of GDP (V_{GPD}) relative to the rate of glucose-6-phosphatase (V_{GLP}) can be calculated by setting isotopic inflow into C-1 of glucose-6-P equal to outflow:

$$\text{Inflow} = (2/3)\,V_{GPD}\,\text{C-2} = \text{outflow} = V_{GLP}\,\text{C-1} + V_{GPD}\,\text{C-1} \tag{4}$$

where C-1 and C-2 represent the ^{13}C enrichments at the first and second carbons of glucose-6-P (or glucose). Rearrangement of Eq. (4) gives

$$V_{GPD}/V_{GLP} = \text{C-1}/[(2/3)\text{C-2} - \text{C-1}] \tag{5}$$

[12] J. Katz, B. R. Landau, and G. E. Bartsch, *J. Biol. Chem.* **241**, 727 (1966).

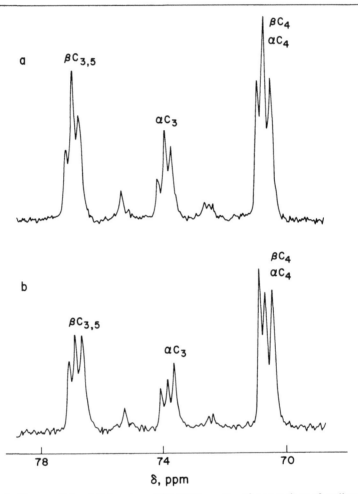

FIG. 1. Glucose C-3 and C-4 region of ^{13}C NMR spectra of suspensions of rat liver cells. (a) Cells incubated with [1,3-^{13}C]glycerol plus unlabeled fructose. (b) Cells incubated with [1,3-^{13}C]glycerol alone.[8]

As measured by ^{13}C spectra of rat liver cells and by use of Eq. (5), the liver glucose-6-P dehydrogenase reaction rate proceeds at 17% of the gluconeogenic rate.[11]

Exploitation of [13]C–[13]C J Splitting. Additional, complementary information on pentose cycle fluxes is obtained by administration of a different labeled glycerol, [1,3-^{13}C]glycerol. As expected with this labeled substrate, carbons 1, 3, 4, and 6 of glucose are intensely and approximately equally labeled via the major gluconeogenic pathway. The peaks labeled as arising from C-3 and C-4 of the α and β anomers of glucose in Fig. 1b are apparent triplets. The outer doublet of each apparent triplet is due to the J coupling that arises whenever a ^{13}C is next to another ^{13}C in the same molecule, as would occur for C-3 and C-4 in glucose formed from condensing two [1,3-^{13}C]trioses. However, if only one of the trioses were labeled, there would be only a single unsplit peak—the center singlet. Comparison of the intensity of the center singlet (S) with the intensity of the outer doublet (D) allows measurement of contributions from *both* labeled and unlabeled triose sources to the total glucose produced. As an example, from the S/D ratio measured in the ^{13}C spectrum of liver cells incubated with [1,3-^{13}C]glycerol and unlabeled fructose shown in Fig. 1a and the theoretical expression for this ratio, derived by taking into account the ^{13}C distribution at the triose level, it was estimated that 22% of the flux into glucose came from unlabeled fructose.[8]

^{13}C–^{13}C splitting patterns can be further exploited when [1,3-^{13}C]glycerol is the gluconeogenic substrate to probe the nonoxidative pentose branch. Consider two alternative pathways for [1,3,4,6-^{13}C]hexose to follow in the pentose cycle. (1) The three [2,3,5-^{13}C]pentoses formed in the oxidative pathway go through the reactions of Eqs. (1)–(3) irreversibly and form two hexoses, [2,4,6-^{13}C]fructose-6-P and [2,3,4,6-^{13}C]fructose-6-P. (2) We assume that the TK reactions of Eqs. (1)–(3) are reversible, with a rate on the order of the pentose cycle, whereas the TA reaction is undirectional. As a consequence, label distribution in the hexoses is affected in two ways: (a) TK exchange can convert [2,3,5-^{13}C]pentoses into hexoses and (b) it must return an equivalent amount of hexoses to the pentose pool. The first of these (a) will create [2,3,4,6-^{13}C]hexose. The second (b) will create [1,3,5-^{13}C]pentoses, three of which must be consumed via Eqs. (1)–(3) to form one [1,3,4,6-^{13}C]fructose-6-P and one [1,4,6-^{13}C]fructose-6-P. Thus, fitting the intensity distribution observed at the C-2 multiplet of glucose[8] provides a stringent test of the model pathway, specifically, for the relative forward and reverse flow through TK in the nonoxidative pentose branch.

^{13}C NMR Estimates of Relative Fluxes through Krebs Cycle

Competition of Pyruvate Dehydrogenase (PDH) with Pyruvate Carboxylase (PC) for Entry of Pyruvate into Krebs Cycle. The relative proportion of pyruvate entering the Krebs cycle by these two routes (Fig. 2) can be estimated from the ^{13}C enrichments at the individual carbons of glutamate in perfused rat liver when [3-^{13}C]alanine is the only exogenous substrate present. As shown schematically in Fig. 2, under these conditions the proportion of [3-^{13}C]pyruvate (the result of direct conversion of [3-^{13}C]alanine) entering the Krebs cycle by decarboxylation to [2-^{13}C]-acetyl-CoA relative to the proportion entering by carboxylation to

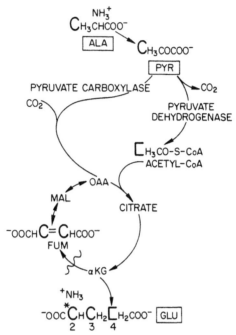

FIG. 2. Simplified model pathway. The original label at C-3 of alanine (boldface C) is followed through pyruvate carboxylase into the Krebs cycle, where in the symmetrical molecule fumarate (FUM) the label is found with equal probability at either of the two middle carbons (boldface Cs). The two middle carbons of oxaloacetate (OAA) label glutamate C-2 and C-3; the asterisk follows the position of the original, unscrambled label through pyruvate carboxylase and into glutamate C-2. The original label at C-3 of alanine is also followed through PDH into acetyl-CoA, which is labeled only at C-2 (boldface square C). C-2 of acetyl-CoA is followed into citrate and α-ketoglutarate (αKG) and, hence, into glutamate C-4 (boldface square C).

[3-^{13}C]oxaloacetate can be estimated from the ^{13}C label at glutamate C-4 relative to the enrichments at glutamate C-2 and C-3. The three ^{13}C spectra shown in Fig. 3 follow changes in concentrations of several metabolites over 10-min intervals commencing 10, 50, or 80 min after the initial administration of [3-^{13}C]alanine to a perfused liver. To focus only on glutamate C-2, C-3, and C-4, we note that the ^{13}C label at glutamate C-4 achieved its steady-state level by 30 min postsubstrate and that the relative ^{13}C enrichments at glutamate C-2 and C-3 are in a ratio of 70:30 (corrected for T_1 and NOE).

This approach was tested by examining three different liver preparations that were selected to provide large gradations in relative flux through PDH and to demonstrate the utility of ^{13}C NMR in evaluating factors responsible for directing the pathway chosen. The relative ^{13}C enrichments of glutamate C-4, C-3, and C-2 were 39 ± 3%, 30 ± 1%, and

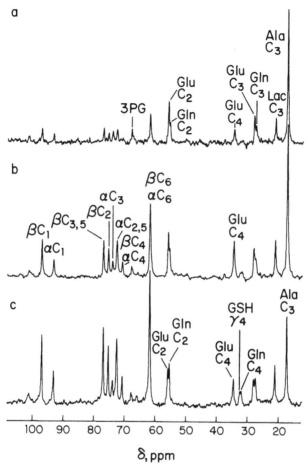

FIG. 3. ^{13}C spectra of perfused liver from a streptozotocin-induced diabetic rat at 35 ± 1°. (a) Spectrum was measured 10–20 min after the addition of [3-^{13}C]alanine, which was maintained at about 10 mM, (b) spectrum was measured 50–60 min after the addition of [3-^{13}C]alanine, and (c) spectrum was measured 80–90 min after the addition of [3-^{13}C]alanine. Labeled ^{13}C peaks include those of the α and β anomers of glucose: βC-1 and αC-1 through βC-6 and αC-6. Other abbreviations: Glu C-2, glutamate C-2; Gln C-2, glutamine C-2; GSHγ_4, C-4 of the glutamyl moiety of reduced glutathione; Lac C-3, lactate C-3; and Ala C-3, alanine C-3.[2]

70 ± 1%, respectively, in streptozotocin-induced diabetic liver (n = 7); 85, 29, and 71% in liver from fed control rats (n = 3); and 13, 30, and 70% in liver from 24-hr starved control rats.[13] From these ^{13}C enrichments we estimate that in liver from fed normal rats, the proportion of pyruvate

[13] S. M. Cohen, *Biochemistry* **26**, 581 (1987).

entering the Krebs cycle by the PDH route relative to the pyruvate carboxylase route was $1:(1.2 \pm 0.1)$; in liver from 24-hr starved controls this ratio was $1:(7.7 \pm 2)$ and in liver from streptozotocin-induced diabetic rats this ratio was $1:(2.6 \pm 0.3)$. Thus, the PC route is estimated to be about 6.4-fold more active than the PDH route in liver from starved controls compared with liver from fed controls, but only 2.2-fold more active in liver from chronically diabetic rats than in liver from fed normal control rats. This determination of the relative proportion of pyruvate entering the Krebs cycle by the competing PDH and PC routes is the ^{13}C NMR analog of the pioneering *in vivo* ^{14}C tracer methods of Freeman and Graff[14] and Koeppe *et al.*,[15] who carried out the demanding isolation and complete degradation of glutamate extracted from rat liver after the administration *in vivo* of either [2-^{14}C]alanine or [2-^{14}C]pyruvate.

Note also that the degree of randomization measured at glutamate C-2 and C-3 for these perfused liver preparations is consistent with the composite rate of mitochondrial malate dehydrogenase and fumarase (fumarate hydratase) exchange being 1.5 times the net rate of the Krebs cycle (Fig. 2).

Relative Fluxes through Krebs Cycle and into Glucose: Exploitation of ^{13}C–^{13}C J Splitting. A useful stratagem for following metabolic fluxes is the coadministration of two different ^{13}C-labeled substrates, one of which is labeled at two adjacent carbons. In this way, a well-defined multiplet structure, due to ^{13}C–^{13}C J coupling, is introduced into the NMR peaks of several key metabolites. In a manner analogous to the apparent triplets observed in glucose synthesized from [1,3-^{13}C]glycerol with respect to fluxes at the triose level, multiplet structures observed when [2-^{13}C]pyruvate and [1,2-^{13}C]ethanol, for example, are coadministered to perfused rat liver are diagnostic of fluxes through the Krebs cycle.[5] That is, beyond giving the percentage label at a particular carbon, ^{13}C NMR measurements also give the distribution of labeled carbons in the *same* molecule from the ^{13}C–^{13}C J splittings. The spectra shown in Figs. 4 and 5 suggest the sort of information contained in the ^{13}C multiplets.

Two labeled fluxes into the mitochondrial acetyl-CoA pool are possible. One flux arises from the conversion of [2-^{13}C]pyruvate to [1-^{13}C]acetyl-CoA through the activity of PDH. The second arises from the production of [1,2-^{13}C]acetyl-CoA via oxidation of [1,2-^{13}C]ethanol as shown in Fig. 4.[5] The multiplet structure observed at C-3 of β-hydroxybutyrate in spectra of perfusates (e.g., Fig. 5) gives the relative contributions of [1,2-^{13}C]acetyl-CoA and [1-^{13}C]acetyl-CoA directly. The center singlet is due to C-3 in a ^{13}C-3–^{12}C-4 moiety derived from [1-^{13}C]acetyl-CoA. Flank-

[14] A. D. Freeman and S. Graff, *J. Biol. Chem.* **233**, 292 (1958).
[15] R. E. Koeppe, G. A. Mourkides, and R. J. Hill, *J. Biol. Chem.* **234**, 2219 (1959).

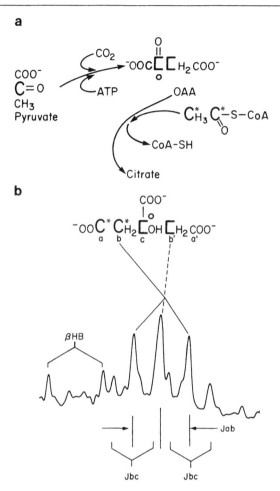

FIG. 4. (a) Simplified pathway from [2-¹³C]pyruvate and [1,2-¹³C]ethanol into citrate. The main flux of pyruvate into the Krebs cycle is via pyruvate carboxylase. Randomization of the ¹³C label in mitochondrial OAA occurred as shown in Fig. 2. The small open circle follows the original label at pyruvate C-2. The main flux into the mitochondrial acetyl-CoA pool is via the oxidation of [1,2-¹³C]ethanol (boldface Cs with asterisk). (b) The citrate methylene region of the ¹³C spectrum of perchloric acid extract of a freeze-clamped liver. This liver had been perfused with 9 mM [2-¹³C]pyruvate, 7 mM [1,2-¹³C]ethanol, and 3.6 mM NH$_4$Cl under steady-state conditions.[5]

ing this singlet is a large doublet, with splitting $J_{\text{C-3,C-4}}$ = 39.06 Hz, which arises from C-3 in a ¹³C-3–¹³C-4 moiety derived from [1,2-¹³C]acetyl-CoA. Each member of the $J_{\text{C-3,C-4}}$ doublet is further split by the less probable interaction with a ¹³C-1–¹³C-2 moiety; this splitting is $J_{\text{C-3,C-2}}$ = 36.62 Hz. The center singlet also showed $J_{\text{C-3,C-2}}$ splitting in convolution difference

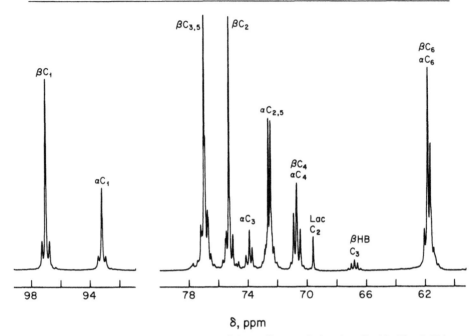

FIG. 5. [13]C spectrum of the perfusate of the rat liver perfusion described in Fig. 4. This spectrum represents 10,800 scans of 45° FIDs, with a 3.5-sec delay between pulses. Only the region between 60 and 100 ppm is shown. Abbreviations are as given for Fig. 3, and βHB C-3 is β-hydroxybutyrate C-3.[5]

spectra. The ratio of the total intensity of C-3 from [13]C-3–[13]C-4 moieties to that from [13]C-3–[12]C-4 moieties is 2.4. This value is consistent with the relative contributions of [1,2-^{13}C]acetyl-CoA and [1-^{13}C]acetyl-CoA to the mitochondrial pool being in a ratio of 71 : 29. The further $J_{C-3,C-2}$ splitting observed is consistent with a negligible unlabeled contribution to the pool. Similarly, the citrate C_b to C_b' intensity ratio (Fig. 4) does not support a significant endogenous acetyl-CoA contribution.

Thus, the ^{13}C multiplets measured for citrate, glutamate, β-hydroxybutyrate, and glucose can be used to unravel the pathways followed by the carbon backbones of these metabolites through the Krebs cycle. These [13]C–[13]C multiplets contain, in effect, an almost complete history of a metabolite with respect to its labeled sources of oxaloacetate and acetyl-CoA and the extent of contributions from unlabeled endogenous sources. First-order analytical expressions[5] describing the distribution of ^{13}C intensity within the multiplets of these metabolites can be solved to give a quantitative estimate of the fluxes. Fluxes are defined unambiguously by analytical expressions, to the accuracy of the model used. Because ^{13}C peaks are broadened in spectra of perfused liver, intensities and splittings within multiplets must usually be measured in spectra of the perfusates or of perchloric acid extracts of liver that had been freeze-clamped at the end of the perfusion period. Although this procedure gives a composite picture of the labeling that existed at that time, the results are representative when perfusions are carried out under steady-state conditions.

One example from the set of analytical expressions that describe the observed intensity ratios will serve to illustrate this approach. The intensity of the glucose C-4 doublet : singlet ratio (Fig. 5) can be written as

$$D/S = y/[y(1 - z) + (1 - y) + (1 - y)(1 - z)] \qquad (6)$$

where y equals $f([1,2-^{13}C]$acetyl-CoA); $(1 - y)$ equals $f([1-^{13}C]$acetyl-CoA); and $(1 - z)$ equals f(methylene ^{13}C in oxalacetate derived from $[2-^{13}C]$pyruvate) (see Fig. 4). Probability is denoted by f. In this way, a set of simultaneous equations can be generated and solved in general form. When the measured intensity ratios are substituted into these expressions, relative fluxes under the conditions of the experiment can be estimated. Because a redundancy of information is available by this approach, checks on self-consistency are built into the estimated fluxes.

^{13}C NMR Monitor of Biosynthesis of Fatty Acids: Kinetics and Pathways

As mentioned above, the relative proportion of pyruvate entering the Krebs cycle by the PDH route was greatest in perfused liver from fed control donor rats. This observation and the well-recognized maximal contribution of acetyl-CoA to the cytosolic process of fatty acid synthesis in liver of fed normal rats both suggest the use of ^{13}C NMR to monitor the incorporation of ^{13}C label into fatty acids in this liver preparation. To demonstrate this application of ^{13}C NMR, consider livers from fed normal control rats perfused under conditions of metabolic and isotopic steady state with specifically labeled substrates that are converted to either $[2-^{13}C]$acetyl-CoA or $[1-^{13}C]$acetyl-CoA, which in the *de novo* synthesis pathway label alternate carbons in fatty acids: $[2-^{13}C]$acetyl-CoA is incorporated into fatty acids according to

$$-OOC-^{13}CH_2-CH_2-(^{13}CH_2-CH_2)_{n/2}-^{13}CH_2-CH_2-^{13}CH_3$$
$$\;\;\; z \;\;\; f \;\;\;\;\; h \;\;\;\;\;\;\;\;\; c,d \;\;\;\;\;\;\; e \;\;\;\;\; g \;\;\;\;\; a$$

1

whereas $[1-^{13}C]$acetyl-CoA is incorporated into fatty acids according to

$$-OO^{13}C-CH_2-^{13}CH_2-(CH_2-^{13}CH_2)_{n/2}-CH_2-^{13}CH_2-CH_3$$
$$\;\;\; z \;\;\; f \;\;\;\;\; h \;\;\;\;\;\;\;\;\; c,d \;\;\;\;\;\;\; e \;\;\;\;\; g \;\;\;\;\; a$$

2

where c, d refers to the resonances arising from the methylene carbons. With the exception of the repeating methylene carbons, fatty acyl carbons labeled by [1-^{13}C]acetyl-CoA (from [2-^{13}C]pyruvate) gave rise to resonances distinguishable on the basis of chemical shift from those observed when label was introduced by [3-^{13}C]alanine plus [2-^{13}C]ethanol, which are converted to [2-^{13}C]acetyl-CoA. Thus, measurement of ^{13}C enrichment at several specific sites in the fatty acyl chains in time-resolved spectra of perfused liver offers a novel way of monitoring the kinetics of the biosynthesis of fatty acids (see Fig. 6.) The sensitivity of the method permitted serial observations with a 10-min time resolution.[13]

In addition to measurement of the rate of synthesis of fatty acids, several inferences about pathways can be drawn from these NMR measurements.

1. It is possible to distinguish the contributions of chain elongation from those of the *de novo* synthesis pathway. Under conditions for **2**, incorporation of ^{13}C label into the carbonyl carbon, z, occurs by both *de novo* and chain elongation pathways, whereas incorporation of label at carbon g (–^{13}CH$_2$–CH$_3$) in the terminal acetyl unit occurs only by the *de novo* synthesis pathway. It follows directly that measurement of the ratio of ^{13}C enrichment at carbon z to that at carbon g under conditions for **2** gives the relative contribution of chain elongation of endogenous fatty acids with labeled acetyl-CoA units.

2. Average chain length for the ^{13}C-labeled fatty acids produced can be estimated from the ratio of ^{13}C enrichment at the repeating methylene carbons, peaks $c + d$, to the enrichment at either the terminal –^{13}CH$_3$ (carbon a for **1**) or –^{13}CH$_2$–CH$_3$ (carbon g for **2**). That is, the average chain length is $2(c + d)/a + 6$ (**1**) or $2(c + d)/g + 6$ (**2**). None of these ^{13}C resonances is obscured by interfering signals from other labeled metabolites under our conditions.[13]

3. Finally, the proportion of the product of the *de novo* synthesis pathway that served as precursor for monoenoic acid can be estimated from the time course for ^{13}C enrichment at the olefinic fatty acyl carbon. As shown in Fig. 6, under conditions for **1**, the ratio of ^{13}C enrichment at –$(CH_2)_n$^{13}C= to that at carbon a (^{13}CH$_3$–CH$_2$–) provides the necessary information.

^{13}C NMR Assay of Pyruvate Kinase Flux

The rate of pyruvate kinase flux under steady-state conditions of active gluconeogenesis in perfused liver can be determined by ^{13}C NMR. In the PEP cycle (pyruvate → PEP → pyruvate) in rat liver, the path in the gluconeogenic direction from pyruvate to PEP is catalyzed by a complex

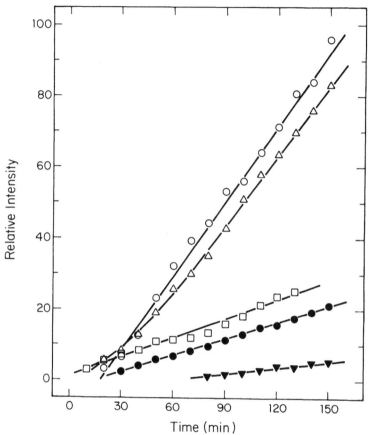

FIG. 6. Time course of ^{13}C enrichment in fatty acyl carbons in perfused liver from fed normal control rats. Livers were perfused under conditions of metabolic and isotopic steady state. Enrichment was measured at $-(^{*}CH_2)_n-$ (peak d) for livers perfused with 10 mM [3-^{13}C]alanine and 7.3 mM [2-^{13}C]ethanol ($n = 2$) ($\triangle$); with 9.1 mM [2-^{13}C]pyruvate, 5.5 mM NH$_4$Cl, and 7.3 mM unlabeled ethanol ($n = 2$) ($\square$); or with 9.1 mM [2-^{13}C]pyruvate, 5.5 mM NH$_4$Cl, and 7.3 mM [2-^{13}C]ethanol ($n = 2$) ($\bigcirc$). ^{13}C enrichments at $^{*}CH_3-CH_2-$ ($\bullet$) (peak a) and at $-(CH_2)_n-^{*}C=$ ($\blacktriangledown$) are each given for the average of four liver perfusions, two with the [3-^{13}C]alanine and [2-^{13}C]ethanol substrate and two with the [2-^{13}C]pyruvate, NH$_4$Cl, and [2-^{13}C]ethanol substrate. Symbol size, in general, covers the range of normalized intensity values.[13]

sequence of enzymes, whereas in the glycolytic direction a single enzyme, pyruvate kinase, catalyzes the reaction PEP → pyruvate (Fig. 7). Although activity is greater in the fed control state, there is considerable flux through pyruvate kinase (PK) under conditions of active gluconeogenesis in the starved state, i.e., under the conditions of this assay.

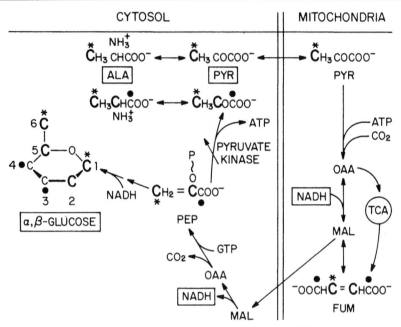

FIG. 7. Simplified model gluconeogenic pathway. The original ^{13}C label at C-3 of alanine (boldface C with asterisk) is followed into the Krebs cycle, where randomization occurs at fumarate (FUM) as described in Fig. 2. Further randomization in the Krebs cycle introduces a lesser amount of label into the terminal carbons of FUM (●). MAL, Malate.

In this ^{13}C NMR determination, a ^{13}C-labeled gluconeogenic substrate ([3-^{13}C]alanine is shown in Fig. 7) that enters the pathway as specifically labeled pyruvate is administered. As Fig. 7 shows schematically, the label that is randomized in the Krebs cycle becomes incorporated into PEP prior to its appearance in either glucose by the usual route or into pyruvate by the action of PK. As noted in Fig. 7, the label distribution in PEP is reflected by ^{13}C enrichments at glucose C-4, C-5, and C-6. Pyruvate bearing the randomized label (at C-1 and C-2 as well as C-3) of its PEP precursor is interconverted to alanine. This randomized label will be effectively trapped in alanine if a sufficiently large pool of alanine with the original label ([3-^{13}C]alanine in this example) is present. The 5–10 mM [3-^{13}C]alanine in the 65 ml of perfusate typically used provides an efficient trap. Under conditions of metabolic[16] and isotopic steady state, an analyt-

[16] By symbolizing the PEP cycle (P $\xrightarrow{k_1}$ P* $\xrightarrow{k_{-1}}$ P) and the reactions from PEP to glucose (2P* $\xrightarrow{k_2}$ G) and applying the metabolic steady-state assumption, we have $dp^*/dt = k_1p - k_{-1}p^* - k_2p^{*2} = 0$ and $dg/dt = k_2p^{*2}$. That is, $-dp/dt = 2V_G = k_1p - k_{-1}p^*$, which can be compared with $V_{PK} = k_{-1}p^*$. Substrate levels must be kept fairly constant.

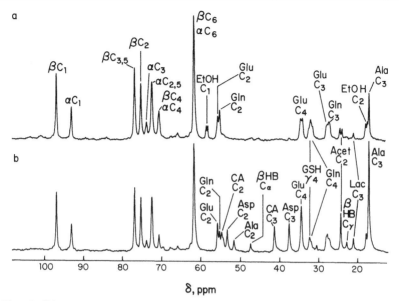

FIG. 8. ^{13}C spectra of livers from streptozotocin-induced diabetic rats. Livers were perfused in the absence (a) or presence of insulin (b). (a) Spectrum is part of a time sequence of spectra and was accumulated 170–180 min after the initial addition of 10 mM [3-^{13}C]-alanine and 7.3 mM [1,2-^{13}C]ethanol. (b) Spectrum is taken from a similar series of spectra taken of another diabetic liver and was measured 170–180 min after initial addition of 10 mM [3-^{13}C]alanine and 7.3 mM [2-^{13}C]ethanol. This liver was treated exactly the same as the liver in a except that insulin was maintained in the perfusate at 7 nM. Abbreviations as given in Figs. 3 and 5, including the following: ETOH, ethanol; CA, N-carbamoylaspartate; Asp, aspartate; Acet, acetate.[17]

ical expression can be written for the flux of PEP through PK relative to the flux of PEP → glucose:

$$\frac{\text{Pyruvate kinase flux}}{(2)(\text{gluconeogenic flux})} =$$

$$\frac{\text{Ala(C-2/C-3)}[1 + (1 - \phi)\text{Glc(C-6/C-5)} + (1 - \phi')\text{Glc(C-4/C-5)}]}{1 - [\text{Glc(C-6/C-5)}][\text{Ala(C-2/C-3)}]} \qquad (7)$$

in which all components are readily measured in ^{13}C NMR spectra of perfused rat liver (see Fig. 8) and in spectra of the corresponding perfusate.[17] That is, Ala(C-2/C-3) is the ratio of ^{13}C enrichment at C-2 to that at C-3 of alanine; Glc(C-6/C-5) is the corresponding ratio for glucose C-6 and C-5; and $\phi(\phi')$ is the fraction of the ^{13}C-labeled alanine pool in which both C-2 and C-3 (both C-1 and C-2) are labeled in the same molecule. $\phi(\phi')$ is

[17] S. M. Cohen, *Biochemistry* **26,** 573 (1987).

conveniently measured in the ^{13}C spectrum of the corresponding perfusate from the ratio D/(D + S) for C-1 (C-4) of the glucose produced. Substrate levels must be kept fairly constant. Continuous monitoring of the substrate ^{13}C peak is used to prompt addition of small increments of substrate at frequent (~15 min) intervals. Substrate levels should not be unphysiologically high, e.g., >10 mM; that is, a rate-limiting step may then cause the steady-state assumption,[16] $dp^*/dt = 0$, to be invalid. The assay includes a check on the reuse of pyruvate with the randomized label (that is, on the adequacy of the trapping pool); this check uses the ^{13}C enrichment measured at glutamate C-4 and C-5 to estimate the flux from recycled pyruvate into the mitochondrial acetyl-CoA pool.[17]

It is useful to use liver preparations from variously conditioned donor rats to provide gradations of pyruvate kinase flux within the assay's requirement of active gluconeogenesis. For example, the rate of PK flux was 0.74 ± 0.04 of the gluconeogenic flux in liver from 24-hr starved controls; in liver from 12-hr starved controls, the relative PK flux increased to 1.0 ± 0.2, and in streptozotocin-induced diabetic liver, PK flux was undetectable. When streptozotocin-induced diabetic liver was treated with 7 nM insulin *in vitro,* a partial reversal of several of the differences noted between control and diabetic liver was demonstrated (Fig. 8). The major reversal caused by insulin was the induction of a large increase in both relative and absolute fluxes through pyruvate kinase in diabetic liver. [Measurement of the rate of gluconeogenesis[2] gives the absolute rate of PK flux using Eq. (7).]

Activity of the malic enzyme, which interconverts malate and pyruvate, is known to be low under conditions favoring gluconeogenesis and should therefore be a negligible source of error in this assay when liver from streptozotocin-induced diabetic or starved control rats is examined. However, it has been demonstrated by ^{13}C NMR that in liver from the starved *ob/ob* mouse, both the gluconeogenic pathway and the *de novo* synthesis pathway for fatty acid production are active simultaneously.[10] Because malic enzyme activity may increase under conditions favorable to lipogenesis, it may be necessary to correct the large apparent flux through pyruvate kinase measured by ^{13}C NMR for the malic enzyme contribution. One way to test this possibility is by use of the malic enzyme inhibitor 2,4-dihydroxybutyrate.[1]

As these several examples may illustrate, there is no single ^{13}C NMR approach to the regulation of metabolic flux by enzymes, but rather an expanding repertoire of approaches based on unique features of ^{13}C NMR and the model system studied.

[23] Monitoring Intracellular Metabolism by Nuclear Magnetic Resonance

By JACK S. COHEN, ROBBE C. LYON, and PETER F. DALY

Introduction

With the growth of magnetic resonance imaging (MRI) as a recognized tool in clinical diagnosis has come the hope that magnetic resonance spectroscopy (MRS or NMR) cannot be far behind. But there is a great deal of confusion in some quarters regarding the delineation of these two related techniques, and how they differ from other methods that provide spatial or metabolic information. MRS and positron emission tomography (PET) scanning are the only two methods employed in radiology that are basically metabolic in their mode of operation. Because MRS is the only metabolic method that is also noninvasive, it can be seen that MRS could potentially provide the physician with a valuable source of localized metabolic information that is not currently available.

On the basis of published results, the potential range of clinical applications of MRS is great. However, to date few examples of clinical diagnosis based on metabolic information have been forthcoming. Radda has focused on skeletal muscle metabolism using ^{31}P NMR spectroscopy,[1] and this is a fortunate choice because skeletal muscle is very accessible and homogeneous and is rich in high-energy phosphates. Using surface coils,[2] Radda and co-workers have carried out diagnostic studies on muscle pathologies, showing the potential of the modality.[3]

Two major problems arise in expanding the scope of the MRS method as a general clinical tool. First is the problem of spatial localization of the spectroscopic information, and second is the meaning of the signals obtained. There are three basic approaches to the problem of localization: (1) the use of a surface coil, (2) the use of purely radiofrequency pulse methods to define the region of interest (ROI), and (3) the use of magnetic field gradients that are adapted from MRI methods.[4] Comparisons of the

[1] G. K. Radda, *Science* **233**, 640 (1986).
[2] J. J. H. Ackerman, T. H. Grove, G. G. Wong, D. G. Gadian, and G. K. Radda, *Nature (London)* **283**, 167 (1980).
[3] B. D. Ross, G. K. Radda, D. D. Gadian, G. Rocker, M. Esiri, and J. Falconer-Smith, *Lancet* **1**, 1338 (1981).
[4] P. A. Bottomley, *in* "Magnetic Resonance Imaging and Spectroscopy" (T. F. Budinger and A. R. Margulis, eds.), p. 81. Soc. Magn. Reson. Med., Berkeley, California, 1986.

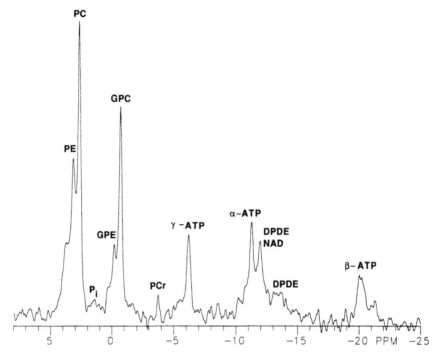

FIG. 1. Representative ^{31}P NMR spectrum at 162 MHz of wild-type MCF-7 human breast cancer cells ($\sim 10^8$/ml) perfused with IMEM media (P$_i$-free) in agarose gel thread (0.5 mm); 200 scans were accumulated with a recycle time of 40 sec and a 90° pulse. The peak assignments are denoted: PE, phosphoethanolamine; PC, phosphocholine; P$_i$, inorganic phosphate; GPE, glycerylphosphoethanolamine; GPC, glycerylphosphocholine; PCr, phosphocreatine; ATP, adenosine triphosphate; DPDE, diphosphodiesters; NAD, nicotine adenine dinucleotide.

advantages and disadvantages of these techniques are beyond the scope of this chapter, which focuses on the second problem, the actual metabolic information obtainable.

In spectra obtained by MRS methods these are two essential pieces of information apart from the spatial origin of the signals, namely, the chemical shift and the relative area of the signals observed. The chemical shift provides identification of the metabolite observed and its peak area provides its concentration, both by comparisons with known standards (Fig. 1). Determining these important parameters *in vivo* is not always easy. Diamagnetic susceptibility effects resulting from the heterogeneity of the body can result in unusual local shift effects,[5] as well as the effects of

---

[5] I. R. Young, S. Khenia, D. G. Thomas, C. H. Davis, and D. G. Gadian, *J. Comput. Assist. Tomogr.* **11**, 2 (1987).

metabolite compartmentation.[6] In addition, several metabolites may have overlapping signals that may not be resolvable even in extracts.[7] In order to identify unambiguously signals *in vivo,* a specific enzymatic process known to be associated with that metabolite should be manipulated.

In order to measure intracellular metabolism, it is necessary to define some terms; in particular, usage of the terms *in vivo* and *in vitro* is not always clear. It should be clear that it is possible to measure intracellular metabolism in intact, functioning mammalian cells both *in vivo* and *in vitro,* where *in vitro* literally means "in glass." Because this terminology was coined well before the advent of noninvasive methods such as NMR, and *in vitro* is generally taken to refer to purified noncellular enzyme systems, this terminology can be confusing. Consequently, wherever possible, we refer to the use of NMR to study metabolism in intact cells as *intracellular* studies, with a further designation of whether the cells are embedded in a matrix, contained in an organ or tissue, or observed in the whole animal. In doing so we affirm the reductive hypothesis, that because the whole animal is extremely complex, and because results obtained *in vivo* are often difficult to understand and analyze, it is essential that any such NMR studies be supported by measurements in a perfused cellular system.

Methods for NMR Studies of Intracellular Metabolism

Studies on Perfused Cells

Many studies of dense cell suspensions have been described,[8] but in this chapter we will refer only to studies in which a system has been used that approximates the *in vivo* cellular situation, which requires an efficient perfusion method. Although several perfusion setups involving dialysis fibers and beads[8,9] have been described, no metabolic studies have thus far been forthcoming from the application of these methods. We choose to describe here in detail the methods that are in principle both the simplest and the best, in that they most closely approximate the biological situation. These methods, collectively known as "gel thread" methods, should enable any investigator to grow any cell line in culture and subsequently

[6] S. J. Busby, D. G. Gadian, G. K. Radda, R. E. Richards, and P. J. Seeley, *Biochem. J.* **170,** 103 (1978).

[7] G. Navon, R. Navon, R. G. Shulman, and T. Yamane, *Proc. Natl. Acad. Sci. U.S.A.* **75,** 891 (1978).

[8] R. J. Gillies, T. J. Chresand, D. D. Drury, and B. E. Dale, *Rev. Magn. Reson. Med.* **1,** 155 (1986).

[9] W. Egan, *in* "Phosphorus NMR in Biology" (C. T. Burt, ed.), Vol. 1, p. 135. CRC Press, Boca Raton, Florida, 1987.

study a particular metabolic pathway or enzymatic process intracellularly in a standard 10-mm NMR probe.[10-12] We will then also describe the basic requirements for carrying out similar studies in perfused organs and in whole animals, with appropriate cautions.

To obtain sufficient signal-to-noise ratio in cellular systems generally requires a high density of cells (~10^8 per ml). Methods to perfuse cells in such densities have been reviewed.[8,9] Here we will describe the methods that have been the most widely applied. It is possible to utilize gels to entrap cells in other ways, including the use of gel beads obtained by a variety of procedures.[8] However, the methods described here are the simplest and most convenient to prepare a sterile preparation of cells suitable for NMR studies of intracellular metabolism.

The two methods described in this section allow different kinds of cells to be studied, notably anchorage-dependent and -independent (cancer) cells. They are not exactly equivalent methods as far as the extracellular environment is concerned, and these criteria will be discussed after the methods are described in detail.

Agarose Gel Thread Method

Preparations of Gel Threads. The fabrication of the agarose gel thread matrix containing cells (50% volume) is carried out as follows.[11] A solution (1.8%, w/v) of low-gelling-temperature agarose (SeaPlaque agarose, FMC Bioproducts, Rockland, Maine) is prepared in a balanced salt solution [0.4 mM Ca(NO$_3$)$_2$; 0.4 mM MgSO$_4$; 5.4 mM KCl; 103 mM NaCl; 24 mM NaHCO$_3$; 5.6 mM Na$_2$HPO$_4$; and 20 mM HEPES, pH 7.4], heated to 55°, and stored in a liquid state at 37° until required. Following harvest and washings, the cells are pelleted at 1640 g and mixed gently with an equal volume of the agarose solution at 37°. The gel matrix is then extruded under mild pressure through a coil of Teflon tubing (0.5-mm i.d.) that is chilled in an ice bath as shown in Fig. 2. The gel threads containing cells are then collected directly into media in a screw-cap NMR tube (Wilmad Inc., Buena, New Jersey). The threads are carefully compacted by the perfusion insert.

Perfusion Apparatus. The perfusion assembly is shown schematically in Fig. 3.[13] A custom-designed plastic insert confines the gel threads to a 2-ml volume at the bottom of the NMR tube. It also directs the inlet Teflon perfusion tubing and supports a spherical capillary bulb (18-μl

[10] D. Foxall and J. S. Cohen, *J. Magn. Reson.* **52,** 346 (1983).
[11] D. Foxall, J. S. Cohen, and J. B. Mitchell, *Exp. Cell Res.* **154,** 521 (1986).
[12] P. F. Daly, R. C. Lyon, E. Straka, and J. S. Cohen, *FASEB J.* **2,** 2596 (1988).
[13] R. C. Lyon, P. J. Faustino, and J. S. Cohen, *Magn. Reson. Med.* **3,** 663 (1986).

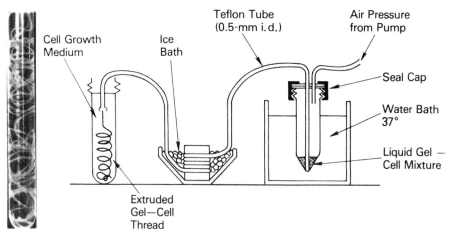

FIG. 2. Diagram of the apparatus used to embed cells within agarose gel threads. A mixture of cells in medium is extruded through a fine Teflon capillary in chilled ice. The gel thread is then extruded directly into medium in the 10-mm screw-cap NMR tube. An actual example is shown on the left.

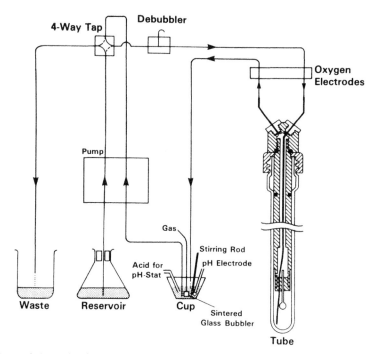

FIG. 3. Schematic of the perfusion system, showing the arrangement of the polyethylene insert.

capacity) containing dioxane-d_6 for adjusting the magnetic field offset. A peristaltic pump was used with a flow rate of 0.5–2.5 ml/min, depending on the density of the threads.

Basement Membrane Gel Method

Preparation of Gel Threads. Cells are embedded in the gel by mixing 0.1 ml of cells with 2 ml of 4° liquid basement membrane gel (Matrigel, Collaborative Research, Bedford, Massachusetts) in a test tube using precooled pipets.[12] The mixture is then pulled into 0.5-mm i.d. Teflon tubing attached to a syringe and allowed to gel at room temperature; 0.3-mm i.d. tubing can also be used. This is then extruded into a petri dish containing media and cells are allowed to grow in an incubator until the desired density. The density can be measured by dissolving a known volume of the threads with dispase and quantitating the cells in a counting chamber. The threads are transferred by removing the media and pouring 2 to 3 ml of the threads through a glass funnel into the screw-cap NMR tube, where they settle under their own weight. A custom-designed cap, with 0.5-mm i.d. inlet and outlet perfusion tubing (Fig. 4), is placed on the tube, and the cells are perfused at 0.5 ml/min with media at 37°.

Perfusion System. Figure 4 illustrates a simple system for maintaining the cells perfused and sterile.[12] The filters have a capacity of 20 liters and have been used for up to 1 month without problems. Filters up to 1000-liter capacity are available. The beaker containing media and 10% fetal calf serum is kept refrigerated under a 95% O_2/5% CO_2 or 95% air/5% CO_2 environment. The measured pH of the media remains between 7.35 and 7.45, which is also the measured pH of the waste fluid. A peristaltic pump from which the line can be easily removed enables the entire system to be moved to an autoclave or a cell culture hood. The use of filters rather than making the system a closed loop to maintain sterility makes it much simpler to change the media or to place additives. The media can be recirculated if desired. All of the components are commercially available except for the cap on the NMR tube. This is a Teflon screw cap that is machined with outlet and inlet openings for 0.5-mm i.d. Teflon tubing and is fitted with O rings to make airtight seals. The spectrometer temperature is set at 37°, and the medium in the Wilmad tube was measured to be 37°, indicating that its temperature equilibrates as the inlet tubing traverses the spectrometer stack. The cells continue to proliferate in the gel threads under these conditions. Because there is no interface between water and plastic or metal in the volume within the NMR coil using the Matrigel thread (as there are in other systems), distortion of the magnetic field is minimized and shimming is improved. Shimming was done on the free

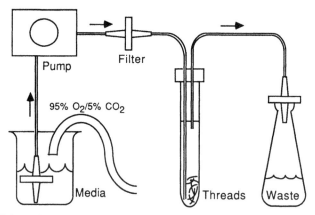

FIG. 4. Schematic representation of perfusion system for studying cells in basement membrane gel threads.

induction decay of ^1H, and a width at half-height of 8 Hz could be routinely obtained for H_2O as compared to 20 Hz by the agarose method.

NMR Studies of Perfused Organs

The apparatus to be described has been utilized for studying perfused rat ovaries in an NMR tube (S. Chen, R. C. Lyon, F. Haseltine, and J. S. Cohen, unpublished results), although the method is generally applicable for a variety of organs. The ovaries are surgically removed by the method of Koos et al.[14] Both ovaries are perfused through the left and right ovarian arteries by cannulating the aorta with a Teflon catheter and ligating all the intermediate arteries connecting with the aorta. The catheter is connected to the inlet line from a pump as shown in Fig. 5. A plastic insert with O rings fits into the NMR tube and forms a watertight seal. By screwing down the top of the insert and compressing the small O rings, the inlet and outlet lines are held tightly in place and will support the tube for placement in the magnet. The level of the perfusate in the tube is regulated by attaching the outlet line to a pump.

A study utilizing this approach has recently been reported by Haseltine et al.[15] ^{31}P spectroscopy was used to monitor adenosine induction of

[14] R. D. Koos, F. J. Jaccarino, R. A. Magaril, and W. J. LeMaire, *Biol. Reprod.* **30,** 1135 (1984).

[15] F. P. Haseltine, F. Arias-Mendoza, A. M. Kaye, and H. Degani, *Magn. Reson. Med.* **3,** 796 (1986).

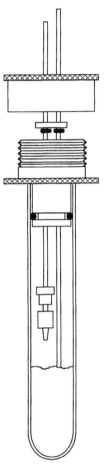

FIG. 5. Adapter for standard 10-mm NMR tube for small-organ perfusion. The inlet tube is connected to the organ for infusion, while the outlet tube bathes the entire organ; perfusion liquid enters and exits at the same rate. This arrangement allows easy adjustment of the height of the perfusion liquid.

FIG. 6. A close-up view of a sedated mouse in the plastic cradle with the subcutaneous tumor protruding into the coil. The grounded copper holder on the left positions the coil, supported by the copper bracket. The mouse is supported in the cradle by Velcro strips, and the body tissues are shielded by a Parafilm-coated thin copper foil. The tubing in the foreground provides a gas inlet.

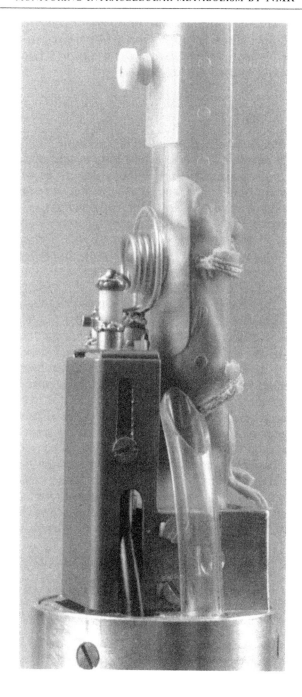

ATP synthesis in luteinized immature rat ovaries. Similar systems have been described for collecting NMR spectra of perfused liver,[16] perfused brain,[17] perfused kidney,[18] superfused skeletal muscle,[19] Langendorff perfused heart,[20] and perfused working heart.[21]

NMR Studies in Vivo

In Vivo Probe. A detailed description of a versatile multinuclear probe for *in vivo* NMR spectroscopy in a vertical-bore magnet of tissues in small animals has recently been reported.[22] Although this probe was specifically designed for monitoring ^{31}P, ^{13}C, and ^1H spectra of tumors subcutaneously implanted in mice, it could be adapted for studying a variety of nuclei and tissues. As shown in Fig. 6, the mouse is positioned vertically in a removable cylindrical plastic cradle and secured by thin Velcro strips laced through small holes in the cradle. The tumor protrudes through an adjustable slot centered on the resonance coil, while the nontumorous tissues are shielded from the coil. The shield consisted of a rectangular piece of copper foil (0.003 inches thick) with a hole in the center and covered with Parafilm for insulation. This is placed around the tumor between the mouse and the plastic cradle. The coils and resonance circuitry (probe inserts) are mounted on interchangeable copper brackets which are supported by a copper holder grounded to the aluminum probe body. The copper bracket can be adjusted vertically while the copper holder can be adjusted horizontally. The bracket assembly is positioned in the stack to center the resonance coil along the z axis for proper alignment of the B_1 field with the area of interest. Because the B_1 field is greatest within the coil, signal from the tumor tissue can be maximized by using a solenoid configuration. Variable capacitors are mounted through the bracket assembly immediately adjacent to the coil and are aligned vertically to allow external circuit tuning while the probe is positioned in the magnet. Examples of different probe inserts, including circuit diagrams, have been described in detail.[22] The most versatile probe was a concentric coil arrange-

[16] S. M. Cohen, R. G. Shulman, and A. C. McLaughlin, *Proc. Natl. Acad. Sci. U.S.A.* **76**, 4808 (1979).

[17] W. I. Norwood, C. R. Norwood, J. S. Ingwall, A. R. Castaneda, and E. T. Fossel, *J. Thorac. Cardiovasc. Surg.* **78**, 823 (1979).

[18] P. A. Sehr, P. J. Bore, J. Papatheofanis, and G. K. Radda, *Br. J. Exp. Pathol.* **60**, 632 (1979).

[19] M. J. Dawson, D. G. Gadian, and D. R. Wilkie, *J. Physiol. (London)* **267**, 703 (1977).

[20] P. B. Garlick, G. K. Radda, and P. J. Seeley, *Biochem. J.* **184**, 547 (1979).

[21] W. E. Jacobus, G. J. Taylor IV, D. P. Hollis, and R. L. Nunnally, *Nature (London)* **265**, 756 (1977).

[22] R. C. Lyon, R. G. Tschudin, P. F. Daly, and J. S. Cohen, *Magn. Reson. Med.* **6**, 1 (1988).

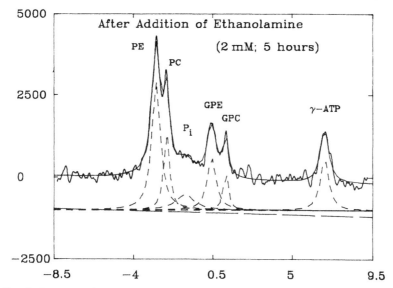

FIG. 7. Deconvolution of a ³¹P NMR spectrum of perfused MB231 human breast cancer cells using a summation of Lorentzian-shaped peaks. The results of the relative areas obtained for a series of closely related spectra after the addition of ethanolamine (2 mM, 5 hr) showed inconsistency for the overlapped peaks of PE and PC.

ment (two coaxial planar coils) used to observe over a broad range (¹³C to ³¹P) with the inner coil, and to decouple ¹H with the outer coil.

Analysis of Spectral Data

To obtain the concentrations of the metabolites, it is necessary to measure the relative areas of the signals. In practice these can depend on the relative relaxation times of the spins giving rise to the signals. It is necessary that these values be known so that either sufficiently long delay times can be used, or appropriate correction factors applied, in order to obtain quantitatively reliable data. An example of this might be the determination of an ATP : P_i ratio, wherein the ATP phosphates have relaxation times in the range of 0.3 to 1 sec and that of the P_i can be up to 3 sec, thus requiring a recycle time for multiple scans of ~15 sec to ensure complete relaxation of the P_i in order to obtain a quantitatively accurate ratio. It should also be noted that, strictly speaking, one should measure the relative areas of peaks in the spectrum, rather than measuring peak heights, particularly because the peaks vary in linewidth and because much of the area of NMR peaks lies in their base. Deconvolution using Lorentzian-shaped peaks is usually difficult with many overlapping signals (Fig. 7), and consequently the use of peak heights to measure relative

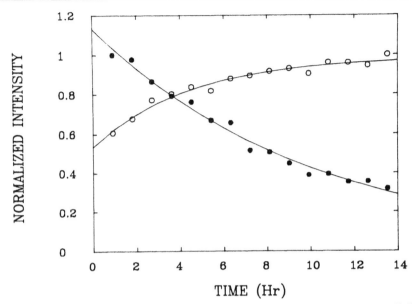

FIG. 8. Exponential fits to the intensity values (above corrected baseline) for the PE (○) and PC (●) ^{31}P NMR peaks following addition of ethanolamine (2 mM) to the perfusate of MB231 cells. Similar data generated by the deconvolution method (Fig. 7) were much more scattered.

metabolite concentrations has become routine, although this is only strictly correct when one is comparing equivalent samples or measuring a change as a function of a variable, such as time, when no change in linewidth is occurring (Fig. 8). Several mathematical manipulations of *in vivo* NMR data have been suggested to overcome the poor signal-to-noise ratio and broad overlapping peaks that are characteristic of these results,[23,24] but detailed discussion of these methods is beyond the scope of this chapter.

Applications

^{31}P and ^{13}C NMR spectroscopy have been used to monitor bioenergetics and phospholipid metabolism of several human tumor cell lines, both embedded in gel threads and subcutaneously implanted in nude mice. It is important to note that the ^{31}P spectra obtained for the same cell line in both of these two environments, namely, perfused cells embedded in

[23] F. Ni, G. C. Levy, and H. A. Scheraga, *J. Magn. Reson.* **66**, 385 (1986).
[24] S. J. Nelson and T. R. Brown, *J. Magn. Reson.* **75**, 229 (1987).

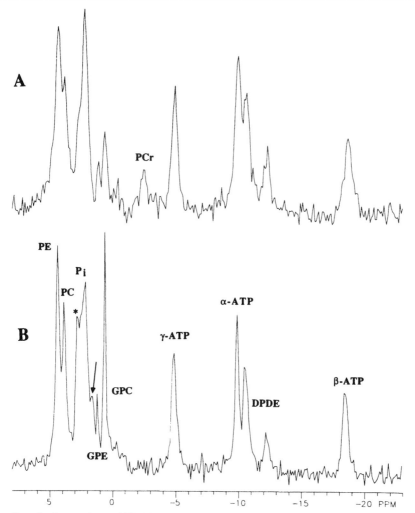

FIG. 9. Comparison of ^{31}P NMR spectra at 162 MHz of (A) *in vivo* and (B) *in vitro* perfused MB231 cells, both with a 5-Hz line broadening. The *in vivo* spectrum was obtained with 100 scans and a 5-sec recycle time. The *in vitro* spectrum is the sum of 56 hr of perfusion with a 5-sec recycle time.

basement membrane gel thread and tumor in a nude mouse, are very similar (Fig. 9). This to some extent justifies the use of cell perfusion to study intracellular metabolism as being relevant to the *in vivo* situation.[12] The applications that follow will focus on the studies of bioenergetics and phospholipid metabolism.

Bioenergetics

Perfused Cells. Energy metabolism in perfused MCF-7 breast cancer cells was investigated by monitoring the levels of ATP and P_i using ^{31}P NMR, and the utilization of glucose and the production of lactate were assessed by ^{13}C NMR.[25] It is well known that the bioenergetics of tumor cells is dominated by glycolysis, the substrate phosphorylation of glucose giving rise to ATP, rather than by the much more efficient oxidative phosphorylation.[26] The ability of cancer cells to rely on anaerobic glycolysis may be a survival mechanism that enables them to proliferate under adverse conditions, particularly in tumors in which dense masses of cells lack normal levels of oxygen. Development of drug resistance during exposure to antitumor agents results in further alterations in carbohydrate metabolism. MCF-7 cells embedded in agarose gel threads were incubated with 5 mM [1-^{13}C]glucose at 4° in a 20-ml closed perfused system. After jumping the temperature of the cells to 37°, the rate of loss of C-1 glucose signal (94.7 and 98.5 ppm) and the rate of increase in C-3 lactate signal (22.9 ppm) were measured with time by ^{13}C spectra. Both processes could be described by simple exponential equations. The rate constants for glucose utilization and lactate production were threefold greater for the adriamycin-resistant cell line (AdrR) compared to the wild-type (WT) parent cell line. These observations were consistent with results obtained by enzymatic assay of lactate production in MCF-7 cell extracts, where it was also found that the rate of glycolysis was three times faster in the AdrR cell line.[27] A maximum of 68% (WT) and 72% (AdrR) of the ^{13}C label from glucose appeared in [3-^{13}C]lactate. Most of the remaining label was incorporated into metabolites of pyruvate oxidation (utilization of lactate). Inhibition of pyruvate oxidation by 20 mM sodium azide (an inhibitor of respiration via cytochrome oxidase) enhanced the incorporation of label into lactate, with an increase to 88% in WT cells and 92% in AdrR cells, and totally blocked the reutilization of lactate. This inhibition of oxidative metabolism had no effect on ATP levels or the rates of glycolysis for either cell line. During a 60-min perfusion with a glucose-free buffer at 37°, ATP levels in both cell lines declined rapidly (20 min) and leveled off at 40% of their initial values, while the P_i values increased over twofold. Within 5 min after the addition of glucose the ATP levels fully

[25] R. C. Lyon, J. S. Cohen, P. Faustino, F. Megnin, and C. E. Myers, *Cancer Res.* **48**, 870 (1988).

[26] E. Eigenbrodt, P. Fister, and M. Reinacher, *in* "Regulation of Carbohydrate Metabolism" (R. Beitner, ed.), Vol. 2, p. 141. CRC Press, Boca Raton, Florida, 1985.

[27] R. J. F. Muindi, M. Shoemaker, and K. C. Cowan, *Proc. Am. Assoc. Cancer Res.* **27**, 272 (1986).

recovered. The addition of 20 mM azide during glucose depletion further reduced the ATP levels to 20% for WT and 5% for AdrR cells. The effect of azide on ATP levels was not evident until the cells were deprived of glucose. Although, ATP production in both cell lines was dominated by active glucose metabolism and substrate-level phosphorylation, in the absence of glucose a lower energy state is maintained by oxidative phosphorylation.

Bioenergetics in Vivo. Levels of ATP in breast cancer cells *in vivo* are approximately the same in cells undergoing perfusion, but there is a smaller P_i peak in cells perfused in agarose gel threads.[22] By contrast, cells grown in basement membrane gel exhibit both higher P_i and PDE peaks, more similar to the spectra of the same cells in *in vivo* tumors in nude mice.[12]

The natural abundance ^{13}C spectra of tumor tissue is characterized by a large envelope of methyl and methylene carbons (14–46 ppm), with a peak at 32 ppm that interferes with the clear observation of [3-^{13}C]lactate. In order to monitor glucose utilization and lactate production in tumor tissues by direct ^{13}C observation, [2-^{13}C]glucose (80 mg in 0.9% saline solution) was injected subcutaneously in nude mice bearing subcutaneously implanted LS-174T human colon carcinomas.[22] [13]C spectra (10-min accumulations) were collected from 30 to 300 min postinjection, and representative spectra are shown in Fig. 10. The [2-^{13}C]glucose signals (73.9 and 76.6 ppm) and [2-^{13}C]lactate signal (71.0 ppm), the only product from glucose metabolism, were clearly detectable and well resolved.

Phospholipid Metabolism

^{31}P NMR offers a unique opportunity not afforded by other methods to study noninvasively the control of the *de novo* pathways for the biosynthesis of phosphatidylcholine and phosphatidylethanolamine.[28] Although the pathways for synthesis of phosphatidylcholine and phosphatidylethanolamine were elucidated in the 1950s,[29] their control has been studied only in the past 10 years. The three-step pathway from choline to phosphocholine to CDP-choline to phosphatidylcholine is catalyzed by choline kinase (EC 2.7.1.32), choline-phosphate cytidylyltransferase (EC 2.7.7.15), and cholinephosphotransferase (EC 2.7.8.2), respectively. Phosphatidylethanolamine is catalyzed by an analogous three steps requiring ethanolamine kinase (EC 2.7.1.82), ethanolamine-phosphate cytidylyltransferase (EC 2.7.7.14), and ethanolaminephosphotransferase

[28] P. F. Daly, R. C. Lyon, P. J. Faustino, and J. S. Cohen, *J. Biol. Chem.* **262**, 14875 (1987).
[29] S. L. Pelech and D. E. Vance, *Biochim. Biophys. Acta* **779**, 217 (1984).

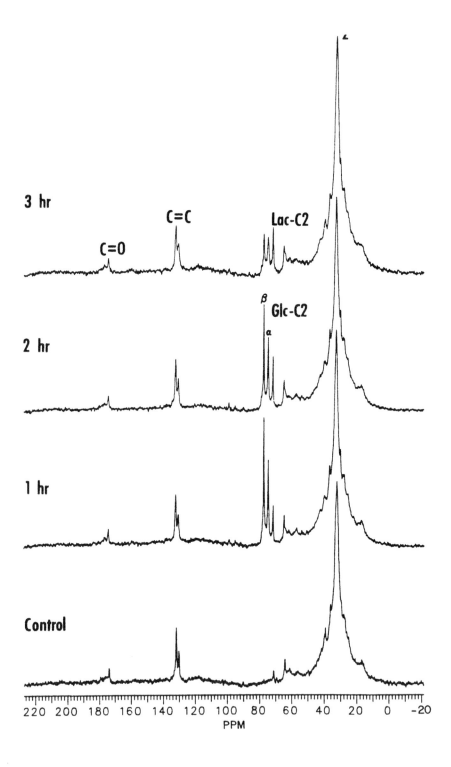

(EC 2.7.8.1).[30] In these pathways the cytidylyltransferase enzymes and phosphotransferase enzymes are membrane bound.[29] The end product of the pathways is the membrane, with net synthesis resulting in growth and phospholipid turnover regulated so that synthesis is coordinated with the rate of degradation.[31] For these reasons the standard methods of destroying the membrane and extracting purified enzymes for study in solutions *in vitro* may not be valid.[32] Ideally, a method such as NMR, in which pathways can be observed noninvasively in growing viable cells, is needed to study these pathways.[32]

We have recently established that the phosphomonoesters observed in tumor cells are phosphoethanolamine (PE) and phosphocholine (PC) and that they are the product of ethanolamine and choline kinase[28] and not degradation products via a phospholipase C, as has been postulated.[33,34] The visibility of PE and PC is particularly fortunate, because they are the products of the kinase enzymes and the substrates of the cytidylyltransferase enzymes. The cytidylyltransferase enzymes are rate limiting and the kinase enzymes are believed to have a modulating effect by controlling the PE and PC pool sizes available to the rate-limiting enzyme.[35] This latter point is controversial, however, because some evidence *in vitro* indicates that PE and PC concentrations are above the saturation points of the cytidylyltransferases.[29]

Data obtained in agarose threads using 2 mM choline or ethanolamine in the perfusate demonstrate that the function of the kinase enzymes can be observed by increases in the PE and PC pool sizes. Alternatively, the inhibition of choline kinase by hemicholinium-3 or ethanolamine resulted in an exponential decrease (e^{-rt}) of PC as it was metabolized by the cytidylyltransferase enzyme. The half-life of this decay was 7.1 hr for ethanolamine and 8.0 hr for hemicholinium-3.[28] The exponential (rather

[30] G. B. Ansell and S. Spanner, in "Phospholipids" (J. N. Hawthorne and G. B. Ansell, eds.), p. 8. Elsevier/North-Holland, New York, 1982.
[31] H. Van Den Bosch, *Biochim. Biophys. Acta* **604**, 191 (1980).
[32] J. D. Esko and C. R. H. Raetz, *Enzymes* **16**, 207 (1983).
[33] G. R. Guy and A. W. Murray, *Cancer Research* **42**, 1980 (1982).
[34] W. T. Evanochko, T. T. Sakai, T. C. Ng, *et al.*, *Biochim. Biophys. Acta* **805**, 104 (1984).
[35] K. W. Ko, H. W. Cook, and D. E. Vance, *J. Biol. Chem.* **261**, 7846 (1986).

FIG. 10. ^{13}C spectra at 100 MHz of subcutaneous human colon carcinoma (LS-17T) in a nude mouse. Control is natural abundance spectrum before injection, and the subsequent spectra are 1, 2, and 3 hr after injection of [2-^{13}C]glucose. Each spectrum was 600 scans at a 1-sec repetition time using a 15-μsec pulsewidth, 2 W of ^{1}H WALTZ decoupling, and a 15-Hz line broadening. The peaks for [2-^{13}C]lactate (Lac-C2) and the α and β anomer of [2-^{13}C]glucose (Glc-C2) are labeled.

than linear) decay implies that the PE and PC levels do influence the forward reaction rates of the cytidylyltransferase enzymes intracellularly. Furthermore, when perfused with choline and ethanolamine, rapid large increases in PE and PC occurred, whereas no appearance of a CDP-choline or CDP-ethanolamine peak was seen. This confirms previous observations that the cytidyltransferase enzymes are rate limiting. In addition, choline and ethanolamine were shown to have an inhibitory effect on GPC phosphodiesterase, which is the last enzyme of the degradation pathway for phosphatidylcholine and phosphatidylethanolamine, and to cause a linear increase in the amount of glycerophosphorylcholine and glycerophosphorylethanolamine. This enzyme and its importance to the phospholipid turnover cycle has not been well studied.[31] With the use of basement membrane gel threads, in which the cells proliferate,[12] a careful study of the synthetic and degradative pathways for these two major phospholipids in growing viable cells is now possible. Preliminary results in our laboratory show that ^{13}C NMR will allow measurement of flux through these pathways by pulse-chase experiments using ^{13}C-labeled choline and ethanolamine.

We have also observed that these pathways are difficult to manipulate *in vivo* in nude mice by using normal or ^{13}C-labeled choline or ethanolamine. This is most likely due to the fact that serum levels of choline and presumably ethanolamine are rapidly regulated by the liver, making it very difficult to change their extracellular levels. This is an example wherein detailed studies of metabolic pathways by NMR methods *are* possible using a perfused mammalian cell system, in which the extracellular environment can be precisely controlled, but *are not* possible *in vivo*, even though PE and PC are clearly visible in *in vivo* spectra of tumors from the same cell line.

The latest development in this area is the ability to obtain localized proton spectra of small volumes (1 ml) of tissue *in vivo* with effective water suppression using strong shielded gradients.[36]

---

[36] P. Van Zijl, C. W. T. Moonen, J. R. Alger, J. S. Cohen, and S. A. Chesnick, *Magn. Res. Med.* **10**, 256 (1989).

Appendix

Computer Programs Related to Nuclear Magnetic Resonance: Availability, Summaries, and Critiques

This Appendix provides a brief overview of selected computer programs discussed in Volumes 176 and 177 and information regarding their acquisition, operation, computer compatibility, maintenance, etc. By virtue of the dynamic nature of the field, any such listing will be incomplete; there are obviously many more computer programs devoted to all aspects of NMR spectroscopy than are listed here. Nonetheless, we hope that the information on the programs included will serve as a resource that will aid investigators in the design of experiments and in the selection of methods for data analysis.

Determination of Solution Structure of Proteins

PROTEAN: Available from Dr. Oleg Jardetzky, Stanford University, Stanford Magnetic Resonance Laboratory, Stanford, California 94305-5055.

PROTEAN is a computer program designed to sample systematically the conformational space accessible to a protein and to determine the entire set of positions for each atom that are compatible with the given set of experimentally derived constraints. It solves a protein structure in a series of hierarchical steps starting with the calculation of a coarse topology of the folded structures and culminating in an iterative calculation of the original data to verify the correctness of the resulting family of structures.

The program is written primarily in LISP with a number of subroutines in C and FORTRAN. The coarse sampling can be run on microcomputers including VAX, μVAX, SUN, and even a Macintosh II (the latter for small proteins with a backbone containing <50 Cα's). The full refinement requires a CRAY or a comparable minisupercomputer. See Chapter [11] by Altman and Jardetzky in Volume 177 of this series.

EMBED and VEMBED, Distance Geometry Calculations of Solution Structures: Available from Dr. I. D. Kuntz, Department of Pharmaceutical Chemistry, University of California, San Francisco, San Francisco, California 94143-0446.

EMBED and VEMBED are programs designed to produce one or more molecular structures that meet a set of constraints that can be derived from a wide variety of experimental or theoretical sources. It provides a means for a relatively high speed and wide ranging search of conformational space to supply starting structures as input for more sophisticated refinement methods.

EMBED is written in FORTRAN and designed for use in the UNIX or VMS operating system environment. VEMBED is the vectorized FORTRAN version designed to run on a Cray or Convex minisupercomputer. See Chapter [9] by Kuntz, Thomason, and Oshiro in Volume 177 of this series.

GROMOS, Molecular Dynamics Calculations: Available from Dr. Wilfred F. van Gunsteren, Laboratory of Physical Chemistry, University of Groningen, 9747 AG Groningen, The Netherlands.

The molecular dynamics approach, based on the molecular dynamics program GROMOS, determines solution structures of some initial set of coordinates, e.g., coordinates generated by a Distance Geometry analysis of the experimental data, by the sequential application of an energy minimization routine followed by a molecular dynamics calculation. The first operation is the Distance bounds Driven Dynamics (DDD) calculation which increases the variety of structures that satisfy the set of constraints, e.g., by removing "obvious" steric inconsistencies that would otherwise cause the molecular dynamics calculations to fail. The Restrained Molecular Dynamics calculations on the resulting structures provide high quality molecular structures consistent with the NMR data and other constraints.

GROMOS is written in FORTRAN 66 and currently runs under the VMS operating system (a UNIX-based version is being developed). The program contains special optimization routines for modern vector supercomputers. Practical considerations regarding computer time limit the number of atoms to <20,000. No graphics programs are provided, however, interfacing to graphics packages is straightforward. See Chapter [10] by Scheek, van Gunsteren, and Kaptein in Volume 177 of this series.

Basic NMR Programs

These programs cover the basic operations needed for analyzing one- and two-dimensional spectra.

NMR1, NMR2: Available from New Methods Research, Inc., 719 East Genesee Street, Syracuse, New York 13210. It is written in FORTRAN 77 and designed for use with VAX, μVAX(VMS and ULTRIX), SUN-(UNIX), Tektronics, and similar color graphics work stations. This package is supported and has a software tool kit including programs for maximum entropy and linear prediction methods and spectral simulations.

FTCGI and FELIX: Available from Hare Research, Inc., 14810 216th Ave. N.E., Woodinville, Washington 98072. These programs are written in FORTRAN (VMS or UNIX) and designed to run on microcomputers including VAX and μVAX as well as SUN and comparable workstations.

Sequence Assignments

SEQASSIGN: Available from Dr. Martin Billeter, Institut für Molekular-biologie und Biophysik, Eidgenössiche Technische Hockschule-Höngger-berg, CH-8093 Zurich, Switzerland.

SEQASSIGN is a program designed for the automated assignment of 2D NMR spectra of proteins. It has two underlying concepts, first that all assignments consistent with currently available data are retained until unambiguously excluded. Second, the process of assignment is split into formal steps with "unambiguous" data, and the interpretation of "ambiguous" NMR data. The latter involves an interactive dialog with the user which can indicate what kind of data would be most important for further assignments. The program is written in Pascal and has been implemented on VAX (using VMS and Berkeley UNIX) and SUN (UNIX). See M. Billeter, V. J. Basus, and I. D. Kuntz, *J. Magn. Reson.* **76,** 400 (1988); Chapter [8] by Billeter in Volume 177 of this series.

Spectra Simulations

SPHINX: Available from Dr. Kurt Wüthrich, Institut für Molekularbiolo-gie und Biophysik, Eidgenössiche Technische Hockschule-Hönggerberg, CH-8093 Zurich, Switzerland.

SPHINX is a program capable of calculating homonuclear and hetero-nuclear 2D NMR experiments using a wide variety of pulse sequences and application of common phase-cycling and data-handling routines. It can also handle strongly coupled spin systems with up to six spins $I = 1/2$. An additional program is available called LINSHA that creates lineshapes and allows graphical presentation of the simulated 2D spectra.

The programs, as well as a stacked plot program, have been written in FORTRAN 77 and a contour plot program has been written in Pascal. The computations have been run on a DEC 10, working with the TOPS 10 system. See H. Widmer and K. Wüthrich, *J. Magn. Reson.* **70,** 270–279 (1986).

MARS: Available from Dr. Paul Rösch, Max-Planck-Institut für Medi-zinische Forschung, Abteilung Biophysik, Jahnstrasse 29, D-6900 Heidel-berg, Federal Republic of Germany.

MARS provides for the simulation of 2D spectra of amino acids including glutamate with 8 coupled spins and leucine with 10 spins (computer memory may impose some limitations for spin systems of more than eight coupled spins). The program is written in Pascal and has been adopted to UNIX and MS-DOS operating systems. See I. Bock and P. Rösch, *J. Magn. Reson.* **74**, 177–183 (1987).

Relaxation and Exchange Rates

Lineshape Analysis and Determination of Exchange Rates: Available from Dr. B. D. Nageswara Rao, Department of Physics, Indiana University–Purdue University at Indianapolis, 1125 East 38 Street, P.O. Box 647, Indianapolis, Indiana 46223.

This program will calculate lineshapes for NMR spin systems undergoing exchange between two magnetically distinct environments. Input parameters include all the individual chemical shifts (expressed as Larmor frequencies ω_A, ω_{BS2A}, etc.), spin–spin coupling constants, linewidths of all the transitions of the spin systems in the absence of exchange, and the exchange rates in either direction. The lineshapes are then calculated for different values of the exchange rate and compared with experiment. The program is written in FORTRAN and requires the availability of an external computer program for the diagonalization of complex unsymmetrical matrices (usually available at most computer centers, e.g., from IMSL). See Chapter [14] by Rao in Volume 176 of this series.

NMR50 and NMR55AUTO: Available from Dr. Peter P. Chuknyisky, Laboratory of Cellular and Molecular Biology, National Institutes of Health, Gerontology Research Center, Baltimore, Maryland 21224.

It is written in FORTRAN and designed for use with VAX 11/780 under VMS. These programs allow for the semiautomatic calculation of the parameters involved in the procedure for distance determination in NMR experiments. The programs are quite general and allow several different types of input. It is important that the number of frequencies used to measure T_{1p} data be at least one more than the number of parameters chosen for computer searching for minimization. The programs can be used to compute the probe-nucleus distance for various combinations of probes and NMR nuclei (C, F, H, P) or to search for a preselected number of parameters (τ_m, B, τ_v, τ_r, r) among those that are important in paramagnetic relaxation phenomena.

CORMA COmplete Relaxation Matrix Analysis: Available from Dr. Thomas L. James, Department of Pharmaceutical Chemistry, University of California, San Francisco, San Francisco, California 94143-0446.

CORMA is a FORTRAN program for calculating expected 2D NOE intensities (actually mixing coefficients) for any user-supplied molecular structure. CORMA requires as input: Cartesian coordinates supplied in Protein Data Bank (PDB) format, a correlation time, and (optionally) experimental intensities. A routine for generating proton coordinates from all other atomic coordinates is also provided. This information is supplied in two input files: model.PDB and model.INT.

The program calculates dipolar relaxation rates based on the proton–proton distances and the supplied correlation time. The correlation time can be a single effective isotropic correlation time for the whole molecule. Alternatively, correlation times are calculated for each proton–proton pair based on a local diffusion time assigned to each residue in the molecule, i.e., $\tau_{ij} = (\tau_i \tau_j)/(\tau_i + \tau_j)$. Correlation times involving methyl groups are further modified by a user-supplied multiplier (≤ 1.0) in order to approximate their shorter effective correlation times. The matrix of relaxation rates is then diagonalized and the exponential function $a = e^{-R\tau_m}$ is evaluated by left- and right-multiplying the exponentiated eigenvalue matrix by the unitary matrix of orthonormal eigenvectors: $a = \chi e^{-\lambda \tau_m} \chi$.

A summary of the calculated intensities (and RMS errors between calculated and experimental intensities if supplied), and a listing of cross-peaks (in the same format as the model.INT file) including the calculated intensity, the distance between protons, and the relaxation rate constants (σ_{ij} or ρ_i), is generated. Additionally, PostScript files are created for plotting (on a laser printer such as the Apple LaserWriter) a representation of the intensity matrix in a schematic gray-scale plot and optionally a logarithmic number plot (with intensities represented by a numeral 0–9). Additional software for drawing contour plots of the intensities on true chemical shift axes for greater ease in making visual comparisons with experimental data will also shortly become available.

The program can do a 100 proton problem in about 15 sec on a VAX 8650 (UNIX f77 compiler), or about 100 sec on a VAX 11/750. Versions of CORMA are available for both UNIX and VMS compilers.

MTFIT for Calculation of Exchange Rates: Available from Dr. Jens J. Led, Department of Chemical Physics, University of Copenhagen, The H. C. Ørsted Institute, 5 Universitetsparken, DK-2100 Copenhagen Ø, Denmark.

MTFIT is a program that calculates the $n(n - 1)/2$ exchange rates and the n longitudinal relaxation rates from the experimental data obtained in an n-site magnetization transfer experiment, consisting of n complementary experiments.

The recovery curves are described by the modified Bloch equations for exchange between n sites

$$\frac{d\overline{M}}{dt} = \overline{\overline{K}}\,\overline{M} + \overline{M}^e \tag{1}$$

where M_i is the time-dependent spin magnetization along the magnetic field, B_0, corresponding to the ith site and $M_i^e = R_{1i}M_i^\infty$, M_i^∞ being the equilibrium value of M_i, and R_{1i} the longitudinal relaxation rate in the ith site. $\overline{\overline{K}}$ is an $n \times n$ matrix where the off-diagonal elements, K_{ij}, denote the pseudo-first-order exchange rate k_{ij} from the jth to the ith site while the diagonal elements are given by

$$K_{ii} = -\left(R_{1i} + \sum_{j \neq i} k_{ji}\right) \tag{2}$$

The solution of Eq. (1) is straightforward and can be written as

$$\overline{M} = \overline{\overline{Q}} \begin{bmatrix} b_1 e^{\lambda_1 t} \\ b_2 e^{\lambda_2 t} \\ \cdot \\ \cdot \\ \cdot \\ b_n e^{\lambda_n t} \end{bmatrix} + \overline{M}^\infty \tag{3}$$

where $\overline{\overline{Q}}$ diagonalizes $\overline{\overline{K}}$. The b_i's are determined from the initial condition through the relation

$$\overline{b} = \overline{\overline{Q}}^{-1}(\overline{M}^0 - \overline{M}^\infty) \tag{4}$$

M_i^0 being the initial value of M_i.

An iterative least-squares procedure that fits a number of functions of the form given in Eq. (2), only differing in $\overline{M}^0$ and $\overline{M}^\infty$, is applied to the n^2 recovery curves. The columns of $\overline{\overline{Q}}$ are calculated as the eigenvalues to the $\overline{\overline{K}}$ matrix, resulting from the latest iteration or from the initial guess. The first-order derivatives, with respect to the rate constants and relaxation rates, needed to calculate the next step in the iteration process, are evaluated numerically since $\overline{\overline{Q}}$ has no analytical form. The remaining derivatives, however, are evaluated analytically. Since only first-order derivatives are applied, the so-called Hartley's modification [H. O. Hartley, *Technometrics* **3**, 269 (1961)] was implemented to assure convergence of the iteration.

Input to the program is the experimental data defining the n^2 recovery curves, and an initial guess of the exchange rates and the longitudinal relaxation rates. Output from the calculation is the rate constants including an estimate of the uncertainties of the parameters.

The program, written in ALGOL and VAX PASCAL/FORTRAN, is available on request from the authors.

Relaxation Parameter Calculations: Available from Dr. Thomas Schleich, Department of Chemistry, University of California, Santa Cruz, Santa Cruz, California 95064.

A family of computer programs was developed to examine the behavior of the theoretical spectral intensity ratio, R, and the spin-lattice relaxation times, T_1 and $T_{1\rho}^{off}$, under different experimental conditions. The mathematical expressions, furnished in Chapter [20] by Schleich et al. in Volume 176 of this series, were coded in Borland TURBO PASCAL, version 5.0, assuming the presence of a math coprocessor chip, and run on IBM Model 70, 80 or AT computers.

These programs allow the effects of isotropic and anisotropic tumbling of rigid hydrodynamic particles, including those of internal motion, on magnetic relaxation times to be studied as a function of axial ratio, rotational correlation time, or precessional frequency in the effective magnetic field. The variety of input parameters, in addition to NMR constants, include protein molecular weight, hydration, partial specific volume, protein X-ray coordinates, Euler angles, rotational correlation time, and off-resonance rf irradiation power and frequency. Depending on the simulation, a particular program requires only a subset of the parameters listed above that are specific to that case.

Program listings will be supplied on request, and can be made available on either 5.25 or 3.5" diskettes (IBM format) or sent electronically via E-mail.

Data Processing and Analysis

NMRFIT: Available from Dr. Jens J. Led, Department of Chemical Physics, University of Copenhagen, The H. C. Ørsted Institute, 5 Universitetsparken, DK-2100 Copenhagen Ø, Denmark.

NMRFIT is a program that extracts the parameters characterizing a Fourier transform NMR spectrum, i.e., the resonance frequency, the intensity, the linewidth, and the phase for a specified number of signals. The program takes into account the effect of nonideal sampling of the free induction decay (FID) and the consequences of processing the FID by a discrete Fourier transformation. Also, extensively overlapping spectral lines can be handled by the program. The output of the program includes an estimate of the uncertainties of the parameters.

The spectral parameters are obtained by a nonlinear least-squares fit of the expression for the discrete Fourier transform of a sum of exponentially damped sinusoids

$$S(\nu_j) = \sum_{l=1}^{L} A_{0,l} T e^{i\varphi_l} \frac{1 - e^{(i2\pi(\nu_{0,l} - \nu_j) - R_{2,l})NT}}{1 - e^{(i2\pi(\nu_{0,l} - \nu_j) - R_{2,l})T}}$$

where $\nu_{0,l}$, $A_{0,l}$, $R_{2,l}$, and φ_l are the resonance frequency, the intensity, the transverse relaxation rate, and the phase, respectively, of the lth signal in the spectrum, consisting of L lines. Further, T is the time between two consecutive measurements, and N is the number of points used in the Fourier transformation. The real and imaginary part of the frequency spectrum are fitted simultaneously using a Newton-type minimization procedure. In this procedure the first- and second-order derivatives are calculated analytically. In the input to the program the predicted number of lines must be specified together with an initial guess of the four parameters for each line.

The program is especially suited for complicated, nonideal spectra with a limited number of signals, or spectra that can be divided into minor parts containing isolated groups of lines.

The program written in ALGOL and VAX PASCAL/FORTRAN is available on request from the authors.

Processing, postprocessing, and storage of 2D NMR data: Available from Operations Assistant, National Magnetic Resonance Facility at Madison, Biochemistry Department, 420 Henry Mall, Madison, Wisconsin 53706.

Software for the Bruker Aspect 1000/3000

The following programs have been written by Zsolt Zolnai to be run on the Aspect computer.

"MAKEUP"[1] is a program that removes noise in phase-sensitive or absolute-value 2D NMR spectra. The relationship between noise and signal is linear in phase-sensitive spectra. This means that the average noise level (dc offset) can be subtracted directly from line intensities. The average noise level is calculated from a reference spectral strip selected to contain no spectral lines along the full length of the f_2 dimension. The standard deviation of the noise also is calculated. The calculations pro-

---

[1] Zs. Zolnai, S. Macura, and J. L. Markley, *Comput. Enhanced Spectrosc.* **3**, 141 (1986).

duce two profiles along the f_2 dimension, along with the dc offset and the rms noise level. Each point of the 2D spectrum is corrected for the corresponding dc offset. The correction is constant within an f_1 column but varies from column to column. In the next step, each spectral point is compared with a number related to the rms noise level: points that fall in the range, ± (rms noise) × (weighting factor), are zeroed; points outside this range are left intact. The overall effect is selective elimination of background t_1 noise irrespective of the variability of its value from column to column. The horizontal ridges (t_2 noise) can be eliminated by carrying out the same treatment rowwise. In practice, all steps are performed simultaneously. Because a linear relationship between signal and noise is required, absolute value spectra must be converted first into the power mode. MAKEUP enables the simultaneous observation of all cross-peaks and diagonal peaks from a single plot, even if very strong noise bands are present in the original spectrum. Spectra processed in this way are amenable to automatic computer analysis and can be compressed into a smaller storage size without loss of information.

"STANLIO" and "OLIO"[2] are programs for the compression and decompression of 2D NMR spectra after MAKEUP. Since MAKEUP replaces noise-bearing points by zeros, compression can be achieved by storing nonzero points normally while representing strings of zeroes by two numbers: the first being the leading zero itself, and the second the number of zeros in the sequence. Depending on the spectra, compression ratios between 5 and 100 can be realized. Reduction in the size of 2D NMR spectra speeds up data storage, retrieval and transmission, and reduces the cost of long-term data storage. Data retrieval, i.e., expansion of compressed data into a format that can be recognized by DISNMR (the Bruker NMR program for the Aspect computer) is performed by OLIO.

"SPELOG"[2] is a program for logarithmic scaling of 2D NMR spectra. The relative error in the final, Fourier-transformed spectrum rarely is less than 1%. Therefore, a large number of bits is not necessary to represent the data, and compression can be achieved without appreciable loss of information by squeezing more than one data point into a single computer word. A convenient way of achieving this is to scale the data logarithmically with a suitably chosen base, b. For example, a 24-bit two's complement word can be compressed into 12 bits via logarithmic scaling by using $b = 1.007815$ (and log 0 = 0), with the maximum relative error due to compression of only 0.8%. Data retrieval, into a format that can be recognized by the DISNMR program, is performed in one or two steps, de-

---

[2] Zs. Zolnai, S. Macura, and J. L. Markley, *J. Magn. Reson.* **80**, 60 (1988).

pending on whether the data are to be viewed in a linear or logarithmic mode.

"SPEMAN"[3] is a routine for making a linear combination of two 2D NMR spectra. It multiplies the intensity of one 2D spectrum by a specified positive or negative factor and adds it to the intensity of another 2D spectrum.

"DIST"[4] is a program for calculating distances between protons from NOEs and ROEs. The program consists of several modules. (1) A module for entering coordinates for each cross-peak to be analyzed. The coordinates are stored in a file and can be used to identify the corresponding two cross-peaks and two diagonal peaks each time they are needed. (2) A module for finding intensities. This module uses the file with cross-peak coordinates, and creates a file containing the intensities of the two cross-peaks and corresponding two diagonal peaks from each spectrum recorded with different mixing times. All intensities are refined by searching for local maximum around actual cross or diagonal peak—local minimum for ROESY cross-peaks. (3) A module that calculates linewidth at half height for each selected cross-peak in both dimensions. (4) A module for distance calculations. This module calculates build-up rates and distances on the basis of intensities or volume integrals, intensity $\times$ width$_1$ $\times$ width$_2$, of cross and diagonal peaks. (5) A module for inspection and printout of files created in the other modules. Files are not text files because of space considerations, and some information is coded. (6) A module for visual checking of intensities or volume integrals versus mixing times and the fit of experimental data to calculated build-up rates.

"CHOPPER"[3] is a program which allows one to "chop out" any part of a 2D NMR spectrum whose size is a power of two in both dimensions. Spectra may be in the absolute-value or pure-phase mode. The "chopped" part is in a format that is recognized by DISNMR and can be treated like any other 2D spectra. Often only some parts of 2D spectra carry relevant information (e.g., spectra obtained with the carrier at one end). CHOPPER retains the imaginary parts of phase-sensitive spectra; thus time and space can be saved by phasing only the important part of the spectrum.

"ROTATE"[3] is a program for transposition of various types of 2D NMR spectra: phase-sensitive, absolute-value, and square and nonsquare spectra. Since the transpose is done in place, no additional disk space is required. It also is useful for faster plotting. In DISNMR, plotting of horizontal strips is much faster than plotting of vertical strips (because of

[3] Zs. Zolnai, S. Macura, and J. L. Markley, to be published.
[4] Zs. Zolnai, S. Macura, and J. L. Markley, *J. Magn. Reson.*, in press (1989).

the way the contour plot routine is written). Therefore, time can be saved if one transposes a vertical strip and plots it as a horizontal strip instead of directly as a vertical strip. After plotting, the data can be transposed back.

"STROLLER"[4] is a program for baseplane correction. Baseplane distortions arise from instrumental artifacts, aliasing of dispersive tails of large peaks (i.e., solvent), and/or t_1 noise. Baseplane correction [26] is an essential step before many kinds of postprocessing: intensity measurement for distance calculation, measurement of volume integrals, automatic linewidth measurement, and peak picking for assignment purposes. In 2D plots it is common to find that some peaks are "lost" because they lie on the negative tail of a large peak whose negative intensity is greater than the positive intensity of the "lost" cross-peak. For baseplane correction, one must first select a grid of control points that best determine the true baseplane. On the basis of this grid, the baseplane is calculated for each point in the spectrum by using the "Cardinal bicubic surfaces algorithm" and is subtracted from this point. The determination of each baseplane point, which requires 20 floating point multiplications and 12 floating point additions, is slow since floating point operations are not efficient on the Aspect computer. Because of this, and because it is often enough to correct the baseplane only in selected regions, the program allows the procedure to be performed on just a subset of the 2D spectrum.

"INTENZY"[3] is a program that gives as output an integer value (stored on the disk) of any point in 2D NMR spectra. This routine is necessary for SPEMAN since DISNMR reports only relative intensities. This program is useful as a subroutine in many other programs. It is embedded in STROLLER and DIST, for example.

Software for Silicon Graphics Iris

Prashanth Darba has rewritten preliminary versions of the programs "PROC2D" and "ANALYS2D"[5] to run under the UNIX operating system on the SGI Iris 2400T and SGI 4D-Series workstations.[6] The main new software package is called "MADNMR".[7] The package supports the usual 2D NMR processing operations plus user-friendly graphics features for interactive analysis of 2D data that take advantage of the graphics engine on the Iris workstations.

Spectral processing can be carried out in either an interactive or batch mode. The package supports both the TPPI[8] and States–Haberkorn–

[5] P. Darba and L. R. Brown, unpublished.
[6] P. Darba and J. L. Markley, to be published.
[7] B. H. Oh, W. M. Westler, P. Darba, and J. L. Markley, *Science* 240, 908 (1988).
[8] D. Marion and K. Wüthrich, *Biochem. Biophys. Res. Commun.* 113, 967 (1983).

Ruben[9] method for generating phase-sensitive spectra and linear combination procedures for efficient generation of multiple quantum filtered 2D NMR spectra.[10] Symmetrization and filtering of noise can be carried out with respect to a user-defined S/N threshold. Peaks above the threshold that do not have a counterpart (of the proper sign and above the specified threshold) in a region related by the symmetry specified are zeroed; all other signals are left as is. Interactive graphics features support the creation of spin-system connectivity databases; peak picking and inking of connectivities is carried out in multiple viewport modes designed to facilitate the analysis of different classes of 2D NMR data. Spin-system connectivities chosen for one data set can be overlaid upon another data set. Inking of spin connectivities is updated automatically in all windows during peak picking procedures; all user-selected portions of previously defined spin-system databases can be inked in all windows in colors of the user's choice. Two 2D spectra can be overlaid and zoomed to resolve higher detail. Specified sections of a 2D spectrum can be plotted along with the spin-system connectivities identified.

Another program for the Silicon Graphics Iris, "GENINT," has been developed to assist in refinement and intensity measurement of the NOESY peaks picked interactively in MADNMR. The user picks peaks interactively by placing the crosswire cursor approximately at the center of a contour and pressing a button on the mouse; the coordinates of the crudely picked peaks are stored in a spin-system database. GENINT uses these coordinates as the starting coordinates for searching the extremes of peaks. With a 2D matrix containing ~200 peaks, the routine takes only a few seconds to create a database of precise chemical shift coordinates and cross-peak intensities.

The Rowland NMR Toolkit: Available from Dr. Jeffrey C. Hoch, Rowland Institute for Science, 100 Cambridge Parkway, Cambridge, Massachusetts 02142.

The implementations of the Burg MEM and LPZ methods are used in the present work are part of the Rowland NMR Toolkit. The Toolkit consists of a suite of programs for processing one- and two-dimensional NMR data on Digital Equipment Corporation VAX computers running the VMS operating system. Although designed as a "workbench" for developing NMR data processing methods, the Toolkit may also be used for routine processing and plotting of NMR data. The Toolkit accommodates data from a variety of sources and supports a wide variety of

[9] D. J. States, R. A. Haberkorn, and D. J. Ruben, *J. Magn. Reson.* **48,** 286 (1982).
[10] R. Ramachandran, P. Darba, and L. R. Brown, *J. Magn. Reson.* **73,** 349 (1987).

graphic display devices. The Toolkit is available to academic institutions for a nominal handling fee. For commercial enterprises the fee is $4000.00 U.S. (version 2.0).

Linear Prediction Singular Value Decomposition and Hankel Singular Value Decomposition: Available from Dr. Ron de Beer, University of Technology Delft, Department of Applied Physics, P.O. Box 5046, 2600 GA Delft, The Netherlands.

The implementations of the LPSVD and HSVD methods were written by H. Barkhuijsen and R. de Beer.

Maximum Entropy Reconstruction: Available from Maximum Entropy Data Consultants, Ltd., 33 North End, Meldreth, Royston SG8 6NR, England.

The implementation of maximum entropy reconstruction used was written by John Skilling.

Author Index

Numbers in parentheses are footnote reference numbers and indicate that an author's work is referred to although the name is not cited in the text.

Subject Index

A

N

Neurofilament proteins, phosphoserine and phosphothreonine residues present simultaneously, 270

Nickel(II), ^{1}H NMR
electron spin relaxation rate, 248
nuclear line broadening effect, 248

Nitrogen-15 labeling, 41
of proteins, 44–73
determination of level of isotopic enrichment, 61–62
selective, by residue type, 52–60
uniform, 50–52
of protein side chains, 59
of T4 lysozyme, selective, protocol for, 60–61

Nitrogen-15 nuclear magnetic resonance
expression and nitrogen-15 labeling of proteins for, 44–73
mechanistic studies utilizing ^{18}O, 376–389
^{18}O isotope effect in, magnitudes of, 378–379
^{18}O isotope shifts in, 376
sensitivity of, 44
use of, 44

Nitrogen-15 nucleus
properties of, 264
in proteins, NMR, 62–65

Nitroxide-induced relaxation of proton resonances, structural information obtained via, 88–89

Nitroxides
affinity chromatography with antinitroxide antibodies, 106
chemical properties of, 104–105

NMRFIT (program), 461–462

NMR1/NMR2 (program), 151–152, 157–158

NMR1 (program), 456–457

NMR2 (program), 456–457

NMR50 (program), 458–459

NMR55AUTO (program), 458–459

NOESY. See Nuclear Overhauser effect spectroscopy; Relayed-NOESY

N-ras p21, selective labeling with ^{15}N-labeled amino acids, 53–54

Nuclear magnetic resonance dispersion spectroscopy, 247

Nuclear magnetic resonance spectroscopy
computer programs for. See Computer programs
determination of equilibrium constants by, 359–375
determination of molecular structures by, 204–218
input routines, 162–163
interpretation of
adjustment paradigm, 219–221
exclusion method, 219–221
methods for, 219–221
monitoring intracellular metabolism by, 435–452
protein structure determination by, 125, 125–131
methods, 127–131
structural data, reporting, 200
structure determination from
comparison of methods, 164–165
general features of, 162–164
two-dimensional, 29, 127
data processing, postprocessing and storage, computer programs for, 462–467

Nuclear Overhauser effect
detection of, 207
from ^{1}H to ^{31}P, for phosphoproteins, 274
measurements, 204
in exchanging systems, 335
structural information obtained from, 161
time dependence of, 336–337
and exchange rate of ligand, 338–339

Nuclear Overhauser effect spectroscopy, 150, 170–171
boundary conditions, 171–172
cross-peaks
assignment of, in structure determination, 149
classification of, 141
data, qualitative interpretation of, 348
data collection, 339–340, 343–344
for elucidation of molecular structure, 333–334
in exchanging systems, 10
future developments, 358
intermolecular protein–ligand, 355–357
ligand conformations studied by, 333–358

Printed and bound by CPI Group (UK) Ltd, Croydon, CR0 4YY

03/10/2024

01040419-0006